Environmental Assessment

Environmental Assessment

R. K. Jain, Ph.D., P.E.
Dean, School of Engineering
University of the Pacific

L. V. Urban, Ph.D., P.E.
Director
Water Resources Center
Professor of Civil Engineering
Texas Tech University

G. S. Stacey, Ph.D.
Former Senior Economist
Battelle-Europe Centres de recherché de Geneve

H. E. Balbach, Ph.D.
Senior Research Scientist
U.S. Army Engineer Research and Development Center

M. D. Webb M.L.A.
Group Leader, Ecology Group
Los Alamos National Laboratory

Second Edition

McGraw-Hill

New York   Chicago   San Francisco   Lisbon   London   Madrid
Mexico City   Milan   New Delhi   San Juan   Seoul
Singapore   Sydney   Toronto

Library of Congress Cataloging-in-Publication Data

Environmental assessment / R. K. Jain...[et al.].—2nd ed.
  p.   cm.
  Includes bibliographical references and index.
  ISBN 0-07-137008-0
  1. Environmental impact analysis—United States   2. Environmental
policy—United States.   I. Jain, R. K. (Ravinder Kumar), 1935-
TD194.65 .E58   2001
333.7'14—dc21   2001034501

McGraw-Hill

A Division of The McGraw·Hill Companies

1 2 3 4 5 6 7 8 9   AGM/AGM   0 9 8 7 6 5 4 3 2 1

ISBN 0-07-137008-0

*The sponsoring editor for this book was Ken McCombs, the editing
supervisor was Daina Penikas, and the production supervisor was
Sherri Souffrance. This book was set by McGraw-Hill Professional's
composition unit, Hightstown, N.J.*

Printed and bound by Quebecor / Martinsburg.

McGraw-Hill books are available at special quantity discounts to use as
premiums and sales promotions, or for use in corporate training
programs. For more information, please write to the Director of Special
Sales, McGraw-Hill, Two Penn Plaza, New York, NY 10121-2298. Or
contact your local bookstore.

 This book is printed on recycled, acid-free paper containing a
minimum of 50 percent recycled, de-inked fiber.

Contents

Preface   ix

Chapter 1.   Environmental Assessment   1

1.1   What Is Environmental Assessment?   4
1.2   Why Environmental Assessment Is Needed   6
1.3   Who Prepares Environmental Assessment Documents?   7
1.4   Integrating Art and Science   8
1.5   Discussion and Study Questions   10
1.6   Further Readings   11

Chapter 2.   Environmental Laws and Regulations   13

2.1   Rationale for Environmental Legislation and Regulations   14
2.2   Shortcomings of Environmental Legislation and Regulations   15
2.3   Legislative Data Systems   15
2.4   Overview of Federal Environmental Legislation   19
2.5   Trends in Environmental Legislation and Regulations   54
2.6   Discussion and Study Questions   57
2.7   Further Readings   58

Chapter 3.   National Environmental Policy Act   61

3.1   Elements of NEPA   62
3.2   Judicial Review   63
3.3   Effects of NEPA   68
3.4   Implementation of NEPA   68
3.5   Council on Environmental Quality   70
3.6   Executive Orders and Agency Response   70
3.7   State Environmental Policy Acts   73
3.8   NEPA and Agency Planning   74
3.9   Discussion and Study Questions   83
3.10 Further Readings   84

Chapter 4.   Environmental Documents and CEQ Regulations   85

4.1   Function and Purpose of the NEPA Assessment Process   85
4.2   NEPA Regulations   88
4.3   Environmental Documents   90
4.4   Application of Environmental Documentation Process   92
4.5   EIS Format and Content   97
4.6   Preparing and Processing of the EA or EIS   104
4.7   Timing of Agency Action   113
4.8   Tiering   115
4.9   Mitigation   115
4.10   General Considerations in EIS Preparation   116
4.11   Case Studies   116
4.12   Discussion and Study Questions   122
4.13   Further Readings   123

Chapter 5.   Elements of Environmental Assessment   125

5.1   Agency Activities   125
5.2   Environmental Attributes   126
5.3   Determining Environmental Impact   148
5.4   Report Findings   157
5.5   Using Information Technology to Aid in the NEPA Process   158
5.6   Discussion and Study Questions   161
5.7   Further Readings   162

Chapter 6.   Environmental Assessment Methodologies   163

6.1   Choosing a Methodology   163
6.2   Categorizing Methodologies   166
6.3   Review Criteria   167
6.4   Methodology Descriptions   176
6.5   Methodology Review   176
6.6   Future Directions   190
6.7   Discussion and Study Questions   191
6.8   Further Readings   192

Chapter 7.   Generalized Approach for Environmental Assessment   193

7.1   Agency Activities   193
7.2   Environmental Attributes   197
7.3   Institutional Constraints   200
7.4   Environmental Setting   201
7.5   System   201
7.6   Output   201
7.7   Rationale for a Computer-Based System   201
7.8   Discussion and Study Questions   205
7.9   Further Readings   206

Chapter 8.   Procedure for Reviewing Environmental Impact Statements 207

8.1   Types of EIS Review   208
8.2   General Considerations in EIS Review   212
8.3   EIS Review Procedures   213
8.4   Approaches to Systematic EIS Review   214
8.5   Summary   223
8.6   Discussion and Study Questions   223
8.7   Further Readings   224

Chapter 9.   International Perspectives on Environmental Assessment   225

9.1   International Implications of NEPA   226
9.2   Future NEPA Trends   228
9.3   Environmental Impact Assessment in Other Countries   229
9.4   EIA in Developing Countries   230
9.5   Limitations to EIA Effectiveness in Developing Countries   231
9.6   EIA in Asia and the Pacific   232
9.7   EIA in Latin America   232
9.8   EIA in Canada   232
9.9   EIA in Europe   233
9.10   International Aid Organizations   235
9.11   Discussion and Study Questions   236
9.12   Further Readings   237

Chapter 10.   Economic and Social Impact Assessment   239

10.1   Socioeconomic Assessment within NEPA   240
10.2   Economic Impact Analysis   242
10.3   Economic Models   243
10.4   Future Direction for Economic Impact Analysis   251
10.5   Social Impact Assessment   251
10.6   Social Impact Analysis Methods   255
10.7   Assembling the Socioeconomic Impact Assessment   259
10.8   Problem Areas   267
10.9   Environmental Justice   269
10.10   Conclusions   278
10.11   Discussion and Study Questions   278
10.12   Further Readings   279

Chapter 11.   Public Participation   281

11.1   Beginnings   281
11.2   Early American Experiences in Public Participation   282
11.3   Modern Applications of the Participation Concept   284
11.4   Public Involvement Requirements within NEPA   284
11.5   What Is a Public?   285
11.6   What Is Participation?   287

11.7   Contemporary Experience in Public Participation   288
11.8   Getting Input from the Public   290
11.9   Commenting on the Draft EIS—A Special Case of Public Participation   291
11.10   Case Study Example 1   292
11.11   Case Study Example 2   294
11.12   Application of Public Input   295
11.13   Participation in Developing Regulations   296
11.14   An Effective Public Participation Program   296
11.15   Benefits from an Effective Public Participation Program   302
11.16   Response to Public Participation Format Variations   303
11.17   Public Participation and the Internet Revolution   304
11.18   Internet Capability to Support Public Participation   305
11.19   Discussion and Study Questions   306
11.20   Further Readings   307

Chapter 12.   Energy and Environmental Assessment   309

12.1   Energy as a Resource   310
12.2   Fuel Alternatives and Development of Supplies—Environmental
        Considerations   316
12.3   Energy Costs of Pollution Control   319
12.4   Energy Aspects of Recycling Materials   319
12.5   Discussion and Study Questions   325
12.6   Further Readings   326

Chapter 13.   Contemporary Issues in Environmental Assessment   327

13.1   Global Warming   327
13.2   Acid Rain   340
13.3   Deforestation   347
13.4   Endangered Species and NEPA   350
13.5   Biodiversity   360
13.6   Cultural Resources   366
13.7   Ecorisk   374
13.8   How Are These Contemporary Issues Different?   391
13.9   Discussion and Study Questions   391
13.10   Further Readings   392

Epilogue   395

Appendix A.   National Environmental Policy Act   397

Appendix B.   Attribute Descriptor Package   407

Appendix C.   Step-by-Step Procedure for Preparing Environmental
Assessments and Statements   567

Appendix D.   Regulations for Implementing Procedural Provisions of
the National Environmental Policy Act   585

Abbreviations and Acronyms   619
Bibliography   625
Index   639

Preface

The signing of the National Environmental Policy Act (NEPA) on January 1, 1970, ushered in a decade of sweeping change in the way the U.S. government did business. By requiring environmental assessments and impact statements for those federal projects significantly affecting the quality of the human environment, NEPA launched a flurry of activities including (1) the development of methodologies for accomplishing the mandated tasks, (2) the establishment of an infrastructure of expertise within various federal agencies and consulting firms to do the actual work, and (3) a seemingly endless string of legal actions (injunctions, court cases, etc.) that helped to interpret and further shape the law. The guidelines and subsequent regulations that were promulgated by the President's Council on Environmental Quality (CEQ) to provide direction to agencies in complying with NEPA were fine-tuned and the assessment and documentation process evolved accordingly.

By the 1980s, many other countries and state governments within the United States had adopted similar requirements, and the NEPA process had become a relatively routine activity, with only the most controversial projects receiving media attention, and hence public notice. Toward the end of the decade, however, a resurgence of environmental awareness and concern occurred, and the 1990s have been referred to as the "new environmental decade." In addition to the more or less traditional concerns for air, water, and ecology, contemporary issues include concerns about biodiversity, global warming, acid rain, deforestation, and other topics scarcely contemplated in the 1960s.

Just as the NEPA process has evolved, so have the authors' presentations of ways to assist it. The beginnings may be traced back to the early 1970s with environmental research conducted at the U.S. Army Construction Engineering Research Laboratory in Champaign, Illinois. In one project, the objective was the development and production of a handbook to assist U.S. Army personnel in addressing the

NEPA requirement to prepare documentation of the environmental impacts of their proposed actions. In response to this challenge, the authors developed procedures to include both the natural and physical environment and the relationship of humans with that environment. This sweeping incorporation of the sum of the biophysical and socioeconomic environment was controversial at the time and was considered to go far beyond that which NEPA "really required." Among the impacts so included were those associated with ecologic, aesthetic, historic, cultural, economic, social, and health effects, whether direct, indirect, or cumulative. We believed (and still hold) that NEPA requires agencies to accomplish environmental impact assessment utilizing a systematic, interdisciplinary approach, and this book continues to present that belief.

To attempt in any single book to guide the accomplishment of all NEPA-related requirements is indeed a tall order! Because of the complexity of the environment, no one individual can be proficient in all areas, nor can a single handbook or text provide every item of information requisite to completely encompass all areas and topics. It should, however, provide readers with insight into each complex area of the environment and furnish guidance in the systematic and interdisciplinary approach to environmental impact analysis. A four-step procedure was developed, consisting of (1) defining and detailing the proposed action, (2) identifying and understanding the affected environment, (3) determining the possible impacts, and (4) reporting the results in an appropriate manner.

The first edition (1993) of this text retained the basic premise but significantly expanded upon the wide range of previously developed concepts involved in environmental assessment. This Second Edition further updates and elaborates on the concepts presented in the earlier edition with a discussion of new requirements and emerging contemporary issues. As with the first edition, the purpose of the book remains to present in a thoughtful and systematic manner a synthesis of ideas and professional experience to address the complex area of environmental assessment. In keeping with the approach outlined in federal law—NEPA—and its implementing regulations, the book provides a comprehensive, systematic approach to analyzing the effects that a project or action may have on the human environment. To replicate effectively the interdisciplinary approach required by the NEPA process, the authors represent many disciplines (civil and environmental engineering, water resources, public policy, economics, ecology, social science and planning) and diverse and rich experiences in research, academia, federal and state agencies, national labs, and industry. It presents or otherwise incorporates a wide range of contemporary issues and other recent developments into the discussion.

Chapter 1 introduces basic concepts associated with environmental assessment and lays the groundwork for the remainder of the text. *Chapter 2* provides an overview of environmental legislation and regulations in sufficient detail to gain an understanding of both the basic provisions and requirements of the acts and their relationship to NEPA and the environmental assessment process. *Chapter 3* discusses the various elements of NEPA and its impact on and importance to federal agencies, gives illustrative examples of executive orders, and provides commentary on state environmental policy acts. *Chapter 4* covers the function and purpose of environmental assessment, NEPA regulations promulgated by CEQ, and the preparation and processing of various documents that are generated as a result of the NEPA process. *Chapter 5* details the four basic elements of environmental assessment. *Chapter 6* provides a review of impact assessment methodologies. *Chapter 7* suggests a framework for a generalized approach for developing an overall impact assessment system. Because of the complexity of the problem and the vast amount of information involved, the rationale for a computer-based analysis system is included. Since persons representing various organizations and management levels review documents generated by the NEPA process, *Chapter 8* presents a general procedure for reviewing environmental impact statements. *Chapter 9* examines environmental assessment from an international perspective. Economic and social impact assessment is presented in *Chapter 10,* supported with a discussion of economic models and social impact methods. Because of the integral role played by the public, *Chapter 11* addresses methods for assuring effective public participation in all phases of the NEPA process, and *Chapter 12* considers energy and environmental assessment. *Chapter 13* examines several contemporary issues in environmental assessment—global warming, acid rain, deforestation, endangered species, biodiversity, cultural resources, and ecorisk.

Four appendices are included: Appendix A reproduces the National Environmental Policy Act (PL 91-190); Appendix B provides a description of environmental attributes that provide the essence of basic information in the eight environmental categories utilized to describe the biophysical and socioeconomic environment; Appendix C provides an illustrative example of a matrix-based, step-by-step procedure for preparing an environmental assessment; and Appendix D reproduces the CEQ (NEPA) Regulations (40 C.F.R. 1500-1508) for implementation of procedural provisions of NEPA.

With discussion questions and suggestions for further reading at the conclusion of all chapters, the result intended by the authors is a book which will be useful to persons ranging from students to practitioners, from concerned citizens to government agency staff, and those whose background ranges from technically-oriented and highly informed to the untrained but environmentally conscious and curious. It may easily be used as a sole or supplemental text in college courses on impact assessment or as an outside assignment in graduate business and law schools. The corporate employee may use it as a handbook, and the government employee as a guide to action.

Many individuals have assisted us along our way. Special credit and our gratitude must go to Ms. Courtney Stovall, a Graduate Research Assistant, who provided extraordinary support in updating many tables, figures, references, and textual material. Likewise, the support and encouragement provided by our many colleagues at the following agencies and institutions is gratefully acknowledged:

University of the Pacific
U.S. Army Engineer Research and Development Center
Texas Tech University
Los Alamos National Laboratory
University of Cincinnati

R. K. Jain
L. V. Urban
G. S. Stacey
H. E. Balbach
M. D. Webb

Environmental Assessment

Environmental considerations were largely ignored for almost 200 years in the development of the United States. Only in the last third of the twentieth century did environmental factors begin to play a significant role in the speed and direction of our national progress. These factors have developed in us a new concern and recognition of the dependence that we, as human beings, have on the long-term viability of the environment for sustaining life. The new "ethic" of conservation of resources has also grown as concern for the environment has grown, because much of our environmental quality is itself a nonrenewable resource.

Human development, especially in the twentieth century, represents an intrusion into the overall balance that maintains the earth as a habitable place in the universe. We are recognizing this fact in our concern for the environment, but most of us are also reluctant to give up the profligate consumption of resources which characterizes the modern lifestyle. Thus, it is incumbent upon the human species to examine its actions and to attune to ensuring the long-term viability of earth as a habitable planet. The development of environmental impact analysis, or assessments, is a logical first step in this process. It represents an opportunity for us to consider, in decision making, the effects of actions that are not otherwise accounted for in the normal market exchange of goods and services. The adverse effects discovered in the assessment process then need to be weighed against the social, economic, and other advantages derived from a given action. The art and science of identifying and quantifying the potential *benefits* from a proposed action has become finely tuned. We must develop the belief that an equally clear exposition of the associated *problems* is equally deserving of careful study and consideration.

Blind adherence to the theory and practice of a pure economic exchange for decision making has possible long-term adverse consequences for the planet Earth. There *are* elements which cannot be accurately represented as monetary values. Economic guidelines for decision making were adequate as long as the effects of societal activities were insignificant when compared to the long-term suitability of the planet as a place to reside. One traditional analogy would compare the swing toward concern for environmental considerations to a pendulum that is on the verge of swinging back toward economic (i.e., cost) dominance. Is this unavoidable? This type of trade-off is essential and is one that will always be made, but humans must be aware that sacrificing long-term viability for short-term expediency is less than a bad solution; it is no solution. Serious environmental problems that surfaced following the collapse of the totalitarian regimes of eastern Europe are vivid examples.

As *glasnost* opened the eastern European and Soviet countries to the west during the late 1980s, it also revealed a region suffering extreme environmental degradation. In previous decades, the area had focused on centrally planned industrial development with disregard for the environmental consequences of this development. Industrialization had been the foremost priority, and production targets were to be met to the exclusion of other goals. Industries had been heavily subsidized, particularly for energy and natural resource needs, and allotments of resources and budgets had been made based on past use and expenditures. Although some countries may have had stringent environmental regulations on their books, these regulations were not enforced. Pollution fines levied by the government were small and easily paid with government subsidies. With the presence of production targets and subsidies and the absence of open markets and a realistic price structure, industries had no incentive to conserve resources, avoid pollution fines, or invest in efficient production technologies.

As a result, environmental conditions are now seriously degraded; air pollution, water pollution, hazardous wastes, and extensive impairment of agricultural land and forests are at extreme levels and among the highest in the world. Air in the region is polluted by exceptionally high levels of sulfur dioxide, due to dependence on coal burning for energy, few pollution controls, and extremely inefficient use of energy (Schultz, 1990). Rivers, lakes, and seashores are heavily polluted by industrial waste discharge and agricultural runoff; 95 percent of Polish rivers are so badly polluted that their water cannot be used directly, even for industrial purposes, because it is corrosive (Hallstrom, 1999).

Indiscriminate dumping of hazardous wastes and the use of substandard landfills have contaminated groundwater sources in the region. In addition, the withdrawal of the Soviet Union from previously occupied territories left behind substantial environmental degradation; 6 percent of Czechoslovakian territories were damaged by toxic wastes, oil, and lead (Renner, 1991). In some instances there has been enough spilled fuel available in the soil for private individuals to dig oil wells (Carter and Turnock, 1997). The Chernobyl accident of 1986 released 1000 times the radioactivity of the Three Mile Island accident, and the radiation was widely dispersed over the northern hemisphere (Flavin, 1987). Many nuclear plants in the region are of the Chernobyl type and present the danger of such an accident recurring at any time.

Inappropriate agricultural practices have eroded soils, and industrial pollution has contaminated large land areas. The land around Glubokoe, a nonferrous metallurgical center in northern Belorus, has 22 times the permitted level of lead, 10 times the permitted level of cobalt, and 100 times the permitted level of zinc (French, 1990). An average of 77 percent of Polish and Czech forests show signs of acid rain damage, most likely as a result of huge amounts of highly toxic dust released into the atmosphere throughout Bulgaria, Romania, Hungary, and Poland from industrial smelter releases and brown coal combustion (Hallstrom, 1999).

The cost of this pollution to human health can be seen in lower life expectancies, higher infant mortality, and higher incidence of respiratory diseases, cancers, birth defects, and other illnesses. Nearly 60 percent of children in inner Budapest show dangerously high levels of lead in their blood (Hallstrom, 1999). Life expectancies for some regions are recognized to be 3 to 5 years less than in cleaner areas (Schultz, 1990).

But this is not the only cost of environmental degradation in the region; without a base of functioning water, land, and air resources, industrial productivity and growth are hampered. The decline in forestry and tourism industries due to damaged forests, the falling crop yields, the damage to historic buildings due to acid deposition, and the corrosion of pipes by polluted water are a few examples of real costs incurred by industrialization without separate regard for environmental consequences. It is estimated that the present state of environmental degradation, rather than providing a cheap avenue to industrial development, is costing Poland 10 to 20 percent of the gross national product (GNP) annually, and Czechoslovakia 5 to 7 percent annually. An estimated 11 percent of GNP has been expended annually in the former Soviet countries toward health costs from pollution alone (French, 1990).

The issues of economic growth, poverty, and environmental protection are intertwined in a perplexing way in today's business climate (*Business Week*, 1990). Lasting economic growth is based on managing natural resources in a sustainable manner. Poverty is both a cause and an effect of environmental problems. Sustainable economic growth provides both the means to address world poverty and the means to solve environmental questions. Industrialization and economic development are essential to provide basic amenities of life and to sustain and improve our standard of living. The challenge is: How to determine the direction and level of development that is not limited by what is most expedient for the present, but will benefit future generations as well as provide for the immediate needs of society.

During the past decade, the business world has become increasingly aware that sustainable development and production can, indeed, be good for business. With the passage of the Pollution Prevention Act (PPA) of 1990, pollution prevention was declared to be the nation's primary pollution control strategy, and a hierarchical system for pollution management was developed, with source reduction at the top of the hierarchy, followed by recycling, treatment, and disposal. Increased support for pollution prevention practices has allowed industry to realize that waste reduction, recycling, conservation, and pollution control can also be tied to lower production costs. Furthermore, a public image as an environmentally responsible company can be essential in gaining community acceptance, attracting top employees, and securing the trust of investors. This "corporate environmentalism," as it has been termed by Edgar S. Woolard, Jr., the CEO of Du Pont, when coupled with the managerial skills and productive capacity commanded by business, appropriately places corporations in a position of leadership in moving toward sustainable use of earth's resources (*Business Week,* 1990).

1.1   What Is Environmental Assessment?

In order to incorporate environmental considerations into a decision or a decision-making process, it is necessary to develop a complete understanding of the possible and probable consequences of a proposed action. However, prior to this development, a clear definition of the environment must be constructed.

The word *environment* means many different things to different people. To some, the word conjures up thoughts of woodland scenes with fresh, clean air and pristine waters. To others, it means a pleasant suburban neighborhood or a quiet campus. Still others relate environment to ecology and think of plant-animal interrelationships, food chains, threatened species, and other recently recognized issues.

Actually, the environment is a combination of all these concepts plus many, many more. It includes not only the areas of air, water, plants, and animals, but also other natural and human-modified features which constitute the totality of our surroundings. Beauty, as well as environmental values, is very much in the "eye of the beholder." Thus, transportation systems, land-use characteristics, community structure, and economic stability all have one thing in common with carbon monoxide levels, dissolved solids in water, and natural land vegetation—they are all characteristics of the environment. In other words, the environment is made up of a combination of our natural and physical surroundings and the relationship of people with these surroundings. It must also include aesthetic, historic, cultural, economic, and social aspects. Thus, in environmental assessment, all these elements should be considered. The ultimate selection of what is "really important" in any one case is very much an art, or at least a refined judgment. Approaches which firmly lay down rules in this area will prove to be too rigid and inflexible for regular use. We seek to develop a feeling for what ought to be emphasized, as well as pointing out ways in which each situation is different.

Environmental assessment implies the determination of the environmental consequences, or impact, of proposed projects or activities. In this context, *impact* means change—any change, positive or negative—from a desirability standpoint. An environmental assessment is, therefore, a study of the probable changes in the various socioeconomic and biophysical characteristics of the environment which may result from a proposed or impending action. Of course, some proposed actions will result in no change at all for one aspect or another of the environment. In these cases, the impact is really one of "no effect." Some proposed actions may also have no *change*, but the present status may be environmentally unacceptable at the start! The terms *environmental effects* and *consequences* are generally interchangeable with *impact*, especially since the latter has come to have solely negative connotations in many circles. Remember, of course, that *some* proposed projects and actions may well have many, or even mostly, positive effects in many sectors of the environment. One should never be afraid to discover them! Environmental assessment need not, in fact should not, *always* be an adversarial activity.

In order to perform the assessment, it is first necessary to develop a complete understanding, and clear definition, of the *proposed action*. What is to be done? Where? What kinds of materials, labor, and/or resources are involved? Are there different ways to accomplish the original purpose? Surprisingly, it is often very difficult to obtain a clear description of these factors, especially at early stages of planning. The project planners may not have a clear idea themselves, or may be unwilling to make the details known.

Second, it is necessary to gain a complete understanding of the *affected environment*. What is the nature of the biophysical and/or socioeconomic characteristics that may be changed by the action? How widely might some effects be felt? The boundary of the work site? A mile? The next state? All are possible.

Third, it is necessary to envision the implementation of the proposed action into that setting and to determine the *possible impacts* on the environmental characteristics, quantifying these changes whenever possible. An interdisciplinary analysis of these effects is encouraged, many say mandated, by current federal law.

Fourth, it is necessary to *report the results* of the study in a manner such that the analysis of probable environmental consequences of the proposed action may be used in the decision-making process. For federal government agencies, this process has been extensively codified. For other entities, the steps vary widely.

The exact procedures to be followed in the accomplishment of each environmental assessment are by no means simple or straightforward. This is due primarily to the fact that many and varied projects are proposed for equally numerous and varied environmental settings. Each combination results in a unique cause-condition-effect relationship, and each combination must be studied individually in order to accomplish a comprehensive analysis. For the project manager, selecting which aspects of a particular environment to emphasize, and which effects to elucidate, is a highly skilled decision-making process. It is potentially as difficult as developing the plan for the project itself. Generalized procedures for conducting an analysis in the manner indicated by the four steps outlined above [(1) define proposed action; (2) define affected environment; (3) determine possible impacts; and (4) report the results] have been developed. These procedures will be explained in subsequent chapters of this book.

1.2   Why Environmental Assessment Is Needed

The necessity for preparing an environmental assessment may vary with individual projects or proposed actions. For many actions, there is a legal basis for requiring such an analysis. Occasionally, Congress may require preparation of environmental documentation as a condition of passing legislation for a particular project, even though other law and regulations may not normally require it. For other types of projects, the environmental analysis may be undertaken simply for incorporation of environmental considerations into planning and design, recognizing the merit of such amenities on an economic, aesthetic, or otherwise desirable basis. Good professional practice may require this analysis even if law or regulation does not. The incorporation of environmental considerations in business practices is an extremely important aspect of environmental assessment.

In the United States, enactment of the National Environmental Policy Act (NEPA), on January 1, 1970, mandated that federal agencies assess the environmental impact of actions "which may have an impact on man's environment" [NEPA, Title I, Sec 102(2)(A)]. Other nations and states within the United States have enacted legislation patterned after NEPA requiring environmental assessment of major actions within their jurisdictions. Chapter 3 further discusses NEPA, and Chapter 4 describes the content and format of documents such as the Environmental Impact Statement (EIS) and the Environmental Assessment (EA).

1.3 Who Prepares Environmental Assessment Documents?

Within the federal government, the *responsible official* of the federal agency which is proposing the action is required to prepare environmental documents and is called the *proponent* of the action. The preparation of these documents, naturally, requires input by a multidisciplinary team of engineers and scientists representing disciplines related to the major potential environmental impacts. In fact, Section 102(2)(A) of NEPA requires that a "systematic and interdisciplinary approach" be used in preparing environmental documentation.

Many times, more than one federal agency is involved in a project due to

1. Sharing of project leadership

2. Joint funding of projects

3. Functional interdependence

In such a case, one federal agency needs to be designated as the "lead agency" and, consequently, the proponent of the project or action. Any other agencies are termed "cooperating agencies."

At times, private industry is undertaking major resource development projects (e.g., offshore oil exploration), and the federal agency is merely issuing a permit, license, lease, or other entitlement for use. The question becomes: "Who should prepare the required EA or EIS?"

In such a case, the federal agency issuing the permit or other entitlement normally relies on the applicant to submit much of the environmental information needed for documentation and analysis. The applicant may be required to submit an essentially complete study. The agency should at least assist the applicant by outlining the types of information required. It is permitted for the agency to prepare the EA or EIS itself, and some have done so. In all cases, the agency granting the permit must make an independent evaluation of the environmental issues involved and must take full responsibility for the scope and content of the environmental documentation actually prepared.

As a result of NEPA-mandated environmental assessment, a number of separate documents may be required at different phases of the effort. Some examples are: Notice of Intent; Scoping Summary; Environmental Assessment; Finding of No Significant Impact; Environmental Impact Statement; and Record of Decision. The place of each of these documents in the assessment process is described in Chapter 4.

Figure 1.1 provides a summary of all EISs filed between 1973 and 1999. In practice, there are many more filed documents than major proposed actions. Each action requires at least a draft and a final EIS, and many have one or more supplements in later years as well. Some draft EISs never result in an action. The 27-year total of documents filed thus may represent less than half as many "major actions." Figure 1.2 details the total EISs filed by selected agencies during the years 1992 to 1998.

1.4   Integrating Art and Science

Environmental assessment, in common with most other complex processes, has elements which represent rigorous scientific endeavor. Some examples might be the analysis of soil or water samples, or the design of a plan to acquire these samples. The selection of instrumentation to measure soil loss or air quality is equally complex, with numerous references, formulas, and guidelines from handbooks and rule books from regulatory agencies. These examples are related to a knowledge of the scientific principles involved. A skillful project manager will be knowledgeable about the basic principles of a dozen or more sciences, from civil engineering through biology, or will seek the advice of persons trained in these areas.

Just as skilled is the art of knowing that soil nutrients, water, air quality, lichen productivity, or aesthetic effects will be relevant and will require examination. This can be taught only to a degree. Through use of real-life examples, we hope to illustrate many ways in which judgment may be developed in this area. In this, the area of analysis which we have termed an art, there are few hard-and-fast rules. One must learn what has been proven desirable in practice, just as one must be aware of what has been considered inadequate. What *are* the elements of a good artistic composition? One may learn a few rules, but that, in itself, is insufficient to qualify one as an artist. We will present those rules, but one must rely on experience, both one's own and that gained through extensive reading in relevant areas. The suggested readings associated with each chapter, in addition to the specific references, are a good start in this direction.

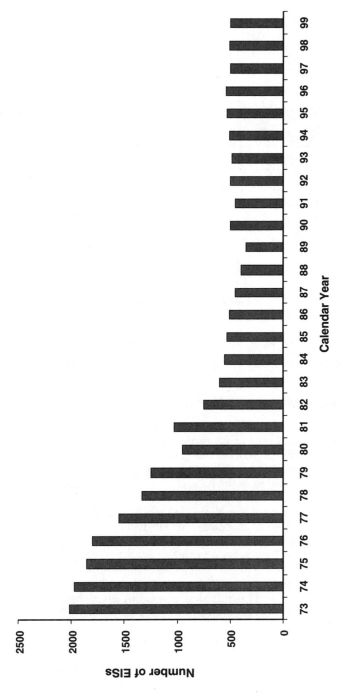

Information Source: http://ceq.eh.doe.gov/nepa/nepanet.htm

Figure 1.1 All EISs filed 1973–1999.

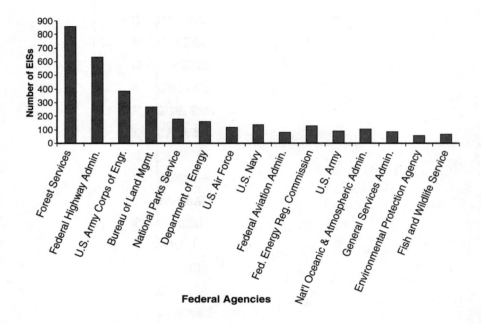

Information Source: http://ceq.eh.doe.gov/nepa/nepanet.htm

Figure 1.2 Total EISs filed by selected agencies for the years 1992–1998.

1.5   Discussion and Study Questions

1   Consider the history of the United States. In its first 200 years, what were the significant federal actions taken with respect to conservation and environmental preservation? Who were the individuals most responsible for these actions and what were their motives? What contemporary federal agencies resulted from some of these actions? How have the roles of these agencies changed with time?

2   Many believe that, historically, Native Americans had a model "environmental ethic" and that we should have patterned our behavior after theirs. Did they have such an ethic and, if so, how widely was it accepted and practiced? How does it differ from that generally practiced today?

3   Discuss the trade-offs between economic development and environmental concerns. How do factors such as inflation, economic conditions, political power, and international concerns affect our environmental "conscience"?

4   Define the term *environment*. Distinguish between (1) the natural and the built environment and (2) the biophysical and the socioeconomic environment. Describe how these environments may be affected by human activities. Are the effects always negative or positive? What kinds of trade-offs may become significant? Is it likely that all these types of considerations would enter into the decision-making process unless mandated by law?

5 How does *interdisciplinary* differ from *multidisciplinary*? Is it possible to thoroughly and adequately evaluate the environmental consequences without utilizing an interdisciplinary approach?

1.6 Further Readings

The following books and articles examine further many of the questions and issues raised in this introductory chapter. Several focus on the questions of maintaining economic competitiveness while considering the environment in business ventures.

Bernstam, Mikhail S. *The Wealth of Nations and the Environment*. London: Institute of Economic Affairs, 1991.

Blackburn, Anne M. *Pieces of the Global Puzzle: International Approaches to Environmental Concerns*. Golden, Colo.: Fulcrum, 1986.

Carter, F. W., and David Turnock. *Environmental Problems in Eastern Europe*. New York: Routledge, 1997.

Cole, Matthew A. "Limits to Growth, Sustainable Development and Environmental Kuznets Curves: An Examination of the Environmental Impact of Economic Development." *Sustainable Development*, 7:87–97, 1999.

Costi, Alterto. "Environmental Justice and Sustainable Development in Central and Eastern Europe." *European Environment*, 8:107–112, 1998.

Council on Environmental Quality. The 1997 report of the Council on Environmental Quality.

Council on Environmental Quality. Unpublished data—CEQ: All EISs filed 1973–1999. http://ceq.eh.doe.gov/nepa/nepanet.htm.

Council on Environmental Quality. Unpublished data—CEQ: Total EISs filed by year and selected agencies, 1992–1998. http://ceq.eh.doe.gov/nepa/nepanet.htm

Davis, John. *Greening Business: Managing for Sustainable Development*. London: Blackwell, 1991.

Dryzek, J. S. *Rational Ecology, Environmental and Political Economy*. Oxford, England: Blackwell, 1987.

Gilbreath, Kent, ed. *Business and the Environment: Toward Common Ground*. Washington, D.C.: The Conservation Foundation, 1984.

Gilpin, Alan. *Environmental Economics, A Critical Overview*. New York: Wiley, 2000.

Hallstrom, Lars K. "Industry Versus Ecology: Environment in the New Europe." *Futures*, 31:25–38, 1999.

Hoffman, Michael, Robert Frederick, and Edward S. Petry, Jr. *The Corporation, Ethics and the Environment*. New York: Quorum Books, 1990.

Kassiola, Joel Jay. *The Death of Industrial Civilization: The Limits to Economic Growth and the Repoliticization of Advanced Industrial Society*. Albany, N.Y.: State University of New York Press, 1990.

Khozin, Grigori. *Talking About the Future: Can We Develop Without Disaster?* Moscow: Progress Publishers, 1988.

Organization for Economic Cooperation and Development. *International Conference on Environment and Economics*. Paris, 1985.

Rao, P. K. *Sustainable Development*. Malden, Mass.: Blackwell, 2000.

Riddell, Robert. *Ecodevelopment: Economics, Ecology, and Development—An Alternative to Growth Imperative Models*. New York: St. Martin's Press, 1981.

Schramm, Gunter, and Jeremy J. Warford, eds. *Environmental Management and Economic Development*. Baltimore: Johns Hopkins University Press, 1989.

Silver, Cheryl Simon, with Ruth S. DeFries. *One Earth, One Future: Our Changing Global Environment.* Washington, D.C.: National Academy Press, 1990.

World Bank. *Striking a Balance: The Environmental Challenge of Development.* Washington, D.C.: World Bank, 1989.

World Commission on Environment and Development. *Our Common Future.* New York: Oxford University Press, 1987.

World Resources Institute. *Multinational Corporations, Environment, and the Third World: Business Matters.* Durham, N.C.: Duke University Press, 1987.

Environmental Laws and Regulations

Much of the environmental legislation in the United States was initiated at the federal government level. Some states have enacted environmental legislation to protect unique environments within their jurisdiction (e.g., coastal areas, wetlands, and cultural and historic sites). Environmental regulations, which form an action-forcing mechanism for implementing the intent of the enabling legislation, are then issued by the regulatory agencies of the government. With the emphasis on giving states the responsibilities for enforcing such regulations, increasingly states are issuing and are responsible for enforcing many of the environmental regulations.

Environmental legislation, and resulting regulations, is continually evolving. Consequently, information presented here is designed to provide a broad perspective on environmental legislation. Clearly, environmental regulations can have a profound effect on economic activity, and these effects should be included in assessment of the implementation of these regulations. To provide an understanding of the purpose and function of these requirements, topics covered in this chapter will be

- Rationale for environmental legislation and regulations
- Shortcomings of environmental legislation and regulations
- Legislative data systems
- An overview of federal environmental legislation

2.1   Rationale for Environmental Legislation and Regulations

The following discussion of the basis for promulgating environmental legislation and regulations focuses on the role of the market economy, the problem of the commons, and long-term viability of the environment. Since labor and capital are scarce resources, their consumption is minimized by industry. Since the environment is, or rather has been in the past, an essentially free resource, its consumption has typically been ignored. Consequently, there has been considerable environmental degradation, with attendant economic and social costs. Simply put, some economic and social costs are ignored in the ordinary marketplace exchange of goods and services. Also, one cannot ignore the third-party interests when looking at two-party transactions of the buyer and the seller (existence of externalities). This, in fact, is the case for many environmental control problems, and thus the transaction results in "market failure." Basically, market failure could result from high transaction costs, large uncertainty, high information costs, and existence of externalities (Schultz, 1977). In order to correct market failure, two choices exist. One can try to isolate the causes of the failure and restore, as nearly as possible, an efficient market process (process-oriented) or alternatively bypass the market process and promulgate regulations to achieve a certain degree of environmental protection (output-oriented).

Some environmental legislation and regulation is needed to protect the health and welfare of society, and market incentives alone will probably never work. For example, it would be very difficult to put a dollar value on discharge of toxic materials, such as polychlorinated biphenyls (PCBs) or mercury, to the environment. Another reason for environmental legislation and regulations is that long-term protection of the life-support systems is important for sustained economic development. Investment decisions can rarely be made to take into account long-term protection of the life-support systems which belong to everyone—a property ultimately leading to the problems of the commons.

Some projects involve exploitation of energy and other natural resources at an unprecedented rate. A question of temporal optimality of market allocations arises. In such cases, a market economy is unable to properly account for all long-term economic and social benefits and costs. As Solow has pointed out, "there are reasons to expect market interest rates to exceed the social interest rate of time preference..." (Solow, 1977). As a result, the market will tend to encourage consumption of exhaustible resources too fast. Consequently, a corrective public intervention—or regulations aimed at slowing down this consumption—needs to be structured. This can be accomplished through compulsory conservation, subsidies, or a system of graduated severance taxes (Solow, 1977).

2.2 Shortcomings of Environmental Legislation and Regulations

Many public administrators, engineers, planners, industrialists, and other decision makers recognize the need for environmental legislation and related regulations to protect the environment. They also recognize the importance of economic efficiency and utility. There are, indeed, a number of concerns regarding many environmental regulations. These concerns are shared by many who feel that environmental regulations can be structured so that they minimally affect the efficiency and productivity of the industry, minimally interfere with essential federal programs such as national defense, and still achieve reasonable environmental protection goals. Some of the concerns related to environmental regulations are

- Regulations seem to be structured in such a way that the costs are often excessive as compared to the benefits they generate.
- In general, the regulations are command-and-control type (i.e., they contain few or no economic incentives for compliance). Consequently, in a free-market economy, they are ineffective and do not preserve elements of voluntary choice.
- Regulations are ineffective because they lack properly structured incentives for achieving social goals.
- It is widely believed that command-and-control regulations generate inefficiencies, at both the micro- and macroeconomic levels.
- Some environmental regulations require unnecessary paperwork and cause unnecessary delays in completion schedules, which, in turn, create additional costs.
- Many regulations at different government levels, such as federal, state, and local, are duplicative and, at times, incompatible with each other; consequently, they create unnecessary work and inefficiencies.

2.3 Legislative Data Systems

The U.S. Congress is continually enacting new legislation and amendments to existing environmental legislation; similarly, the Office of the President periodically issues new Executive Orders regarding the environment. Federal agencies continuously modify environmental regulations pertaining to these laws and executive orders. Because of this, those interested in the current legislative climate must ensure that they are working under the current legal regime. The advent of Internet access and electronic data retrieval systems has greatly aided this process. Described here are some of the current Internet resources for existing environmental legislation which readers may want to use, depending on their specific needs. Because Internet access addresses and content often change, the information given below should be checked to see if it has been updated.

Federal agencies: http://www.firstgov.gov

This web site provides information about the federal government and its branches. It includes links to federal agencies, a list of interesting topics, and a search window.

Council on Environmental Quality: http://www.whitehouse.gov/ceq/

The Council on Environmental Quality (CEQ) is part of the Executive Office of the President, and so is included under the White House web site. The CEQ was established in 1970 under the National Environmental Policy Act (NEPA). The CEQ homepage gives information about the Council and includes a link to its "NEPAnet" site. The NEPAnet site, found at http://ceq.eh.doe.gov/nepa/nepanet.htm, gives the text of NEPA, the CEQ regulations, CEQ guidance, and recent CEQ documents, including the CEQ annual reports. One useful feature is a link to case law (interpretation of statues by courts) that helps to define specific aspects of NEPA. The site also provides links to federal agency NEPA web sites and points of contact. These sites provide information relevant to the environmental activities of the administration and allow users to access large volumes of information concerning NEPA and other environmental issues.

United States Environmental Protection Agency: http://www.epa.gov

The U.S. Environmental Protection Agency (EPA) web site offers a direct link to information on laws and regulations. The user can choose to search major environmental laws, current legislation in Congress, U.S. code, regulations and proposed rules, or Code of Federal Regulations. Each of these sites can be searched through a keyword search option and allows the user to download documents directly.

Federal Register and the Code of Federal Regulations: http://www.nara.gov/fedreg

The text of U.S. federal laws, regulations, and notices can be accessed and downloaded through the National Archives and Records Administration site, http://www.nara.gov. The *Federal Register* is a daily publication that provides notices of federal activities for all federal agencies, including notices about NEPA documents. The *Federal Register* web site also gives information on how to write and submit notices to the *Federal Register* for publication. The Code of Federal Regulations, updated annually, provides the text of all official regulations of all federal agencies. New regulations and updates to existing regulations are printed in the *Federal Register,* so both documents must be consulted to understand the current regulatory situation. The web site also gives access to public laws, Executive Orders, and other federal documents of interest. In addition, the *Federal Register,* United States Code, Code of Federal Regulations, and many other documents can be accessed through the Government Printing Office web site at http://www.access.gpo.gov. This web site also gives requirements for printing government documents such as environmental impact statements, and gives access to the Government Printing Office Style Guide and other documents of interest.

Advisory Council on Historic Preservation: http://www.achp/gov

The Advisory Council on Historic Preservation is part of the Executive Office of the President, and can also be accessed through www.whitehouse.gov. Of interest to the NEPA practitioner are the requirements for consultation on historic properties under Section 106 of the National Historic Preservation Act (16 USC 470f). The Section 106 regulations, Protection of Historic Properties (36 CFR Part 800), went into effect on January 11, 2001. The full text of the revised regulations and their preamble can be found at 65 F.R. 77698–77739, which is linked to this web site. The site also links to other information about Section 106 consultations.

U.S. Fish and Wildlife Service: http://fws.gov

The U.S. Fish and Wildlife Service of the Department of the Interior is responsible for the administration of several laws of interest to the NEPA preparer. Of particular note are the web site devoted to threatened and endangered species and the site regarding migratory birds. The Fish and Wildlife Service in the Department of the Interior and the National Marine Fisheries Service of the Department of Commerce share responsibility for administration of the Endangered Species Act. The two agencies sponsor the web site at http://endangered.fws.gov, which provides information and links to laws, regulations, notices, and species lists regarding endangered species management. The U.S. Fish and Wildlife Services, Division of Migratory Bird Management, web site can be found at http://migratorybirds.fws.gov. The site provides a link to Executive Order 13186, "Responsibilities of Federal Agencies to Protect Migratory Birds," January 11, 2001, which provides guidance to federal agencies regarding actions that may have an adverse effect on migratory birds. This Executive Order updates the requirements of the Migratory Bird Treaty Act of 1918, and provides that federal agencies are to consider habitat and conservation for migratory birds in agency plans and actions. The web site also provides links to other relevant laws and regulations, and species lists.

LEXIS®-NEXIS®R: http://web.lexis-nexis.com/universe/

In 1973, the LEXIS service became the first commercial, full-text legal information system, developed to aid legal professionals in researching the law. The addition of the NEXIS service, in 1979, provided additional references to recent and past news and financial information. Today, the LEXIS-NEXIS organization helps legal, business, and government professionals collect, manage, and use information more efficiently. Perhaps the most useful application of LEXIS-NEXIS for the reader is LEXIS-NEXIS Academic Universe. Associated with this service is a Legal Research option. The Legal Research preference allows the user to search documents under headings such as secondary literature, case law, codes and regulations, and patent research. LEXIS-NEXIS, which is an extremely powerful tool and offers many useful services to the user, is a proprietary tool.

Westlaw: http://www.westlaw.com

Westlaw is a research tool for both legal and business professionals. Several services are available through Westlaw, but perhaps those most useful to the reader include information on cases, statutes and administrative materials, public records and court dockets, law reviews and legal newsletters, practice-area treatises, and legal forms. Users are able to search for documents using numerous fields such as keyword, subject, and date. Documents can be downloaded directly from the site. This site is designed for use by legal personnel and requires a subscription; therefore, it may be better suited for use by corporations, government agencies, law firms, and other similar institutions.

**Federal Legal Information
Through Electronics:
http://www.fedworld.gov/supcourt**

The Federal Legal Information Through Electronics (FLITE) system is an information retrieval and analysis service that provides the text of U.S. Supreme Court decisions from 1937 through 1975. More recent Court decisions can be accessed through proprietary systems such as LEXIS-NEXIS and Westlaw, discussed above.

2.4   Overview of Federal Environmental Legislation

An overview of federal environmental legislation is provided in this section. State environmental legislation and regulations have been patterned after the federal programs. Information on the selected major federal environmental laws is organized under the headings of "Basic objective" and "Key provisions."

Clean Air Act (42 U.S.C. 7401 et seq.)

Basic objective.   The Clean Air Act of 1970, which amended the Air Quality Act of 1967, was established "to protect and enhance the quality of the Nation's air resources so as to promote public health and welfare and the productive capacity of its population." Since 1970, the basic act has been significantly amended to reflect national concern over air quality. Support for cleaner air has come from both environmentalists and the general public, although legislation has been politically controversial because of its impact on industry and economic growth.

The major provisions of the act are intended to set a goal for cleaner air by setting national primary and secondary ambient air quality standards. Primary standards define levels of air quality necessary to protect public health, while secondary standards define levels necessary to protect the public welfare from any known or anticipated adverse effects of a pollutant.

The basic objectives of the Clean Air Act Amendments of 1977 were to define issues related to significant deterioration and nonattainment areas, to implement a concept of emission offset, to encourage usage of innovative control technologies, to prevent industries from benefiting economically from noncompliance with air pollution control requirements, to state that using tall stacks to disperse air pollutants is not considered a permanent solution to the air pollution problem, to state that federal facilities must comply with both procedural and substantive state pollution control requirements, and to establish guidelines for future EPA standard setting in a number of areas.

The 1990 Clean Air Act Amendments (CA 90) represent another major effort by the U.S. Congress to address many complex and controversial issues related to clean air legislation. CA 90 is expected to have profound and far-reaching effects on federal facilities and industry. One indication of the magnitude of the efforts commanded by these amendments is their estimated cost. Expenditures to meet these requirements are projected to reach $100 billion per year, with an annual compliance cost of over $30 billion per year.

Basic objectives of CA 90 are to overhaul the nonattainment provisions, to create an elaborate technology-based control program for toxic air pollutants, to address acid precipitation and the power plant emissions, to mandate the phase-out of chlorofluorocarbons (CFCs), and to greatly strengthen enforcement powers of regulatory agencies.

Key provisions.   Key provisions of the seven titles of the act are summarized as follows:

Title I: Attainment and Maintenance of the National Ambient Air Quality Standards. This title describes air pollution control requirements for geographic areas in the United States which have failed to meet the National Ambient Air Quality Standards (NAAQS). These areas are known as nonattainment areas. Ozone is currently the most pervasive nonattainment pollutant in the United States, and this title is directed at controlling the pollutants (volatile organic compounds and nitrogen oxides) which contribute to ground-level ozone formation. Title VI of this act discusses stratospheric ozone issues.

Title II: Mobile Sources. This title deals with revised tailpipe emission standards for motor vehicles. Requirements under this title compel automobile manufacturers to improve design standards to limit carbon monoxide, hydrocarbon, and nitrogen oxide emissions. Manufacturers must also investigate the feasibility of controlling refueling emissions. For the worst ozone and carbon monoxide nonattainment areas, reformulated and oxygenated gasolines will be required.

Title III: Hazardous Air Pollutants. This title deals with control of hazardous air pollutant emissions and contingency planning for accidental release of these pollutants. Requirements of this title are, perhaps, the most costly aspects of CA 90.

Title IV: Acid Deposition Control. The amendments establish a totally new control scheme for addressing the acid rain problem. The exclusive focus is on power plant emissions of sulfur dioxide and nitrogen oxide. Sulfur dioxide emissions are to be reduced by approximately 10 million tons annually in two phases—the first to take effect in 1995, the second in 2000. It is important to note that these reductions are to be achieved through a new market-based system under which power plants are to be allocated "emissions allowances" that will require plants to reduce their emissions or acquire allowances from others to achieve compliance. The target for the reduction of nitrogen oxide is established at 2 million tons per year.

Title V: Permits. This title provides for the states to issue federally enforceable operating permits to applicable stationary sources. The permits are designed to enforce the ability of the federal EPA, state regulatory agencies, and private citizens to enforce the requirements of CA 90. These permits will also be used to clarify operating and control requirements for stationary sources.

Title VI: Stratospheric Ozone Protection. This title limits emissions of CFCs, halons, and other halogenic chemicals which contribute to the destruction of stratospheric ozone. Provisions of this title closely follow the control strategies recommended in June 1990 by the second meeting of parties to the Montreal protocol.

Title VII: Enforcement. Requirements of this title completely replace existing enforcement provisions in the Clean Air Act Amendments of 1977. New enforcement actions include higher maximum fines and terms of imprisonment. Seriousness of violations has been upgraded and liabilities are now targeted at senior management rather than on-site operators.

Enforcement responsibilities; federal-state relationship. A major provision of the Clean Air Act establishes the concept that the state accepts the primary issue. Under the enforcement responsibilities established in the act, the EPA sets certain federal minimum standards and procedures. The states must then pass their own regulatory programs based upon these minimum standards. State programs must be submitted to the EPA for approval before the state can accept enforcement responsibilities. In lieu of an approved state program, the federal program will be in force. State regulatory programs must address the issue of how to improve air quality in areas not meeting NAAQS and protecting areas that meet NAAQS from deterioration of air quality.

Since the 1990 amendments, states have been passing and reviewing their regulatory programs to reflect deadlines mandated by the act. The impact of these regulatory programs has been enormous. The EPA must review and approve or disapprove programs for 50 states, each of which must incorporate all the key provisions of the act into the program.

Accomplishments and impacts. Although significant strides have been made in improving air quality since the Clean Air Act was originally passed in 1970, the nation's concern with air pollution and its impact is still evolving. Some politically unpopular control strategies in the area of land use regulations and transportation controls have been modifies or eliminated. Many statutory deadlines have been postponed. In order to provide for continued economic growth, and recognizing the energy needs of the nation, many air pollution control requirements continue to be modified.

Noise Control Act (42 U.S.C. 4901 et seq.)

Basic objective.   Noise pollution is one of the most pervasive environmental problems. A report to the President and Congress on noise indicates that between 80 and 100 million people are bothered by environmental noise on a daily basis, and approximately 40 million people are adversely affected (Report to the President, 1971).

Since noise is a by-product of human activity, the extent of exposure increases as a function of population growth, population density, mobility, and industrial activities. Acts such as NEPA also have an effect on noise control requirements and related land uses.

In congressional hearings regarding federal aviation noise policy (Federal Noise Policy, 1990), it was pointed out that aviation noise is a serious environmental problem for those who live near airports. The Federal Aviation Administration has authority to regulate aircraft noise emissions, and classifies aircraft into three categories based on their noise levels. Stage 1 aircraft, with the highest emissions, are planes manufactured in the 1960s and 1970s. The original 707 and DC-8 are examples. Stage 2 aircraft represent newer designs, such as the 737 and later models of the 727. Stage 3 aircraft are the newest designs, mostly of mid-1980s production, such as the MD-80 and 767, and are notably quieter than older designs.

Since 1988, operation of Stage 1 aircraft has been flatly prohibited at many urban airports, which has reduced the number of persons seriously affected by noise from an estimated 7 million in the mid-1970s to 3.2 million in 1990 (Federal Aviation Noise Policy, 1990). Stage 2 aircraft may continue to be operated, though their proportion in the fleet is decreasing through natural attrition, and all were expected to drop out of use after the year 2000. Many citizens' groups and airport authorities are requesting even faster phase-out of Stage 2 aircraft. The European Community prohibits the purchase of new Stage 2 aircraft, even as replacements, and plans to phase out their use well before 2000. The business and economic implications of this regulation of aircraft type are serious. The mix of Stage 2 and Stage 3 aircraft varies widely among airline companies, with some of the highest proportions of older aircraft being held by companies that are in relatively poor financial condition and may not be able to afford the purchase of new aircraft.

The Noise Control Act has four basic objectives:

1. New product noise emission standards directed principally at surface transportation and construction noise sources
2. The utilization of "in-use" controls directed principally at aviation, interstate motor carriers, and railroad noise sources
3. The labeling of products for protection against voluntary high-level individual exposure
4. The development of state and local programs to control noise

Key provisions.   The act mandates the EPA to promulgate standards for noise emissions from the following new products:

1. Portable air compressors
2. Medium- and heavy-duty trucks
3. Earth-moving machinery
4. Buses
5. Truck-mounted solid waste compactors
6. Motorcycles
7. Jackhammers
8. Lawn mowers

Additionally, the act specifies that the following sources will be regulated via performance standards:

1. Construction equipment
2. Transportation equipment (with the aid of the Department of Transportation)
3. Any motor or engine
4. Electrical or electronic equipment
5. Any other source which can feasibly be regulated

Section 7 of the act also amends the Federal Aviation Act and regulates aircraft noise and sonic booms. The Federal Aviation Administration (FAA) is given the authority to regulate such noise after consultation with and review by the EPA.

In 1978, the Noise Control Act was amended by the Quiet Communities Act. This amendment provided for greater involvement by state and local authorities in controlling noise. Its objectives are

1. The dissemination of information concerning noise pollution
2. The conducting or financing of research on noise pollution
3. The administration of the quiet communities program, which involves grants to local communities, the monitoring of noise emissions, studies on noise pollution, and the education and training of the public concerning the hazards of noise pollution
4. The development and implementation of a national noise environmental assessment program to
   a. Identify trends in noise exposure
   b. Set ambient levels of noise
   c. Set compliance data
   d. Assess the effectiveness of noise abatement
5. The establishment of regional technical assistance centers.

The EPA is further given the authority to certify a product as acceptable for low noise emission levels. These certified products are to be used by federal agencies in lieu of a like product that is not certified.

Enforcement responsibilities; federal-state relationship.   The EPA has enforcement responsibilities under the act, as indicated in the key provisions, and is mandated to promulgate noise emission standards. The FAA controls noise from aircraft and sonic booms.

The 1978 amendments (Quiet Communities Act) were an attempt to recognize that noise pollution is very often a local community problem and needs to be regulated at that level. Thus, many noise regulations are promulgated at the local level, with support from the state and national level in the form of grants and research results.

Accomplishments and impacts.   The effects of the law include:
1. The establishment of noise emission standards for
   a. Construction equipment
   b. Interstate motor vehicles (40 CFR, Part 202)
   c. Railroads (40 CFR, Part 201)
   d. Portable air compressors (40 CFR, Part 204)
   e. Aircraft noise and sonic booms
2. The establishment of labeling requirements for certain types of equipment (40 CFR, Part 211).
3. The establishment of the quiet communities program, which has encouraged more involvement by state and local agencies in the setting of more stringent noise levels and the enforcement of those levels.
4. The requirement for federal agencies to purchase equipment certified by the EPA as having low noise emissions in lieu of like products not having a certificate.

On a more fundamental level, the act has served to increase noise pollution awareness on the part of the public and has validated concerns over this often overlooked type of pollution. It has stimulated more and better research into the effects of noise on the quality of life and the health hazard aspects.

Safe Drinking Water Act (42 U.S.C. 300f et seq.)

Basic objective.   The primary objectives of the act are twofold: (1) to protect the nation's sources of drinking water and (2) to protect public health to the maximum extent possible, using proper water treatment techniques. The act establishes the need to set contaminant levels to protect public health. These levels were established in regulations issued pursuant to the act, which requires the EPA to develop regulations for the protection of underground sources of drinking water. Any underground injection of wastewater must be authorized by a permit. Such a permit will not be issued until the applicant can prove that such disposal will not affect drinking water sources. Finally, the act requires procedures for inspection, monitoring, record keeping, and reporting.

Key provisions.   Key provisions of the act can be summarized as

1. The establishment of national primary drinking water standards based upon maximum contaminant levels.
2. The establishment of treatment techniques to meet the standards.
3. The establishment of secondary drinking water standards.
4. The establishment of those contaminants for which standards are set, based on studies conducted by the National Academy of Sciences. The EPA shall request comments from the Science Advisory Board, established under the Environmental Research, Development, and Demonstration Act of 1978, prior to proposals on new or revised maximum contaminant levels.
5. The establishment of state management programs for enforcement responsibilities. States must submit regulatory programs to the EPA for approval. These programs must set primary and secondary drinking water standards which meet or better the national standards. They must also regulate by permit facilities which treat drinking water supplies.
6. The protection of underground sources of drinking water.
7. The establishment of procedures for development, implementation, and assessment of demonstration programs designed to protect

critical aquifer protection areas located within areas designated as sole or principal source aquifers.

8. The requirements for state programs to protect wellhead areas from contaminants which may have any adverse effects on public health.

9. Originally, the EPA was required to regulate 25 additional drinking water contaminants each year. The 1996 amendments changed this requirement and instead mandated that the EPA regulate the contaminants that pose the greatest risk and are most likely to occur in water systems.

10. The 1996 amendments created a fund that aids water systems. The fund provides assistance for infrastructure upgrades and source water protection programs.

Enforcement responsibilities; federal-state relationship. The passage of the Drinking Water Act in December 1974, and amendments passed through 1996, have broadened the EPA's authority and responsibility to regulate the quality of the nation's drinking water regulations, with the states having the major responsibility for enforcing these regulations.

States must submit drinking water programs to the EPA for approval. These programs must meet, at a minimum, the federal standards for drinking water quality. They must also include procedural aspects of inspection and monitoring, as well as control technology and emergency procedures for noncompliance to protect the public health. States are also given enforcement responsibilities for the control of underground sources of water supply. These responsibilities must include permitting procedures.

Accomplishments and impacts. There are more than 240,000 public water supply systems serving over 200 million people. Many of these systems are not using the most effective equipment and techniques to collect, treat, and deliver potable water to the public. According to the EPA (EPA, 1979), more than half of these systems are out of compliance because of

1. Inadequate treatment techniques

2. Inadequately trained operators

3. Poor system design

4. Inadequate monitoring procedures

The only variations from state to state are procedural, such as record keeping. Issues involving other legislation are also closely tied to safe drinking water; for example, the protection of the nation's waterways under the Clean Water Act affects the ultimate protection of the water supply for potable water. Similarly, the leaching of hazardous wastes into groundwater can affect underground water quality. Thus, the quality of sources of drinking water is closely tied by other major legislation to the control of pollution.

Clean Water Act (33 U.S.C. 1251 et seq.)

Basic objective. The Clean Water Act is the primary authority for the water pollution control programs. The objective of these programs is to "restore and maintain the chemical, physical, and biological integrity of the Nation's waters." It sets national goals to

1. Eliminate the discharge of pollutants into navigable waters by 1985

2. Set interim goals of water quality which will protect fish and wildlife and will provide for recreation by July 1, 1983

3. Prohibit the discharge of toxic pollutants in quantities that might adversely affect the environment

4. Construct publicly owned waste-treatment facilities with federal financial assistance

5. Establish waste-treatment management plans within each state

6. Establish the technology necessary to eliminate the discharge of pollutants

7. Develop and implement programs for the control of nonpoint sources of pollution to enable the goals of the act to be met.

The goals are to be achieved by a legislative program which includes permits under the National Pollutant Discharge Elimination System (NPDES). Effluent limitations imposed under the initial legislation required the existing sources of pollution to use the "best practicable" treatment technology by 1977 and the "best available" technology by 1983; amendments provided means for modification of compliance dates. It requires an independent set of effluent limitations for new sources.

Key provisions. Development of effluent standards and permit systems and state and local responsibilities are key provisions of the act.

Effluent standards for existing and new sources of water pollution are established. These are source-specific limitations. Also, the act lists categories of point sources for which the EPA must issue standards of performance for new sources. States must develop and submit to the EPA a procedure for applying and enforcing these standards.

The EPA may establish a list of toxic pollutants and establish effluent limitations based on the best available technology economically achievable for point sources designated by the EPA. The EPA has also issued pretreatment standards for toxic pollutants.

Anyone conducting an activity, including construction or operation of a facility, which may result in any discharge into navigable waters must first obtain a permit. Permit applications must include a certification that the discharge will meet applicable provisions of the act, under NPDES. Permits for a discharge into ocean water will be issued under separate guidelines from the EPA. The Corps of Engineers will issue permits for the discharge of dredged or fill material in ocean water, based on criteria established by the Corps.

The act makes provision for direct grants to states to help them in administering pollution control programs. It also provides grants to assist in the development and implementation of waste-treatment management programs, including the construction of waste-treatment facilities. The federal share of construction costs will be no more than 55 percent after October 1, 1984.

To be eligible for these grants, states must develop waste-treatment management plans that are based upon federally issued guidelines. These programs must be approved by the EPA and must include

1. Regulatory programs to assure that the treatment facilities will include applicable pretreatment requirements

2. The identification of sources of pollution and the process by which control will be achieved

3. A process to control sources of groundwater pollution

4. The control of pollution from dredged or fill material into navigable waters. This must meet Section 404 requirements of this act.

Waste-treatment management shall be on an areawide basis, providing for the control of pollution from all point and nonpoint sources. In addition, the states must develop implementation plans for EPA approval to meet minimum water quality standards established by the EPA.

Other provisions of the act state that federal facilities must comply with all federal, state, and local requirements for the abatement and control of pollution. Also, the act provides grants to conduct a national wetlands inventory.

In November 1990, the EPA issued regulations setting forth the NPDES permit application requirements for stormwater discharges associated with industrial activities, discharge from municipal storm sewer systems which serve urban areas of 250,000 population or greater, and discharges from municipal storm sewer systems serving populations between 100,000 and 250,000.

Enforcement responsibilities; federal-state relationship. Except for issuing permits for the discharge of dredged or fill material, the EPA has no enforcement responsibilities for the act. The Corps of Engineers has the responsibility of issuing permits for specific categories of activities involving discharge of dredged or fill materials if the discharge will cause only minimal adverse effects. Sites for the discharge of dredged or fill material shall be specified by EPA guidelines.

Like many other major environmental statutes, the Clean Water Act emphasizes eventual state primacy and enforcement responsibilities. When the state has plans for preserving or restoring water quality and the EPA has approved those programs, the state will then assume enforcement responsibilities. Both these programs are based upon a minimal federal regulatory involvement. The federal role is also one of providing grants to states for the implementation of these programs.

Accomplishments. The Clean Water Act is enforced through two major interrelated strategies—a statutory program for the improvement of water quality and a related program of federal grants for the construction and expansion of wastewater treatment works.

A national clean water goal, initially to have been achieved by 1983, first provided a statutory guideline for a legislative program intended to eliminate all pollution in national waters. Discharge permits were then required for all water effluent discharges into national waters, and these permits may not be granted unless the source of the discharge utilizes the effluent treatment technology required by the act. These discharge permits are granted and administered according to the NPDES, initially to be administered by the EPA but which may be transferred for administration to the states subject to their compliance with detailed criteria contained in the federal law.

Effluent limitations imposed under the act generally require that existing sources of pollution make use of the "best practicable" treatment technology by 1977 and the "best available" technology by 1983, and the statute also imposes an independent set of effluent limitations on new sources of water pollution. Discharges from wastewater treatment plants also require a discharge permit under the NPDES system.

Water quality standards established under the earlier water quality act are also continued. Standards must be established by a state if it has not done so previously. The EPA and the states must establish more stringent effluent limitations than those otherwise required by the act if needed to meet water quality standards. As in the Clean Air Act, the water quality and discharge permit requirements of the Clean

Water Act can be expected to have a major impact on land development patterns through the influence they exert on the location of water pollution sources.

As in the Clean Air Act, the water pollution control requirements are applied principally to point sources of pollution. Where the Clean Water Act differs from the Clean Air Act is in its specific statutory requirements for a water quality planning program that includes specific land development control authority.

As a result of amendments (Water Quality Act of 1987), the Clean Water Act also addresses problems caused by the diffusion of water from nonpoint areawide sources of pollution, such as stormwater runoff and water runoff from on-site construction activities. Controls over nonpoint sources are also required by the act. They are first required in the "regulatory program," which must be a part of the areawide waste treatment planning process. This program must include "procedures and methods," including "land use requirements," to control nonpoint pollution sources.

The dredge and fill program under Section 404 of the 1972 Water Pollution Control Act authorizes a permit program for dredge and fill activities in "waters of the United States" to be administered by the U.S. Army Corps of Engineers. Deliberate congressional selection of the language defining the jurisdiction of the Corps led to an expansion of the program to include coastal and freshwater wetlands as well as navigable waters. This extension of jurisdiction makes a federal dredge and fill permit necessary for residential and other development in wetlands areas. The Corps is authorized to issue permits for dredge and fill activities and disposal sites specified by the Corps under regulations jointly developed by it and the EPA. The required review covers analysis similar to the environmental assessments or environmental impact statements. The Corps is to consider the need for the permit, alternative locations and methods, beneficial and detrimental effects, and cumulative impacts.

The act also authorizes the EPA to veto dredge and fill permits issued by the Corps of Engineers if they have an "unacceptable adverse effect" on municipal water supplies or shellfish beds or on fishery, wildlife, or recreational areas. With the 1977 amendments, Congress preserved the broad jurisdiction of the dredge and fill program over all waters, but authorized a delegation to the states of the authority to issue permits for waters not classed as navigable and shifted control over nonpoint sources of pollution to Section 208. An amendment to Section 204 exempts from the dredge and fill permit requirement a series of earthmoving activities such as normal farming and construction sites, as well as any nonpoint sources subject to control under a state nonpoint sources control program approved under Section 208.

The Water Pollution Control Amendments of 1981 (33 U.S.C. 1251 et seq.) legalized oxidation ponds, lagoons and ditches, and trickling filters as the equivalent of secondary treatment if water is not adversely affected.

Under this act, the EPA administers programs that provide financial grants to local agencies for the planning of wastewater management facilities. The Corps of Engineers participates in the planning of wastewater facilities or systems as follows:

1. The Corps of Engineers may perform a single-purpose wastewater management study in response to a congressional resolution or an act of Congress.

2. The Corps of Engineers may engage in wastewater management planning as part of an urban study.

3. The Corps of Engineers may provide advisory assistance to local or state agencies engaged in areawide waste treatment planning at the request of such agency.

Water quality planning under Section 208, also referred to as "208 Planning," was initiated under this act. A substantial number of 208 plans were developed.

The prevailing trend in water pollution control regulation and research has been in the direction of technology-based rather than water-quality-based, causing some point-source pollution control projects to become excessively costly by providing treatment beyond the levels required by receiving waters. On the other hand, pressing problems like surface runoff, combined sewer overflows, operation and maintenance, and toxic and hazardous waste disposal remained unresolved. Many scientists and engineers recommend that the facilities to treat point-source pollutants should be developed in concert with measures that may be needed for control of nonpoint sources.

Resource Conservation and Recovery Act (42 U.S.C. 6901 et seq.)

Basic objective.   The Resource Conservation and Recovery Act (RCRA), as it exists now, is the culmination of a long series of pieces of legislation, dating back to the passage of the Solid Waste Disposal Act of 1965, which address the problem of waste disposal. It began with the attempt to control solid waste disposal and eventually evolved into an expression of the national concern with the safe and proper disposal of hazardous waste. Establishing alternatives to existing methods of land disposal and to conversion of solid wastes into energy are two important needs noted by the act.

RCRA gives the EPA broad authority to regulate the disposal of hazardous wastes; encourages the development of solid waste management plans and nonhazardous waste regulatory programs by states; prohibits open dumping of wastes; regulates underground storage tanks; and provides for a national research, development, and demonstration program for improved solid waste management and resource conservation techniques.

The control of hazardous wastes will be undertaken by identifying and tracking hazardous wastes as they are generated, ensuring that hazardous wastes are properly contained and transported, and regulating the storage, disposal, or treatment of hazardous wastes.

A major objective of the RCRA is to protect the environment and conserve resources through the development and implementation of solid waste management plans by the states. The act recognizes the need to develop and demonstrate waste management practices that not only are environmentally sound and economically viable but also conserve resources. The act requires the EPA to undertake a number of special studies on subjects such as resource recovery from glass and plastic waste and managing the disposal of sludge and tires. An Interagency Resource Conservation Committee has been established to report to the President and the Congress on the economic, social, and environmental consequences of present and alternative resource conservation and resource recovery techniques.

Key provisions. Some significant elements of the act follow. Hazardous wastes are identified by definition and publication. Four classes of definitions of hazardous waste have been identified—ignitability, reactivity, corrosivity, and toxicity. The chemicals that fall into these classes are regulated primarily because of the dangerous situations they can cause when landfilled with typical municipal refuse. Four lists, containing approximately 1000 distinct chemical compounds, have been published. (These lists are revised as new chemicals become available.) These lists include waste chemicals from nonspecific sources, by-products of specific industrial processes, and pure or off-specification commercial chemical products. These classes of chemicals are regulated primarily to protect groundwater from contamination by toxic products and by-products.

The act requires tracking of hazardous wastes from generation, to transportation, to storage, to disposal or treatment. Generators, transporters, and operators of facilities that dispose of solid wastes must comply with a system of record keeping, labeling, and manufacturing to ensure that all hazardous waste is designated only for authorized treatment, storage, or disposal facilities. The EPA must issue permits for these facilities, and they must comply with the standards issued by the EPA.

The states must develop hazardous waste management plans, which must be EPA-approved. These programs will regulate hazardous wastes in the states and will control the issuance of permits. If a state does not develop such a program, the EPA, based on the federal program, will do so.

Solid waste disposal sites are to be inventoried to determined compliance with the sanitary landfill regulations issued by the EPA. Open dumps are to be closed or upgraded within 5 years of the inventory. As with hazardous waste management, states must develop management plans to control the disposal of solid waste and to regulate disposal sites. The EPA has issued guidelines to assist states in developing their programs.

As of 1983, experience and a variety of studies dating back to the initial passage of the RCRA legislation found that an estimated 40 million metric tons of hazardous waste escaped control annually through loopholes in the legislative and regulatory framework. Subsequently, Congress was forced to reevaluate RCRA, and in doing so found that RCRA fell short of its legislative intent by failing to regulate a significant number of small-quantity generators, regulate waste oil, ensure environmentally sound operation of land disposal facilities, and realize the need to control the contamination of groundwater caused by leaking underground storage tanks.

Major amendments were enacted in 1984 in order to address the shortcomings of RCRA. Key provisions of the 1984 amendments include

- Notification of underground tank data and regulations for detection, prevention, and correction of releases
- Incorporation of small-quantity generators (which generate between 100 and 1000 kg of hazardous waste per month) into the regulatory scheme
- Restriction of land disposal of a variety of wastes unless the EPA determines that land disposal is safe from human health and environmental points of views
- Requirement of corrective action by treatment, storage, and disposal facilities for all releases of hazardous waste regardless of when the waste was placed in the unit
- Requirement that the EPA inspect government-owned facilities (which handle hazardous waste) annually, and other permitted hazardous waste facilities at least every other year
- Regulation of facilities which burn wastes and oils in boilers and industrial furnaces

Enforcement responsibilities; federal-state relationship. Subtitle C of the Solid Waste Disposal Act, as amended by RCRA of 1976, directs the EPA to promulgate regulations for the management of hazardous wastes.

The hazardous waste regulations, initially published in May 1980, control the treatment, storage, transport, and disposal of waste chemicals that may be hazardous if landfilled in the traditional way. These regulations (40 CFR 261–265) identify hazardous chemicals in two ways—by listing and by definition. A chemical substance that appears on any of the lists or meets any one of the definitions must be handled as a hazardous waste.

Like other environmental legislation, RCRA enforcement responsibilities for hazardous waste management will eventually be handled by each state, with federal approval. Each state must submit a program for the control of hazardous waste. These programs must be approved by the EPA before the state can accept enforcement responsibilities.

The state programs will pass through three phases before final approval will be given. The first phase is the interim phase, during which the federal program will be in effect. The states will then begin submitting their programs for the control of hazardous wastes. The second-phase programs will address permitting procedures. A final phase will provide federal guidance for design and operation of hazardous waste disposal facilities. Many states have chosen to allow the federal programs to suffice as the state program to avoid the expense of designing and enforcing the program.

It should also be noted that the Department of Transportation has enforcement responsibilities for the transportation of hazardous wastes and for the manifest system involved in transporting.

Accomplishments and impacts. The 1980 regulations for the control of hazardous wastes were a response to the national concern over hazardous waste disposal. States have begun to discover their own "Love Canals" and the impacts of unregulated disposal of hazardous wastes on their communities. While the "Superfund" legislation provides funds for the cleanup of such sites, RCRA attempts to avoid future Love Canals.

Comprehensive Environmental Response, Compensation and Liability Act (42 U.S.C. 9601 et seq.)

Basic objective. The Comprehensive Environmental Response Compensation and Liability Act (CERCLA), also known as "Superfund," has four objectives. These are

1. To give the enforcement agency the authority to respond to the releases of hazardous wastes (as defined in the Federal Water

Pollution Control Act, Clean Air Act, Toxic Substances Control Act, and Solid Waste Disposal Act, and by the administrator of the enforcement agency) from "inactive" hazardous waste sites which endanger public health and the environment

2. To establish a Hazardous Substance Superfund

3. To establish regulations controlling inactive hazardous waste sites

4. To provide liability for releases of hazardous wastes from such inactive sites

The act amends the Solid Waste Disposal Act. It provides for an inventory of inactive hazardous waste sites and for the appropriate action to protect the public from the dangers possible from such sites. It is a response to the concern for the dangers of negligent hazardous waste disposal practices.

Key provisions

1. The establishment of a Hazardous Substance Superfund based on fees from industry and federal appropriations to finance response actions.

2. The establishment of liability to recover costs of response from liable parties and to induce the cleanup of sites by responsible persons.

3. The determination of the number of inactive hazardous waste sites by conducting a national inventory. This inventory shall include coordination by the Agency for Toxic Substances and Disease Registry with the Public Health Service for the purpose of implementing the health-related authorities in the act.

4. The provision of the authority for the EPA to act when there is a release or threat of release of a pollutant from a site which may endanger public health. Such action may include "removal, remedy, and remedial action."

5. The revision, within 180 days of enactment of the act, of the National Contingency Plan for the Removal of Oil and Hazardous Substances (40 CFR, Part 300). This plan must include a section to establish procedures and standards for responding to releases of hazardous substances, pollutants, and contaminants and abatement actions necessary to offset imminent dangers.

CERCLA requires that federal agencies assess injury or damage to natural resources caused by spills of oil or hazardous substances; these requirements are called the Natural Resource Damage Assessment (NRDA) provisions of CERCLA. The Department of the Interior regulations (43 CFR 11) explain how to conduct damage assessments under

NRDA and calculate the monetary cost of restoring five types of natural resources—air, surface water, groundwater, biotic, and geologic—from this type of injury. Under the CERCLA National Contingency Plan regulations (40 CFR 300), the Departments of Agriculture, Commerce, Defense, Energy, and Interior, states, and Native American governments have specific trust responsibilities over natural resources and can claim injury in the event of resource damage.

Enforcement responsibilities; federal-state relationship. The EPA has responsibility for enforcement of CERCLA as it pertains to the inventory, liability, and response provisions. The EPA is also responsible for claims against the Hazardous Substance Superfund, which is administered by the President. The EPA is responsible for promulgating regulations to designate hazardous substances, reportable quantities, and procedures for response. The National Response Center, established by the Clean Water Act, is responsible for notifying the appropriate government agencies of any release.

The following Department of Transportation agencies also have responsibilities under the act:

1. U.S. Coast Guard—response to releases from vessels
2. Federal Aviation Administration—responses to releases from aircraft
3. Federal Highway Administration—responses to releases from motor carriers
4. Federal Railway Administration—responses to releases from rolling stock

States are encouraged by the act to participate in response actions. The act authorizes the EPA to enter into contracts or cooperative agreements with states to take response actions. The fund can be used to defray costs to the states. The EPA must first approve an agreement with the state, based on the commitment by the state to provide funding for remedial implementation. Before undertaking any remedial action as part of a response, the EPA must consult with the affected states.

Accomplishments and impacts. On July 16, 1982, the EPA published the final regulations pursuant to Section 105 of the act, revising the National Contingency Plan for Oil and Hazardous Substances under the Clean Water Act, reflecting new responsibilities and powers created by CERCLA. The plan established an effective response program. Because the act requires a national inventory of inactive hazardous waste sites, the intent is to identify potential danger areas and effect a cleanup or remedial actions to avoid or mitigate public health and environmental dangers. In studying a sampling of these sites, the House Committee on Interstate and Foreign Commerce (House Report No. 96-1016) found four dangerous characteristics common to all the sites. These characteristics are

1. Large quantities of hazardous wastes

2. Unsafe design of the sites and unsafe disposal practices

3. Substantial environmental danger from the wastes

4. The potential for major health problems for people living and working in the area of the sites.

The intent of the act is to eliminate the above problems by dealing with the vast quantities of hazardous and toxic wastes in unsafe disposal sites in the country. The immediate impact of the act has been the identification of the worst sites, where the environmental and health dangers are imminent. This priority list will be used to spend the money available in the Hazardous Waste Response Fund in the most effective way to eliminate the imminent dangers. The long-term impact of the act will be to clean up all the identified inactive sites and develop practices and procedures to prevent future hazards in such sites, whether active or inactive. Another accomplishment of the act is to establish liability for the cost of cleanup to discourage unsafe design and disposal practices.

Superfund Amendments and Reauthorization Act (42 U.S.C. 11001 et seq.)

Basic objective.   The Superfund Amendments and Reauthorization Act (SARA) revises and extends CERCLA (Superfund authorization). CERCLA is extended by the addition of new authorities known as the Emergency Planning and Community Right-to-Know Act of 1986 (also known as Title III of SARA). Title III of SARA provides for "emergency planning and preparedness, community right-to-know reporting, and toxic chemical release reporting." This act also establishes a special program within the Department of Defense for restoration of contaminated lands, somewhat similar to the Superfund under CERCLA.

Key provisions.   There are key provisions which apply when a hazardous substance is handled and when an actual release has occurred. Even before any emergency has arisen, certain information must be made available to state and local authorities and to the general public upon request. Facility owners and operators are obligated to provide information pertaining to any regulated substance present on the facility to the appropriate state or local authorities (Subtitle A). Three types of information are to be reported to the appropriate state and local authorities (Subtitle B):

1. Material safety data sheets (MSDSs), which are prepared by the manufacturer of any hazardous chemical and are retained by the facility owner or operator (or if confidentiality is a concern, a list of hazardous chemicals for which MSDSs are retained can be made available). These sheets contain general information on a hazardous chemical and provide an initial notice to the state and local authorities.

2. Emergency and hazardous chemical inventory forms, which are submitted annually to the state and local authorities. Tier I information includes the maximum amount of a hazardous chemical which may be present at any time during the reporting year, and the average daily amount present during the year prior to the reporting year. Also included is the "general location of hazardous chemicals in each category." This information is available to the general public upon request. Tier II information is reported only if requested by an emergency entity or fire department. This information provides a more detailed description of the chemicals, the average amounts handled, the precise location, storage procedures, and whether the information is to be made available to the general public (allowing for the protection of confidential information).

3. Toxic chemical release reporting, which releases general information about effluents and emissions of any "toxic chemicals."

In the event that a release of a hazardous substance does occur, a facility owner or operator must notify the authorities. This notification must identify the hazardous chemical involved; amounts released; time, duration, and environmental fate; and suggested action.

A multilayer emergency planning and response network on the state and local government levels is to be established (also providing a notification scheme in the event of a release).

Enforcement responsibilities; federal-state relationship.   Local emergency planning committees or an emergency response commission appointed by the governor of the state is responsible for the response scheme. The primary drafters of the local response plans are local committees, which are also responsible for initiating the response procedure in the event of an emergency. Each state commission will supervise the local activities.

Accomplishments and impacts. SARA legislation to promote emergency planning and to provide citizen information at the local level was a response to the 1984 disaster in Bhopal, India. A major intent is to reassure U.S. citizens that a similar tragedy will not occur in this country, and thus have a calming effect. The standardization of reporting and record keeping should produce long-term benefits and well-designed response plans. Whether a high-quality emergency response involvement can be maintained indefinitely at the local level remains a question.

Federal Insecticide, Fungicide, and Rodenticide Act (7 U.S.C. 136 et seq.)

Basic objective. The Federal Insecticide, Fungicide, and Rodenticide Act (FIFRA) is designed to regulate the use and safety of pesticide products within the United States (which is in excess of one billion pounds). The 1972 amendments (a major restructuring which established the contemporary regulatory structure) are intended to ensure that the environmental harm resulting from the use of pesticides does not outweigh the benefits.

Key provisions. Key provisions of FIFRA include

- The evaluation of risks posed by pesticides (requiring registration with the EPA)
- The classification and certification of pesticides by specific use (as a way to control exposure)
- The restriction of the use of pesticides which are harmful to the environment (or suspending or canceling the use of the pesticide)
- The enforcement of the above requirements through inspections, labeling, notices, and state regulation

Enforcement responsibilities; federal-state relationship. The EPA is allowed to establish regulations concerning registration, inspection, fines, and criminal penalties, and to stop the sales of pesticides. Primary enforcement responsibility, however, has been assumed by almost every state. Federal law only specifies that each state must have adequate law and enforcement procedures to assume primary authority.

As in the case of almost any federal law, FIFRA preempts state law to the extent that it addresses the pesticide problem. Thus, a state cannot adopt a law or regulation that counters a provision of FIFRA, but can be more stringent.

Accomplishments and impacts. While the volume of pesticides and related information is enormous, FIFRA has enabled the EPA to acquire much information for analysis of risk and environmental degradation that results from the use of pesticides. This information has been, and will continue to be, generally invaluable in such analysis. However, Congress continues to struggle with the balancing of benefits and detriments of the use of pesticides in its attempt to deal with the economic, scientific, and environmental issues that are involved in the regulation of pesticides.

Marine Protection, Research, and Sanctuaries Act (33 U.S.C. 1401 et seq.)

Basic objective. This act regulates the dumping of all types of materials into the ocean. It prevents, or severely restricts, the dumping of materials adversely affecting human welfare, the marine environment, ecological systems, or economic potentialities. It provides for a permitting process to control the ocean dumping of dredged material.

The act also establishes the marine sanctuaries program, which designates certain areas of the ocean waters as sanctuaries when such designation is necessary to preserve or restore these areas for their conservation, recreation, ecology, or aesthetic values. States are involved in the program through veto powers to prohibit a designation.

Key provisions. The EPA is responsible for issuing permits for the dumping of materials in ocean waters except for dredged material (regulated by the Corps of Engineers), radiological, chemical, and biological warfare agents, and high-level radioactive waste, for which no permits will be issued.

The EPA has established criteria for reviewing and evaluating permit applications (40 CFR, Subchapter H). These criteria shall consider

1. The need for the proposed dumping

2. The effect of such dumping on human health and welfare, including economic, aesthetic, and recreational values

3. The effect of such dumping on marine ecosystems

4. The persistence and permanence of the effects of the dumping

5. The effect of dumping particular volumes and concentrations of such materials

6. Locations and methods of disposal or recycling, including land-based alternatives

7. The effect on alternate uses of oceans such as scientific study, fishing, and other living resource exploitation

The Secretary of the Army is responsible for issuing permits for the transportation and disposal of dredged material in ocean waters. The

Secretary shall apply the same criteria for the issuance of permits as the EPA uses and will issue permits in consultation with the EPA. Permits issued by the EPA or the Corps of Engineers shall designate

1. The type of material authorized to be transported for dumping or to be dumped

2. The amount of material authorized to be transported for dumping or to be dumped

3. The location where such transport for dumping will be terminated or where such dumping will occur

4. The length of time for which the permits are valid

5. Any special provisions

The other major provision of the act is the establishment of the Marine Sanctuaries Program.

Enforcement responsibilities; federal-state relationship. The EPA has responsibility for issuing and administering permits for the dumping of all materials (except for dredged material) into ocean waters. The Corps of Engineers has responsibility for permits for the dumping of dredged or fill material in ocean waters. Each agency has issued regulations to control ocean dumping. The National Oceanic and Atmospheric Administration (NOAA) in the Department of Commerce is responsible for administering the Marine Sanctuaries Program and issuing regulations to implement it. The states in which a sanctuary is designated can stop the designation by certifying that the terms are unacceptable to the state.

Accomplishments and impacts. In January 1982, the Department of Commerce released the "Program Development Plan" for the national Marine Sanctuaries Program. In this program, emphasis is on the use of marine sanctuaries for both public and private concerns. This will be particularly evident in the exploitation of the areas for mineral resources. A greater participation by those states in which the sanctuaries are located is being fostered. This greater involvement on the part of affected states will also extend to the permitting process for the dumping of wastes and dredged material into ocean waters.

Toxic Substances Control Act (15 U.S.C. 2601 et seq.)

Basic objective. The Toxic Substances Control Act (TSCA) sets up the toxic substances program, which is administered by the EPA. If the EPA finds that a chemical substance may present an unreasonable risk to health or to the environment and that there are insufficient data to predict the effects of the substance, manufacturers may be required to conduct tests to evaluate the characteristics of the substance, such as persistence, acute toxicity, or carcinogenic effects. Also, the act establishes a committee to develop a priority list of chemical substances to be tested. The committee may list up to 50 chemicals which must be tested within 1 year. However, the EPA may require testing for chemicals not on the priority list.

Manufacturers must notify the EPA of the intention to manufacture a new chemical substance. The EPA may then determine if the data available are inadequate to assess the health and environmental effects of the new chemical. If the data are determined to be inadequate, the EPA will require testing. Most importantly, the EPA may prohibit the manufacture, sale, use, or disposal of a new or existing chemical substance if it finds the chemical presents an unreasonable risk to health or the environment. The EPA can also limit the amount of the chemical that can be manufactured and used and the manner in which the chemical can be used.

The act also regulates the labeling and disposal of polychlorinated biphenyls (PCBs) and prohibits their production and distribution after July 1979.

In 1986, Title II, "Asbestos Hazard Emergency Response," was added to address issues of inspection and removal of asbestos products in public schools and to study the extent of (and response to) the public health danger posed by asbestos in public and commercial buildings.

Key provisions.   Testing is required on chemical substances meeting certain criteria to develop data with respect to the health and environmental effects for which there are insufficient data relevant to the determination that the chemical substance does or does not present an unreasonable risk of injury to health or the environment.

Testing shall include identification of the chemical and standards for test data. Testing is required from the following:

1. Manufacturers of a chemical meeting certain criteria

2. Processors of a chemical meeting certain criteria

3. Distributors or persons involved in disposal of chemicals meeting certain criteria

Test data required by the act must be submitted to the EPA, identifying the chemical, listing the uses or intended uses, and listing the information required by the applicable standards for the development of test data.

The EPA will establish a priority list of chemical substances for regulation. Priority is given to substances known to cause or contribute to cancer, gene mutations, or birth defects. The list is revised and updated as needed.

A new chemical may not be manufactured without notifying the EPA at least 90 days before manufacturing begins. The notification must include test data showing that the manufacture, processing, use, and disposal of the chemical will not present an unreasonable risk of injury to health or the environment. Chemical manufacturers must keep records for submission to the EPA as required. The EPA will use these reports to compile an inventory of chemical substances manufactured or processed in the United States.

The EPA can prohibit the manufacture of a chemical found to present an unreasonable risk of injury to health or the environment or otherwise restrict a chemical. The act also regulates the disposal and use and prohibits the future manufacture of PCBs, and requires the EPA to engage, through various means, in research, development, collection, dissemination, and utilization of data relevant to chemical substances.

Enforcement responsibilities; federal-state relationship. The EPA has enforcement responsibilities for the act, but the act makes provision for consultation with other federal agencies involved in health and environmental issues, such as the Occupational Safety and Health Administration and the Department of Health and Human Services. Initially, the states could receive EPA grants to aid them in establishing programs at the state level to prevent or eliminate unreasonable risks to health or the environment related to chemical substances.

Accomplishments and impacts. TSCA has provided a framework for establishing that chemical manufacturers take responsibility for the testing of chemical substances as related to their health and environmental effects. It places the burden of proof on the manufacturer to establish the safety of a chemical, yet still gives the EPA the final authority to prohibit or severely restrict chemicals in commerce. Thus, it is an attempt at the introduction of a chemical to prevent significant health and environmental problems that may surface later on. The fact that, when this legislation was initially passed, PCB effects were such an issue because of their widespread and uncontrolled use is reflective of public concerns over the number of other possible chemicals commonly used which could be carcinogenic. Public concern was so visible that an immediate need was perceived to regulate PCBs. Thus, PCBs are controlled as specifically prohibited by TSCA rather than RCRA.

National Environmental Policy Act (42 U.S.C. 4341 et seq.)

NEPA is considered the cornerstone of environmental legislation in that it establishes a national policy regarding protection of the environment. The complete text of this legislation is presented in Appendix A. Basic objectives and key provisions of the act are well defined in the language of this legislation. The CEQ has the main responsibility for overseeing federal efforts to comply with NEPA. In 1978, the CEQ issued regulations to comply with the procedural provisions of NEPA (40 CFR 1500–1508, which appears in Appendix D). Other provisions of NEPA apply to major federal actions significantly affecting the quality of the human environment.

This act requires federal agencies to assess the environmental impact of implementing their major programs and actions early in the planning process. For those projects or actions which either are expected to have a significant effect on the quality of the human environment or are expected to be controversial on environmental grounds, the proponent agency is required to file a formal environmental impact statement (EIS).

Enforcement responsibilities; federal-state relationship. The CEQ has responsibility for overseeing federal efforts to comply with NEPA. Each federal agency has the responsibility to comply with NEPA, and most agencies have developed agency-specific regulations, guidelines, or requirements for complying with NEPA. Some states have enacted state laws similar to NEPA. Occasionally, an action with both a federal and state component may fall under both laws.

Accomplishments and impacts. This act has added a new dimension to the planning and decision-making process of federal agencies in the United States. This act requires federal agencies to assess the environmental impact of implementing their major programs and actions early in the planning process. Other accomplishments and impacts of NEPA are

1. It has provided a systematic means of dealing with environmental concerns and including environmental costs in the decision-making process.

2. It has opened governmental activities and projects to public scrutiny and public participation.

3. Some projects have been delayed because of the time required to comply with the NEPA requirements.

4. Many projects have been modified or abandoned to balance environmental costs with other benefits.

5. It has served to accomplish the four purposes of the act as stated in its text.

National Historic Preservation Act (16 U.S.C. 470–470t)

Basic objective. The act, first passed in 1966 and amended several times since, declares a national policy of preserving, restoring, and maintaining cultural resources—broadly defined as historic, tribal, or archaeological properties (King, 1998). The President's Advisory Council on Historic Preservation is given responsibility under the act to implement this national policy. The law authorizes the Secretary of the Interior to maintain a National Register of Historic Places; amendments to the act in the 1970s gave the National Park Service, U.S. Department of the Interior, the responsibility for determining the eligibility of sites for inclusion on the National Register. Federal agencies cannot undertake projects that would affect properties listed, or eligible for listing, on the Register without considering the effect on those properties. Under Section 106 of the National Historic Preservation Act, federal agencies must consult with the Advisory Council or the State Historic Preservation Officer if a project will affect, or is likely to affect, either a listed site or an eligible site. The section's latest 106 regulations, Protection of Historic Properties (36 CFR Part 800), effective January 11, 2001, provide specific requirements for the consultation process. (In addition to the National Historic Preservation Act, there are many other laws related to cultural resource protection and preservation that must be considered during a NEPA review; see Chapter 13.)

Key provisions.   In summary, major provisions of the act are

1. Regulations for determination of eligibility for the National Register of Historic Places.

2. A federal agency must take into account the effect of a project on any property included in or eligible for inclusion on the National Register.

3. The Advisory Council must be given an opportunity to comment on a federal project.

4. Federal agencies must inventory all property and nominate any eligible properties to the National Register.

5. Federal agencies must provide for the maintenance of federally owned registered sites.

6. Agencies must coordinate projects with the state historic preservation officer of the state in which the project is located.

7. States can qualify for federal grants for the protection, restoration, and maintenance on properties on the National Register.

Enforcement responsibilities; federal-state relationship. Enforcement responsibilities involve a triad of agencies. The Advisory Council on Historic Preservation is given the ultimate authority to comment on a federal project that may affect a property on or eligible for inclusion on the National Register. The National Park Service has responsibility for making determinations on eligibility. The state historic preservation officer has the final responsibility for protecting and maintaining eligible properties.

Accomplishments and impacts. The greatest impact of the act has been the inclusion of cultural concerns in the environmental area. Federal agencies are including cultural assessments as part of the environmental assessment process. The act has served to highlight the national concern to preserve its cultural heritage in the form of the protection of historic sites and properties.

The major accomplishment has been the publication of a list of protected sites on the National Register and the provision of funds to restore and maintain those sites for future generations. Many new projects in urban areas proposed to be located in a historic district may be opposed by the community on the grounds of their adverse effects in terms of character, scale, or style of the historic district.

Wild and Scenic Rivers Act (16 U.S.C. 1271 et seq.)

Basic objective. This act establishes the Wild and Scenic River System. It protects rivers designated for their wild and scenic values from activities which may adversely affect those values. It provides for a mechanism to determine if a river can meet certain eligibility requirements for protection as a wild and/or scenic river.

Key provisions. In planning for the use and development of water and land resources, federal agencies must give consideration to potential wild and scenic river areas. This potential must be discussed in all river basin and project plans submitted to Congress. No federal agency is allowed to assist in any way in the construction of a water resources project having a direct and adverse effect on the values of a river designated as part of the Wild and Scenic River System.

Likewise, no agency is allowed to recommend authorization or request appropriations to begin construction of a project on a designated river without informing the administering secretary (Secretary of the U.S. Department of the Interior or Agriculture) in writing, 60 days in advance, and without specifically reporting to Congress on how construction would conflict with the act and affect values of the river being protected by the act.

No agency is permitted to recommend authorization of, or request appropriations to initiate, construction of a project on or directly affecting a river designated for potential addition to the system during the full 3 fiscal years after the designation, plus 3 more years for congressional consideration, unless the Secretary of the Interior or Agriculture advises against including the segment in the system in a report that lies before Congress for 180 in-session days. The comparable time limit for state-promised additions is 1 year.

Agencies must inform the secretary of any proceedings, studies, or other activities which would affect a river that is designated as a potential addition to the system. Agencies having jurisdiction over lands which include, border upon, or are adjacent to any river within or under consideration for the system shall protect the river with management policies and plans for the lands as necessary.

Enforcement responsibilities; federal-state relationship. The Department of the Interior has ultimate authority for administering the program, but the states can designate rivers for inclusion in the system. The Department of Agriculture administers and designates rivers in national forests.

Accomplishments and impacts. As of July 1996, 160 rivers or river segments had been designated wild, scenic, or recreational, as part of the act. The act has attempted to preserve designated rivers and their values from adverse impacts.

Coastal Zone Management Act (16 U.S.C. 141 et seq.)

Basic objective. The act was passed in response to the public's concern for balanced preservation and development activities in coastal areas. It was designed to help states manage these competing demands and provided funding to states participating in the federal program.

The legislation emphasized the state leadership in the program, and allowed states to participate in the federal program by submitting their own coastal zone management proposals. The purpose of these state programs, which are federally approved, is to increase protection of coastal areas while better managing development and government activities at all levels.

The act established the Office of Coastal Zone Management (OCZM) in the NOAA. Once the OCZM has approved a state program, federal agency activities within a coastal zone must be consistent "to the maximum extent practicable" with the program.

Key provisions. Federal agencies must assess whether their activities will directly affect the coastal zone of a state having an approved program.

The 1980 amendments included, as part of coastal areas, wetlands, flood plains, estuaries, beaches, dunes, barrier islands, coral reefs, and fish and wildlife and their habitats. The act also provides public access to the coast for recreational purposes.

States are encouraged to prepare special area management plans addressing such issues as natural resources, coastal-dependent economic growth, and protection of life and property in hazardous areas. Federal grants are available to the states to cover 80 percent of the costs of administering their federally approved coastal zone management programs. They may use 30 percent of their grants to implement the 1980 amendment provisions.

The states are also encouraged to inventory coastal resources, designate those of "national significance," and establish standards to protect those so designated.

Enforcement responsibilities; federal-state relationships. The act is administered by the OCZM as part of the NOAA. However, the underlying objective of the act is to involve agencies at the state and local levels in the administering process. While the act does not require states to submit a coastal zone management program for approval, it does provide two major incentives for states to join the federal program. One incentive is financial assistance to administer the program, and the other is that any federal activity in a coastal zone must include the consistency determination process, which involves consultation with the state.

Accomplishments and impacts. Because the consistency determination is a major factor or incentive in encouraging states to participate in the coastal zone management program, it is imperative to clearly define when such a consistency determination is required. The act states that this determination is necessary when a federal activity will "directly affect" the coastal zone. Since 1979, the NOAA has been attempting to define "directly affecting." The latest attempt, in January 1982, was withdrawn in May 1982. Thus there is not a clear definition of this term.

The central issue is whether off-coast survey (OCS) activities by the Department of the Interior are subject to consistency determinations. The recent extensive off-shore tracts opened for lease by the Secretary of the Interior serve to highlight the conflict between the federal government and affected states. At the present time, the NOAA is awaiting the outcome of litigation involving OCS activities before attempting a further definition of "directly affecting," although at the appeals court level, the court ruled in favor of including a specific lease in the consistency determination process. Thus, the two major incentives for encouraging states to participate in the program are currently in jeopardy.

Endangered Species Act (16 U.S.C. 1531–1542)

Basic objective. The Fish and Wildlife Service of the Department of the Interior and the National Marine Fisheries Service of the Department of Commerce share responsibility for administration of the Endangered Species Act. This act seeks to conserve endangered and threatened species. It directs the Fish and Wildlife Service to promulgate a list of endangered and threatened species and designate critical habitat for those species. Amendments also created the Endangered Species Committee to grant exemptions to the act.

Federal agencies must carry out programs for the conservation of listed species and must take actions to ensure that projects they authorize, fund, or carry out are not likely to jeopardize the existence of the listed species or result in the destruction or modification of habitat declared to be critical.

The act divides procedures for those projects begun before and after November 10, 1978. For those not under construction before November 10, 1978, agencies must request the Fish and Wildlife Service to furnish information as to whether any species listed, or proposed to be listed, are in the area. If such species are present, a biological assessment must be completed by the proponent agency within 180 days.

If the biological assessment or other project information reveals that a listed species may be affected, the agency must consult with the Fish and Wildlife Service (or National Marine Fisheries Service). Consultation must be completed by the service within a 90-day period. The Department of the Interior shall provide the agency with an opinion as to how the action will affect the species or its critical habitat, and suggest reasonable alternatives. The agency may apply for an exemption to the act to the Endangered Species Committee.

Key provisions. Of major significance is the promulgation of a list of species which have been found to be either threatened or endangered and the protection of species on the list from activities which may affect their continued protection and survival. Also, the act provides for the designation of habitat to be protected from activities which may harm the delicate ecological balance necessary for the existence of a listed species.

Federal agencies are required to perform a biological assessment before undertaking a project to determine the impact of a project on a listed species or its habitat. If that impact is negative, the agency must undertake mitigation procedures or the project must be halted. An important provision of the act is the establishment of an Endangered Species Committee to grant exemptions from the act.

A federal agency must consult with the Fish and Wildlife Service if the results of the biological assessment show a listed species may be affected by a project. The Fish and Wildlife Service will suggest alternatives to the agency.

A process is established whereby a species can be determined to be threatened or endangered, and thus eligible for the list, or can't be removed from the list.

Enforcement responsibilities; federal-state relationship. The Fish and Wildlife Service of the Department of the Interior has enforcement responsibilities under the act and must ultimately decide on all biological assessments and mitigation procedures. While states can compile their own lists of species and the degrees of protection required, species on the federal list are under the jurisdiction and protection of the federal government, and a violation of the act caries federal penalties.

Accomplishments and impacts. The Endangered Species Act has served to stop the rapid rate of extinction of many species. Perhaps the greatest success has been with the bald eagle, which is making a successful return, largely due to its protection under the act. Perhaps the most visible of its impacts was the halting of a major water project, the Tellico Dam, in the 1970s due to its impact on a listed species. The result of that action and the result of the Supreme Court decision was the 1978 amendment establishing the Endangered Species Committee, which can grant exemptions from the act.

For many of the species listed, it is too late to prevent ultimate extinction, but for others, such as the bald eagle, the grizzly bear, and the alligator, the act has protected the species and its habitat to allow for its survival.

Fish and Wildlife Coordination Act (16 U.S.C. 661 et seq.)

Basic objective. This act provides that wildlife conservation be given equal consideration and be coordinated with other aspects of water resource development programs. It establishes the need to coordinate activities of federal, state, and private agencies in the development, protection, and stocking of wildlife resources and their habitat. Also, the act provides procedures for consultation between agencies with the purpose of preventing loss of and damage to wildlife resources from any water resource–related project. Any such consultation shall include the Fish and Wildlife Service, the head of the agency having administrative control of state wildlife resources, and the agency conducting the project.

Key provisions. The act requires officers of the agency conducting the project to give full consideration to Fish and Wildlife Service recommendations or recommendations of the state agency. "Full consideration" includes mitigation measures.

Any report recommending authorization of a new project must contain an estimate of wildlife benefits and losses and the costs and amount of reimbursement. Adequate provision must be made for the use of project lands and water for the conservation, maintenance, and management of wildlife resources, including their development and improvement.

Lands to be measured by a state for the conservation of wildlife must be managed in accordance with a plan which must be jointly approved by the federal agency exercising primary administrative responsibility, the Secretary of the Interior, and the administering state agency.

In addition to this law, the federal government has passed dozens of other laws pertaining to fish, birds, and other animals. For example, the Migratory Bird Treaty Act of 1918 (16 U.S.C. 703) and its implementing Executive Order 13186, "Responsibilities of Federal Agencies to Protect Migratory Birds," January 11, 2001, provide guidance for the requirement that federal agencies consider migratory bird habitat and conservation in agency plans and actions, such as those considered under NEPA.

Pollution Prevention Act of 1990 (42 U.S.C. 13101 et seq.)

Basic objective. Traditionally, environmental legislation in the United States has focused on an end-of-pipe-control approach for minimizing discharge of pollutants to the environment. By using this approach, considerable progress has been made in reducing the total discharge of pollutants to the environment. However, this often has resulted in transferring pollutants from one medium to another and in many cases is not cost effective. The basic objective of the Pollution Prevention Act is to establish a national policy of preventing or reducing pollution at the source wherever feasible, and it directs the federal EPA to undertake certain steps in that regard. Prior to this act, RCRA

Hazardous and Solid Waste Amendments of 1984 had established a program of waste minimization. This law has many provisions, including requiring large-quantity generators to certify on their waste manifests that they have a program in place to minimize the amount and toxicity of wastes generated to the extent economically feasible.

Key provisions.   The Pollution Prevention Act of 1990 established as national policy the following waste management hierarchy:

1. *Prevention.*   The waste management priority is to prevent or reduce pollution at the source whenever feasible.
2. *Recycling.*   Where pollution cannot be prevented, it should be recycled in an environmentally safe manner whenever feasible.
3. *Treatment.*   In the absence of feasible prevention and recycling, pollution should be treated to applicable standards prior to release or transfers.
4. *Disposal.*   Only as a last resort are wastes to be disposed of safely.

The Pollution Prevention Act further directed the EPA to

1. Establish a prevention office independent of the agency's single-medium program offices (the EPA added pollution prevention to the existing function of Assistant Administrator for Pesticides and Toxic Substances). Congress appropriated $8 million for each of the fiscal years 1991, 1992, and 1993 for the new office to fulfill the function delineated in the act.
2. Facilitate the adoption by business of source-reduction techniques by establishing a source-reduction clearinghouse and a state matching grants program. Congress further appropriated $58 million for each of the fiscal years 1991, 1992, and 1993 for state grants, with a 50 percent state match requirement.
3. Establish a training program on source-reduction opportunities for state and federal officials working in all agency program offices.
4. Identify opportunities to use federal procurement to encourage source reduction.
5. Establish an annual award program to recognize companies that operate outstanding or innovative source reduction programs.
6. Issue a biennial status report to Congress.
7. Require an annual toxic chemicals source reduction and recycling report for each owner or operator of a facility already required to file an annual toxic chemical release form under Section 313 of SARA.

The EPA is pursuing the integration of pollution prevention into all its programs and activities and has developed unique voluntary reduction programs with the public and private sectors. The EPA 33/50 Program, through voluntary enrollment and direct action by industry, sought to reduce the generation of high-priority wastes from a target group of 17 toxic chemicals by 50 percent by 1995, with an interim goal of 33 percent reduction by 1992, as measured against a 1988 baseline. The 33/50 Program achieved its goal in 1994.

The executive branch of the federal government has sought to apply pollution prevention requirements broadly throughout the government. Under Executive Order 13148, "Greening the Government Through Leadership in Environmental Management," April 21, 2000, federal agencies became responsible for integrating environmental accountability and more stringent pollution prevention considerations into their day-to-day decisions and long-term planning. The executive order is administered by the EPA, with certain responsibilities delegated to the CEQ.

Enforcement responsibilities; federal-state relationship. The EPA conducts a yearly audit of major users of toxic substances and producers of toxic wastes.

"The purpose of the audits is to determine:

1. whether there are better and less environmentally damaging ways to complete the task without use of toxic substances,

2. whether there are ways to minimize the production of toxic wastes,

3. whether there are ways to recycle the toxic substances,

4. and who is regulated" (Trudeau and Olexa, 1994).

Again, the federal government's statutes take precedence over state statutes. The act also pledges federal assistance to states (up to 50 percent) with pollution prevention programs under the act.

Accomplishments and impacts. "The primary purpose of the Pollution Prevention Act is to discourage the disposal of recyclable toxic substances" (Trudeau and Olexa, 1994). The act focuses on industry, government, and public attention on reducing the amount of pollution through cost-effective changes in production, operation, and raw materials use (EPA, 1997). "Opportunities for source reduction are often not realized because of existing regulations, and the industrial resources required for compliance, focus on treatment and disposal. Source reduction is fundamentally different and more desirable than waste management or pollution control" (EPA, 1997).

"Pollution prevention also includes other practices that increase efficiency in the use of energy, water or other natural resources, and protect our resource base through conservation. Practices include recycling, source reduction, and sustainable agriculture" (EPA, 1997).

2.5 Trends in Environmental Legislation and Regulations

The 1970s were a decade of extensive new federal legislation covering all spheres of environmental concerns. In the 1980s, emphasis was directed toward refining existing legislation and fine-tuning current regulatory and enforcement policies. During the 1990s, emphasis shifted toward balancing economic and environmental costs and toward pollution prevention. In the initial decades of the twenty-first century, the pace of new environmental legislation seems certain to slow considerably, barring a significant crisis or disaster that might spark new legislative initiatives.

New legislation

Concern for protecting the quality of groundwater resources, casually expressed in the Clean Water Act and more forcibly articulated by RCRA, is likely to be the focus of new environmental legislation in the near future. These resources supply all or a part of the drinking water for about one-half of our population. Furthermore, we are accustomed to withdrawing water from the ground and using it without extensive treatment, except perhaps for softening. Recently, it has become widely known that groundwater supplies are extremely vulnerable to permanent damage due to seepage from chemical waste disposal sites and other forms of contamination. Often pollutants are persistent trace organics that defy treatment with conventional technology at affordable costs. Experience and expertise developed in response to RCRA groundwater requirements will have to be expanded greatly to provide the degree of protection and capability for corrective action that is likely to be called for.

Another major initiative stemming from concern for the disposal of hazardous wastes will revolve around limitations on land disposal of such wastes. States have begun the processes that will likely ban disposal of certain kinds of wastes that can be shown to be able to be treated and handled by alternative methods. Federal initiatives in the form of an amendment to RCRA are likely to establish national limitations on land disposal of certain kinds of waste.

Another high-profile environmental issue is the protection of wetlands in the near term and possible restoration and creation of wetlands in the future. Key issues for forthcoming legislation will be changes in Section

404 of the Clean Water Act, designation of a single, lead federal agency, and delegation procedures for states with approved plans to have primary responsibility for planning and permitting wetland protection. For the most part, protection refers to actions to prevent destruction of wetlands. Unless human-made for wastewater treatment, wetlands in the United States are protected from pollution damage by the Clean Water Act.

In other areas, legislation and regulations will continue to evolve to address issues related to air pollution, such as global warming, ozone depletion, acid rain, and indoor air quality. Water supply and water pollution issues that will become important in the future include non-point (and stormwater) controls and effective use of water resources management practices (i.e., allocations for withdrawal and for waste assimilation). Medical and infectious wastes are newer public issues in the management of solid and hazardous wastes.

Nuclear waste management, always a controversial and emotional issue, is likely to create major environmental and economic problems for society. Regulations for effective nuclear waste management are likely to be made more stringent.

Balancing federal and nonfederal roles

The legislation of the 1970s and the implementing regulations were structured largely on the basis of a dominant federal role in environmental protection. This balance shifted in the 1980s as part of an overall change in federal government policy transferring much of the regulatory enforcement responsibilities to the states.

For a number of years, popular rhetoric of state and local agencies expressed a desire for more say in environmental affairs. Along with reduced federal direction, fewer federal dollars are being earmarked for the federal share in implementing environmental protection programs. In fact, it is the desire to lower federal expenditures that is driving a decreasing federal involvement in environmental programs and not a basic philosophical shift in how government affairs can best be conducted on behalf of the populace.

Reduced federal financial support, however, is not part of the package previously espoused by state and local politicians. Several states have voiced objections to having to assume the burden of administration and enforcement of certain environmental programs if federal financial support drops below a certain threshold level.

Balancing economic and environmental costs

The common theme of the environmental movement is that good environmental quality is good for the economy in the long run. The short-run economic dislocation problems with this philosophy were largely ignored in the 1970s. Corporations and municipalities were expected to pay whatever was needed to correct past environmental problems and to provide future environmental protection, no matter what the price. Federal laws and regulations established ambitious compliance schedules, which were occasionally relaxed, but which for the most part committed industry and public to considerable expenditures.

Opposition to spending what it takes was often stated ineffectively, mostly because the arguments advanced tended to overstate the problem. Too often, decisions to close companies or shut down plants were attributed solely to the cost of environmental regulations. No doubt these were important factors and may have been the sole factor in some instances, but not to the extent that was claimed.

The national priority now is continuous improvement and strengthening of the economy in harmony with the environment and a trend toward cost-effective regulations. There is, and will be, a requirement on regulators and enforcers to collect facts before imposing major and costly requirements. The philosophy of the 1970s was that all potential problems imaginable had to be prevented. Now, it is recognized that the possibilities that could be safeguarded against are too numerous for this approach to be affordable. Another manifestation of the recognition that priority must be given to the economy will be reduced paperwork requirements for the industry.

In summary, the trends toward environmental regulations and environmental protection can be stated as follows:

1. Adjustments in the federal and nonfederal roles are likely to increase state participation in the enforcement and administration of environmental regulations.

2. Balancing of economic and environmental goals is likely to take the form of moderation in achieving some environmental goals that adversely affect economic activities.

3. Public support for environmental protection and related life-support systems is expected to continue, especially in the industrialized countries.

4. In the United States, midcourse correction to major environmental legislation is expected to be made by the legislative bodies. This midcourse correction will be based upon benefits (environmental protection and enhancement) and costs associated with environmental requirements.

5. To the extent possible, regulations will move away from the command-and-control type of approach presently used in most cases because in a free-market economy, these regulations are inefficient

and do not preserve elements of voluntary choice. To the extent technologically practical, future regulatory approaches will focus on the use of economic incentives, such as marketable discharge licenses or permits and effluent charges.

6. With increasing experience in the pollution control technology areas, regulatory controls will move away from the "hothouse" types of control technologies that deteriorate rather quickly and end up contributing large amounts of pollutants and incurring high operation and maintenance costs during the life cycle of the control devices. Instead, more practical emission standards, with built-in economic incentives, will be established so that cost-effective pollution control technology that provides overall lower pollutants during the life cycle of the equipment could be used.

7. More emphasis will be placed on new concepts such as pollution prevention, industrial ecology, and sustainable development.

8. Many problems of the global commons, such as acid rain, global warming, deforestation, and biodiversity, will become issues of international concern.

9. Industrialization in developing countries and continued population increases will further adversely affect environmental quality, especially in developing countries.

10. Concern for the environment and support for environmental protection and sustainable development internationally, including among developing countries, will increase.

11. International agreements to protect the global commons and to address issues such as global warming will face significant difficulties. Factors contributing to this will be the disproportionate economic burden borne by industrialized countries as compared to developing countries, disparity of political and economic power among countries involved, and the historical parochial nature of some political leaders in industrialized countries.

12. Vigorous public support for incorporating environmental concerns into decision-making process, as embodied in the provisions of legislation such as NEPA, is expected to continue.

2.6   Discussion and Study Questions

1   One interpretation of trends in environmental legislation has been presented above. What are other ways in which this sequence of laws and regulations could be interpreted? Provide some evidence for this alternative point of view.

2 Discuss whether the changing relationships between the federal government and the states, especially in enforcement, may properly be referred to as a trend. Is the direction of movement consistent? Where does your state fit in this relationship?

3 What problems do you see with U.S. environmental laws and regulations? Many contend that the United States is overburdened with laws and regulations, and point to the environmental arena as an example. Are such laws and regulations really necessary? What are the consequences of reducing them? Of increasing them?

4 What is the net effect of pollution control regulations with respect to employment? Are more jobs lost than are gained? How is the economy affected on balance? How does one factor in the effects of intangibles such as cleaner air and water?

5 Do environmental regulations result in U.S. companies relocating to foreign countries? What are these companies' environmental responsibilities if they do choose to relocate?

6 Review relevant environmental laws and regulations for other countries and compare them with U.S. requirements in an area which interests you. How do they differ? In what ways are they similar?

7 Obtain copies of your state's environmental code. How do the rules compare with the corresponding federal regulations? Are they more or less stringent? Is there a relationship between these laws and the economic activity within your state?

8 Which agencies in your state administer environmental regulations? Can you develop a comprehensive list? Consider such areas as air and water quality, solid and hazardous waste, and noise. What about administration of resources such as parks, public lands, wildlife, soil conservation, and similar topics? Are the administrators appointed or elected? Are there oversight boards? Are there questions of conflict of interest? What suggestions have been made for improvement of their operation?

2.7 Further Readings

Corbitt, Robert A. *Standard Handbook of Environmental Engineering.* New York: McGraw-Hill, 1999.

Environmental Statutes—Annual Edition. Rockville, Md.: Government Institutes, Inc.

EPA, 1997. From EPA web page; maintained by Jeff Kelley, Office of Public Affairs.

EPA, 1999. 33/55 Program: the final record, March 1999. EPA-745-R-99-004. Washington, D.C.: Environmental Protection Agency, Office of Pollution Prevention and Toxics.

King, Thomas F. *Cultural Resource Laws and Practice: An Introductory Guide.* Walnut Creek, Calif.: AltaMira Press, Rowman and Littlefield Publishers, Inc., 1998.

Legislation, Programs and Organization, U.S. Environmental Protection Agency, January 1979, p. 28.

McEldowney, John F., and Sharron McEldowney. *Environmental Law and Regulation.* London: Blackstone, 2000.

Percival, Robert V., et al. (eds.). *Environmental Regulation: Law, Science, and Policy.* Gaithersburg, Md.: Aspen Law & Business, 2000.

Trudeau, Rebecca L., and M. T. Olexa, "The Pollution Prevention Act," SS-FRE-13, Food and Resource Economics, Florida Cooperative Extension Service, Institute of Food and Agricultural Sciences, University of Florida. September 1994. Taken from the web page, 1 page.

National Environmental Policy Act

On January 1, 1970, the President of the United States signed the National Environmental Policy Act (NEPA), PL 91-190, into law (NEPA, 1969). The enactment of this legislation established a national policy of encouraging productive and enjoyable harmony between us and our environment. The full text of this act is in Appendix A. The symbolism of the timing of this law did not go unnoted by the President and other concerned Americans, who heralded the 1970s as a decade of environmental concern. Enactment of NEPA and concern regarding the environment and quality of life among people around the world have generated significant environmental protection legislation and regulations in many industrialized nations besides the United States. Provisions and policies set forth in NEPA are being emulated by many states within the United States and within other nations as well.

The main purposes of this legislation, as set forth in the act, are "to declare a national policy which will encourage productive and enjoyable harmony between man and his environment; to promote efforts which will prevent or eliminate damage to the environment and biosphere and stimulate the health and welfare of man; to enrich the understanding of the ecological systems and natural resources important to the Nation; and to establish a Council on Environmental Quality."

3.1 Elements of NEPA

There are two titles under this act: Title I, Declaration of National Environmental Policy, and Title II, Council on Environmental Quality (CEQ).

Title I

Title I sets forth the national policy on restoration and protection of environmental quality. The relevant sections under this title are summarized as follows.

Section 101. Requirements of Section 101 are of a substantive nature. Under this section, the federal government has a continuing responsibility "consistent with other essential considerations of national policy" to minimize adverse environmental impact and to preserve and enhance the environment as a result of implementing federal plans and programs.

Section 102. Section 102 requirements are of a procedural nature. Under this section, the proponent federal agency is required to make a full and adequate analysis of all environmental effects of implementing its programs or actions.

In Section 102(1), Congress directs that policies, regulations, and public laws shall be interpreted and administered in accordance with the policies of NEPA; Section 102(2) directs all federal agencies to follow a series of steps to ensure that the goals of the act will be met.

The first requirements are found in Section 102(2)(A), where it is stipulated that "a systematic and interdisciplinary approach" be used to ensure the integrated use of social, natural, and environmental sciences in planning and decision making.

Section 102(2)(B) states that federal agencies shall, in consultation with CEQ, identify and develop procedures and methods such that "presently unquantified environmental amenities and values may be given appropriate consideration in decision making…" along with traditional economic and technical considerations.

Section 102(2)(C) sets forth the requirements and guidelines for preparing the environmental impact statement (EIS). This section requires all federal agencies to include in every recommendation or report on legislative proposals and other major federal actions significantly affecting the quality of the human environment, a detailed statement by the responsible official covering the following elements.

1. The environmental impact of the proposed actions
2. Any adverse environmental effects which cannot be avoided should the proposal be implemented

3. The alternatives to the proposed actions

4. The relationship between local short-term uses of our environment and the maintenance and enhancement of long-term productivity

5. Any irreversible and irretrievable commitments of resources which would be involved in the proposed action should it be implemented

Specific EIS format, coordination, instruction, approval, and review hierarchy are established by each federal agency within the NEPA regulation promulgated (since 1978) by the President's Council on Environmental Quality (40 CFR 1500-1508). Persons preparing an EIS should first follow the instructions of their organizations, then the content of the NEPA regulations, and finally, if not explicit elsewhere, the letter and spirit of the law itself. Chapter 4 provides further information regarding the detailed content of an EIS and other environmental documents required by the NEPA regulations.

Section 103. This section requires all federal agencies to review their regulations and procedures "for the purpose of determining whether there are any deficiencies or inconsistencies therein which prohibit full compliance with the purposes and provisions of this Act and shall propose to the President...such measures as may be necessary to bring their authority and policies into conformity with...this Act."

Title II

Title II establishes the President's Council on Environmental Quality (CEQ) as an environmental advisory body for the Executive Office of the President. In addition, the President is required to submit to the Congress an annual "Environmental Quality Report." This yearly summary sets forth (1) the status and condition of the major natural, human-made, or altered environmental classes of the nation, (2) current and foreseeable trends in the quality, management, and utilization of such environments and socioeconomic impacts of these trends, (3) the adequacy of available natural resources, (4) a review of governmental and nongovernmental activities on the environmental and natural resources, and (5) a program for remedying the deficiencies and recommending appropriate legislation.

3.2 Judicial Review

Initially, the court cases resulting from NEPA dealt primarily with procedural requirements of the act. Most of these basic procedural questions were settled early. Litigation in 1972 and 1973 dealt with the content of statements and, more recently, with the substantive requirements of NEPA and the agency decisions made after statements are completed.

In one case (*Sierra Club v. Froehlke,* February 1973), a Federal District Court enjoined the U.S. Army Corps of Engineers from proceeding with the Wallisville Dam Project because of the inadequacy of the EIS content (*Environmental Quality,* 1973). The court concluded that (1) the statement did not adequately disclose its relationship to the much larger project (Trinity River Project), (2) the statement lacked the requisite detail to satisfy the act's full disclosure requirement, (3) alternatives to the project were inadequately considered, and (4) there was no indication that genuine efforts had been made to mitigate any of the major impacts on the environment.

In a more recent example (*City of Carmel-By-The-Sea v. U.S. DOT,* 1996), the U.S. Court of Appeals for the 9th Circuit reviewed a plan to expand California Highway 1 and found the EIS inadequate for several reasons. First, the EIS relied on wetlands studies that were several years old. The court stated that "reliance on stale scientific evidence is sufficient to require re-examination of an EIS." Moreover, the EIS did not sufficiently address the potential cumulative impacts associated with the project, nor did it adequately consider all relevant alternatives (Findlay and Farber, 1999).

In addition, several cases have confirmed the role of the judicial branch of the U.S. government in reviewing the substance of the agency decisions. Affirmation of this judicial role came in the Gillham Dam case, in which the Court of Appeals concluded that there is a judicial responsibility to make sure that an agency has not acted "arbitrarily and capriciously" in making decisions affected by NEPA (*Environmental Quality,* 1973).

One case involving a program or a comprehensive EIS was *Swain v. Brinegar.* In this case, the U.S. Court of Appeals for the 7th Circuit rejected the Federal Highway Administration's plan to prepare an EIS for a 15-mile segment of a 42-mile highway project in Illinois and reaffirmed the necessity to prepare a program EIS, since the individual action (the 15-mile segment) was an integral part of the 42-mile project (*Environmental Quality,* 1977). Completion of the first segment of the project would, for all practical purposes, foreclose later project alternatives. The court cited the long-standing NEPA goal of eliminating potentially disastrous errors that may result when the cumulative impacts of individual parts of a major program are ignored. However, if a particular action is substantially independent of other actions to be included in a comprehensive EIS, interim activity may begin on the single action if an adequate EIS is prepared for it, if a decision on the interim action would not prejudice the outcome of the larger programmatic review, and if the effects of the interim action are analyzed cumulatively in the final comprehensive EIS.

It is important to note that courts usually give great deference to agency expertise and do not set aside agency decisions unless there are significant procedural or substantive reasons. In other words, in appropriate cases, courts reserve judicial authority to review agency decisions in light of NEPA's substantive goals, along with any deficiencies which may be present in the agency's compliance with procedural requirements (*Environmental Quality,* 1978).

The clearest decision on the subject of a substantive requirement came in the case of *Burger v. County of Mendocino* (California) (*Environmental Quality,* 1977). In this case, a developer had applied for a permit to build a motel complex in an environmentally fragile forest. An EIS concluded that of seven possible alternatives, the applicant's was the worst environmentally; however, a local agency approved the application as submitted. The California State Court of Appeals reversed the decision, because sufficient evidence was not provided in the EIS to make the case for overriding environmental impacts if the project were to be approved. The court held that the agency had illegally approved the project. This case clearly involved review of a decision based upon substantive requirements of NEPA, not simply the procedural aspects of an EIS.

The substantive requirements of NEPA and the extent to which courts can require agencies to consider them are discussed in *Environmental Law,* by Professor William H. Rodgers of Georgetown Law School (Rodgers, 1977). In this document, it is pointed out that Section 101 of NEPA contains sufficiently clear substantive standards to permit meaningful judicial review. Professor Rodgers points out that the NEPA requirement "to use all practicable means" to carry out environmental policies is consistent with traditional court decisions on the "nuisance doctrine." This doctrine measures the actions of people accused of creating nuisances against a standard which considers the extent and degree to which best efforts were made to mitigate the nuisance.

Congressional purpose in NEPA was clearly to limit the statute's oversight to federal decision making and to leave private decision making subject to specific regulatory constraints. Because its intent is to improve governmental decision making by requiring federal agencies to consider the environmental consequences of their actions, the scope of what constitutes federal action has received substantial judicial review.

According to regulations adopted by the CEQ, a private action may become a federal action (1) if the project is funded by a federal agency or (2) if it involves an activity which legally requires a permit, license, or other federal approval as a precondition.

In three 1987 cases, the definition of "federal action" has been clarified as being based on the jurisdiction of the federal agency. A case brought against the EPA by the Natural Resources Defense Council tested the authority of the EPA to prohibit the construction of new plants until any required pollutant discharge permits are obtained. The court ruled that in the absence of federal funding, and because pollutant discharge permits are not a legal precondition of construction, the construction itself did not constitute a federal action. Therefore, because the EPA has jurisdiction only over the issuance of permits, it had no NEPA authority to prohibit the construction (Ellis, 1988).

Also in 1987, the Army Corps of Engineers proposed to amend the regulations that required it to produce EISs covering entire private construction projects when only a portion of the project required a Corps permit. The CEQ approved this amendment, agreeing that Corps permit requirements for a small portion of a project did not constitute sufficient federal involvement to make the entire project a federal action (Ellis, 1988).

In *Ringsted v. Duluth,* a Native American tribe purchased a building to be used as a bingo parlor and transferred it to federal trust as an addition to its reservation. The city of Duluth purchased adjacent land to construct a parking ramp serving the parlor and other users. The Secretary of the Interior produced an EIS which addressed the parlor, but excluded the parking ramp. The court rejected the complainant's contentions that the ramp was part of a federal action or a secondary effect of the federal action, on the basis that no federal action is required as a legal precondition to the construction of a parking ramp (Ellis, 1988).

In *Robertson v. Methow Valley Citizen Council,* 1989, the Supreme Court ruled that a worst-case analysis on the possible impacts of air pollution at a ski resort was not required by NEPA, overturning a previous appellate court decision. In the same case, the court found that the Forest Service was not required to formulate and adopt a plan to mitigate the adverse effects of air pollution on mule deer. The basis for this decision was the court's view that consideration of mitigation possibilities is a procedural requirement of NEPA; NEPA does not substantively require that a plan be developed and formally adopted (CEQ, 1990).

The question as to whether NEPA applies outside of the United States has also been addressed by the courts. In *Greenpeace v. Stone* (9th Cir. 1991), plaintiffs argued to enjoin the transport of previously stockpiled U.S. Army artillery shells filled with nerve gas through Germany to Johnston Atoll on the grounds that the U.S. Army had not complied with NEPA (Clark and Canter, 1997). Transport within Germany was planned and supervised by the German government, and plans for safety and hazard management were prepared by the German federal authorities, but they were not made public for security reasons.

The U.S. Army had previously prepared EISs for both the construction and the operation of the incinerator on Johnston Atoll (which is U.S. territory, though not within any state). For this action, an EIS was prepared for the receipt of the new shipment of chemical munitions into U.S. territory and their placement into storage on Johnston Atoll. An EA was prepared examining the environmental impact and risk to human populations of transoceanic transport of the munitions from a German North Sea port to Johnston Atoll. Army and Department of Defense (DOD) regulations required that the effects be assessed in a manner similar to NEPA, but noted that the procedural requirements of actions totally outside the U.S. did not require that all EIS processes be met. Plaintiffs filed suit against the Department of the Army to prohibit movement of the munitions from Germany to Johnston Atoll, partly on the grounds that a comprehensive EIS covering all aspects of the transportation and disposal of the stockpile was required by NEPA. The court concluded that applying NEPA requirements to the transport within Germany would infringe upon its jurisdiction. In addition, the court found that transoceanic transport of the munitions was a necessary consequence of the project and involved the same foreign policy considerations. (Additional allegations that the effects of an accident at sea were not considered fully were rejected following review by the court.) The court interpreted that NEPA "intended to encourage federal agencies to consider the global impact of domestic actions and may have intended under certain circumstances for NEPA to apply extraterritorially; [however]...that action should be taken 'consistent with the foreign policy of the United States'" (Clark and Canter, 1997).

The 1993 case of *Environmental Defense Fund v. Massey* (D.C. Cir. 1993) focused on the National Science Foundation's (NSF) food waste disposal practices at a research facility in Antarctica. NSF decided to stop burning food wastes in an open landfill and develop an alternative method of disposal. During the interim period, NSF resumed burning in a temporary incinerator until a state-of-the-art incinerator could be delivered. The Environmental Defense Fund (EDF) objected, arguing that the proposed incineration might generate toxic pollutants that could be hazardous to the environment. The EDF filed suit, claiming that the NSF did not adequately consider the environmental impacts under NEPA. The court noted that NEPA applicability to federal actions is not limited to actions occurring in the United States and that the primary purpose of considering extraterritoriality is "to protect against the unintended clashes between our laws and those of other nations." The court found, therefore, that the presumption against extraterritoriality did not apply in this case (CEQ, 1993).

Additional discussion of the application of NEPA outside the United States may be found in Chapter 9.

3.3   Effects of NEPA

Effects of NEPA have been far-reaching. This act, in many instances, has been instrumental in requiring reassessment of many federal programs (and programs where federal participation, approval, or license is involved)—both newly proposed programs and ongoing programs in various stages of completion and implementation. In the *re*assessment process, federal agencies have been required to consider not only the economic and mission requirements but also both the positive and negative environmental impacts.

When the environmental costs, as surfaced because of the requirements of NEPA (i.e., documentation of an EIS), are made known to the decision makers at various official levels and to the public, modification or abandonment of the project will be made at the federal agency's own initiative; however, in most cases, strong pressure from the public, environmental groups, and court actions will be the driving forces.

As a result of court cases and issuance of CEQ regulations, and in order to comply with the requirements of NEPA, the agencies should

1. Satisfy the act's full disclosure requirement with adequate detail
2. Adequately consider all reasonable alternatives to the project
3. Make genuine efforts to mitigate any major impacts on the environment due to implementation of the project
4. Prepare comprehensive program-level environmental impact statements where there is a clear interdependence of various phases of the project.
5. Consider substantive requirements of NEPA and properly weigh environmental matters relative to other considerations

3.4   Implementation of NEPA

It must be noted that NEPA and its implementation have not been without their critics [as perhaps typified by Paul Ehrlich's article, entitled "Dodging the Crisis" (Ehrlich, 1970)]. Considerable litigation has developed concerning compliance with (or, in the view of some, circumvention of) the provisions of the act. Notable among these was the Calvert Cliffs case, in which the courts held that compliance with established environmental standards did not relieve a governmental agency from the NEPA requirement of considering all environmental factors when assessing impact. In this case, the Atomic Energy Commission had sought to exclude water quality considerations from its assessment of the impact of a nuclear power plant, on the grounds that a state had certified compliance with water quality standards under the relevant federal water pollution control legislation (Calvert Cliffs, 1971).

Among the frequently voiced concerns about the implementation of NEPA are

1. Impact statements are not available in time to accompany proposals through review procedures.
2. Statements are prepared in "mechanical compliance" with NEPA.
3. Impact statements are biased to meet the needs of predetermined program plans.
4. Agencies may disregard the conclusions of adverse impact statements.
5. The CEQ lacks authority to enforce the intent of NEPA.
6. Intangible environmental amenities are being ignored.
7. Secondary effects are being ignored.
8. Inadequate opportunity is available for public participation and reaction.

Perhaps the most severe of these reservations concerning NEPA was summarized by Roger C. Crampton, who testified that "the agencies must guard against a natural but unfortunate tendency to let the writing of impact statements become a form of bureaucratic gamesmanship, in which the newly acquired expertise is devoted not so much to shaping the project to meet the needs of the environment, as to the shaping of the impact statement to meet the needs of the agency's preconceived program and the threat of judicial review" (Crampton, 1972).

Perhaps the point is best made that an impact statement for a project should not be used as a justification for a preconceived program, but rather it should be used as a vehicle for a full disclosure of the potential environmental impacts involved. Also, it should be used as a tool for adequately considering the environmental amenities in decision making and for allowing participation in the project by other federal and state agencies and the public, to provide proper consideration of the environment, along with economic and project objective requirements.

Industry concerns about NEPA are important and should not be ignored. Some of these concerns are

- Costs are excessive for the benefits derived.
- There are already too many government regulations.
- EIS/EA preparation causes project delays.
- The paperwork represents wasted effort.
- Untoward concern for the environment stifles economic development.

3.5   Council on Environmental Quality

Title II of NEPA created in the Executive Office of the President of the United States a Council on Environmental Quality (CEQ). This council is composed of three members, who are appointed by the President with the advice and consent of the Senate. The President designates one of the members of the council to serve as chair. In addition, the council employs environmental lawyers, professional scientists, and other employees to carry out its functions as required under NEPA. Duties and functions of CEQ may be summarized as follows (NEPA, 1969):

1. Assist and advise the President in the preparation of the Environmental Quality Report as required by NEPA.
2. Gather, analyze, and interpret, on a timely basis, information concerning the conditions and trends in the quality of the environment, both current and prospective.
3. Review and appraise the various programs and activities of the federal government in light of the policy of environmental protection and enhancement, as set forth under Title I of NEPA.
4. Develop and recommend to the President national policies to foster and promote improvement of environmental quality to meet many goals of the nation.
5. Conduct research and investigations related to ecological systems and environmental quality.
6. Accumulate necessary data and other information for a continuing analysis of changes in the national environment and interpretation of the underlying causes.
7. Report at least once a year to the President on the state and condition of the environment.
8. Conduct such studies and furnish such reports and recommendations as the President may request.

A significant feature to note is that both the charter assigned and the responsibilities delegated to the CEQ are quite extensive. The CEQ has proven to be highly influential in its advisory capacity, although it does not have any regulatory or policing responsibilities.

3.6   Executive Orders and Agency Response

To further enhance and explain NEPA and other environmental legislation, several executive orders have been issued by Presidents, and the federal agencies have responded with appropriate guidelines and directives. As an illustration, brief descriptions of some of the executive orders and agency responses follow.

Executive Order 11602, "Providing for Administration of the Clean Air Act with Respect to Federal Contracts, Grants, or Loans," June 30, 1971. This order sets the policy with respect to federal contracts, grants, or loans for the procurement of goods, materials, or services as being undertaken in such a manner that will result in effective enforcement of the Clean Air Act Amendments of 1970.

Executive Order 11514, "Protection and Enhancement of Environmental Quality," March 5, 1970, as amended by EO 11991, May 24, 1977. The federal government shall provide the leadership in protecting and enhancing the quality of the nation's environment to sustain and enrich human life. Federal agencies shall initiate measures needed to direct their policies, plans, and programs so as to meet national environmental goals. The Council on Environmental Quality, through the chairperson, shall advise and assist the President in leading this national effort.

Also, the heads of federal agencies are required to monitor, evaluate, and control, on a continuing basis, their agencies' activities so as to protect and enhance the quality of the environment.

The May 1977 amendment required the CEQ to issue regulations for implementation of procedural provisions of NEPA. These regulations are designed to make the environmental impact statement process more useful to decision makers and the public and to reduce paperwork and unnecessary delays. It is this amendment which provided the authority under which the CEQ researched the need for clear regulations, and finally issued the NEPA regulations in November 1978. (See Chapter 4.)

Executive Order 11990, "Protection of Wetlands," May 24, 1977. States that each agency shall provide leadership and shall take action to minimize the destruction, loss, or degradation of wetlands, and to preserve and enhance the natural and beneficial values of wetlands in carrying out the agency's responsibilities....It further states that agencies should avoid undertaking or providing assistance for new construction located in wetlands unless there is no practicable alternative to the project, and that, when such a project is necessary, all reasonable measures to minimize environmental damage should be implemented.

Executive Order 12088, "Federal Compliance with Pollution Control Standards," October 13, 1977. Directs that the head of each executive agency "...is responsible for ensuring that all necessary actions are taken for the prevention, control, and abatement of environmental pollution with respect to Federal facilities and activities under the control of the agency"; "...is responsible for compliance with applicable pollution control standards..."; and "...shall submit...an annual plan for the control of environmental pollution."

Executive Order 12114, "Environmental Effects Abroad of Major Federal Actions," January 9, 1979. Directs federal government agencies to assess the consequences of actions which take place outside U.S. jurisdiction. NEPA itself does not clearly address the issue, and the CEQ regulations address only the consequences across borders of an action taking place *within* the United States. That the order exists at all represents a compromise between those executive agencies with overseas activities and the CEQ. In general, domestic law does not apply outside the United States without clear wording from Congress to the contrary. The usual principle is that the United States should not infringe upon the sovereignty of other nations, and the separation of powers doctrine normally requires that domestic law not limit the President's conduct of foreign affairs. It is acknowledged, however, that many overseas programs, such as military bases, pipelines, and water development projects, have the potential to result in environmental problems in the host country, just as they would if performed inside the United States. This order provides for the preparation of environmental documentation, either with or without the active participation of the host country, to cover such projects. It also provides for examination of activities taking place in the "global commons" (i.e., not within any nation's territory). The high seas, outer space, and Antarctica may be examples of the global commons.

Executive Order 12898, "Federal Actions to Address Environmental Justice in Minority Populations and Low-Income Populations," February 11, 1994. Directs that each Federal agency make achieving environmental justice part of its mission by identifying and addressing areas of disproportionately high adverse human health or environmental effects on minority and low-income populations within its programs and policies. It further states that "each Federal agency shall conduct its programs, policies, and activities...in a manner that ensures that such programs, policies, and activities do not have the effect of excluding persons from participation in, denying persons the benefits of, or subjecting persons to discrimination under, such programs, policies, and activities because of their race, color, or national origin." Human health and environmental research and analysis "shall include diverse segments of the populations in epidemiological and clinical studies, including segments at high risk from environmental hazards, such as minority populations, low-income populations, and workers who may be exposed to substantial environmental hazards."

Executive Order 13148, "Greening the Environment Through Leadership in Environmental Management," April 21, 2000.   States that the head of each federal agency is responsible for ensuring that all necessary actions are taken to integrate environmental accountability into agency day-to-day decision making and long-term planning processes, across all agency missions, activities, and functions. Consequently, environmental management considerations must be a fundamental and integral component of federal government policies, operations, planning, and management. The head of each federal agency is responsible for meeting the goals and requirements of this order. These goals include environmental management, environmental compliance, right-to-know and pollution prevention, toxic chemical release reduction, toxic chemicals and hazardous substances use reduction, reductions in ozone-depleting substances, and environmentally and economically beneficial landscaping.

Agency responses

Nearly all federal agencies have issued directives, guidelines, circulars, and other appropriate documents in response to executive orders and NEPA. Because of the changing nature of these documents, it would be infeasible to include extensive information about them. It is suggested that the current appropriate agency information be consulted prior to embarking upon an environmental impact analysis.

3.7   State Environmental Policy Acts

Because of the concern for environmental protection and enhancement, NEPA was enacted by Congress and signed into law by the President on January 1, 1970. NEPA applied directly only to the activities and programs of the *federal* agencies and to those activities and programs supported by federal funds and/or federally issued permits and licenses. Many states felt that, in many instances, the problems and concerns at the state level were different from those at the federal level and that they varied from one state to another. Since many state-supported projects were not covered by the requirements of NEPA, some of the states enacted their own state environmental policy acts or guidelines, sometimes referred to as State Environmental Policy Acts (SEPAs) or "little NEPAs."

NEPA was drafted with at least some expectation that it would serve as a model for similar programs at the state level. Sixteen states, the District of Columbia, and Puerto Rico have adopted NEPA-like systems requiring environmental assessment. Eighteen states and the District of Columbia have limited environmental review requirements established by statute, executive order, or other administrative directives (CEQ, 1992). The legal basis, administration, and requirements of the state systems vary; California, New York, and Washington are examples of systems with comprehensive legislation and judicial enforcement, while others are more restricted in scope (CEQ, 1990).

Although most state systems were initiated in the 1970s, there is significant recent interest in state NEPA programs. Montana and New York have held conferences on the EA process, Washington and New Jersey have revised their regulations, and state environmental quality agencies are being considered by Michigan and Maine. Several cities have also adopted environmental assessment procedures. The New York City program was established as part of its responsibilities under the state's Environmental Quality Act.

Since state environmental policy acts are patterned after NEPA, discussion and procedures presented in this text can be used to address impact analysis requirements set forth by the states. We note, however, that many state acts require preparation of NEPA-like documents by *private* applicants for a state-granted permit. This differs somewhat from the general federal practice of *agency* preparation of the documentation.

3.8 NEPA and Agency Planning

NEPA is at heart a planning tool. The law requires federal agencies to consider environmental consequences along with other types of issues (such as financial, political, social, or technical) when making decisions, and evaluate alternative courses of action. Although NEPA does not dictate an environmentally benign outcome from federal decisions, as a matter of national policy the law asks that agencies act as stewards of the environment and try to protect it from harm. NEPA requires that if a proposed action is expected to cause adverse consequences, the agency must fully disclose these adverse consequences and must identify mitigation actions and put these into place over time to ameliorate the adverse consequences of federal action. A federal agency must review its proposed projects to establish NEPA compliance. Such a "NEPA review" may result in any one of several types of documents, such as an EA or EIS. In order to most effectively do this, the agency must plan ahead.

The CEQ regulations address the relationship between NEPA and agency planning (see 40 CFR 1501). The regulations emphasize that integration of NEPA early in the agency planning process will be the most effective way to avoid conflicts and delays in seeing a project through to completion. Adoption of formal agency plans is one of the four main types of federal actions that trigger a NEPA review (see 40 CFR 1508.18(b)(2)). NEPA reviews prepared on plans, broad programs, or closely related proposals are often referred to as "programmatic" NEPA reviews (see 40 CFR 1502.4). While some federal agencies are continuing to increase the use of NEPA as an agency strategic planning tool, this use of NEPA is still growing (CEQ, 1997).

The planning process

Many books and articles have been written on the planning process, which may be of interest to the NEPA practitioner from both the management perspective and the physical, or land-use, perspective. (See, for example, Drucker, 1973; Goodman, 1968; Faludi, 1973; Lynch and Hack, 1984; McHarg, 1991.) Broadly, the approach to planning often used is the "rational comprehensive" (or "synoptic") approach, although other approaches exist (Hudson, 1979). Synoptic planning has four elements: (1) establish goals, (2) identify alternatives, (3) evaluate options, and (4) implement decisions. These elements can be further refined into a series of cyclic steps, as shown in Fig. 3.1.

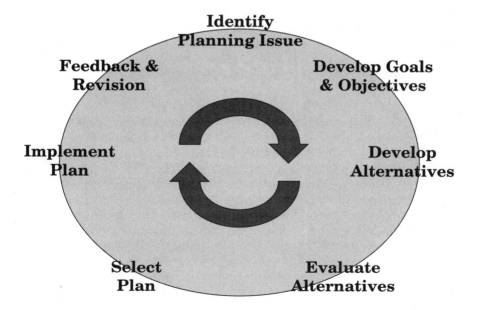

Figure 3.1 Generalized planning process.

First, the community, agency, executive, or planner must decide what is "at issue," in other words, what problem needs to be solved. For the planner or the executive some care must be given to defining this question (Drucker, 1973); if the central question, or issue, is not thoughtfully parsed, the answer, or solution, will be inadequate, ineffective, or irrelevant. "The first step—the most difficult and most often bungled step—is to ask what the problem is" (Lynch and Hack, 1984).

An agency plan is generally developed to answer a specific question: What course of action would be optimal to address a specific set of issues? A related question may be *why* the action is proposed. Is it required by specific legislation? Is it necessary to correct a violation of law or regulation? Is it clearly part of the agency mission? Is it in response to some change in the environment? All these factors will affect the development of the plan. A good plan is structured to address how an agency will meet certain *goals,* or end points. The plan may outline *objectives,* which are means (operational actions) to reach the goals (Drucker, 1973). Sometimes a planner will confuse "goals" and "objectives," or combine them into one term of "goals and objectives," but the two are quite distinct. For example, in a football game, the *goal* would be to win the game, while an *objective* might be to make a first down. A *strategy* is a way to achieve an objective, and consists of implementing a series of *actions.* There may be many strategies pertaining to a single objective, and more than one series of actions can be deployed at the same time. Going back to the football analogy, one action to achieve a first down might be for the quarterback to throw a forward pass, while at the same time a teammate might execute a second action and block a defensive player. Both of these would be actions taken by the team to implement the objective (first down) and reach the ultimate goal (win the game). Goals and objectives may be set by the agency, or may be established with input from interested parties, the scientific community, or the public at large.

Once there is consensus on what is to be addressed, the planner, with input from appropriate parties, can develop options, or alternative means, to address the goals and objectives that need to be met. Through the planning process, the options are sifted and examined, compared and analyzed. The results of this evaluation are presented to the decision maker, and a course of action is selected. The planning process does not stop there, however; the executive or agency must decide how to implement the plan and take the appropriate actions to put the plan into place. Effective plans are often conceived as "living documents"; that is, the results of the plan are monitored over time, and the plan's effectiveness evaluated.

An agency should not "plan to plan," but should "plan to *do.*" A plan is prepared to give an agency a means to weigh options and address uncertainty over time. By developing a plan, and following it, an agency can ensure that its near-term actions will be lined up toward achieving a common long-term goal. The agency can avoid taking actions that are mutually counterproductive and avoid squandering its resources on unneeded or incompatible actions. For federal agencies, the plan implementation should give guidance on what NEPA or other environmental reviews would be required at each step. Plan implementation may additionally specify what permits or licenses would be needed to carry out each step. If the NEPA review indicated specific mitigation measures to ameliorate adverse impacts, these may be appropriately addressed during the plan implementation.

The last step of the planning cycle is feedback: Did the plan, when implemented, effectively address or solve the problem that was at issue? Perhaps the plan missed the mark and needs to be revised or fine-tuned to better address the planning issue. Even if the plan was a good one, the conditions leading to the original issue may have changed over time, leading to the need for planning revisions.

Relationship between NEPA and planning

There are strong parallels between the planning process and the NEPA review process. Just as a plan addresses a "planning question," a NEPA review addresses a decision that must be made. A plan responds to specific goals and objectives; a NEPA review responds to a purpose and need for action. A plan sifts through alternative courses of action to compare options and determine the "best" or optimal approach; a NEPA review analyzes alternatives to compare the impacts of different courses of action. A plan may come up with ways to address or soften adverse consequences; a NEPA review identifies ways to mitigate adverse impacts from the course of action chosen. A plan may include a way to conduct plan monitoring, or provide a feedback loop; a NEPA review may include provisions to monitor the outcome of the proposed action.

Like a NEPA review, a plan may benefit from public involvement at many points of the process. A planner may solicit public input to help decide what is at issue and to help determine alternatives. This is analogous to the public scoping process of a NEPA review. A planner may seek additional information about various options or input on the analysis or weighing of alternative courses of action. The addition of public members to the planning team is highly recommended as an excellent public involvement activity (see Chapter 11). This is analogous to the public review of a draft environmental impact statement. The planner may put a draft plan out for review and comment, or may check with interested parties before the decision maker implements a plan. This is analogous to public review of a decision or mitigation commitments following a NEPA review, but prior to taking action.

Types of plans

There are many types of plans. Except for things such as financial plans (budgets) or personnel management (hiring and dismissing employees), most federal plans are subject to NEPA reviews. A few examples of various types of plans follow.

A *programmatic plan* spells out how an agency will carry out a set of related, or "programmatic," actions over time to achieve an overall objective, or "program." These types of plans are often broad-scale, visionary documents. They may define long-term goals and include objectives as to how the program will respond to national mandates. They may address specific strategies, such as hiring a skilled workforce. They may discuss how existing facilities can be used to further the agency program or if new facilities would need to be built over time. These programmatic plans generally have a longer planning horizon, 10 years or more, and are designed as "roadmaps" to carry an agency into the future. They generally are stand-alone documents, and provide a path forward to meet a specific goal independently of other agency goals.

A *resource management plan* addresses how to manage a given set of natural, cultural, or economic resources to meet a specific agency or national goal. This may, for example, address how to improve wildlife habitat, develop a logging plan to address silvicultural needs, or manage archaeological sites to meet specified legal requirements. These types of plans generally weigh trade-offs among competing resource uses, and select among options to manage for one set of resource values, possibly at the expense of other resource uses. These plans are sometimes seen as "living documents" where the agency agrees to reconsider the trade-offs among resource management issues at a given interval, often 5 or 10 years.

A *land use plan* may need to be developed to determine future growth patterns, optimize land use over time, sort out competing or conflicting uses of federal lands, lay out options for site development for a new park or building, or plan for new utility or infrastructure corridors. A land use plan generally covers a large area, such as an entire military base, a complete national park, a city, a county, or a district or region. These types of plans are geographic in nature, and generally result in some type of map or blueprint to lay out physical actions to be taken over some set time. The period of time considered is often called the "planning horizon," and typically may be 5, 10, or 25 years. As the planning horizon increases, the uncertainly level also increases;

it is much easier to declare with some certainty what is intended to be carried out during the next year (planning horizon of 1 year) than to foresee and second guess what might be needed in two decades (planning horizon of 20 years).

A *facility plan* defines a path for managing a specific facility. This may be a single building (such as a wastewater treatment plant) or a cluster of buildings and related infrastructure (such as a set of barracks and their related utilities and parking lots). For example, a facility plan may identify when a building will need a roof replacement, schedule how utilities will be phased to complement expected growth patterns, determine how parking structures may be sized, lay out an emergency evacuation route for employees or residents, or determine an authorization basis to safely operate machinery or radiological equipment. A facility plan may have a relatively short planning horizon and a fairly detailed set of strategies. The plan may include routine maintenance as well as plans for future expansion or additions.

Timing of NEPA review

NEPA and agency planning can come together at three junctures. In the first case, the agency may develop a plan and then perform NEPA reviews on the planning options after the plan is finished. Alternatively, the agency may prepare a NEPA review of a programmatic set of actions, and then prepare a plan to determine how to implement the programmatic decisions. Thirdly, the agency may use the NEPA review process as a basis for developing a plan, including a consideration of alternative courses of action (the "heart" of the NEPA analysis), using an interdisciplinary approach to look at a wide range of environmental facets and involving the public or other interested parties in the planning process.

A discussion of the pros and cons of these three approaches follows. Each has its place in the agency planner's toolbox.

Plan first, NEPA review second. This approach is useful when a plan itself is not subject to NEPA, but the implementing strategies would require a NEPA review. At times an agency may wish to develop a high-level strategy or program to provide long-term or mission-specific direction, such as a plan to document a national energy policy. Because this type of plan looks far into the future or does not define specific actions, it may be premature to pursue a NEPA review until a set of actions emerge that are "ripe for decision." Once a course of direction is established through the policy or plan, the agency may then want to pursue NEPA review of the discrete actions that would emanate from the plan.

Pros. This approach allows an agency the flexibility to sketch out a broad plan of action without detailing large suites of "reasonable" alternatives for analysis in a NEPA review. Often the possible courses of action are too uncertain or too amorphous early in the planning process to be able to sustain a meaningful NEPA analysis. Once a plan is established, the agency can then appropriately focus its attention on the specific decisions that must be made to carry out the planned actions. This sharpens the agency focus. The follow-on NEPA reviews can be staged to allow the agency to sequentially focus on issues that are "ripe for decision" within the context of the larger course of action developed through the plan.

Cons. It is easy for an agency to focus its attention on a preferred course of action too soon, thereby abbreviating or sharply narrowing the spectrum of "reasonable" alternatives to be considered in the NEPA review. The agency may come to rely on its plan as if it had made a decision through the NEPA process, and perform NEPA too late in the process, after all implementing decisions have been made. Officials may develop "ego commitment" to one course of action, and feel they cannot entertain modifications because of the risk of appearing weak or indecisive within the agency. If the agency has used the planning process to divide an entire course of action into smaller pieces, it may be guilty of "segmentation," which is the inappropriate use of NEPA to look at one small part of a larger proposal apart from its broader context. If the plan is too broad or too high-level, there may be little substance to use as the subject of a NEPA review, resulting in a large number of analysis assumptions that may have little grounding in fact.

NEPA review first, plan second. This approach is useful to determine if there would be unacceptable environmental impacts from potential courses of action, and to use this information as a starting point to develop a plan. The NEPA review would be followed by a plan on how to carry out the decisions derived from the NEPA review. For example, an agency may use the NEPA process with public involvement to reach an early decision to select a site or technology for a major new facility, then develop project-specific plans at only the selected site or focus only on the selected technology. If needed, the agency may engage in a site-specific or "tiered" NEPA review of the impacts of constructing and operating the facility on the selected site.

Pros. The agency can use the NEPA process to narrow the field so that it does not have to develop detailed plans for a number of "reasonable" alternatives. The agency can look at "connected actions" through the NEPA process without fear of improper "segmentation" of the suite of actions that need to be taken.

Cons. The agency will have to develop a reasoned set of analysis assumptions in order to make an informed choice among NEPA alternatives prior to engaging in the planning process. The scope of the programmatic NEPA review, if too narrow, might inadvertently limit the scope of the subsequent plan.

Plan developed through a NEPA review. This approach is useful when the development of the plan and NEPA review of implementing strategies are intertwined. Many federal land managing agencies, such as the U.S. Forest Service, the National Park Service, or the Bureau of Land Management, prepare land-use plans or resource management plans to guide their stewardship of the nation's forests, parks, and public lands. Typically, these plans are developed by dovetailing the planning process with the NEPA review. The agency prepares one document, which is jointly a plan and an environmental analysis.

The Bureau of Land Management, for example, follows a planning process that was established in the 1970s by law and related regulations. The Bureau administers vast acreages of the nation's public lands, primarily in 12 western states. The Bureau must balance many different types of resources under the principles of "multiple use" and "sustained yield." These uses include such things as cattle grazing, wildlife management, oil exploration, mining, recreation, and paleontological preserves; although often not all of these resource uses take place on the same tract of land, in some cases they do. The land manager must decide how to resolve conflicts among different types of land uses on a given tract of land. For example, if cattle need to use a water source, wildlife using the same source might be driven away. Lands that are being used to extract minerals may be unsafe for hikers, or active hard rock mining operations may conflict with oil and gas exploration.

To assist the agency, in 1976 Congress passed the Federal Land Policy and Management Act (43 U.S.C. 1701 et seq.). Sometimes called an "organic act," the law established many ways to organize the agency and its processes. Title II of that law specifically provides that the agency inventory public lands to determine their resource values, and prepare land use plans to prioritize and allocate resource management. Through agency regulations (found at 43 CFR 1600), the planning process combines the elements required by Title II with the elements required by NEPA. The resource management plan for a given Bureau resource area, which may cover several thousand acres, is developed in conjunction with the required NEPA review (BLM, 2000). The environmental impact statement analyzes alternative resource uses and includes a draft plan as a preferred alternative; the record of decision presents the plan finally decided upon and explains the trade-offs among competing land and resource uses.

Similar to the NEPA review process, the Bureau envisions its planning process as tiered. At the highest level, the agency has a strategic plan that outlines broad mission goals. At the next tier, the agency prepares resource management plans to weigh resource uses within a given area. At the lowest tier, the agency may prepare site-specific implementation plans to determine how a given activity will be carried out (BLM, 2000). By using a tiered approach, a Bureau manager can focus resources and attention on questions that are "at issue."

Pros. The agency can use the NEPA process to add value to its plans by simultaneously weighing resource value trade-offs while disclosing environmental impacts of the various options considered. The agency can streamline its work by completing two types of reviews at the same time. The planning process can narrow the scope of the "reasonable" alternatives considered in the NEPA review and eliminate unnecessary or spurious analyses; at the same time the NEPA process can provide needed information on the environmental impacts of possible approaches and allow the plan to concentrate on those that would have lesser environmental impacts. This approach also allows planners to develop mitigation measures to offset adverse impacts.

Cons. Combining two similar, but different, types of reviews in one joint document can be confusing to the agency, the document preparers, and the general public. The agency planners must be able to conduct the two similar, but different, processes at the same time.

Conclusion

There are many similarities between NEPA and planning, and the federal agency planner can use the NEPA process to strengthen and improve the agency planning process. Similarly, the NEPA practitioner will consider NEPA early in the planning process. Because both of these processes are flexible, the planner has many avenues to perform a NEPA review in conjunction with the plan. Sometimes the type of plan to be developed will dictate or influence the timing of the related NEPA review. Plan implementation is important to ensure that the agency objectives are carried out and its long-term goals are met. The agency must recognize that the plan is not an end in itself, but, like NEPA, a guide for "excellent action": "Ultimately, of course, it is not better documents but better decisions that count. NEPA's purpose is not to generate paperwork—even excellent paperwork—but to foster excellent action" [CEQ Regulations, 40 CFR 1500.1(c)].

3.9 Discussion and Study Questions

1 Which of the major provisions (sections) of NEPA do you think was believed at the time to be the most important and far-reaching? Explain why you are led to this conclusion. Which provision do you feel was actually proven to be the most important over the next 30 years? Why? If this is not the same as your answer to the first part, explain why.

2 Discuss the issue of molding the document to fit the needs of the project versus molding the project to fit the environmental problems found at the site. Is this a major defect? May public needs be met in either situation? Which needs or values may be compromised (assuming the EA or EIS is accurate and truthful in both cases)?

3 NEPA is directed toward federal actions and agencies. Should it also be directed to industries and other nonfederal agencies and individuals? If not, why not?

4 What is specifically required to be included in an EIS, according to Section 102(2)(c) of NEPA? What additions, deletions, or modifications to these requirements would you suggest to make the purpose of the act more easily attainable?

5 Since NEPA does not transfer to the CEQ authority for directing or overruling agency decisions, how does the act purport to improve decision making? Is this effective (i.e., is NEPA "working")?

6 How are the members of the CEQ selected? Who are the current members and what are their qualifications to serve?

7 Obtain copies of CEQ annual reports. After reviewing them in light of NEPA requirements, do these documents meet your expectations?

8 Does your state have a SEPA that requires impact statements on state-funded or private projects? If it does, compare/contrast it and its requirements with NEPA. If not, discuss the pros and cons of initiating one. Examine, especially, who is required to prepare an EIS (or equivalent document), who reviews it, and who approves it.

9 Why are some types of plans subject to NEPA while others are not?

10 How does the "purpose and need" section of an environmental impact statement compare to the "goals and objectives" section of a programmatic plan?

11 When is NEPA an effective planning tool? When might the planning process benefit from the NEPA process?

12 How does public participation aid the planning process? Is this the same as, or different from, public participation in the NEPA process?

13 Some federal agencies conduct a NEPA review on planning actions, when not otherwise required, "to further the purposes of NEPA." An example would be the sitewide environmental impact statements or assessments prepared by the Department of Energy to address cumulative impacts on its large, multifunctional sites. Why would this be advantageous to the agency? What are the pitfalls?

3.10   Further Readings

American Planning Association (APA), Chapter Presidents Council. 1990. A Study Manual for the Comprehensive Planning Examination of the American Institute of Certified Planners (AICP). Memphis, Tennessee: Graduate Program in City and Regional Planning, Memphis State University.

Bauch, Carl. "Achieving NEPA's Purpose in the 1990's." *The Environmental Professional,* 13:95–99, 1991.

Bureau of Land Management (BLM). 2000. Manual 1601 - Land Use Planning. Rel. 1-1666, November 22, 2000. Washington, DC: Bureau of Land Management.

Caldwell, Lynton K. "A Constitutional Law for the Environment: 20 Years with NEPA Indicates the Need." *Environment,* 31(10):6–28, 1989.

Canter, Larry, and Ray Clark. "NEPA Effectiveness—A Survey of Academics." *Environmental Impact Assessment Review,* 17:313–327, 1997.

Clark, Ray, and Larry Canter, eds. *Environmental Policy and NEPA: Past, Present, and Future.* Boca Raton, Florida, CRC Press LLC. 1997.

Council on Environmental Quality (CEQ). 1997. The National Environmental Policy Act: A Study of Its Effectiveness after Twenty-Five Years, January 1997. Washington, DC: Executive Office of the President, Council on Environmental Quality.

Drucker, Peter F. *Management: Tasks, Responsibilities, Practices.* New York: Harper & Row, 1973.

Faludi, Andreas. *Planning Theory.* Oxford: Pergamon Press, 1973.

Findley, Roger W., and Daniel A. Farber, Cases and Materials on Environmental Law. St. Paul, Minn.: West Group, 1999.

Goodman, William I., editor, and Eric C. Freund, associate editor. 1968. Principles and Practice of Urban Planning. Washington, DC: Institute for Training in Municipal Administration, International City Managers' Association.

Hudson, Barclay M. "Comparison of Current Planning Theories: Counterparts and Contradictions," *Journal of the American Planning Association,* 45 (4):387, 1979. Chicago: American Planning Association, Journal of the American Planning Association.

Lee, Jessee. "The National Environmental Policy Act Net (NEPAnet) and DOE NEPA Web: What They Bring to Environmental Impact Assessment." *Environmental Impact Assessment Review.* 18:73–82, 1998.

Lynch, Kevin, and Gary Hack. *Site Planning.* Cambridge, Massachusetts: The MIT Press, 1984.

McHarg, Ian L. *Design With Nature.* New York: John Wiley & Sons, 1991 (originally published in 1969).

Rees, William E. "A Role for Environmental Assessment in Achieving Sustainable Development." *Environmental Impact Assessment Review,* 8:273–291, 1988.

U.S. Executive Office of the President, Council on Environmental Quality (CEQ). The National Environmental Policy Act: A Study of Its Effectiveness after 25 Years, Washington, DC: Executive Office of the President. 1997. http://ceq.eh.doe.gov/nepa/nepanet.htm.

Environmental Documents and CEQ Regulations

The President's Council on Environmental Quality (CEQ) issued a set of regulations in 1978 to direct federal agencies how to comply with the National Environmental Policy Act (NEPA)—these regulations are found at 40 CFR 1500 to 1508. The regulations address both the substantive requirements of NEPA—the "what"—and the procedural requirements—the "how." In addition to describing the analytic process and documentation requirements, the CEQ regulations brought many new terms into the vocabulary and identified many then-new documents. The text of the CEQ regulations is included in full in Appendix D. Included in Chapter 4 are a discussion of the evolution of the NEPA regulations, a description of the various environmental documents required by the CEQ regulations and related terms, a detailed discussion of the content of an environmental impact statement (EIS), and other related information. The discussion in this chapter focuses mainly upon NEPA, the CEQ regulations, the EIS process, and related procedural requirements. Many states and local governments, as well as many other countries, have enacted environmental policy requirements that parallel the federal process; therefore the guidance presented in this chapter is relevant to many of these procedures as well.

4.1 Function and Purpose of the NEPA Assessment Process

The environmental analysis process serves to meet the primary goal of Congress in enacting NEPA—to establish a national policy in favor of protecting and restoring the environment. The EIS process was included in NEPA to achieve a unified response from all federal agencies to the policy directives contained within the act. Section 102 (2) (c) of NEPA, which requires that an EIS be prepared for major federal action, was intended as an "action forcing device" to ensure that federal agencies meet their obligations under NEPA.

The primary purpose for preparing an EIS is to make known the environmental consequences of a proposed action. This alerts the agency decision maker, other agencies, states, American Indian tribes, the public, and ultimately Congress and the President to the environmental risks involved. An important and intended consequence of this disclosure is to build into the agency's decision-making process a continuing consciousness of environmental considerations.

Environmental impact assessment should be undertaken for reasons other than to simply conform to the procedural requirements of the law. According to the letter of the law, environmental impact must be assessed for federal activities with significant impact. However, the spirit of the law is founded on the premise that to use resources in an environmentally compatible way and to protect and enhance the environment, it is necessary to know how activities will affect the environment and to consider these effects early enough so that changes in plans can be made if the potential impacts warrant them.

In standard cost-benefit analysis and program evaluation, the intangible impacts on the environment cannot be taken into account. The impact assessment process provides the basis for operating within the spirit of the law by encouraging recognition of impacts early in the planning process and by providing an inventory of potential environmental effects of human activities.

The planning process inevitably involves projecting activities into the future to determine how well the projected activities conform to anticipated alternative functions. The methods for dealing with short-term exigencies and complexities can be identified only with reference to the long-term plan.

Environmental impact analysis fits into the long-term planning process because it provides the vehicle for identifying the potential effects of activities on the environment. While immediate knowledge of these effects is important, the long-term aspects of impact are probably more important, because only on a longer time horizon can adequate, effective, and low-cost alternatives to reduce the impact be identified.

If, for example, the potential for an adverse impact of an activity or program planned for 5 years in the future was identified, adequate time to consider significant mitigation alternatives (including stopping the program) would exist. This is much preferred to finding out about serious impacts only after an activity is half completed and (potentially) millions of dollars expended. In the latter case, modifications to reduce the impact could be very costly, or opposition could force costly delays in completion or even prevent continuation.

NEPA-related documents provide a vehicle for recording anticipated impacts of activities so that concerned institutions or individuals will be aware of possible repercussions of the subject activity. Future projects can look at the impacts of similar, past projects to gain insight into potential environmental impact. Historically, few records have been maintained of the long-term environmental effects of activities. Frankly, reliable records of the pre-action conditions may never have existed, and cannot now be located for many pre-1970 projects.

Another valuable use for the inventory of impacts is to identify the potential cumulative effects of a group or series of activities in an area. Although a single activity might not be likely to cause serious changes in the environment, when its effects are added to those of other projects, the impacts on the environment might be severe. The potential for cumulative impacts must be identified, and in some cases, this may be possible only at the intra-agency level. Thus, to account for cumulative impacts, it might be more desirable to assess the environmental impact at a program level, which covers many projects or activities.

Again, NEPA has the primary goal of incorporating environmental considerations into the decision-making process. NEPA should not be used, nor was it intended to be used, simply to stop unwanted projects, provided the requirements of the act are fulfilled. The prudent course of action for any agency, however, must be to avoid the possibility that such obstructionism is able to utilize deficiencies in NEPA documentation as a tool.

The essence of NEPA is simple: Use a systematic and interdisciplinary approach to evaluate the environmental consequences of the proposed action, include this analysis in environmental documents, give appropriate consideration to environmental accommodation meeting the substantive requirements of the act, and incorporate the results into the decision-making process. If this is done in a complete, honest, and straightforward manner, and if impacts are disclosed to the public, NEPA requirements are satisfied. The project or action ultimately decided upon may have significant environmental effects; however, if the probable consequences are known, fully disclosed, and weighed with other factors related to economic and technical considerations and agency statutory missions, and all reasonable mitigation measures are taken, the letter and the spirit of NEPA have been fulfilled.

As discussed earlier, nothing within NEPA requires that every environmental problem be totally resolved. NEPA does not require a particular outcome or that the most environmentally benign course of action be pursued. Nor does the act require that consideration for the environment be the *primary* factor in the agency decision-making process. What *is* required is that the environment be *included* in the decision process. Typically, it is only when the environmental assessment procedure is looked upon as a "paper exercise," or when the assessment is done in an incomplete or shortsighted manner, that legal difficulties develop.

It is not unknown, of course, for environmental considerations to be used as a lever by persons or groups who simply oppose the mission of the proponent agency, the basic purpose of the proposed action, or the location proposed for it. As with all differences of opinion, greater polarization leads to a more heightened, more adversarial relationship. When a dispute reaches the "anything's fair" stage, it will be difficult to determine whether or not substantive environmental questions exist, or whether allegations of incomplete assessment are being used as a partisan tool. The combination of procedural and substantive requirements found within the NEPA regulations and agency NEPA rules do, however, provide many opportunities for an opponent to identify errors in process or fact. A court will probably not be sympathetic if an agency has not followed its own published and approved procedures.

4.2   NEPA Regulations

The CEQ is responsible for overseeing federal efforts to comply with NEPA. In 1970, the Council issued guidelines for the preparation of EISs under Executive Order 11514 (1970). Until 1979, the 1973 revised guidelines were in effect, but under Executive Order 11991 (1997), the President directed the CEQ to issue regulations to supersede the 1973 guidelines.

Initially proposed in 1977, the CEQ regulations became effective on July 30, 1979. In the executive order, the President directed that the regulations should be "... designed to make the environmental impact statement process more useful to decision makers and the public and to reduce paperwork and the accumulation of extraneous background data, in order to emphasize the need to focus on real environmental issues and alternatives."

The new regulations were developed to achieve three principal goals: reduction of paperwork, reduction of delays, and, most important, production of better decisions which further our national policy to protect and enhance the quality of the human environment.

The executive order was based on the President's constitutional and statutory authority, including NEPA, the Environmental Quality Improvement Act, and Section 309 of the Clean Air Act. The President has a constitutional duty to ensure that the laws are faithfully executed, and this authority may be delegated to appropriate officials. In signing Executive Order 11991, the President delegated this authority to the agency created by NEPA—the CEQ.

In accordance with this directive, the Council's regulations are binding on all federal agencies, and replaced some 70 different sets of earlier agency regulations. The CEQ regulations provide uniform standards applicable throughout the federal government for conducting environmental reviews. The CEQ regulations also provide for agencies to develop and publish their own internal NEPA regulations, tailored to the types of actions which the agency needs. The regulations also establish formal guidance from the Council on the requirements of NEPA for use by the courts in interpreting this law.

In arriving at these regulations, the CEQ used a vigorous process of input by diverse groups and conducted many reviews of its draft regulations issued earlier. In all, the Council sought the views of almost 12,000 private organizations, state and local agencies, and private citizens. The CEQ affirmatively involved critics of NEPA as well as its friends.

There was broad consensus among these diverse witnesses. Incredible as it might seem, all, without exception, expressed the view that NEPA benefited the public. As an example, during one hearing, an official spokesperson for the oil industry said that he adopted in its entirety the presentation of the president of the Sierra Club—a well-known conservation organization.

Information from the hearings was organized into a 38-page "NEPA Hearing Questionnaire" that was sent out to all the witnesses, every state governor, all federal agencies, and everyone who responded to an invitation in the *Federal Register*. More than 300 replies were received. In addition, meetings were held with every federal agency affected by the proposed regulations, which had been circulated for comment to all federal agencies in December 1977. While federal agencies were reviewing the proposed regulations, the CEQ continued to meet with, listen to, and brief members of the public, including representatives of business, labor, state and local governments, environmental groups, and others.

On June 9, 1978, the CEQ regulations were proposed in a draft form and the Council announced that the period for public reviews of and comment on the draft regulations would extend for 2 months, until August 11, 1978. During this period, the Council received almost 500 more written comments on the draft regulations. Most of these comments contained specific and detailed suggestions for improving them.

The CEQ meticulously responded to these comments on November 29, 1978. A written environmental assessment for these regulations was prepared by the CEQ. These regulations were effective for actions proposed after July 30, 1979. NEPA itself continued to apply to actions started before the signing of the act into law (i.e., January 1, 1970).

The CEQ regulations are designed to ensure that the action-forcing procedures of Section 102(2) of NEPA are used by agencies to fulfill the requirement of the congressionally mandated policy set forth in Section 101 of the act. Since these regulations are applied uniformly to all federal agencies, this will minimize misinterpretation, redundancy, and misapplication. Also, the time required to learn these regulations and review these documents will be minimized.

4.3   Environmental Documents

Before a federal agency can undertake a new proposed action, the CEQ regulations require that the agency document its consideration of environmental factors and their bearing on the agency decision-making process. The CEQ regulations recognize the following environmental documents (40 CFR 1508.10):

1. Environmental assessment (EA)
2. Finding of no significant impact (FONSI)
3. Notice of Intent (NOI) to prepare an EIS
4. Draft EIS (DEIS)
5. Final EIS (FEIS)
6. Record of Decision (ROD)

Individual agencies may require or allow other documents as part of their NEPA implementing procedures, but these specialized documents are not included here. This section briefly discusses each of the six NEPA documents identified above, and the differences among them.

Environmental assessment (EA)

If an agency is not certain whether a proposed action would result in significant environmental impacts within the meaning of NEPA, it may prepare an EA (40 CFR 1508.9). An EA provides sufficient information to allow the agency to decide whether the impacts of a proposal or its alternatives would be expected to be significant, in which case an environmental impact statement would be prepared, or whether no significant impact would be expected to occur. An EA can help an agency meet the purpose of NEPA even when no EIS is required.

Finding of no significant impact (FONSI)

The FONSI briefly presents the reasons why the action considered in an EA would not have a significant impact on the human environment, and the rationale for why an EIS would not be required (40 CFR 1508.13). If an agency cannot reach a FONSI for a proposal, it must prepare an EIS before proceeding with the action.

Notice of Intent (NOI)

This document is a formal notice published in the *Federal Register* that informs the public and other agencies that an EIS will be prepared and considered in agency decision making (40 CFR 1508.22). All timing for the EIS process is tied to the publication date of this document. The NOI should state, at a minimum,

- The agency's proposed action, and potential alternatives
- The agency's proposed scoping process, including whether public meetings will be held, and if so, when and where they will be held
- The point of contact for the project, and for the EIS, if different

As an option, although not required, an agency may publish an NOI or similar public notice if it intends to prepare an EA instead of an EIS.

Draft environmental impact statement (DEIS)

A DEIS is the first of the two documents prepared to meet the requirements of Section 102(2)(c) of NEPA (40 CFR 1502.9). The DEIS is circulated for a formal agency and public review and comment process as outlined by the CEQ regulations.

Final environmental impact statement (FEIS)

A FEIS is the second of the two documents prepared to meet the requirements of Section 102(2)(c) of NEPA (40 CFR 1502.9). The FEIS incorporates the results of the formal review of the DEIS, as outlined by the CEQ regulations.

Record of Decision (ROD)

At the time of its decision or, if appropriate, its recommendation to Congress, each agency should prepare a concise public record of its decision. This record may be integrated into any other documentation which is prepared by the agency for a similar purpose. This record (40 CFR 1505.2) should include

- A statement of what the decision is.
- Identification of all alternatives considered by the agency in reaching its decision, including specification of alternatives which were

considered environmentally preferable. An agency may discuss preferences among alternatives based on factors related to economic and technical considerations and agency statutory missions. The agency should identify and discuss all such factors, including any other essential considerations of national policy which were balanced by the agency in making its decision.

▪ A statement of what practicable means to mitigate environmental damage from the selected course of action will be included in implementing the action. If some practicable mitigation techniques were not included, the reasons for their exclusion should be stated. A monitoring and enforcement program designed to carry out the mitigation techniques identified should be summarized. If a monitoring and enforcement program designed to carry out the mitigation techniques was not included, the reasons for its exclusion should be stated.

It should be noted that although the decision made is often either the "preferred alternative" or the "proposed action," there can be modifications. The decision maker can select any of the alternatives analyzed as the final agency course of action. Beyond the alternatives analyzed in the EIS, the decision maker can select a hybrid course of action (some elements from one alternative, some from another), or can select a course of action that was not specifically analyzed in the EIS as long as it can be shown that the impacts of the action selected fall within the bounds of the environmental impact analysis.

4.4 Application of Environmental Documentation Process

The broad spectrum of potential federal (or state or local) action ranges from major to minor, and the associated environmental impacts from highly significant to truly insignificant. Because of the wide spectrum of possibilities, the CEQ regulations recognize three ways to proceed with an environmental analysis of a proposed action, and the regulations encourage federal agencies to identify and provide guidance on the types of actions that would fall under these three classes of actions. These three classes are as follows:

1. The first class of actions are those known or presumed to result in significant environmental impacts, for which an EIS would be required.

2. The second class of actions are those where the agency has sufficient experience in preparing NEPA reviews on similar proposals to be able to accurately predict that no significant impact would occur, and that

the agency has formally identified the actions to the public through a list in the *Federal Register.* The agency may exclude proposed actions in this category from the requirement to prepare either an EIS or an EA; actions of this type are referred to as "categorical exclusions." Some agencies abbreviate this as "Cat-X" or "CX."

3. The third class of actions are those expected to result in impacts that are not significant, or those where the degree of significance cannot be accurately predicted, for which an EA would be the appropriate initial NEPA review.

Figure 4.1 depicts the process used to determine which path of analysis to take.

Although some preparers think of EISs as "big NEPA," EAs as "medium NEPA," and categorical exclusions as "little NEPA," this mindset oversimplifies the situation. It is possible that the degree of significance of impacts from a proposal analyzed under an EIS, one analyzed under an EA, and one covered by a categorical exclusion would be very similar. The point of a NEPA review is not to explain away or minimize the impacts that might occur, but rather to accurately capture the significance of the environmental factors that would bear upon an agency decision. As stated in the CEQ regulations (40 CFR 1500.1(c)), the purpose of NEPA is not to generate paperwork but to lead to "excellent action."

An agency should embark upon an EIS as the initial level of NEPA review for those "major actions" that would "significantly affect" the environment, as those terms are defined in the CEQ regulations. An agency might have identified a proposal as falling within a class of actions that normally require an EIS. Or, using the gift of common sense, an agency may discern without lengthy analysis or a preliminary EA that a proposal would result in significant environmental impacts and proceed with an EIS, such as when considering plans for very large, very expensive facilities, proposals involving transportation or use of large quantities of highly toxic materials, or actions that would obviously adversely affect large areas of critical habitat for endangered species or infringe upon major archaeological sites. An agency may decide to prepare an EIS as the initial level of NEPA review for one-of-a-kind actions where impacts are unknown or highly uncertain without going through an initial EA process. Lastly, an agency may decide to prepare an EIS on any action "to further the purposes of NEPA," even if environmental impacts would not be significant. The EIS process includes the NOI, the draft EIS (DEIS), the final EIS (FEIS), and the ROD. These documents should be formatted in accordance with the CEQ regulations and the requirements of the proponent agency.

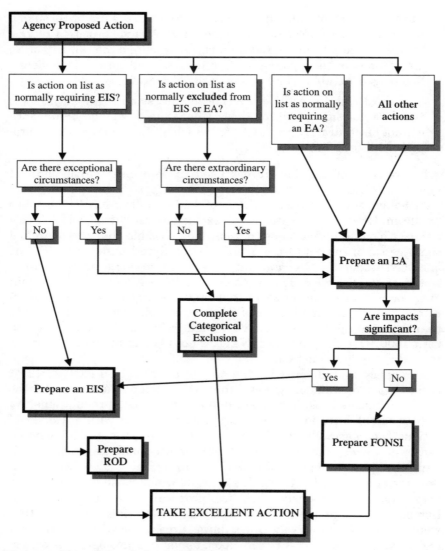

Figure 4.1 Determining the level of NEPA review.

In theory, if an agency prepares an EIS and the analysis indicates that no significant impact would result, the agency could prepare a FONSI instead of an ROD and proceed with the action. In practice, this is rarely (if ever) done because it is difficult to explain or justify to other agencies or the public why the EIS is being abandoned. Furthermore, there is no procedural advantage in terms of paperwork or timing (by the time the agency discovers that there would not be a significant impact, it would generally be just as quick to complete the EIS process and prepare an ROD as to abandon the EIS process and prepare a FONSI). This course of action (abandoning the EIS and preparing a FONSI) would lead to a greater risk of litigation, and higher probability of litigation success against the agency, than the option of completing the EIS process.

If an action is listed by the proponent agency as one that normally would require an EIS, but in a specific case the agency has reason to believe that impact might not be significant, the agency may prepare an EA as an initial level of NEPA review. Depending on the agency, this approach may require several levels of agency approval before being initiated; check with agency requirements before seriously considering this path.

If it is clear that an EIS would not be the initial level of NEPA review, the NEPA preparer should then look to see if a proposed action could be categorically excluded from preparation of either an EIS or an EA. A categorical exclusion is not always reserved for actions with very minor impacts. While the impacts of a categorically excluded action may be minimal, in other situations they may be essentially the same as the impacts of an action addressed by an EA and FONSI. Regardless of how minor the environmental impact may be, an action *must* appear on the agency's list of classes of actions that can be categorically excluded in order to pursue this approach. Categorical exclusions are specific to the issuing agency: An action that is categorically excluded by one federal agency may not be on the list of exclusions of another agency. If the agency knows that in a specific extraordinary instance the impacts from a listed action might be significant, such as if the action were to take place in a designated wetland, the CEQ regulations provide that the categorical exclusion not be applied and that another type of NEPA review (EA or EIS) be performed instead. Requirements for documenting categorical exclusions vary from agency to agency; many agencies do not require any documentation for all or some types of categorical exclusions, and others use some type of checklist or short memorandum. Check with the agency procedures for documentation and notification requirements. As a word of warning, an agency may have reason to believe that a proposed action falls within its listing of categorical exclusions, and fail to perceive that other parties, including regulatory agencies, do not share that opinion. Great care should be taken to try not to stretch the listed definitions, or to "force fit" an action into the desired categorical exclusion, if a substantial segment of the public does not agree.

Between those actions for which the agency will prepare an EIS and those actions that are categorically listed as excluded from the requirement to prepare either an EIS or an EA lies the broad spectrum of actions where either it is thought that impacts would not be significant or it is difficult to determine whether impacts would be significant or not. In these cases, an EA is prepared. The EA analysis is very similar to the EIS analysis, but generally there are fewer alternatives to the proposed action, the alternatives do not have to be analyzed to the same degree (as is the case with an EIS), and the analysis of the proposed action and alternatives needs only to be sufficiently detailed to demonstrate that no significant impacts would occur (leading to a FONSI) or to demonstrate that they would occur (leading to an EIS). In some cases the analysis is inconclusive and it cannot be determined if impacts would be significant or not. In these cases an EIS would be prepared instead of a FONSI. In some cases, the EA analysis might demonstrate that while the proposed action would not be expected to have significant impacts, a reasonable alternative course of action *would* be expected to have significant impacts. In this event, an EIS should be prepared so that the decision maker will have full disclosure of the significance of the options available.

The EA review process includes preparing an EA and, if warranted, a FONSI. Most agencies prepare an EA using a format modeled on that required for an EIS. However, there are other approaches and other formats in use among federal agencies. Some agencies have developed a checklist approach used for simple EAs. An alternative format for a simple EA is described later in this chapter.

While the content of an EA is similar to that of an EIS, the process is simpler. An agency is not obligated to provide notification to the public or other agencies that an EA is being prepared, does not have to prepare a draft EA for circulation and comment, and does not have to consider public input in the preparation of the document, although many agencies take some or all of these steps. The agency does not have to disclose in any detail its final decision or its decision factors, although it does have to disclose its FONSI and the EA upon which it is based. To facilitate agency action and public access to information, some agencies prepare a decision record along with the FONSI and make it available, similar to the ROD following an EIS.

As a word of caution, it is not prudent to attempt to avoid preparing an EIS by intentionally understating the possible impact of the action, trying to "explain away" all potential impacts, or selecting alternatives that will make the proposal look relatively benign. One common avenue for citizen action against the federal government is the challenge before a court that an agency prepared an EA and FONSI when an EIS was required; since NEPA is a procedural law and the determination of significance is somewhat subjective, judges are often sympathetic to this argument and remand the documents to the agency with instructions to prepare an EIS before taking further action. These types of challenges have caused lengthy and expensive project delays.

4.5   EIS Format and Content

The CEQ NEPA regulations provide a standard format for all federal agencies to follow in preparing an EIS (40 CFR 1502.10) (see Appendix D). If an agency determines that there is a compelling reason to do otherwise, it may use an alternative format, but this is discouraged. A generic outline for a prescribed EIS format is provided in Fig. 4.2. Each required item is discussed briefly, below.

The CEQ regulations state that an EIS should normally be less than 150 pages, or 300 pages for unusually complex analyses, but many agencies routinely exceed these limits. Some agencies have additional formatting requirements, and agency guidance should always be consulted prior to beginning an EIS.

Although this section discusses each part of the EIS document in turn, the parts of the EIS are interdependent. Preparers may need to work on more than one section at a time, and make iterative changes as alternatives are refined, mitigation measures developed, and impacts analyzed. The document manager should develop a schedule for completing each different part of the document and ensure that the members of the EIS team communicate frequently so that the analysis is internally consistent.

Cover sheet

The cover sheet should be one page. See Fig. 4.2 for required information. (See also 40 CFR 1502.11.)

Summary

Each EIS is required to contain a summary; this is often called an Executive Summary. It is suggested that the format of the summary follow the general outline of the main body of the EIS. The summary should outline the decision to be made, and should stress the major points of the analysis, including alternatives considered, conclusions, areas of controversy (including topics raised by other agencies or the public), and the issues to be resolved. For a long EIS, the summary may be published as a separate volume of the document.

Table of contents

The table of contents should include, at a minimum, the headings for all chapters and major sections, and appendices. A list of figures and

Format for an EIS

1. **Cover sheet (1 page)**
    Title of the proposed action
    Location of project
    EIS designation (agency number): Draft, Final, or Supplemental
    Lead agency, and cooperating agencies if any
    Agency point of contact (name, address, phone number, and e-mail if available)
    Date by which comments must be received
    Abstract (one paragraph)

2. **Summary**
    Proposed action and alternatives considered
    Summary of EIS content (suggested to follow EIS format)
    Conclusion
    Areas of controversy
    Issues to be resolved

3. **Table of contents (and other front matter)**
    Chapters and section headings, including Appendices
    List of figures
    List of tables
    List of abbreviations and acronyms
    List of scientific or foreign symbols (explanations of scientific notation may be helpful)

4. **Purpose and need for agency action**
    Underlying need (goal) for agency action
    Purpose (objectives) of the proposed action

5. **Proposed action and alternatives**
    Description of the proposed action
    Description of the preferred alternative
    Description of each alternative considered and analyzed (including No Action)
    Description of alternatives considered and dismissed
    Comparative summary of environmental consequences of alternatives
    Mitigation measures identified

6. **Affected environment**
    Description of each affected environmental attribute and baseline condition, including
      natural, cultural, social, economic, and aesthetic environments
    Reference to material summarized, appended, or incorporated by reference

7. **Environmental consequences**
    Direct effects
    Indirect effects
    Cumulative effects
    Consultation requirements with other agencies, if applicable
    Conflicts with plans of other federal, state, local, or tribal agencies
    Irreversible or irretrievable consequences
    Energy requirements and conservation potential
    Natural resource requirements and conservation potential
    Need for mitigation measures, and an analysis of their potential effectiveness

Figure 4.2 CEQ-prescribed outline for EIS content.

8. **List of preparers**
   Name and qualifications (to demonstrate interdisciplinary approach), and area of expertise in the document (reference sections if possible)

9. **Distribution list**
   List of agencies from whom official comment is requested (including state clearinghouses)
   Identify other agencies, officials, and organizations from whom comment is solicited
   Other parties and individuals receiving a copy of the EIS (optional)
   Locations where EIS copies are available to the public for review

10. **Index**
    At a minimum, generate index by major environmental topics, such as "wildlife"

11. **Appendices**
    Material prepared in support of EIS
    Analysis to support effects
    Analytic methodologies
    Analytic computations relevant to the analysis
    Classified, proprietary, or confidential matter may be placed in an appendix and reserved
      from public review

Figure 4.2   CEQ-prescribed outline for EIS content. (*Continued*)

tables should also be included. If the preparer takes a bit of extra care in organizing the table of contents, it will help the document readers considerably.

In addition to the table of contents, the front matter of the document often includes other helpful information that applies to the entire document. This often includes lists of abbreviations and acronyms, a short explanation of scientific notation, notation of Greek or other foreign symbols used, and other similar material. A glossary of technical terms is desirable in many cases, especially in cases where the topics are not generally familiar to the general public or other agencies (such as, for example, medical, military, or geologic terms). Sometimes a glossary is placed at the back of the document near the index instead of with the front matter.

Purpose and need

Briefly describe the purpose and need to which the agency is responding. The "need" is the underlying goal of the proposed action, while the "purpose" of the proposal is to meet the stated need. The "proposed action" is how the agency plans to meet the purpose and need. For example, a need may be to "enhance national security posture," the purpose may be to "provide laser capability," and the proposed action to "construct a new laser facility." For another example, a need may be to "augment the intrastate transportation system," a purpose to "provide a cross-city transportation route," and the proposed action to "construct a freeway across the city." Agencies sometimes confuse the purpose and need with the proposed action—in the example just given, an agency may state that the purpose of the proposed project is to "construct a freeway across the city" and then go on to also state that the proposed action is to "construct a freeway across the city." This is circular logic, and will not allow for a reasoned development of alternatives to the proposed action.

Some agencies combine purpose and need and do not distinguish between the two; other agencies consider them separately. In either case, the section should clearly state the underlying problem to which the agency is responding. While the proposed action and alternatives represent possible *answers* to the problem, the section on purpose and need states *what the problem is*. It is crucial to accurately define the purpose and need, because in the event that the EIS is reviewed by a court, the judge may well define the adequacy of the analysis by how well the proposed action and its alternatives respond to the agency's statement of purpose and need. At times, the need to take agency action is required by law, the wording of the authorizing legislation for a new project, a congressional budget line item, a court order, or other legal or judicial considerations. In these cases it is helpful to make it very clear that the underlying purpose and need for agency action is in response to these drivers that are beyond the agency's control. At times an EIS is written in response to a suggestion in congressional debate language; although perhaps this purpose does not carry the weight of law, it should not be overlooked.

For any action, the agency may decide to prepare an EIS where one is not otherwise required to "further the purposes of NEPA." This may be done in order to provide a fuller public disclosure of environmental impacts, to increase the degree of public participation in the environmental review process, to accommodate the requests of another agency, or to assess the impacts of ongoing activities where there is no proposal for change.

An early format for an EIS, used in the early 1970s, called for the proponent to state, as the first point, what it wished to do and why. The advantage of this earlier process was that a clear, logical, unequivocal statement of purpose was integral to the development and success of the succeeding analysis. The public is rarely accepting of unclear statements of purpose and need, or justification of a decision after the fact. The EIS preparer may want to focus on this aspect of the NEPA analysis to clarify the concepts of the underlying purpose and need for the project before proceeding with developing the remainder of the document.

Alternatives including the proposed action

The CEQ considers this section to be "the heart" of the EIS. In this section the agency describes the proposed action and identifies the alternatives that it considered. The proposed action and the other alternatives should be responsive to the problem stated in the section on purpose and need. The lead agency is expected to consider a range of reasonable alternatives, including alternatives under the jurisdiction of another agency if relevant.

A NEPA analysis is a comparative analysis—it compares the environmental impacts of taking a proposed course of action against the impacts if the action were not taken (the so-called no-action alternative) and the impacts that would occur if an alternative course of action were taken. The impacts are calculated over time and use a common time frame: For example, what would be the impact in 10 years if a facility were to be constructed at site A compared to the impact in the same 10-year time frame if the facility were to be constructed at site B? The proposed action and the alternatives to be analyzed must be carefully defined, or else potential impacts may be overlooked (if the description is too sketchy) or exaggerated (if the description includes overly conservative assumptions or broad parameters). The EIS preparer may find that it is more difficult to define and describe the proposed action than one would think, and may have to work closely with the project engineers to define details before the analysis can proceed.

Often there are several alternatives that are considered initially. Many may prove to be unreasonable due to cost, schedule, or agency mission constraints; are unresponsive to the agency purpose and need; or are very similar to other alternatives and so would not provide a range of alternatives. Alternatives that are considered and dismissed should be briefly discussed, along with the rationale for their dismissal. Through the agency and public scoping process, the EIS preparers will narrow the list of alternatives to those reasonable alternatives that will assist the decision maker in focusing on the issues ripe for decision. These alternatives will be analyzed in the EIS, and must be considered to a comparable level of detail. Usually they include things such as alternative locations, technologies, timing, or construction techniques. Most EISs fully analyze between three and six alternatives.

At such time as it knows its preference, the agency should identify its preferred alternative. Usually this is the proposed action and is known at the time of the Notice of Intent or publication of the draft EIS, but occasionally the agency may not select a preferred course of action until the final EIS is prepared and the agency has had a chance to consider public input on its draft analysis. The EIS should include a rationale for why this is the agency's preferred course of action.

The EIS is required, by law and regulation, to include, describe, and analyze the no-action alternatives. Sometimes the EIS preparer will hear the argument that the no-action alternative is not reasonable, or else is not responsive to the purpose and need for action and therefore should not be included in the analysis. It is true that if the agency had no problem with the status quo, it would not be seeking an alternative course of action through the initial proposal. However, it must be understood that the no-action alternative is used in the EIS as a baseline for the comparative analysis: What would be the environmental impact if the agency took a proposed action *versus* the environmental impact if the agency did not take the proposed action? The no-action alternative might be a continuation of current actions over time, or it might be the condition of the ambient environment over time if a new facility were not constructed. It would not be the cessation of existing activities, or pretending that a proposed construction site is a pristine meadow instead of a heavily disturbed site that has known past industrial use. The impacts of the proposed action are defined by the comparison to the no-action alternative. The impacts of other alternatives may be compared to the impacts of either the proposed action or the no-action, but the document preparer must be careful to explain which analytical technique is used and be internally consistent within the document. See also Chapter 5.

Under the CEQ regulations, this section of an EIS must include a comparative description of the environmental impacts of each alternative, including the no-action alternative. This is drawn from the full impact analysis prepared under section 7, environmental consequences, below. In most EISs this information is presented in tabular form, although it can be presented as a comparative summary.

Whether these are considered as part of an alternative or identified through the impact analysis, the EIS must identify means to mitigate adverse environmental impacts. A summary of mitigation measures and their ameliorating effects on the adverse impacts can be included in this section, but are often included in section 7 instead. Mitigation measures are determined not only for the preferred alternative, but for any or all alternatives.

Affected environment

This section describes those attributes of the environment that would be affected by the proposed action or one of the alternatives analyzed. The section is also commonly called the environmental setting, or the environmental baseline. The different aspects of the affected environment should be described succinctly and the information should be relevant to the impacts discussed. The description should not be so verbose that the agency could be accused of "hiding" significant effects among many insignificant items. This section should emphasize the aspects of the environment that could change (be affected) due to the proposed action or one of the alternatives analyzed, in order to help the decision maker focus sharply on the environmental issues that distinguish among the alternative courses of action. Lengthy analyses should be incorporated by reference or moved to an appendix; completed documents should be incorporated by reference instead of being appended. It is important to note that the baseline is not necessarily static, but would be expected to change over time.

Environmental consequences

This section of the EIS provides the scientific and analytic basis for the comparison of alternatives analyzed in the document. Environmental consequences of impacts to be considered should include direct, indirect, cumulative, and induced impacts in the biological, physical, social, economic, cultural, and aesthetic environments. The discussion should include adverse environmental impacts which cannot be avoided, the relationship between short-term uses of the human environment and long-term productivity, and any irreversible or irretrievable commitments of resources which would be involved should the action be implemented.

Determining cumulative impacts can be particularly challenging. Analysts must predict the cause-and-effect relationships between the proposed and alternative actions and the resources, ecosystems, and social environments of concern. They must then describe the consequences of the action and alternatives using mathematical modeling, trend analysis, and scenario building. The CEQ has provided additional instruction on this issue in its 1997 guidance document *Considering Cumulative Impacts* (CEQ, 1997).

The CEQ regulations suggest that an EIS show how the alternatives considered and decisions made upon these alternatives will or will not achieve the requirements of Sections 101 and 102(1) of NEPA and other environmental laws and policies. This information can be included in this section.

List of preparers

The names and qualifications (expertise, experience, professional disciplines, educational background) of those persons primarily responsible for preparing the EIS or developing background papers or analyses should be included in the list of preparers. Where possible, persons responsible for a particular analysis section of the EIS should be so identified. The list of preparers serves three purposes: (1) It provides a basis for evaluating whether a systematic, interdisciplinary approach was actually used in preparing the EIS, (2) it increases the accountability and professional responsibility of those who prepared the different parts of the EIS, and (3) it gives due credit to and enhances the professional standing of the preparers. Although the CEQ regulations suggest a list of about two pages, it is common for the list of prepares to be longer, especially for complex analyses that draw upon many types of expertise.

Index

The document should have an index that allows immediate identification of where EIS elements of particular interest are located. For example, by referring to the table of contents and the index, a reader should be able to readily determine all sections where water quality is discussed. Most word-processing systems allow for easy identification and notation of items for the index, which is relatively simple if thought out before the document is compiled.

Appendices

The EIS may include as many appendices as actually needed; however, the appendices should not become a dumping ground for inclusion of irrelevant or unnecessary paperwork. Only material specifically prepared in support of the EIS and material needed to substantiate analysis in the main body of the EIS should be included. It is often necessary to perform extensive computations to determine some impacts. It would be appropriate to include such analytic computations and their scientific basis in an appendix. The appendices can be circulated with the EIS, or, if they are lengthy or costly, can be made available only upon request. If the analysis relies upon classified or proprietary information, this material can be placed in an appendix that is withheld from public circulation.

4.6   Preparing and Processing the EA or EIS

Preparing and processing the NEPA document depends upon whether it is an EA, FONSI, DEIS, FEIS, or ROD. All NEPA documents are public records that can be accessed under the provisions of the Freedom of Information Act (FOIA) (5 USC 552). Most agencies anticipate these requests and have procedures to make NEPA documents readily available.

Environmental assessment

If an agency is not certain whether a proposed action would result in significant environmental impacts within the meaning of NEPA, it may prepare an EA. An EA is intended to be a concise document that briefly provides sufficient information to allow the agency to decide whether impacts of a proposal or its alternatives would be expected to be significant, in which case an environmental impact statement would be prepared, or whether no significant impact would be expected to occur (40 CFR 1508.9). An agency is not required to prepare an EA to determine significance. In practice, most agencies *do not* prepare an EA if they have already decided to prepare a full environmental impact statement either because it is anticipated, or known, that impacts will be significant, or because the agency has decided that preparing a full environmental impact statement will further the purposes of NEPA regardless of the degree of significance of impacts. Most agencies *do* embark on an EA if the proposed action is on the agency list as normally requiring an EA, if the degree of significance of anticipated impacts is not known, or if impacts are not expected to be significant, because the preparation of an EA is generally seen as quicker and less onerous than preparing an EIS. If the EA indicates that the anticipated environmental impacts of the proposal, or the alternatives considered, *will* be, or may *possibly* be, significant within the meaning of NEPA, the proposed action cannot be taken unless the agency prepares an environmental impact statement. At the point where this becomes obvious, most agencies do not complete the EA but abandon it and begin the EIS process by issuing a formal NOI. For this reason, the student of NEPA will find very few, if any, completed EAs that demonstrate significance.

If an EA is prepared and it is determined that no EIS is required, the documentation, processing, and other follow-up procedures vary both among and within federal agencies. The CEQ NEPA regulations give the agency considerable leeway to determine its own procedures. EAs are sometimes viewed as "mini-EISs" and are prepared and formatted accordingly. Although most agencies use a format for an EA that is similar to that required by regulation for an EIS, some agencies use variations, especially for shorter or simpler analyses.

An alternative format, and comparison to the standard EIS-type format style used for EAs, is given in Fig. 4.3. The alternative format identifies and describes the impacts for each attribute of the environment (such as wildlife habitat or cultural resources) in turn, whereas the standard format identifies and describes the impacts for each alternative in turn. The advantage of the standard format is that it collects impacts as a total picture under each alternative, and so is of greater use to those who want to see the impact of each alternative as a whole. The advantage of the alternative format is that it more sharply defines the impacts to each environmental attribute by describing the differences among the alternatives for each attribute. It

Standard EA Format	Alternative EA Format
Summary	Summary (same)
Purpose and Need	Purpose and Need (same)
Description of Proposed Action and Alternatives	Description of Proposed Action and Alternatives (same)
Affected environment • Existing situation – Attribute A • Existing situation – Attribute B • Existing situation – Attribute C	Affected environment – Attribute A • Existing situation • Environmental consequences, Alt. 1 • Environmental consequences, Alt. 2 • Environmental consequences, Alt. 3
Environmental consequences – Alt. 1 • Attribute A • Attribute B • Attribute C	Affected environment – Attribute B • Existing situation • Environmental consequences, Alt. 1 • Environmental consequences, Alt. 2 • Environmental consequences, Alt. 3
Environmental consequences – Alt. 2 • Attribute A • Attribute B • Attribute C	Affected environment – Attribute C • Existing situation • Environmental consequences, Alt. 1 • Environmental consequences, Alt. 2 • Environmental consequences, Alt. 3
Environmental consequences – Alt. 3 • Attribute A • Attribute B • Attribute C	
Cumulative Impacts	Cumulative Impacts (same)

Figure 4.3 Comparison between standard EA format and alternative EA format.

is also easier for a subject matter expert to prepare or review, since all the salient information about a given environmental attribute is collected in one place, and it is easier to ensure that impacts to the given attribute were projected for each alternative and not overlooked.

In many cases, the proponent of the action is required only to document the assessment and retain a copy in the project files. Some proponent agencies may require that copies be forwarded to offices within the agency, as specified in their own agency guidance. It is common for an agency to require that an EA be prepared prior to agency commitment of funds to carry out the project, and that this fact be recorded with the funding entity. Some agencies circulate draft EAs or completed EAs to other agencies and the general public for review and comment, similar to the EIS process.

Finding of no significant impact

If an EA demonstrates that the impacts from a proposed project or any of the alternatives analyzed would not be significant, the agency may prepare a FONSI. Although the CEQ regulations do not require that a FONSI be formally recorded, it is not uncommon for agency procedures to require notification in local media or the *Federal Register* that a FONSI has been prepared, or even publication of the full text of the document in the *Federal Register.* This is especially true when the circumstances of the action are such that

- The proposed action is on the agency's list as one that would normally require preparation of an EIS, but an EA has been prepared instead.
- The FONSI is based on mitigation of potentially significant impacts.
- The action considered is of nationwide interest.
- The nature of the proposed action is one without precedent.

In certain limited circumstances, which the agency may cover in its own NEPA procedures, the agency might make the FONSI available for public review and comment (including federal, state, and tribal agencies and statewide clearinghouses) for 30 days before the agency makes its final determination whether to prepare an EIS or proceed with the project on the merits of the FONSI. If a federal agency prepares an EA that demonstrates that a proposed action and the alternatives considered would not result in a significant environmental impact within the meaning of NEPA, the agency may issue a FONSI and proceed with the action.

The FONSI briefly presents the reasons why the action considered in an EA would not have a significant impact on the human environment, and the rationale for why an EIS would not be required. Some agencies include the FONSI within the body of the EA, and others prepare the FONSI as a separate document with reference to the EA. Some agencies combine a FONSI with a statement of what the agency decision is, similar to the ROD following an EIS. Some agencies allow a FONSI to include reference to mitigating measures that must occur to enable impacts to remain below the threshold of significance. If an agency cannot reach a FONSI for a proposal, it must prepare an EIS before proceeding with the action.

Scoping process

The term *scoping* refers to the process to determine the range of alternatives and analysis, that is, the *scope* of issues to be addressed in the EIS. For example, what is the definition of the proposed action? What are the reasonable alternatives? Exactly which aspects of the environment are important for this project at this time and in this place? The scoping process has two primary aspects: the *internal* process within the proponent agency, and the *external* process which involves other federal agencies, states, tribes, local governments, and the general public. Through the internal process the agency develops the proposed action; considers, rejects, or accepts alternatives; defines the no-action alternative; checks to see whether the proposal is consistent with agency plans and policies; and identifies the environmental baseline information that will be needed for the analysis. Through the external process, the agency gathers relevant information from other cognizant agencies regarding alternatives or the environmental baseline, gains information from the public that may bear on the proposal, such as who may be affected and whether it will be considered controversial, and engages other interested parties in the information collection process.

The informal (internal) scoping process starts at the time that the agency articulates its proposal; the formal (external) process should start as soon as possible after the proponent agency proposes to take action. The public scoping process starts with the agency's publication in the *Federal Register* of its formal Notice of Intent to prepare an EIS.

As part of the scoping process, internal and external, the lead agency should

1. Invite the participation of affected federal, state, and local agencies, Native American tribal governments, and other potentially interested parties.
2. If required by the agency, or as an option, hold public scoping meetings.
3. Invite, receive, and consider spoken or written comments on the scope of the document and the environmental analysis.
4. Define the proposed action and alternatives, including the no-action alternative.
5. Identify other EISs, EAs, or environmental studies that have been prepared or are under preparation by the proponent agency, other agencies, or other entities (such as a state government) that are related to the proposed action.
6. Identify issues to be analyzed in depth in the EIS.
7. Identify and eliminate issues which are not relevant, or which have been adequately covered by prior environmental review.
8. Determine additional analyses, such as field studies, statistical computations, or siting studies, that will be required to support the EIS.
9. Assemble an interdisciplinary team to prepare the EIS, including personnel from cooperating agencies (where appropriate), and allocate assignments.
10. Develop a timetable for preparing the EIS and agency decision making.

11. As an option, develop and circulate a potential table of contents, including page limits.

The draft EIS itself will attest to the scope of its analysis, and the agency does not need to issue a separate document indicating the results of the scoping process or its determination of scope (although sometimes an agency will do this, particularly if it anticipates a lengthy time to prepare the draft EIS or experiences a delay in the project). The agency is free to change the scope of the proposal, the analysis, the identification of issues, or environmental documentation at any time up to and including preparation of the final EIS.

Draft environmental impact statement

If the agency determines that an EIS is required, either because the proposed action is on the agency's list of actions normally requiring preparation of an EIS or as a result of the EA process, the next step is to prepare and process a draft EIS. The DEIS is prepared in accordance with the CEQ regulations. After undergoing internal review, a process that varies among agencies, the DEIS is circulated for agency and public review as outlined in the CEQ regulations.

The DEIS must be filed with the Office of Federal Activities, EPA, Washington, D.C., in compliance with the CEQ regulations (40 CFR 1506.9). The EPA will publish a notice of availability of the DEIS in the *Federal Register*, and will review the document as described in Chapter 7.

After preparing the DEIS and before proceeding with the FEIS, the agency should solicit comments from the following groups:

- Federal agencies that have jurisdiction over the proposed action or an alternative by law or special expertise

- State and local agency that are authorized to develop and enforce environmental standards

- Native American bands or tribes, when the proposed action or an alternative would affect them, their tribal lands, or their traditional cultural properties

- The applicant, if the federal agency is considering the action of issuing a lease, license, permit, or entitlement

- Any party that has requested a copy of the DEIS

- The general public and nongovernmental organizations affected by, or potentially interested in, the action (it is the agency's responsibility to make a reasonable attempt to identify such parties)

Although the CEQ regulations refer to Office of Management and Budget Circular A-95 clearinghouses as a means to circulate the DEIS to state and local environmental agencies, many of these clearinghouses are inactive. Reliance on this method may be problematic if it is the sole avenue used for the release of time-sensitive material, including a DEIS, because many of these clearinghouses, even where functioning, operate at a low level of activity and may be slow to distribute information or choose not to duplicate large documents because of the expense.

Federal agencies with jurisdiction by law or special expertise in the environmental arena covered by the DEIS are expected to comment on the document. As an example, the U.S. Fish and Wildlife Service, an agency of the U.S. Department of the Interior, has special expertise on endangered species and is required by regulation to comment within a certain time frame on all actions potentially affecting federally listed threatened or endangered species. An agency or other party may respond with the notation that it has no comments to make. In order to reduce unnecessary delays, agencies are requested to provide their comments within the time limits designated by the preparing agency.

The CEQ regulations ask that comments be as specific as possible and that comments be designed to further assist the NEPA process. To be most useful, comments should focus on the following:

- The adequacy of the document, including the merits of the alternatives or of the analysis

- The adequacy of the agency's scientific or predictive methodology, and if in disagreement, a description of the preferred approach and the rationale for the preference

- Additional information held by the commentor that may assist in determining effects

When a cooperating agency is issuing a permit, license, or entitlement, the proponent agency may request additional information if needed to determine site-specific effects. If the cooperating agency has any reservations about the proposal on the grounds of environmental impacts identified in the DEIS, the agency proffering the objections should specify proposed mitigation measures or alternative conditions that it considers necessary for its approval of the proposal.

In practice, the NEPA preparer will find that comments received on the DEIS run the gamut from specific questions about the analysis to broad statements about the agency mission. The preparer should make an attempt to address all comments received rather than to curtly dismiss them as irrelevant.

Final environmental impact statement

The FEIS incorporates all changes to the DEIS that have come about from refinements generated by the proponent agency, comments from other agencies, and comments from the general public. The agency should take advantage of this input and make changes, if necessary, whether in response to an error in the draft or, at the agency's discretion, to incorporate good ideas. The agency must consider comments received on the draft EIS both individually and collectively, and prepare a summary of the comments received and their disposition.

The regulatory charge to "incorporate" comments into the preparation of the FEIS may be handled in many different ways. All substantive comments received on the draft EIS should be considered and addressed. (A comment stating simply that the respondent was opposed to, or in favor of, the proposed action may not be considered "substantive" by many agencies.) If changes have been made to the draft due to substantive comments, the agency can simply annotate that the comment was accepted and the change made. If the comment does not warrant a change in the document text, the agency should write a brief response to state why the comment was not accepted. It should be noted that spoken comments received at a public hearing are expected by the commentor to be given the same weight as a written comment, and the agency should have a way to record and consider these, such as through a verbatim transcript of the hearing.

Agencies have different ways of acknowledging public comments. At a minimum, the agency should summarize comments and agency response. In addition, most agencies attach all comments, or all substantive comments, to the FEIS, along with the agency response to each comment or category of comments. This is often done by placing the comments and responses in a separate volume or appendix of the FEIS. Voluminous comments (such as submission of an entire book) or duplicatory comments (such as a form letter) may be summarized, but are part of the public record and must be made available for public review if requested under FOIA.

When changes made to the FEIS in response to review of the DEIS are truly minor, such as corrections of typographical errors or minor changes to data, the proponent agency can sometimes prepare a simple document including only the comments and responses, and reprint only the pages with changes (or prepare an errata sheet). The FEIS, then, would consist of the DEIS text as modified by those few revised pages.

It is more common to find that "incorporation" of agency and public comments on the draft EIS requires considerable revision to the text of the draft EIS. The agency may need to revise alternatives, complete additional field studies, and discuss areas brought to light through the comment process or raised as a point of dissention. A well-prepared FEIS may add a considerable amount of new material. The newly drafted FEIS, ·including the comment summary, comments, and agency responses, would then completely replace the DEIS. Occasionally the agency may find that the document must be so extensively revised in response to public or regulatory comments that it is necessary to issue a second, revised, DEIS on the basis that the action originally described, the alternatives to meet the need, or the analysis originally made would not be recognizable. Another reason to issue a second draft EIS would be if agency procedures changed, or new information became available that would affect the conclusions of the analysis; examples are a change in the listing of endangered species and implementation of new agency guidance such as designation of critical habitat. If the changes to the draft EIS are extreme, but not so drastic as to warrant preparation and circulation of a second DEIS, the agency may circulate the FEIS for review and comment prior to reaching a final agency decision.

The CEQ regulations require that the final EIS, when completed, together with comments and responses (or a comment/response summary) be filed with the Office of Federal Activities, EPA, in Washington, D.C. The FEIS should be made available to federal, state, tribal, and local government agencies, and the general public, at the same time it is filed with the EPA. To comply with the regulatory requirement to make the FEIS available to the President, the EPA will deliver one copy of the document to the CEQ.

Supplemental reviews

After a NEPA review is completed, and before the action decided upon is implemented in its entirety, an agency may find a need to reopen, or supplement, the initial analysis. This is most often done if there are substantial changes to the proposed action that are relevant to environmental concerns or if there are significant new circumstances that bear on the analysis, such as changes to either the affected environment or changes in the knowledge base concerning the affected environment. In either case—a change in the action or a change in the environment—the agency may prepare a supplemental analysis to assist in its agency decision making (see 40 CFR 1502.9(c)). The agency may prepare some sort of document to record its consideration of whether or not a supplement is needed, and may make that discussion public.

The CEQ regulations provide guidance on supplementing an EIS; in addition, individual agencies may cover this topic in their own NEPA regulations, guidelines, or procedures. If an agency decides to prepare a supplemental EIS, a draft and final EISs are prepared, formatted, circulated, and filed in the same manner as an initial EIS. However, for a supplemental EIS, external scoping may be omitted, unless provided for under the proponent agency NEPA requirements. The regulations are silent on supplementing an EA, but this is often done, usually following the guidance given for supplemental EISs.

4.7   Timing of Agency Action

Time frames for an EIS are calculated from the date that the NOI is published in the *Federal Register*. See Fig. 4.4. The EPA publishes a weekly notice in the *Federal Register* of the EISs filed during the preceding week. The minimum time periods set forth by the CEQ should be calculated from the date of *publication* of this notice, not delivery of the document to the EPA. Additionally, if the EIS is not *delivered* by the time the *Federal Register* is received, commenting agencies and groups may request, and be given, a time extension dating from when the document was constructively available to them. No decision on the proposed action should be made or recorded by a federal agency except as follows:

- Ninety days after publication of the notice for a draft EIS.

- Thirty days after publication of the notice for a final EIS. However, there is an exception for those rule-making actions where an agency may announce a preliminary decision at the same time an EIS is filed. In such cases there must be a real opportunity to alter the decision. This means that the period for appeal of the decision and the 30-day period required for the EIS process may be concurrent. An agency engaged in rule making under the Administrative Procedures Act or other statute for the purpose of protecting the public health or safety may waive this 30-day period.

- The minimum of 90 days required between the draft EIS and the final action (or recording of the decision) and the 30-day waiting period after the EIS can be concurrent. However, a minimum of 45 days must be provided for comments by other agencies and the public. This is often extended considerably.

- The lead agency may extend (but not reduce) these prescribed time periods. Only the EPA may reduce the prescribed time periods at the request of the lead agency due to compelling reasons of national policy. Also, the EPA may extend the time periods at the request of other federal agencies (other than the lead agency) in consultation with the lead agency. The EPA is required to notify the CEQ for any such extension or reduction of the time periods. There are some other restrictions as well, and these are further described in the CEQ regulations.

Total elapsed time – 8 months to 2 years

Act	Issue ROD	Prepare FEIS		Prepare DEIS	
Min. 90 da. After DEIS Or 30 da. After FEIS	Opportunity for response: 30 da. min.	Variable	Write FEIS Variable: 1 – 6 mos.	Comment Period Variable: 45 da. min.	Write DEIS Variable: 2 – 6 mos.	Public Scoping Process Variable: 30 – 90 da.	Variable

Agency Concept for New Action
- Determine that a new action is needed
- Determine level of initial NEPA review
- Develop initial proposed scope for EIS analysis
- Prepare Notice of Intent

Begin Public Scoping Process
- Publish Notice of Intent in *Federal Register*
- Invite comments from other agencies, states, tribes, local governments, and public
- Publish times, dates, and location of public meetings, if held (min. 15 days prior to meeting)
- Hold public meetings (optional)

Complete Public Scoping Process
- Hold public meetings (optional)
- Receive and analyze agency and public comments

Prepare DEIS document
- Conduct field studies, modeling, and analyses
- Write text
- Prepare text and figures, print
- Complete internal agency review

Issue DEIS for review and comment
- Complete text and file with EPA
- EPA issues Notice of Availability
- Circulate DEIS
- Invite comments from other agencies, states, tribes, local governments, and public
- Publish times, dates, and location of public hearings (min. 15 days prior to hearing)
- Hold public hearings

Prepare FEIS document
- Receive and analyze agency and public comments
- Prepare comment response and summary
- Develop additional analyses or mitigation measures
- Revise text as needed, print

Issue FEIS
- Revise text per agency and public comments
- Complete text and file with EPA
- EPA issues Notice of Availability
- Circulate FEIS

Complete ROD (or ROD becomes effective)
- Determine final decision
- Determine mitigation measures
- Prepare ROD
- Issue ROD – make publicly available

Take Action
- Implement action
- Apply mitigation
- Monitor over time

Figure 4.4 Time requirements for processing an EIS.

4.8 Tiering

The ability to "tier" environmental assessments and statements can be very useful. Tiering refers to the coverage of one level of environmental documentation in a broad "programmatic" EIS, followed by more detailed analyses and environmental documentation for a site-specific action or a subset of the broad program. The subsequent EIS or EA need not repeat in full the issues treated in the programmatic document, but may summarize the issues discussed in the broader statement and concentrate on issues specific to the subsequent action. Tiering is often used where an agency has developed a plan or program using the NEPA process (top tier) and uses subsequent EAs or EISs to analyze impacts of specific implementing actions under that plan (bottom tier). Tiering may also be appropriate for large projects where all the details are not available in the earlier stages of the project.

4.9 Mitigation

An important part of the analysis content of an EA or EIS is mitigation-specific statements showing how potentially adverse impacts may be lessened or avoided. If the inclusion of mitigation techniques is to have any meaning, it is essential that those identified in the EIS be carried out. Case law clearly shows that the mitigation procedures incorporated in an EIS are legally binding commitments on the proponent agency. The lead agency, therefore, should provide a framework for implementing mitigation techniques, and the framework should encompass the following:

- Appropriate conditions in grants, permits, or the approvals. (This item would apply when a federal agency is issuing such a grant, permit, or other approval to a nongovernmental agency.)
- Funding of actions conditioned on proper implementation of the mitigation techniques required.
- Upon request, informing cooperating or commenting agencies on the progress in carrying out mitigation procedures that were a part of the EIS.
- Upon request, making available to the public results of relevant monitoring to ensure that mitigation is being carried out.

We note here that the term *mitigation* has developed, over the years, two distinct meanings within the environmental assessment community. As used here, it implies "means taken to minimize damage that would otherwise occur." The alternate meaning, most commonly used when the resources involved include fish and wildlife habitat, is closer to "land which the agency will purchase and allow to be devoted to fish and wildlife use as compensation for habitat damaged or occupied by the agency's project." It would be preferable if the term were restricted to the former definition, but the latter has become widely used in many agencies, and is acceptable *if it is clear what is meant when the word is used.* There are certain agencies where the term is used almost entirely in the latter sense. When discussions take place with those agencies, the use of this term must be made unequivocally clear. We note that the EPA normally rates as "unacceptable" an EIS which shows effects on threatened or endangered species or fish and wildlife habitat, but does not detail the exact mitigation measures proposed.

4.10   General Considerations in EIS Preparation

General comments included in the 1979 NEPA regulations regarding the preparation of EISs can be summarized as follows:

- EISs should be analytic rather than encyclopedic.
- Impacts should be discussed in proportion to their significance.
- EISs should be concise.
- EISs should state how alternatives considered and decisions made based upon the EIS will or will not achieve the requirements of Sections 101 and 102(1) of NEPA and environmental laws and policies.
- Alternatives discussed should be limited to those which are expected to be considered by the agency decision maker.
- The agency should not commit resources prejudicing selection of alternatives before completing the NEPA process.
- EISs should be a means of assessing the environmental impact of the *proposed* action, rather than a means of justifying decisions already made.
- A systematic and interdisciplinary approach should be used to prepare EISs.
- EISs should be written in plain language and appropriate graphics used so that decision makers and the public can readily understand the documents.

4.11   Case Studies

This section presents three case studies of interest to the NEPA student. They illustrate points and potential pitfalls brought out in this chapter.

Case study 1. Scope creep

It is normal for the scope, or details, of a project to change or evolve as designs are finalized, new equipment becomes available, or new technologies are developed. Sometimes, however, many small incremental changes over time can add up to a change in the scope of the project that is large enough to cross a threshold of significance and negate the original NEPA review. A case in point was the Dual Axis Radiographic Hydrodynamic Test (DARHT) Facility, an accelerator-based diagnostic test machine constructed in the 1990s at the Los Alamos National Laboratory, a New Mexico nuclear weapons research and development lab administered by the U.S. Department of Energy (Webb, 1997). In this case study, we see that agency reluctance to reexamine the accretion of environmental impacts due to incremental project changes resulted in a court injunction, expensive project delays, and court direction to prepare an EIS (the course of action the agency initially sought to avoid).

The DARHT Facility was designed to use two linear accelerators to power radiographic equipment that produces x-ray images during hydrodynamic diagnostic tests (DOE, 1995). These types of tests are used to measure material motion and compression. The twin accelerators would allow for two views, or dual axes, for radiographic lines of sight, thereby producing three-dimensional imaging.

As developed in the early 1980s, the original project concept placed a small accelerator on a track-mounted cart, about the size of a semi-truck van, adjacent to an existing small accelerator. Over the next dozen years, however, incremental changes to the evolving project design resulted in a much larger facility. In 1994 construction started on two accelerator halls, 225 feet long and four stories high, adjacent to an open-air explosives firing site. In early 1995, this construction was enjoined in response to suit brought by two citizen organizations, pending completion of an adequate EIS.

The agency initially sought to avoid the time, expense, and public intrusion of preparing an EIS. To support its contention that no EIS was needed, the Department of Energy prepared several small environmental impact reviews of the initial DARHT proposal and its iterations. In 1982 the agency wrote a memorandum to file indicating that no further NEPA review was needed due to the minor nature of the project. Over the next 12 years, the agency prepared four revisions to its original memorandum. However, in so doing, the agency did not take into account whether the overall environmental review of the total project was still reasonable, given the magnitude of cumulative design changes. The large project under construction in 1994 bore little resemblance to the small project envisioned in 1982, although it responded to the same original purpose and need for a better hydrodynamic diagnostic machine. Each incremental change made to improve upon the original design, incorporate new aspects of evolving accelerator technology, or allow for additional features was compared only to the most-recent prior change. Because each step seemed minor, potential environmental impacts of each increment were dismissed as "similar" to those of the previous step rather than comparing them to the 1982 baseline project first analyzed.

To further complicate the situation, the Department of Energy made several substantial changes to its internal NEPA review process during the period of time that DARHT was designed (see the DOE NEPA regulations at 10 CFR 1021, first issued in 1992). Additionally, during this time the Department reassessed and softened its prior "hands-off" relationship with the general public. Consequently, the agency's requirements for NEPA review at the time DARHT construction started in 1994 were not the same as its requirements at the beginning of the process in the early 1980s. In addition, changes in national policy regarding the nuclear weapons complex had increased the programmatic importance of DARHT in 1994 compared to the low-key project envisioned in 1982. Not only had the project changed, but the procedural expectations of the agency had changed, and the national stakes were higher in regard to the importance of the project.

In response to the action brought by two citizen organizations, in January 1995 the U.S. District Court for the District of New Mexico enjoined further work on the 7-month-old construction project. The Department agreed to prepare an EIS, and met an exceptionally ambitious, expedited schedule of 10 months to complete the EIS. The resultant August 1995 DARHT EIS was reviewed by the court and, in April 1996, found to be "adequate." The injunction was lifted. Construction of DARHT resumed after a 15-month delay, and the facility went into operation in 1999. The Department of Energy and the Laboratory incurred not only the cost of the litigation, but additionally the cost of bringing the construction site to a safe stand-down, maintaining the site, and keeping the construction contractor on standby (approximately $1 million per month); the cost of reassigning technical personnel to other projects; and the cost of preparing an EIS.

In retrospect it is easy to see that 10 to 15 years of incremental design changes at DARHT resulted in starting construction on a project that had never been subject to environmental review in its totality. These changes in design, or "scope creep," caused the project to cross the threshold from a minor modification of an existing building that was not expected to result in significant impact to a hundred-million-dollar construction project with significant impacts. Given the changes in agency procedures and the additional importance of the new facility to national programs, it would have been prudent to have initiated a full-scale EIS review prior to finalizing project design and starting construction. However, given the agency climate of the time, it is also easy to see how project engineers and program managers felt that a comparison of each design change to the prior step was sufficient. The result of the path of NEPA avoidance, however, was court action, costly project delays, and the eventual publication of an EIS.

Case study 2. Defining the no-action alternative: The Camp Shelby, Miss., use permit

Camp Shelby is an Army National Guard installation located near Hattiesburg, Miss. The total land area is about 137,000 acres, and consists of a mix of Department of Defense, State of Mississippi, and DeSoto National Forest lands. The 117,000 acres of Forest Service lands are used under a permit. When, in the 1980s, this permit was up for renewal, the issuance of the permit was challenged by, among others, the Sierra Club, on the basis that no full examination had been made of the effects of off-road military *maneuver* training on these lands. A complicating factor was that the Guard was asking for access to some lands on which they had not previously been allowed to maneuver off-road with tracked vehicles. The State Forester (U.S. Forest Service Supervisor of the DeSoto National Forest) asked the Mississippi Army National Guard to prepare an EIS covering the proposed permit.

During the planning and scoping of the EIS, the definition of no action was questioned. While all parties agreed that the option must be included, there was little initial agreement as to what it meant in this case. Did it mean that all activities would proceed as before? Normally, it was argued, no action implies that the status quo maintains, so the comparisons should be with present activities. Did it mean that the entire installation would close? Did it just mean that no permit would be issued for use of U.S. Forest Service lands and that use of the other 20,000 acres would continue? All these were proposed, in some cases with considerable passion. Broadly speaking, the "sides" were arrayed as follows:

1. The U.S. Forest Service believed that no action meant that no *maneuver* permit of any type would be issued, but that use of other U.S. Forest Service lands was not affected.

2. Many environmental groups believed that lack of a maneuver permit should not affect actions on the federal Department of Defense lands and state lands.

3. The National Guard's position was that the sole function of Camp Shelby in its structure was to provide a place for the heavy brigades

of the Mississippi and Tennessee Army National Guard to maneuver, and that, absent that capability, there was no need for maintaining the facility.

In the end, the proponent's position prevailed. The permit applicant was the Mississippi National Guard, and it was allowed to define the consequences of lack of issuance of the permit sought. Thus, in the draft and final EIS, the consequences of no action were not the maintenance of the status quo, but involved all aspects of the closure of the installation, including effects on local employment. In some ways this is analogous to the situation where a business applies for a permit to continue operation of, for example, a privately owned hydropower dam. If the permit renewal is denied, the business closes and does not continue as before.

This circumstance is slightly unusual, but constitutes a class of exceptions to the belief that no action always means a continuation of the present effects.

Case study 3. Emergency actions

Extraordinary situations call for extraordinary action. The CEQ regulations recognize that in the event of an emergency, an agency may need to take immediate action without benefit of a NEPA review (40 CFR 1506.11). In such cases, the agency is not excused from the NEPA process; rather, the agency is to consult with the Council to determine alternative compliance arrangements. This provision is limited to actions needed to control the immediate impacts of the emergency. A key point of this regulatory clause is that in the face of an emergency, an adequate NEPA review is not *forgiven,* although it might be postponed, modified, or otherwise amended to meet the specifics of the emergency situation.

In these situations, what constitutes an emergency? A reminder of what is *not* an emergency is found in the phrase often seen tacked to office walls: "Failure to take action on your part does not constitute an emergency on my part." Agencies have been known to try to invoke the emergency clause of the CEQ regulations to address their failure to timely prepare NEPA documents because of budget delays, schedule changes, procrastination, poor planning, or discovery of unforeseen environmental conditions. Agencies with national security missions have also tried to use this clause to forgo NEPA reviews when faced with new security postures, legislative changes, or Presidential Decision Directives related to security measures. Although perhaps distressing to the agency, none of these situations constitute an emergency as that word is used in the regulations.

This case study looks at the emergency actions taken during and in the wake of the Cerro Grande Fire, a major forest fire near Los Alamos in northern New Mexico, and how the Council and the cognizant agency (the U.S. Department of Energy) modified the related NEPA review process. (DOE, 2000). In May 2000, the National Park Service set a "controlled burn" to restore a meadow by removing overgrown trees and deadwood. Unexpected high winds whipped the burn into a raging wildfire, named the Cerro Grande Fire, which burned for several weeks. The Cerro Grande Fire ultimately consumed over 47,000 acres of federal and private lands, becoming the largest wildfire ever recorded in New Mexico and resulting in the greatest property loss ever recorded in the state (approaching $1 billion). Over 200 homes and duplexes in the small town of Los Alamos were incinerated in a matter of a few hours, leaving some 400 families homeless. Traditional hunting grounds and fisheries in the nearby American Indian Pueblo of Santa Clara were destroyed. High mountain slopes in Bandelier National Monument and the adjacent Santa Fe National Forest were reduced to blackened stubble. Over 7000 acres of the Department of Energy's Los Alamos National Laboratory burned, and dozens of buildings were lost. In the space of less than a week the extremely severe burn reduced tens of thousands of acres of old growth pine and spruce to charred trunks set in glazed hydrophobic soil. The loss of vegetative cover and soil damage on steep slopes presented a secondary flood hazard that is common in burned areas.

The thickly forested, mountainous federal lands involved in the fire are administered by the U.S. Department of Agriculture Forest Service, the National Park Service, the Department of the Interior Bureau of Indian Affairs, and the U.S. Department of Energy (BAER, 2000). Although the Secretaries of Agriculture and the Interior have signed formal interagency cooperative agreements to address firefighting and wildfire emergency actions, the Department of Energy is not a party to these agreements, which do not cover firefighting actions taken on Energy lands. As is commonly the case during a large wildfire, the Departments of Agriculture and the Interior convened an Interagency Burned Area Emergency Rehabilitation Team to direct actions on National Forest, National Park, and American Indian trust lands following the Cerro Grande Fire (BAER, 2000); although the Department of Energy participated on the team, its lands were not covered by the interagency agreements, and therefore its actions were subject to the emergency provisions of NEPA.

In order for an agency to invoke the emergency provisions of NEPA, it must gain agreement from the Council on Environmental Quality; alternative arrangements to the standard NEPA review are limited to those actions necessary to control the immediate impacts of the emergency.

During the Cerro Grande emergency, the Department of Energy consulted with the Council on Environmental Quality and subsequently published a *Federal Register* notice outlining emergency actions taken in response to the fire and to mitigate flood hazards (65 F.R. 38522, June 21, 2000). Emergency actions taken on Department of Energy lands during the fire included bulldozing several miles of firebreaks and access roads, cutting hazard trees near buildings, lighting backfires, and conducting emergency aircraft flight operations. The Department conducted enhanced environmental sampling to monitor smoke, ash, soils, and contaminant transport. Hundreds of archaeological and historic properties burned on Department of Energy land, and habitat areas of three federally listed (threatened or endangered) bird species were affected. Following the fire, the Department took a variety of actions to mitigate the fire conditions and to alleviate the risk of flash flooding. These included seeding; aerial hydromulching; felling hazard trees; replacing power poles, guard rails, and culverts; removing contaminated soils; building flood control weirs and channels; and stabilizing archaeological sites. The Department acknowledged that the post-fire actions were more likely to result in significant adverse impacts than the actions taken during the fire.

As part of the "alternative arrangements" agreed to with the CEQ, the Department of Energy prepared a special environmental analysis of the known and potential impacts from wildfire and flood control actions (DOE, 2000). The special analysis included public involvement, although the public input was after the fact for the actions taken during the fire. The special analysis describes the actions taken and defines mitigation of adverse impacts of those actions. It is important to note that this analysis does not include the impacts of the fire per se, because while these effects are of scientific and ecological interest, the Department did not have any control over the fire. The Department did have control (exercise choice) over its own firefighting and flood control measures, but because of the emergency conditions, it did not have time to prepare an analysis of environmental impacts prior to taking action, as is normally done in a NEPA review.

This is a classic case of an environmental emergency as envisioned by the Council regulations; under those regulations the NEPA review was postponed until the emergency abated. The Department of Energy EIS-level special analysis fulfills the NEPA requirement to disclose agency action for public scrutiny.

4.12   Discussion and Study Questions

1   Obtain several EISs. Evaluate the statements with respect to the following:

    a.   Format—well-organized or confusing?

    b. Content—easy or difficult to follow? Concise?

    c. Readability—understandable?

    d. Alternatives—viable alternatives identified and equitably treated?

    e. Decision-making tool—usefulness of document in decision making?

    f. Effectiveness—has NEPA been served?

2 The CEQ regulations bring many new terms to the vocabulary, and provide definitions of others that relate specifically to the NEPA process. Define the following terms with respect to the NEPA process:

Categorical exclusion	Cooperating agency	Cumulative impact
Effect	Human environment	Lead agency
Major federal action	Mitigation	Scope
Significantly	Special expertise	Tiering

3 Compare and contrast the following NEPA documents:

Notice of Intent (NOI)	Final environmental impact statement (FEIS)
Environmental assessment (EA)	
Draft environmental impact statement (DEIS)	Finding of no significant impact (FONSI)
	Record of Decision (ROD)

4 Why does the CEQ consider the section on alternatives to be the "heart" of an EIS? Is it considered appropriate for an agency to identify its "preferred" alternative? Why or why not? At what step in the process?

5 An EIS may minimize the effect of adverse impacts by identifying various mitigating measures. What assurance do we have that these measures will indeed be carried out and are not just empty promises?

6 According to CEQ regulations, when and how does the EPA become involved in the NEPA process?

7 How does your state government review federal EISs? Does this process appear to be functioning as intended? Is the availability of EISs for public and agency review, and how to obtain them, a well-known fact in your area?

4.13   Further Readings

Cheney, P., and D. Schleidher. "From Proposal to Decision: Suggestions for Tightening up the NEPA Process." *Environmental Impact Assessment Review*, 9:89–98, 1985.

Council on Environmental Quality. 1997. *Considering Cumulative Effects under the National Environmental Policy Act*, January 1997.

Department of Energy (DOE). 2000. Special Environmental Analysis for the Department of Energy, National Nuclear Security Administration: Actions Taken in Response to the Cerro Grande Fire at Los Alamos National Laboratory, Los Alamos, New Mexico, DOE/SEA-03, September 2000. Los Alamos, NM: Los Alamos Area Office, Department of Energy.

Department of Energy; National Nuclear Security Administration. Emergency Activities Conducted at Los Alamos National Laboratory, Los Alamos County, New Mexico in Response to Major Disaster Conditions Associated with the Cerro Grande Fire, *Federal Register*, Vol. 65, No.120, pp. 38522–38527, June 21, 2000.

Department of Energy (DOE). 1995. Dual Axis Radiographic Hydrodynamic Test Facility Final Environmental Impact Statement, DOE/EIS-0228, August 1995. Los Alamos, NM: Los Alamos Area Office, Department of Energy.

Gilpin, Alan. *Environmental Impact Assessment: Cutting Edge for the Twenty-First Century*. New York: Cambridge University Press, 1995.

Herson, A., and K. M. Bogdan. "Cumulative Impact Analysis Under NEPA: Recent Legal Developments." *The Environmental Professional,* 13:100–106, 1991.

Morgan, Richard K. *Environmental Impact Assessment: A Methodological Perspective*. Boston, Mass.: Kluwer Academic, 1998.

Rau, John G., and David C. Wooten, eds. *Environmental Impact Analysis Handbook*. New York: McGraw-Hill, 1980.

U.S. Environmental Protection Agency. 1989. Filing System Guidance for Implementing 1506.9 and 1506.10 of the CEQ Regulations. March 7, 1989.

U.S. Interagency Burned Area Emergency Rehabilitation (BAER) Team. 2000. Cerro Grande Rehabilitation Report, Cerro Grande Burned Area Emergency Rehabilitation Team, June 2000. Santa Fe, NM: National Park Service.

Webb, M. Diana. 1997. DARHT—an "Adequate" EIS: A NEPA Case Study. In: *Proceedings of the 22nd Annual Conference of the National Association of Environmental Professionals*, National Association of Environmental Professionals, 1997, pp. 1014–1027. Boulder, CO: National Association of Environmental Professionals.

Westman, Walter E. *Ecology, Impact Assessment, and Environmental Planning*. New York: Wiley, 1985.

5

Elements of
Environmental Assessment

As indicated in previous chapters, environmental assessment encompasses varied disciplines, and consequently requires the expertise of personnel knowledgeable in various technical areas. It is superficially easy to lump together an admixture of elements under the heading "problems." While many good assessments have been performed by persons and groups which have used little or no structure in planning or executing the task, the development of a more rigorous structure is highly recommended. At a minimum, a good structure will allow the separation of cause and effect—surely critical to a good study. Therefore, when assessing the environmental impact of a given project, four major elements are involved:

1. Determine agency activities associated with implementing the action or the project.
2. Identify environmental attributes (elements) representing a categorization of the environment such that changes in the attributes reflect impacts.
3. Evaluate environmental impact [i.e., the effects of the *activities* (1, above) on the *attributes* (2, above)].
4. Report findings in a systematic manner.

5.1 Agency Activities

A comprehensive list of activities associated with implementing the project or action throughout its life cycle should be developed.

Necessary levels of detail would depend upon the size and type of project. As an illustration, an example of detailed activities for construction is included in the matrix in Appendix C. It is easy to trivialize this step—isn't it obvious that an agency (or firm) knows *exactly* the activities which are required to complete an action? Well, this must be answered "Yes and no."

Most planners know the *stages* through which a project passes—these are the stuff that project management charts are made of—but exactly when, where, and how do actions such as land clearing, excavation, equipment refueling and maintenance, and pest control take place? These subtasks are those which affect the real environmental consequences, not the stage called "preliminary site preparation," which might have a place in a milestone list. In construction, for example, the input of an experienced site supervisor may be more valuable than that of the engineer or architect in charge. The message here is "Think more detailed!" Think actions rather than concepts.

5.2   Environmental Attributes

Consisting of both natural and human-caused factors, the environment is admittedly difficult to characterize because of its many attributes (elements) and the complex interrelationships among them. Anticipated changes in the attributes of the environment and their interrelationships are defined as potential impacts.

An environmental assessment (EA) or environmental impact statement (EIS) is prepared to characterize the environment and potential changes to be brought about by a specific activity. Such a document is advantageous in that it presents an organized and complete information base for achieving the benefits intended by NEPA. In order for this objective to become fulfilled, it is necessary that a complete description, hence understanding, of the environment to be affected is first achieved. A wide variety of impact assessment methodologies have been developed (see Chapter 6), and virtually all of them employ a categorization of environmental characteristics of some form. This approach is recommended so that aspects of the environment are not overlooked during the analysis phase.

DEFINITION: Variables that represent characteristics of the environment are defined as *attributes,* and changes in environmental attributes provide indicators of changes in the environment.

All lists of environmental attributes are a shorthand method for focusing on important characteristics of the environment. Due to the complex nature of the environment, it should be recognized that any such listing is limited and, consequently, may not capture every potential impact. The more complete the listing is, the more likely it will reflect all important effects on the environment, but this may be expensive and cumbersome to apply.

Figure 5.1 presents a general listing of 49 suggested attributes in eight categories which comprise the biophysical and socioeconomic environment at a generalized level. While it is felt that this list of attributes represents a reasonable, concise breakdown of environmental parameters, it is likely to require modification or supplementation depending upon the type of action to be assessed. Appendix B provides details of these specific attributes, and the following sections provide a general discussion of the eight categories.

Air

When assessing the primary resources that are needed to sustain life, one must consider air as being one of the most, if not the most, critical resources. What makes air quality vulnerable is that air, unlike water or other wastes, cannot, in practice, be reprocessed at some central location and subsequently distributed for reuse. If the air becomes poisonous, the only natural alternative, if it is to sustain life, is for each individual to wear some sort of breathing (life support) system. For normal operating conditions this is unworkable and economically infeasible. When emissions and unfavorable climatic conditions interact to create undesirable air quality, the atmospheric environment may begin to exert adverse effects on humans and their surroundings. Air may be replenished through photosynthetic processes and cleansed through precipitation, but these natural processes are limited in their effectiveness in solving contemporary air pollution problems. Hence, great care must be exercised when assessing and maintaining the quality of air resources. It, therefore, seems self-evident that the protection of our air quality is a vital consideration when assessing the environmental impact of diversified human activities.

To better understand why our air quality has deteriorated—and will probably continue to deteriorate, even if the most advanced technology developed to date is applied—one must recognize the factors responsible for air pollution problems. Air quality is intimately connected to population growth, expansion of industry and technology, and urbanization. In particular, the energy use associated with these activities is increasing. Since energy use and air pollution are very strongly correlated, it seems imperative that we, as a society, examine each of our everyday activities in light of its potential impact on the environment. In effect, we must examine our lifestyle, at both a professional and a personal level, to assure that the precious resource, clean air, is preserved.

Air

1. Diffusion factor
2. Particulates
3. Sulfur oxides
4. Hydrocarbons
5. Nitrogen oxide
6. Carbon monoxide
7. Photochemical oxidants
8. Hazardous toxicants
9. Odors

Water

10. Aquifer safe yield
11. Flow variations
12. Oil
13. Radioactivity
14. Suspended solids
15. Thermal pollution
16. Acid and alkali
17. Biochemical oxygen
    demand (BOD)
18. Dissolved oxygen (DO)
19. Dissolved solids
20. Nutrients
21. Toxic compounds
22. Aquatic life
23. Fecal coliforms

Land

24. Soil stability
25. Natural hazards
26. Land-use patterns

Ecology

27. Large animals (wild and domestic)
28. Predatory birds
29. Small game
30. Fish, shellfish, and waterfowl
31. Field crops
32. Threatened species
33. Natural land vegetation
34. Aquatic plants

Sound

35. Physical effects
36. Psychological effects
37. Communication effects
38. Performance effects
39. Social behavior effects

Human Aspects

40. Lifestyles
41. Psychological needs
42. Physiological systems
43. Community needs

Economics

44. Regional economic stability
45. Public sector review
46. Per capita consumption

Resources

47. Fuel resources
48. Nonfuel resources
49. Aesthetics

Figure 5.1   Examples of environmental attributes.

The Clean Air Act of 1970 was established "to protect and enhance the quality of the nation's air resources so as to promote public health and welfare and the productive capacity of its population." In 1971, the EPA set forth national primary and secondary ambient air quality standards under Section 109 of the Clean Air Act. The primary standards define levels of air quality necessary to protect the public health, while secondary standards define levels necessary to protect the public welfare from any known or anticipated adverse effects of a pollutant. These standards are continually amended as health and environmental risks resulting from exposure to these pollutants are better understood and as monitoring technologies improve.

Air pollution legislation has been politically controversial because of its impact on industry and economic growth. After 3 years of intense legislative effort, the Clean Air Amendments of 1977 were passed.

Additionally, the Clean Air Act Amendments (CAAA) of 1990 moved farther still in the direction of controlling automobile emissions and ground-level ozone. There were 11 major titles to the CAAA, including provisions on ambient air quality standards, mobile source emissions, hazardous air pollutants, acid rain, stratospheric ozone, and enforcement. More recently, consideration has been given to air pollution on a regional and global scale. Effects such as acid deposition, stratospheric ozone depletion, and global warming are all of major concern on a national as well as an international level and may direct the trend for future environmental legislation.

To assess the impact of various activities on air quality, the major elements of the air pollution problem may be examined. These are (1) the presence of a source or "generator" of pollution, (2) a means of transporting the pollutant to a receptor, and (3) the receptor. If any of these elements is removed, the problem ceases to exist. When examining the sources, two types of classifications may be used: particulates, and gases and vapors. Under the particulate category are found smoke, dust, and fumes, as well as liquid mists. To further identify the impact of these particulates, it may be necessary to further subdivide them into chemical and biological classifications. Likewise, for gases and vapors one may consider sulfur oxides, nitrogen oxide, carbon monoxide, and hydrocarbons as hazardous toxicants.

Finally, environmental factors influence the transport mechanism of the pollutant. The pollutant transport, or lack of it, is controlled by the meteorological and topographical conditions. Clearly, a lower ground-level pollutant concentration will occur on a flat, open plain under windy conditions than in a valley under calm conditions. These factors and situations are discussed for the air attributes listed in Appendix B. Table 5.1 describes some of the more serious effects of air pollutants on humans.

Not only humans are affected by air quality. Air pollution has been definitely identified as having deleterious effects on animals, plant life, and materials as well. A drastic reduction in air quality is bound to severely affect the overall ecosystem behavior. Acid rain (actually an air quality characteristic) and global warming are two air quality topics discussed in Chapter 13.

Water

Water of high quality is essential to human life, and water of acceptable quality is essential for agricultural, industrial, domestic, and commercial uses. In addition, much recreation is water-based. Therefore, major activities having potential effects on surface water are certain to be of appreciable concern to the consumers and taxpayers. Additionally,

TABLE 5.1   Effects of Some Major Air Pollutants on Humans

Pollutant	Effect
Carbon monoxide	Combines with hemoglobin in blood, displacing the vital oxygen that hemoglobin normally transports, thereby reducing the oxygen-carrying capacity of the circulatory system. Results in reduced reaction time and increased burden on pulmonary system in cardiac patients.
Photochemical oxidants	Nitrogen oxides react with hydrocarbons in the presence of sunlight to form photochemical oxidants; cause eye, ear, and nose irritation; and adversely affect plant life.
Hydrocarbons	Combine with oxygen and NO_x to form photochemical oxidants.
Nitrogen oxides	Form photochemical smog and are associated with a variety of respiratory diseases.
Sulfur oxides	Associated with respiratory diseases and can form compounds resulting in corrosion and plant damage.
Particulate matter	Injures surface within respiratory system, causes pulmonary disorders and eye irritation, and creates psychological stress. Results in economic loss from surface material damage.

developments of recent years suggest that Americans are far more concerned now about water quality than they were a few decades ago.

Perhaps the political process provides the best barometer to measure the extent of public concern about water quality. The U.S. House of Representatives and Senate overwhelmingly enacted (over presidential veto) the Federal Water Pollution Control Act Amendments of 1972.

The Federal Water Pollution Control Act Amendments of 1972 were further amended in 1977. The amendments to the act were then termed the Clean Water Act of 1977. The 1977 changes to the Federal Water Pollution Control Act were "mid-course corrections" to the previous provisions of the act. The Clean Water Act provides the primary authority for water pollution control programs, and the act is periodically amended by Congress to incorporate contemporary national concerns.

Potential impacts on surface water quality and quantity are certain to be of concern in assessment of the effects of many federal programs. Almost any human activity offers the potential for impact on surface water through generation of waterborne wastes, alteration of the quantity and/or quality of surface runoff, direct alteration of the water body, modification of the exchanges between surface and groundwaters through direct or indirect consumption of surface water, or other causes.

The hydrologic environment is composed of two interrelated phases: groundwater and surface water. Impacts initiated in one phase eventually affect the other. For example, a groundwater system may charge one surface water system and later be recharged by another surface water system. The complete assessment of impact dictates consideration of both groundwater and surface water. Thus, pollution at one point in the system can be passed throughout, and consideration of only one phase does not characterize the entire problem.

Due to the close interrelationship between surface and groundwaters, most environmental attributes inevitably interface. Hence, aside from those aspects dealing specifically with surface or subterranean features, the attributes may be considered as applicable to both. Many attributes of the aquatic environment could be viewed as being physical, chemical, or biological in nature.

Physical attributes of surface water can be categorized as relating to either the physical nature of the water body or the physical properties of the water contained therein. Examples of individual parameters in the former category include the depth, velocity, and rate of discharge of a stream. Features of this type might be influenced by major activities, such as withdrawal of water, dredging, and clearing of shoreline vegetation.

The other category of physical characteristics—those related to the water itself—includes water characteristics such as color, turbidity, temperature, and floating solids. Many types of activities could influence the physical properties of water. A few examples are clearing of land and construction of parking lots, roads, and even rooftops (which concentrate runoff and may accelerate erosion, flooding, and sedimentation), discharge of scale-laden boiler waters, and discharge of cooling waters. Some other quality aspects which could be included in this category are dissolved gases and tastes and odors, which are actually manifestations of chemical properties of water. This serves to illustrate the occasional difficulty in strict categorization of attributes in the water environment.

Chemical attributes could be categorized conveniently as organic or inorganic chemicals. Some inorganic chemicals (like cadmium, lead, and mercury) may have grave consequences to human health; some (notably phosphorus and dissolved oxygen) have severe effects on the water environment, while others (such as calcium, manganese, and chlorides) relate mainly to the economic and aesthetic value of water in commercial, industrial, and domestic uses.

Normal personal use of water increases the concentration of many inorganic chemicals in water. Additionally, almost any type of industrial activity and land drainage is a source of chemicals. Because of the hundreds of thousands of organic (carbon-based) chemicals produced both naturally and by humans, most of the attributes contained in the organic chemical category are "lumped parameters." Examples include biochemical oxygen demand (BOD), oil, and toxic compounds. Some organic compounds are natural constituents of surface drainage and human and animal wastes, while others are unique to industrial activities and industrial products.

Biological attributes of the water environment could be categorized conveniently as either pathogenic agents or normal aquatic life. Pathogenic (disease-causing) agents include viruses, bacteria, protozoa, and other organisms, and they originate almost exclusively from human wastes. Aquatic life refers to the microorganisms and microscopic plants and animals, including fish, which inhabit water bodies. They are affected directly or indirectly by almost any natural or human-made change in a water body.

It is difficult to conceive of an alteration of surface water quantity or quality which is not accompanied by secondary effects. The physical, biological, and chemical factors influencing water quality are so interrelated that a change in any water quality variable triggers other changes in a complex network of interrelated variables. Thus, while individual water quality and quantity parameters may seem far more amenable to quantitative expression than parameters describing the quality of other sectors of the environment, the total effect of a particular impact on surface water may be as intangible as those on any sector of the total environment, because of the complex secondary, tertiary, and higher-order effects.

Land

Considering both the physical makeup and the uses to which it is put, land constitutes another important category of the environment. The soil that mantles the land surface is the sole means of support for virtually all terrestrial life. As this layer is depleted by improper use, so is the buffer between nourishment and starvation destroyed. However, the ability of soil to support life varies from place to place according to the nature of the local climate, the surface configuration of the land, the kind of bedrock, and even the type of vegetation cover. At the same time, the vulnerability of soil to destruction through mismanagement will vary as these factors change. Cultivated soils on slopes greater than 6 percent, or those that developed on limestone, are prone to erosion; soils in arid climates are sensitive to degradation by excessive salt accumulation. On the other hand, those in the tropics may quickly lose their plant nutrients by exposure to the abundant rainfall of those areas.

Soil serves well as an example of an interface between the three great systems that comprise the earth sciences: the lithosphere, the atmosphere, and the hydrosphere. The biosphere also operates in this interface, but it is usually considered to comprise the life sciences. For purposes of this discussion, the lithosphere consists of the various characteristics of landforms (slope, elevation, etc.), landform constituent materials (substratum), and the weathered layer, or soil. In the case of the atmosphere, the main elements are those that describe its state of temperature, moisture, and motion—or, in a word, climate. With regard to the hydrosphere, the principal concern will be with water flowing over the land surface or in streams.

Climate profoundly influences the nature of site characteristics, such as soils and vegetation. Soil stability, to a substantial degree, is the result of the interaction of rainfall and temperature with the local rock types. The rate of soil erosion, other things being equal, will depend upon the amount and intensity of rainfall. The details of the site climate must be known before an adequate environmental impact assessment can be made.

Climates are commonly identified and described by the total annual amount of precipitation and its seasonal distribution and by temperature and its seasonal distribution. Climatic types may be described as warm-humid, cool-humid, cool with summer droughts, arid, semiarid, and so on. There are additional descriptive elements of climate that are important in causing substantial differences within any one climatic type. Some of these include probability of maximum rainfall intensity, probability of drought, length of growing season, wind intensity, and the kind and frequency of storms.

The preparer of an environmental impact assessment or statement should be aware of the local landform type and its constituent materials. This information will enable him or her to more quickly evaluate the potential hazards of his or her activity upon the local physical environment. For example, slope erosion problems should be slight in plains areas with low relief. Areas underlain by soft limestone must always be treated cautiously with respect to groundwater pollution, due to the likelihood of solution channels in the rock strata.

Landform types are based upon only two descriptive characteristics of topography: local relief and slope. Other important properties are pattern, texture, constituent material, and elevation. These, along with local relief and slope, can be used to identify landforms with a considerable degree of precision.

However, the above define the landform system only at a given moment of time. Landforms are not static, but are continually changing (i.e., the landform system is dynamic since landforming processes are continuously at work, although the rate at which they operate varies from place to place). The factors that influence process rate include some of the attributes of landforms, as well as the attributes of climate and biota. Figure 5.2 shows one way of illustrating this complex system.

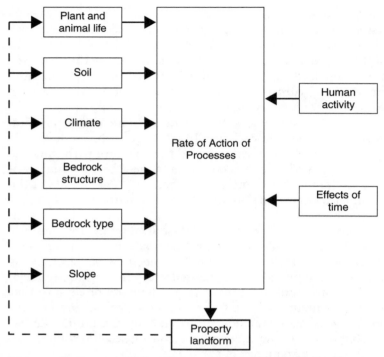

Figure 5.2 Processes, including human activity, which modify landforms.

From the above relationships, it is evident that landform evolution can also be considered important. Among the more important processes are weathering (disintegration and decay of rock), stream and wind erosion (removal of weathered debris by those forces), mass wasting (direct removal of weathered material by gravity), deposition (the cessation of movement of the entrained rock debris), and soil formation (those processes of weathering that give soils their distinctive regional characteristics). It is also evident that human activity is an important factor in changing the rate of process operation. This is done by modifying the land surface—changing the vegetation or destroying it; by plowing or otherwise disturbing the soil; by paving, construction, or otherwise sealing the surface; by changing the chemical or physical equilibrium in the soil; and by other actions. Through these actions, we reduce the natural resistance of the physical environmental system and permit the physical processes to operate at accelerated rates—with respect not only to one attribute, but to many.

As a typical historical example, one might consider the Piedmont region of the southeastern United States, where the interaction of soils, slope, and climate and the introduction of clear field cultivation of cotton and tobacco led to widespread destruction of the physical setting. The bare, gentle to moderate slopes, combined with the clay-rich middle layers of the local soils and the extremely heavy late summer and fall precipitation associated with hurricanes, created circumstances of exceedingly rapid soil erosion. The intense runoff quickly formed gullies in the surface layers which spread laterally, stripping off the soil. Once gullies were eroded through the midlayers, they deepened and lengthened rapidly, the water table was lowered, and the potential plant growth was thereby diminished. The process continued, with damage spreading to all parts of the system, and eventually returning to humans with a vengeance. The vegetation was impoverished, the wildlife destroyed, and the streams were polluted with excessive sediment. This process was advanced enough that land abandonment in the Piedmont began more than 150 years ago, before the middle of the nineteenth century, and continued through the 1950s.

All social and economic activities are located in time and space; therefore all physical problems have a socioeconomic component, just as all socioeconomic problems have biological and physical aspects. The spatial or locational aspects of human activities involve land in some way. Thus, land is a resource (i.e., it is useful in the production of goods and services needed to satisfy human wants and desires).

Land may be used directly, as in agriculture or forestry, where production depends partly on the inherent capability of the soil, and where land serves to locate the activity in space. Or, land may mainly provide the locational base upon which all sorts of commercial structures, transportation, and communication facilities or residential housing are built, and on which every sort of social and economic activity takes place.

Our activities mainly affect the availability or suitability of land for certain uses and thus land-use patterns. The activities may have negative or positive repercussions of varying magnitude on the local or regional economy, on community social or cultural patterns, or on the biophysical characteristics of the land itself, depending on the nature and extent of the activity. For instance, increases in the number of employees, due to a major federal action, may cause shortages of presently available rental housing, followed by rent increases. However, increased housing demand may stimulate residential and related construction requiring more land, thus having some beneficial economic effects. Similarly, increased local consumption of meat, dairy products, or fresh fruit and vegetables, due to the influx of new population, may encourage more intensive grazing and truck farming, with possible resulting beneficial or detrimental changes in land-use patterns. An unplanned, sudden population increase may tax the capacity of local indoor or outdoor recreational facilities beyond design limits, sometimes to the detriment or destruction of these resources, or force the conversion of wild lands to parks, and older parks to more highly developed recreational areas.

Some activities can affect the present or potential suitability of land for certain uses, rather than its availability. For example, the establishment of an industrial complex near a residential area would seriously limit the use of the adjacent land as a school site or for additional housing. On the other hand, where the adjacent land is being used for heavy industry, sanitary landfill, or warehouses, its potential would be much less affected. Thus, the ramifications of the proposed project may reach far beyond the perimeter of the project area in diverse ways.

Ecology

Ecology is the study of the interrelationships among organisms (including people) and their environment. Based upon this definition, all the subject areas discussed in this section would constitute a part of the overall category of ecology. In the context of this discussion, however, the category is utilized to include those considerations covering living animal and plant species.

Interest in plant and animal species, especially those becoming less common, prompted the beginnings of modern environmental concern in the mid-1950s. The general recognition that society was seriously disturbing organisms in the ecosystem without intending to do so caused ripples of concern, disbelief, and protest which are still with us. While it has always been recognized that many species have been crowded out of their habitats, and that others have been deliberately exterminated, the gradual comprehension of the fact that humans were unknowingly killing many entire species, such as by indiscriminate use of broad-spectrum pesticides, came as a distinct shock to the scientific community. Even greater public controversy was generated by citizen groups that actively pressured governmental agencies to enact legislation to prevent recurrence of such widespread detrimental impacts. The present legislation requiring assessment of likely effects before initiation of a project,is an outgrowth of these movements of the 1960s.

It is generally agreed that an aesthetically agreeable environment includes as many species of native plants and animals as possible. In many ways, one may measure the degradation of environments by noting the decrease in these common wildlife species. Since many types of outdoor activities are based directly on wildlife species, there may be economic as well as moral and aesthetic bases for maintaining large, healthy populations. The values derived from hunting and fishing activities are the difference between existence and relative affluence for many persons engaged in services connected with these outdoor recreational pursuits.

In considering the impact of human activities on the biota, it can be determined that there are at least three separable types of interests. The first, *species diversity,* includes the examination of all types and numbers of plants and animals considered as species, whether or not they have been determined to have economic importance or any other special values. The second general area, *system stability,* is basically concerned with the dynamics of relationships among the various organisms within a community.

A third important area, *managed species,* deals with the agricultural species and those nondomesticated species known to have some recreational or economic value. The wild species are usually managed by state or other conservation departments under the category "wildlife management." Agricultural species have economic and cultural value, and their close ties to human needs may cause extraordinarily acute controversy if effects on agriculture are likely, especially if the quality or safety of the human food supply appears threatened.

All the areas in ecology are very difficult to quantify, often being almost impossible to present in familiar terms to scientists of other disciplines. Furthermore, there are literally millions of possible pathways in which interactions among the plants, animals, and environment may proceed. To date, even those scientists knowledgeable in the field have been able to trace and analyze only a small minority of these, although thousands more may be inferred from existing data. Thus, many impacts predicted cannot be absolutely verified. Other interactions are probably correct by comparison with known cases involving similar situations, while many more are simply predicted on the basis of knowledge and experience in a broad range of analogous, although not closely similar, systems.

The question of chance effects is also an important one in ecology. One may be able to say that the likelihood of serious impact following a certain activity is low, based on available experience. This is definitely not the same as saying that the impact, if it develops, is not serious. The impact may be catastrophic, at least on a regional basis, once it develops. When one works with living organisms, the possibility of spread from an area where little chance of damage exists to one in which a greater opportunity for harm is present is itself a very real danger. The vectors of such movement cannot be predicted with any accuracy; however, the basic principles best kept in mind are simple enough. Any decrease in species diversity tends to also decrease the stability of the ecosystem, and any decrease in stability increases the danger of fluctuations in populations of economically important species.

Many other scientific disciplines are often closely related to ecology. When the question of turbidity of water in a stream is examined, for example, it will be found that this effect not only is displeasing to the human observer but has ecological consequences also. The excessive turbidity may cause eggs of many species of fish to fail to develop normally. It may even, in extreme cases, render the water unsuitable for the very existence of several species of fish. The smaller animals and the plant life once characteristic of that watershed may also disappear. Thus, the turbidity of the water, possibly caused by land-clearing operations in the stream watershed, may have effects ramifying far beyond the original, observed ones. Similarly, almost all effects which are observed relating to the quality of water will also have some ecological implications, in addition to those already of interest from a water supply viewpoint.

Since it was the observation of damage to the biological environment that helped to initiate the "environmental awareness" juggernaut of the 1960s, we must recognize that there is almost *no* activity which takes place that does not have some ecological implications. These may be simply aesthetic in nature, damaging the appearance of a favorite view, for example. They may also be symptoms of effects which could possibly be harmful to humans if ignored, such as pesticide accumulation in birds and fish. If we are to view the area of biology, or ecology, in perspective, we must realize that it includes a wide variety of messages to us. These should be interpreted as skillfully as possible, if our future is to be assured.

Sound

Noise is one of the most pervasive environmental problems. The "Report to the President and Congress on Noise" indicates that between 80 and 100 million people are bothered by environmental noise on a daily basis and approximately 40 million are adversely affected in terms of health (U.S. EPA, 1971). Relative to the occupational environment, the National Institute for Occupational Health and Safety (NIOSH) estimated the number of noise-exposed workers in the U.S. to be approximately 26 percent of the total production workforce (Sriwattanatamma and Breysse, 2000).

Since noise is a by-product of human activity, the area of exposure increases as a function of population growth, population density, mobility, and industrial activities. Figure 5.3 shows the range of sound levels for some common noise sources. The most common sources of noise include road traffic, aircraft, construction equipment, industrial activity, and many common appliances.

Road traffic continues to be the largest contributor, and trends indicate that the problem will worsen because traffic is extending into weekend and evening hours and into rural and recreational areas. In 1990, the average passenger car traveled 10,280 miles during the course of the year, and by 1997, the average distance had increased to 11,575 miles

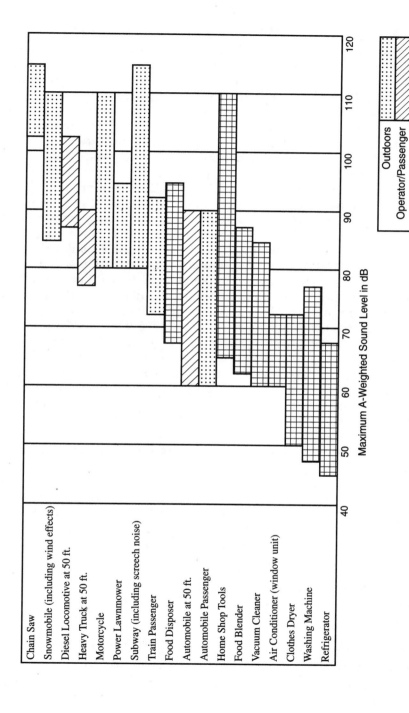

Figure 5.3 Typical range of common sounds (U.S. EPA, 1978).

per year (CEQ, 1997). Additionally, truck transportation has tradition-ally grown at a faster rate than the general population. For example, a total of 33.6 million trucks were registered in the United States in 1980. That number grew to 45.5 million in 1989, an increase of about 35 per-cent (Suter, 1991). Noise from the motors and exhaust systems of large trucks provides the major portion of highway noise impact, and in the city, the main sources of traffic noise include the motors and exhaust systems of automobiles, smaller trucks, buses, and motorcycles.

Options for managing noise pollution include increased restrictions on noise emissions, promotion of quieter products, traffic manage-ment, building insulation and noise barriers, and appropriate land-use planning. In the case of transportation systems, most options for reducing noise pollution are also consistent with energy conservation goals, and careful design planning can resolve conflicts between noise emissions and energy consumption for transportation. Concerning air-craft noise, in 1990 the FAA began a phased elimination of civil, sub-sonic aircraft weighing over 75,000 pounds flying into or out of airports in the United States by December 31, 1999. In 1995, FAA esti-mated a decline of over 75 percent in individuals exposed to day-night noise levels greater than 65 decibels since 1975 (CEQ, 1997).

Research into the physiological effects of noise indicates these con-clusions: The body does not become physiologically accustomed to noise, and even after several years' exposure, the heart remains responsive; an average level of external noise under 45 dB(A) is required to avoid sleep disturbances; a high noise level in residential areas is positively correlated to higher rates of hypertension and con-sumption of sleep medications; long-term exposure to noise over 80 dB(A) presents an increased risk for hypertension; noisy environments interfere with the development of communicative and auditory ability in children (OECD, 1986).

The health effects of noise are substantial. It was reported that 50 to 70 percent of the United States' population is annoyed by noise on a daily basis (U.S. EPA, 1971); the resulting social and psychological stresses are of major concern to the scientists and planners. The impli-cated health-related effects due to noise include

1. Permanent or temporary hearing loss

2. Sleep interference

3. Increased human annoyance

4. Communications interference resulting in reduced efficiency

5. Impairment of mental and creative types of work performance

6. Possible increase in usage of drugs like sleeping pills as a method of adaptation to noise stress (Bragdon, 1972).

Hearing loss is one of the most obvious effects of excessive exposure to noise. The first stages of noise-induced hearing loss, however, are often not recognized because they do not impair speech communication ability, and often the impairment can reach the handicapping stage before an individual is aware of any damage (Berglund and Lindvall, 1995). In addition, as the median age of the population is increasing, the loss of hearing which often accompanies age will be aggravated by higher noise levels. According to the U.S. Public Health Service (U.S. PHS, 1991), approximately 10 million of the estimated 21 million Americans with hearing impairments owe their losses to noise exposure (Suter, 1991).

Noise is also one of the most common forms of sleep disturbance and is regularly reported as a source of annoyance, stress, and dissatisfaction (Job, 1996). Exposure to noise can cause sleep disturbance in terms of difficulty in falling asleep, alterations of sleep patterns or depth, and awakenings. These effects are referred to as primary sleep disturbance effects. Exposure to nighttime noise can also induce secondary effects such as reduced perceived sleep quality, increased fatigue and annoyance, and decreased performance (Berglund and Lindvall, 1995). Although people often believe that they get used to nighttime noise, physiological studies have shown that while the subjective response improves with time, cardiovascular responses remain unchanged (Suter, 1991).

Damage to physical objects is another important consideration. Many natural and human-made features in the environment have become increasingly vulnerable to an ever-expanding technology, of which noise is a by-product. Damages associated with noise exposure include

1. Structural impairment

2. Property devaluation

3. Land-use incompatibility

This concern may be supported by considering the damages which have been sought by various plaintiffs for transportation noise (Bragdon, 1971). Figure 5.4 summarizes these and other impacts on human activity.

It has already been noted that noise may affect human health and land-use integrity. If a noise has an adverse impact on human physical and mental health, it is likely that the ecosystem (specifically animal life in an exposed area) is also being affected. Chronic noise annoyance and distraction may lead to (1) human error in handling and disposal of hazardous materials, thereby potentially affecting land, air, and water quality, as well as (2) disrupting harmonious social interaction by creating minor upheavals and disagreements.

Physiological Effects

1. Vasoconstriction
2. Gastrointestinal modification
3. Endocrine stimulations
4. Respiratory modification
5. Galvanic skin resistance alteration

Hearing Impairment

1. Permanent/temporary hearing loss
2. Recruitment
3. Tinnitus

Communication Interference

1. Aural—face-to-face; telephone
2. Visual—distortion; color blindness

Task Interference

1. Reduced production
2. Increased error rate
3. Extended output

Sleep Interference

1. Electroencephalographic modification (EEG)
2. Sleep stage alteration
3. Awakening
4. Medication

Personal Behavior

1. Annoyance
2. Anxiety—nervousness
3. Fear
4. Misfeasance

Figure 5.4 Human activity impacts resulting from increased noise stress.

On the other hand, because noise restricts the scope of land use, it also tends to depreciate the value of affected property, including undeveloped as well as developed land. Therefore, the impact of noise may be far-reaching, having a potentially significant impact on nearly every other environmental area.

An environmental assessment needs to describe the proposed activities and provide details about possible changes (either adverse or beneficial) in the noise environment. This description can be obtained with the following steps:

1. Classify all land within the area of interest into the following use categories:
   a. Industrial/commercial
   b. Residential
   c. Special—schools, hospitals, churches, parks, etc.
2. Plot the land-use data on an appropriately scaled map. Select acoustic criteria for different land uses.
3. Generate day-night average sound levels (L_{dn}) contours for each source.
4. Overlay a transparent sheet on the land-use map, locate each noise source, and plot its contours using the same scale as the land-use map. Computer-based geographic information systems are commonly used for these calculations and to prepare the contours. The contours should begin at the nearest residence, school, hospital, or other noise-sensitive area and extend outward in 5-dB zones until the affected area is covered.

5. Combine the noise contours for the different sources to obtain a composite contour. Identify affected areas and then compute sound level weighted population based upon the concept that some annoyance begins at 35 L_{dn} values, with increasing reaction as the sound level intensifies. (See Appendix B under Sound for explanation of dB, L_{dn}, and L_{eq}.)

Using existing analytical models and databases, noise levels can be estimated for proposed project activities. Duration and intensity of noise levels generated are important, and so is the population exposed to different levels. Equivalent population response representing population-weighted measures of the severity of the noise impact can then be computed. Details of these computations are beyond the scope of this text; excellent examples are provided by Goff and Novak (1977).

High-amplitude impulse noise (typically less than 1 second) is characteristically associated with a source such as sonic boom, piledrivers, blasts, artillery, and helicopters. Noise level measurement and determining human response and environmental impact due to impulse noise are complex issues. Further information about this is provided in Appendix B.

Human aspects

People everywhere react to situations as they define them, and if one defines a problem as real, then that situation is real in its consequences. This tendency has become a principle of advertising, public and community relations, and "image management." The fact that scientists and engineers think a solution of their own requirements is perfectly rational, economic, and altogether good may be beside the point. If that solution provokes a public controversy because numerous people and organizations believe it threatens a certain quality of life which they value, then the *consequences* will be real. The "facts" depend greatly upon who is perceiving them. Hence, there is the great practical importance of sociopsychological thinking by environment-conscious planners and managers.

Environment is surroundings. Social environment is people surroundings: human beings and their products, their property, their groups, their influence, their heritage. Such are the surroundings of almost any undertaking. There is no one social environment; there are many. Each event—the construction of a major facility, a reservoir, or a power project, proposed legislation, etc., as long as it is at a different place or time—has its own social environment, its own surroundings.

The effects of a project or plan on people and people's responses may be direct and immediate or remote and attenuated. But it is likely that people are somehow, sooner or later, implicated. And this is apt to be the case even if an activity occurs on a deserted island, miles from human habitation, and the action is triggered by electronic push buttons.

Prerequisite to any rational assessment of human impacts and responses is an inventory and depiction of the relevant social environment. It applies equally well to a wide variety of event-environment situations, and some straightforward observation and fact gathering is all that is necessary.

First, the location of the event itself is established. This can be done on a map having lines and boundaries that have been established by law (town, city, county, state). Location can also be described in terms of topography and physical dimensions: near a river, on a hill, two miles from a freeway, etc. Both means of placing the event may be necessary.

A place (with its people) may be a community or a neighborhood; on the other hand, it may be only a settlement, housing people who have so little in common that they constitute neither a neighborhood nor a community. It is important to learn just what kind of place, socially and politically, one is dealing with. To this end, more questions must be investigated.

Having located the place, the next question is: What are the place's resources upon which people have become dependent? What are the hopes and prospects which they hold dear? This part of environmental description calls for some of the same knowledge that is generated by those who analyze biological and physical environments—the conditions and the resources of the earth, water, air, and climate. The student of social environment, however, is concerned with these things only to the extent that people have come to value them, use them, and require them. This extent and its consequences may both be considerable. People are inclined to fear that their way of life will be damaged or disrupted if the resource base is altered. Their fear is quite understandable.

People, place, and resources, each element acting on the others, produce land uses. A land use is literally the activity and the purpose to which a piece of land—a lot, an acreage, an acre—has been put by people. Uses are mapped and analyzed by many environmental scientists, business people, and public officials. Patterns and changes in land use are identified as a basis for locating stores, highways, utilities, and schools. Millions of dollars (or political fortunes) can be lost or made as profits or tax revenues on accurate predictions of land-use trends—from agricultural to residential or from industrial to unused, for example.

Like many things in society, land uses are never completely stable, and they may change very rapidly. It all depends on what is happening to the people—their numbers, their characteristics, their distribution—and to their economy and technology. Therefore, the person assessing environmental impacts, who wants to predict outcomes and weigh alternatives, must know the land-use patterns and population trends of one or more places. At the same time, he or she must figure the economic dimension of the social environment. (In this connection, note the attributes classified as economics.)

So far in this brief account, only what teachers and research scientists call "human ecology" has been introduced. But that is only half of the social environment. Project managers must also assess the political realities of the place in which they would locate their projects and their activities. For engineers, especially, this seems to be difficult. They are used to thinking and working with physical things and with tools from the physical sciences. They strive to identify the "correct" answer, as defined in terms of time and costs. Social considerations are not their forte. Nevertheless, engineering managers and decision makers today, as never before, must reckon with human stubbornness and controversy. This is to say, they must anticipate and calculate the political reactions which their work is bound to produce. And they will engage in social engineering insofar as they act upon these considerations.

Because the essential ingredient of politics is power, and power is generated in organizations of people, the wise planner/manager will ask, "What are the organizations in this place, or with a stake in it, that I must reckon with?" State and local governments, business corporations, property owners' associations, environmental groups, families; these are some of the kinds of organizations that may be present. How big are they? How powerful and influential are they? How is their policy making done—by what persons and what procedures? Have they enacted laws or regulations that could or should affect major projects? Local and state government land use plans, zoning regulations, and building codes are examples.

An organization may react favorably, unfavorably, or neutrally. The position an organization takes, as well as its capacity to generate broader support for its policy and to execute it successfully, will depend upon whether and how its members and its public believe their quality of life will be affected by the proposed new project. Finally, community needs (the overall effects on the local community and public facilities operation) change with changes in the population, human resources, and community facilities. As such, these needs deal with potential effects on local housing, schools, hospitals, and local government operations.

Economics

Measurement of economic impact may be as simple as estimating the change in income in an area, or as complicated as determining the change in the underlying economic structure and distribution of income. Generally, effects may be examined for impact on conditions (income, employment) or structure (output by sector, employment by sector). These effects may be measured as impacts on the stock of certain resources or the flow of an economic parameter. We will discuss briefly the value of assets (stock), employment, income, and output as categories of variables.

Community or regional assets may be affected by project activity, and these assets may or may not be replaceable. The change in value of land and natural resources is an indicator of change in the stock or quantity of certain resources—for example, minerals—which are used in the conduct of social and economic activities. The category of land and natural resources which are not readily replenished by additional economic activities includes coal, a natural resource which, once mined and utilized, cannot be replaced. This category of economic change is important to decision makers because the extent to which the quantity of irreplaceable resources is changed will become increasingly more controversial as real or feared shortages of these resources develop.

The value of structures, equipment, and inventory is an indicator of change in the stock or quantity of resources such as buildings, trucks, or furniture which are used in the conduct of human social and economic activities. This group of resources represents capital stocks that are replaceable by additional economic activity. For example, it might be possible to reconstruct a building elsewhere if it were rendered useless by project activity. If proposed rule making were to make some vehicles obsolete, replacement with other, newer alternatives might be possible.

Total employment effects relate to all full-time and part-time employees in a region, on the payroll of operating establishments or other forms of organization, who worked or received pay for any part of a specified period. Included are persons on paid sick leave, paid holidays, and paid vacations during the pay period. Officers of corporations are included here as employees. Total employment can be affected by direct demand for services to perform a specified task or by indirect demand and secondary and tertiary activities that affect the requirements for goods and services.

Total income for a region refers to the money income of people employed in the conduct of economic activities in the region. This income normally comes from salaries and wages paid to the individuals in return for services performed. Included are incomes from social security, retirement, public assistance, welfare, interest, dividends, and net income from property rental. Incomes are most easily affected by changes in purchasing patterns in the region. The magnitude of a project's potential effect is related to the extent to which purchases of goods and services in the region are significant and will increase or decrease.

Output can be defined as goods and services produced by sectors of the economy in the region. Indicators of regional output are (1) value added to a product as a result of a manufacturing process, (2) gross receipts for service industries, (3) total sales from the trade sector, and (4) values of shipments. Output can be affected by direct and indirect expenditure and employment changes.

Other areas of potential impact relate to income distribution, the distribution of production by sector, governmental expenditures, and revenue collections by governmental units. The possible impact categories are extensive, and this brief introduction touches on a few of the more widely recognized areas.

Resources

The United States entered a new era of its history in the early 1970s. Supplies of many commonplace items, such as meat, building materials, and gasoline, fluctuated from adequacy to virtual nonavailability in many sections of the country. The period beginning in the early 1970s has been termed "the era of shortages" by many commentators surveying the American scene.

The rampant gas and oil shortage came as no surprise to experts in the economic and energy fields, but for the first time, the American public became aware that the question of energy supply could dramatically affect the quality of day-to-day life. Federal agencies experienced cutbacks in allocations for fuel and petroleum products. Interest in energy conservation was stimulated as a result of these shortages; magazine articles, news broadcasts, and newspapers pointed out energy conservation methods, presented information on energy supply, and exposed many groups involved with wasteful practices.

The energy situation was not the only concern resulting from the shortages experienced in 1973 and 1974. Increasing realization of the fact that many of our domestic mineral resources are rapidly approaching depletion has prompted renewed interest in the search for new materials which could be substituted for heavily affected resources. In addition, the question of obtaining raw materials has generated concern. The United States' increasing dependence on foreign sources for petroleum, minerals, and other nonrenewable critical resources, along with concern for the balance of payments and national security, has increased interest in conserving and recycling resources, and has renewed the search for alternative sources of energy. The most apparent example in the search for alternative fuels is that for petroleum-powered vehicles. The U.S. DOE reported that the use of alternative and replacement fuels doubled from 1992 to 1998 (U.S. DOT, 1999).

Environmental quality is directly linked with the use and procurement of energy. The continued degradation of air and water resources, the irrevocable loss of wilderness areas, and land-use planning dilemmas are problems which must be dealt with in the development of resources. Environmental considerations delayed the construction of the Alaskan pipeline and have delayed or totally stopped many offshore drilling projects and power plants. Air pollution resulting from emissions from the combustion of fossil fuels in engines and furnaces is also another cause for concern. Even such "safe" emissions as carbon dioxide have been implicated as "greenhouse gases," possibly contributing to climatic change (see Chapter 13). The necessity of providing a safe and healthful environment is another motive for the development of alternate energy sources which are also nonpolluting.

Another environmental characteristic which may be thought of as a resource is the aesthetic component. Although difficult to measure or quantify, the environment, as apprehended through hearing, sound, sight, smell, and touch, is important to everyone, although each individual perceives and responds to this environment differently. Project planners today are faced with increased pressures not only to incorporate functional engineering and cost aspects but also to include aesthetic considerations in every planning activity.

5.3   Determining Environmental Impact

The distinction between "environmental impact" and "change in an environmental attribute" is that changes in the attributes provide an indication of changes in the environment. In a sense, the set of attributes must provide a model for the prediction of all impacts. The steps in determining environmental impact are

1. Identification of impacts on attributes

2. Measurement of impacts on attributes

3. Aggregation of impacts on attributes to reflect impact on the environment.

With and without the project

The conditions for estimating environmental impact are measurement of attributes with and without the project or activity under consideration at a given point in time. Figure 5.5 indicates the measure of an attribute with and without an activity over time.

Consideration of the potential for impact if no action is taken, that is, maintaining the status quo, is called the no-action alternative. Figure 5.5 shows how the concept of no action is used. The dashed line shows the

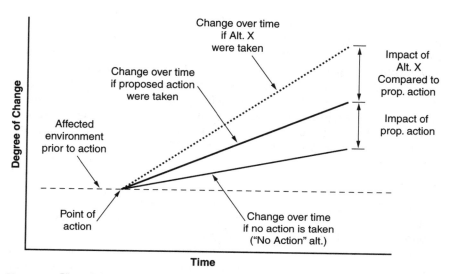

Figure 5.5 Use of the no-action alternative as a basis of comparison for the proposed action.

condition of a hypothetical environmental attribute prior to taking action (i.e., the affected environment). While *affected environment* describes the condition of the environment when the action is proposed to take place, the environment will not remain static over time. For the purposes of this figure, the condition of the environmental attribute at the time of the proposal is projected as the dashed line, although in actuality the line would reflect change over time, or environmental trends. In Fig. 5.5 the bottom solid line shows the degree to which the environment would be expected to change over time if no action were taken. However, if a hypothetical "proposed action" were implemented, shown as the heavy line in Fig. 5.5, the impact would be the degree of change over time if the action were taken, compared to the condition of the environment over the same span of time if the action were *not* taken (the impact would *not* be the comparison between the proposed action over time compared to the ambient environment prior to the point of action). For other alternatives, shown in the figure by the heavy dotted line for Alternative X, either the comparison can be to the impacts of the proposed action (as shown in the figure), or all alternatives can be compared to the no-action alternative. Both approaches are used; the only caution is to be consistent throughout the analysis and explain clearly which approach is used. Note that in some cases legislation or a court may require that an agency pursue some specific action. In this case, the preparer should describe the consequences of not taking the action, and note that this alternative, if implemented, would not fulfill the requirements of the law.

This concept of impact is used to avoid problems of comparing the present measure (without the activity) with the future measure (with the activity). The difficulty is that data for a "with activity" and "without activity" projection of impacts are difficult to obtain, and results are difficult to verify. However, several well-established forecasting techniques are available for establishing the "without" project condition, based on assumptions made for alternative futures. Quantifiable attributes, especially, can be forecast using past data and mathematical trend forecasting techniques (IWR, 1975).

Identifying impacts

The list of environmental attributes that might be evaluated is practically infinite because any characteristic of the environment is an attribute. Therefore, it is necessary to reduce the number of attributes to be examined. Duplicative, redundant, difficult to measure, and obscure attributes may be eliminated in favor of those that are more tractable. This procedure is valid only if the remaining attributes reflect all aspects of the environment. This means that some attributes, even if difficult to measure or conceptualize, may remain to be dealt with. Thus, identification of impacts is based on review of potentially affected attributes to determine whether they will be affected by the subject activity.

Baseline characteristics

Conditions prior to the activity.   The nature of the impact is determined by the conditions of the environment prior to the activity. Base data are information regarding what the measure of the attributes would be (or is) prior to the activity at the project location. Because the measurement and analysis of environmental impact cannot take place without base data, identifying the characteristics of the base is critical.

Geographic characteristics.   There may be significant differences in impact on attributes for a given activity in different areas. Geographical location is, therefore, one of the factors that affects the merit or relative importance of considering a particular attribute. For example, the impact of similar projects on water quality in an area with abundant water supplies and the impact in an area with scarce water resources would differ significantly. The spatial dispersion of different activities introduces one of the difficult elements in comparing one activity and its impact with another.

Temporal characteristics.   Time may also pose problems for the impact analysis. It is essential to ensure that all impacts are examined over the same projected time period. Furthermore, to adequately compare (or combine) activity impacts, it is necessary that the same time period (or periods) apply. An effect which will last 1 month is obviously different in many respects from the same effect projected to last for many years.

Role of the attributes

Although potential effects of impacts can be considered as effects on definite discrete attributes of the environment, the impression must not be created that actual impacts are correspondingly well categorized. That is, nature does not necessarily respect our discrete categories. Rather, actual impacts may be "smears" comprising effects of varying severity on a variety of interrelated attributes. Many of these interrelationships may be handled by noting the attributes primarily affected by activities and by utilizing the descriptions contained in the descriptor package in Appendix B to point out the secondary, tertiary, etc., effects.

Measurement of impact

Identifying the impact of a project on an attribute leads directly to the second step of measuring the impact. Ideally, all impacts should be translatable into common units. This is, however, not possible because of the difficulty in defining impacts in common units (e.g., on income and on rare or endangered species). In addition to the difficulties in quantitatively identifying impacts are the problems that arise because quantification of some impacts may be beyond the state of the art. Thus, the problems of measuring and comparing them with quantitative impacts are introduced.

Quantitative measurements

Some attributes, such as BOD for example, may be measured and changes projected. Quantitative measurements of impact are measures of projected change in the relevant attributes. These measurement units must be based on a technique for projecting the changes into the future. The changes must be projected on the basis of a no-activity alternative. One difficulty in assessing the quantitative change arises from the fact that changes in different attributes may not be in common units. In addition, there are difficulties in assessing the changes in the attributes through the use of projection techniques.

Qualitative measurements

Changes in some attributes of the environment are not amenable to measurement. The attribute may not be defined well enough in its relationship to the overall environment to determine what the most adequate measurable parameter might be. Therefore, instead of a specific measure, a general title and definition may be all that is available. For example, one may project that the aesthetic elements of a view may be degraded, but a quantified measure may not be available. In such cases, it may be necessary to rely on expert judgment to answer the question of how attributes will be affected by the subject project.

Comparison among attributes

In the development of any technique or methodology for environmental impact analysis, inevitably a time will come when someone asks the question "How do you compare all these environmental parameters with one another?" And, as is usually the case, long-lasting and frequently heated arguments follow, with the final result generally being the consensus that there is no single conclusion. Indeed, the question of comparing "apples and alligators" or, even worse, "biochemical oxygen demand and public sector revenue" bears no simple or well-defined solution. There have been some attempts at developing schemes for making numerical comparisons, which will be discussed in more detail in Chapter 6.

Another interesting procedure for developing such information is also available—a modification of the Delphi technique (Jain et al. 1973). The Delphi technique is a procedure developed originally at the Rand Corporation for eliciting and processing the opinions of a group of experts knowledgeable in the various areas involved. A systematic and controlled process of queuing and aggregating the judgments of group members is used, and stress is placed upon iteration with feedback to arrive at a convergent consensus. The weighting system discussed in the following section does not include all the elements of the original Delphi technique. In addition, results of these ranking sessions need further study, feedback, and substantive input from field data before use in your studies. They are a tool, not the answer.

The weighting procedure can be accomplished in a very simple manner. A deck of cards is given to each person participating in the weighting. In this example each card names a different technical specialty. Each of the participants is then requested to rank the technical specialties according to their relative importance to explain changes in the environment that would result from major activities in a particular project. Then each individual is asked to go back through the list, making a pairwise comparison between technical specialties, beginning with the most important one. The most important technical specialty is compared with the next most important by each individual, and the second technical specialty is assigned a percentage. This assignment is to reflect the percentage of importance of the second technical specialty with respect to the first. For example, the first technical area would receive a weight of 100 percent, and the second most important technical area might be considered by a specialist to be only 90 percent as important as the first. Then the second and third most important technical specialties are compared, and the third most important area is assigned a number (for example, 95 percent) as its relative importance compared to the second most important technical specialty. A sample diagram of the comparison is presented in Fig. 5.6.

The formula for weighting the technical specialties is

$$W_{ij} = \frac{V_{ij}}{\sum_{i-1}^{n} V_{ij} P} \qquad (i = 1,2,3,...,n)$$

$$V_{ij} = \begin{cases} 1 & (i = 1) \\ V_{i-1j} X_{ij} & (i = 2,3,...,n) \end{cases}$$

where W_{ij} = weight for the ith technical specialty area by the jth scientist

n = number of technical specialties

P = 1000: total number of points to be distributed among the technical specialties

X_{ij} = the jth scientist's assessment of the ratio of importance of the ith technical specialty in relation to the $(i-1)$th technical specialty

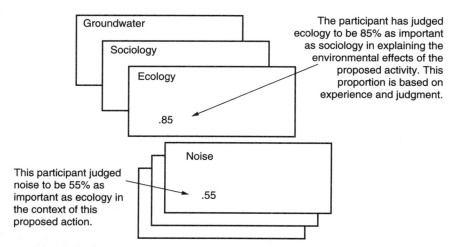

Figure 5.6   Pairwise comparison of environmental attributes (modified Delphi technique).

V_{ij} = measure of relative weight for the ith technical specialty area by the jth scientist

To accomplish the second part of this technique (i.e., to rank attributes within a technical specialty), each scientist independently ranks attributes in his or her own specialty. A group of scientists *within* one area could perform a similar comparison for the attributes. The information from these pairwise comparisons then can be used to calculate the relative importance of each of these technical specialty areas; a fixed number of points (e.g., 1000) is distributed among the technical specialties according to individual relative importance.

After the weights are calculated from one round of this procedure, the information about the relative weights is presented again to the experts, a discussion of the weights is undertaken, and a second round of pairwise comparisons is made. The process is repeated until the results become relatively stable in successive rounds.

In a demonstration of this method, an interdisciplinary group of college graduates with very little interdisciplinary training was asked to rate the following areas according to their relative importance in environmental impact analysis, and to distribute a 1000-point total among the categories:

1. Air quality

2. Ecology

3. Water quality

4. Aesthetics

5. Economics

6. Transportation

7. Earth science

8. Sociology

9. Natural resources and energy

10. Health science

11. Land use

12. Noise

After a thorough group study of all 12 areas, the group was asked to rate the areas again. The results, shown in Table 5.2, indicate that although some relative priorities changed, the points allocated to each category remained essentially the same. Similar ratings may be developed for attributes within each group.

TABLE 5.2   Results of Modified Delphi Procedure for Comparing Environmental Areas (This is an example only.)

Before interdisciplinary study		After interdisciplinary study	
Area	Average point distribution	Area	Average point distribution
Water	125	Water	128
Air	122	Air	126
Natural resources	109	Natural resources	105
Health	100	Ecology	93
Ecology	97	Health	88
Land use	81	Earth science	87
Earth science	79	Land use	78
Economics	62	Sociology	64
Sociology	60	Noise	62
Transportation	56	Economics	62
Aesthetics	54	Transportation	61
Noise	53	Aesthetics	46
	1000		1000

It should be emphasized that this procedure, as described, is only a tool for arriving at group decisions. This was not a group trained in environmental considerations. Please do not apply the numeric values in Table 5.2 to other studies. Different groups would certainly arrive at different decisions, and any application directed toward comparison between attributes should be made in the context of a specific planning situation.

Aggregation

After measuring project impacts on various attributes, two aggregation problems must be addressed. The first problem deals with how to aggregate among the different attributes (quantitative and qualitative) to arrive at a single measure for activity impact. Doing this involves expressing the various impact measures in common units. Then, a method for aggregating the impacts on a specific attribute must be identified. (Some methodologies utilize a weighting procedure to accomplish this.) Finally, the impacts are summed and compared with the impact of an alternative activity. A method for summarizing impacts is discussed in Appendix C.

Secondary impacts

Secondary or indirect consequences for the environment should be addressed, especially as related to infrastructure investments that stimulate or induce secondary effects in the form of associated investments and changed patterns of social and economic activity. These effects may be produced through their impact on existing community facilities and activities, through induced new facilities and activities, or through changes in natural conditions. A specific example calls out possible changes in population patterns and growth that may have secondary and indirect effects upon the resource base, including land use, water, and public services. In the biophysical environment, the secondary impacts can also be important.

To illustrate the nature of interrelationships among environmental attributes, consider, as an example, an activity which involves extensive removal of vegetation in a watershed. The environmental attribute indicated as being affected by this activity would be erosion. The examination of this attribute leads to other potentially affected attributes, such as dissolved oxygen, suspended solids, and nutrient concentration (which may stimulate growth of algae), that can cause a change in community maintenance (the numbers of organisms and composition of aquatic species in the stream). The pH of the stream could be affected by the growth of algae, and this, in turn, could affect the concentration of many of the chemicals in the stream by changing their solubility. Changes in each of the chemical constituents affected could trigger further change in the complex system. Excessive growth of algae could, at some location, result in high BOD values and loss of oxygen from the stream. Clearly, the interrelationships would not be limited to the stream, for evolution of gases from decomposition could create air pollution problems. This and/or the green color of the stream could affect land use and cause adverse social and economic effects.

Cumulative impacts

A single activity may produce a negligible effect on the environment. However, a series of similar activities may produce cumulative effects on certain aspects of the environment. This raises the question of how to deal with these potential cumulative effects. The most obvious solution is to prepare impact assessments on broad programs rather than on a series of component actions. Unfortunately, the definition of activities at the program level may be so vague as to preclude identification of impacts on the attributes of the environment. Nevertheless, review of activities at the program level, requiring enough detail to evaluate impacts, is the best way to handle the problem of cumulative impacts.

In real life, determination of cumulative impacts on an ecosystem is rather complex. Conceptually, cumulative impacts should include impacts on environmental attributes by different activities of the project and incremental stresses placed on the environment as a result of present or planned projects, and degradation which might result due to the interrelationship of affected attributes.

Recognizing the complexity and importance of assessing cumulative impacts, CEQ (1997) developed a handbook, *Considering Cumulative Effects under the National Environmental Policy Act*. Its recommendation, based on considerable research and consultations, is to consider the process of analyzing cumulative effects as "enhancing the traditional components of an environmental impact assessment: (1) scoping, (2) describing the affected environment, and (3) determining the environmental consequences," with the results contributing to the refinement of alternatives and design of mitigations. Table 5.3 illustrates how cumulative effects analysis can be incorporated into NEPA process components. Additional discussion on cumulative impact analysis is presented in Chapter 6.

5.4 Reporting Findings

Results of the impact analysis process are documented in one or more of the following:

1. An assessment
2. A finding of no significant impact

TABLE 5.3   Incorporating Principles of Cumulative Effects Analysis (CEA) into the Components of Environmental Impact Assessment (EIA)

EIA components	CEA principles
Scoping	Include past, present, and future actions. Include all federal, nonfederal, and private actions. Focus on each affected resource, ecosystem, and human community. Focus on truly meaningful effects.
Describing the affected environment	Focus on each affected resource, ecosystem, and human community. Use natural boundaries.
Determining the environmental consequences	Address additive, countervailing, and synergistic effects. Look beyond the life of the action. Address the sustainability of resources, ecosystems, and human communities.

SOURCE: CEQ, 1997.

3. A draft statement

4. A final statement

The content of each of these is discussed in Chapter 4.

It is useful to consider displaying the results in a way that makes it easy to comprehend the total impact from a brief review. One suggested method for doing this is by displaying the impacts on the summary sheet described in Appendix C.

Details of the specific format for an environmental impact analysis documentation are given by individual agency guidelines. These guidelines should be consulted and followed for each analysis.

5.5   Using Information Technology to Aid in the NEPA Process

Since the U.S. government initiative to "reinvent" government began in 1993, there have been fundamental changes in the way federal agencies provide access to information and how information is shared within agencies. Many of these changes have been made possible through the increased development of computerized information technology and the Internet, especially the World Wide Web (WWW). Both federal agencies and private organizations have developed sites on the Internet where one can easily find information on environmental laws, guidance on environmental compliance, and notices on agency activity. In addition, most of these sites contain links to environmental groups, data repositories, and/or electronic environmental journals and reports. However, the Internet addresses (access codes) for these web sites can change suddenly, and older web addresses may no longer be accessible when an agency updates its home page.

Developing and providing information to agencies and the public is specifically mandated by NEPA. Section 102 of NEPA requires that significant environmental data be gathered prior to decision making, and it is stated in Section 102(2)(g) that agencies are required to "make available to states, counties, municipalities, institutions, and individuals, advice and information useful in restoring, maintaining, and enhancing the quality of the environment." Additionally, Section 102(2)(h) further requires agencies to "initiate and utilize ecological information in the planning and development of resource-oriented projects." The Internet is a powerful and convenient means for quickly providing this information; users can access on-line versions of environmental laws and regulations in addition to project information and environmental, spatial, and demographic data.

Access to laws, regulations, and guidance documents

A key starting point for sound environmental decision making is a knowledge and understanding of environmental laws, regulations, and agency procedures. In the past, this information was typically only available through expensive subscription services or traditional law libraries and public reading rooms. The Internet, however, has made the dynamic body of U.S. laws, regulations, Executive Orders, and departmental directives and orders easily accessible to both agency personnel and the public.

In many cases, agencies provide guidance documents on-line to assist agency personnel as well as the public in understanding the necessary processes to be followed under specific environmental regulations. Additionally, agencies often furnish information on environmental impact statements by providing new releases, *Federal Register* notices, announcements, annual reports, and sometimes summaries of EISs on the Internet.

Internet technology can help an agency fulfill certain requirements of many of the U.S. environmental laws, such as involving the public in agency decision-making processes, providing easy access to environmental information, and providing a method for interagency cooperation. The Internet also allows for quick distribution of agency and Executive Office information.

Access to data

Internet technology makes it possible for the CEQ to better fulfill Section 205(2) of NEPA. That section of the law requires the CEQ to "utilize, to the fullest extent possible, the services, facilities and information (including statistical information) of public and private agencies and organizations, and individuals, in order that duplication of effort and expense may be avoided, thus assuring that the Council's activities will not unnecessarily overlap or conflict with similar activities authorized by law and performed by established agencies."

As an example, both the CEQ's home page and its NEPAnet web site provide a link to the U.S. Geological Survey (USGS) Environmental Impact Data Links. This site currently provides on-line access to diverse data sets and data centers such as the U.S. Department of Agriculture's Economics and Statistics System, the USGS Hydro-Climatic Data Network Streamflow Data Set, National Oceanic and Atmospheric Administration National Oceanographic Data Center, U.S. Census demographic data sets, and Earth Resources Observation Systems Data Center. Large amounts of environmental data are also available on-line through the EPA home page (http://www.epa.gov) and the U.S. Department of Energy web site (http://www.energy.gov).

Access to models

Section 102(2)(a) of NEPA requires federal agencies to "utilize a systematic, interdisciplinary approach which will insure the integrated use of the natural and social sciences and the environmental decision-making which may have an impact on man's environment." Computational models that simulate the complex interactions of natural environmental systems are valuable tools in projecting the effects of human activity or natural events on the environment. Computer models have been created to study many aspects of the environment, including ocean circulation, air dispersion, noise propagation, storm water runoff, erosion, groundwater flow, traffic circulation, and human migration. Computer models allow analyses to be both systematic and interdisciplinary by examining complex interactions.

Increasingly, agencies have included brief descriptions of models that they use and the model development process on their web sites. The EPA has identified many media-specific tools available on the Internet and has made these available on the Internet. For example, the ability to forecast travel demand is included in the Bureau of Transportation Statistics Travel Model Improvement Program, its multiagency program created to develop new travel demand modeling procedures (http://www.bts.gov/tmip/tmip.html).

On-line libraries and electronic journals

The Internet has dramatically changed how agencies and researchers access reference works and professional journals. There is a vast amount of environmental information available on the Web from publicly maintained libraries. On-line libraries offer an efficient and low-cost way of providing NEPA documents and reference materials to a wide audience in a timely manner. The EPA National Center for Environmental Publications, accessed through the EPA home page, provides access to the national Environmental Publications Internet Site with over 6000 EPA documents available to browse, view, or print online. The Government Printing Office (GPO) (http://www.access.gpo.gov) provides extensive access to on-line federal databases, including the *Federal Register, Congressional Record,* Code of Federal Regulations, statutes, congressional bills, budgets, and other resources. In addition, the General Services Administration also provides links to environmental libraries (http://www.gsa.gov).

NEPA requirements place a heavy burden on environmental analysts to be knowledgeable about the evolving state of science.

Environmental training, professional associations, and professional journals are all critical to environmental professionals remaining current in their fields. Increasingly, journals related to the environment are available on-line. The Committee for the National Institutes for the Environment maintains a list of environmental journals on the Internet (http://www.cnie.org). This list includes tables of contents, articles, and journals available in full text and those available with abstracts.

The American Association for the Advancement of Science has a summary version of its publication *Science* available on-line. Similarly, one can access summaries of articles in *Nature: An International Journal of Science. Issues in Ecology* is an on-line series designed to deal with major ecological issues and is published by the Pew Scholars in Conservation Biology Program and the Ecological Society of America. An important aspect to on-line publications is that the same information available to environmental professionals is also easily accessed by environmental groups and interested citizens, thus making for a better-informed public (CEQ, 1997).

5.6   Discussion and Study Questions

1   The organization of environmental characteristics presented above is very generalized. Discuss why a particular department or agency might either accept this structure completely or create a very different one altogether.

2   Is the ideal prioritization of attributes of the environment?

3   Is it better to create a set of attributes *before* you begin to prepare environmental documentation? Or is it better to develop such a list *after* you have completed your studies and have better knowledge of local conditions? Discuss which approach seems best to you, and why.

4   Select an ongoing or proposed project in your area (e.g., a reservoir, airport, highway relocation, or prison). Identify local or otherwise easily accessible sources of data that could be used to develop the baseline characteristics of the affected environment. Include relevant federal, state, and local agencies; institutions; associations; organizations; and/or individuals with special knowledge or expertise.

5   For the project selected for question 4, identify data needs beyond those currently available and identified in question 4. Describe the qualitative and quantitative measurements that would be necessary, and estimate the cost and time frame for obtaining the data.

6   In a group setting, discuss the project identified in question 4 and the eight environmental categories outlined in Fig. 5.1. Apply the Delphi technique in response to the question: "What is the relative importance of each of these eight areas in describing the environmental impact of the proposed project?" After averaging the group responses, discuss the results and conduct a second round. Did the group average change significantly?

5.7 Further Readings

Brown, Jennifer, ed. *Environmental Threats: Perception, Analysis, and Management.* London, New York: Belhaven Press, 1989.

Kryter, Karl D. "Aircraft Noise and Social Factors in Psychiatric Hospital Admission Rates: A Re-examination of Some Data." *Psychological Medicine,* 20:395–411, 1990.

McCold, Lance N., and James W. Saulsbury. "Defining the No-Action Alternative for National Environmental Policy Act Analyses of Continuing Actions." *Environmental Impact Assessment Review,* 18:15–37, 1998.

Wathern, Peter, ed. *Environmental Impact Assessment: Theory and Practice.* Winchester, Mass.: Allen & Unwin, 1988.

World Resources 1990–1991. World Resources Institute and United Nations Environment Programme Report, New York: Oxford University Press, 1990.

6

Environmental Assessment Methodologies

Many methodologies have been developed which allow the user to respond in a substantive manner to NEPA regulations when preparing an EA/EIS document. Presented in this chapter is a review and analysis of some of these environmental assessment methodologies. The general categories of methodologies evolved quickly after the passage of NEPA, as researchers and practitioners sought to ensure that a "systematic, interdisciplinary approach" was used in preparing environmental documentation. The purpose of this discussion of assessment methodologies is (1) to acquaint the reader with the different general types of methodologies, and (2) to provide illustrative examples of available methodologies in each category. An initial review and analysis of assessment methodologies was first completed by Warner and Preston (1974). The discussion here draws substantially from their work. Other approaches, such as multiattribute utility theory, systems diagrams, and simulation modeling, provide alternative ways of grouping assessment methodologies. Some of these methodologies were reviewed by Bisset (1988). The Further Readings section provides references to still other approaches and organizational methodologies which may be of interest.

6.1 Choosing a Methodology

Depending upon the specific needs of the user and the type of project being undertaken, one particular methodology may be more useful than another. Each individual must determine which tools best fit a given task. To select the most appropriate tools, the following key considerations may be useful.

Application

Is the analysis primarily a decision, an information, or a regulatory compliance document? (A decision document is vital for determining the best course of action, while an information document primarily reveals implications of the selected choices.) A decision document analysis generally requires greater emphasis on identification of key issues, quantification, and direct comparison of alternatives. An information document requires a more comprehensive analysis and concentrates on interpreting the significance of a broader spectrum of possible impacts. A study whose sole purpose is for regulatory compliance combines the two approaches.

Alternatives

Are alternatives fundamentally or incrementally different? If differences are fundamental (such as preventing flood damage by levee construction as opposed to flood plain zoning), the impact significance should be measured against some absolute standard, since impacts will differ in type as well as size. On the other hand, incrementally different alternative sets permit direct comparison of impacts and a greater degree of quantification. An example might be that of comparing the effects of a four-lane highway to those of highways with six and eight lanes. There should always be a no-action alternative, though in practice it is often hard to define. In some situations, especially those surrounding the *continuation* of an action, the no-action alternative may be the one which brings about large changes in the status quo (e.g., the cessation of the activity). Many agencies have grappled with this paradox, with varying degrees of success. One must, as well, overcome the confusing public relations issues which arise when the no-action alternative is the one which has the more severe consequences.

Public involvement

Does the role of the public in the analysis involve substantive preparation of studies, especially those destined for public review? Substantive preparation allows use of more complex techniques, such as computer or statistical analysis, that might be difficult to explain to a previously uninvolved but highly concerned public. A substantive preparation role will also allow a greater degree of quantification or weighting of impact significance through the direct incorporation of public values. Are regulatory agencies expected to have a high level of interest? If so, not only will detailed data likely be required, but the agency may require use of *its* models and criteria. The issue of managing public involvement will be separately examined in Chapter 11.

Resources

How much effort, skill, money, and data and what computer facilities are available? Generally, embarking on the more quantitative analyses will require more of everything, especially time. Many of the most complex EIS studies have required several years or more than 100 person-years of effort to complete. Is the project sponsor expecting that assessment of a multimillion-dollar action may be completed in a few months and at a cost of less than $25,000? This is not likely to be the case, and realizing the magnitude of an effort *prior* to agency commitment rather than after work is supposed to begin may be vital to eventual success.

Time

Is there an announced project schedule? Have the officials in charge already announced a starting date? A completion date? Are they remotely realistic? Have they allocated at least the minimum preparation and processing times as presented in Chapter 4? All too often, though much less frequently today than in the past, the time for preparation of environmental documentation is severely underestimated—or omitted entirely.

Familiarity

Is the preparer familiar with both the type of action contemplated and the physical site? Greater familiarity will improve the validity of a more subjective analysis of impact significance. This is where the real value of the interdisciplinary team is seen. Together, they may exhibit knowledgeable oversight through understanding of both action and environment, whereas separately, only parts of the picture are clear.

Issue significance

How big is the issue being dealt with? All other things being equal, the bigger the issue, the greater the need to be explicit, to quantify, and to identify key issues. Arbitrary weights or formulas for trading off one type of impact (e.g., environmental) against another (e.g., economic) become less appropriate as the stakes increase.

Controversy

Are the activities known to be controversial? Certain types of actions are inherently controversial, or carry high potential to raise public ire and, in the U.S. tradition, congressional interest and involvement. Some past and present examples are nuclear power, hazardous waste disposal, highway routing, threats to endangered species, and closure of military installations. In some ways, treating such issues makes planning easier, because you know a smooth, rigid assessment process can be ruled out from the start. If the "quick and easy" route is acknowledged by all to be impossible, it is often easier to obtain agency support for more thorough presentations.

Administrative constraints

Are choices limited by agency procedural or format requirements? Specific agency policy or guidelines may rule out some tools by specifying the range of impacts to be addressed, the need for analyzing the trade-offs, or the time frame of analysis. A programmatic EIS may require that all follow-on assessments will have a certain format, content, and methodology. Another aspect of constraint may be the cooperativeness of the planners and decision makers within your own agency. Are *they* willing to accept that the proposed action, or its time schedule, may need to be modified to accommodate environmental assessment activities or findings? The professional assessment staff, whether in-house or contracted, should be able to expect that two-way communication will be allowed. If not, this constraint should be identified as early as possible, and the anticipated problems associated with this lack of cooperation made known.

6.2   Categorizing Methodologies

The various methodologies examined can be divided into six types, based upon the way impacts are identified.

Ad hoc

These methodologies provide minimal guidance for impact assessment beyond suggesting broad areas of possible impacts (e.g., impacts upon flora and fauna, lakes, and forests), rather than defining the specific parameters within the impact area which should be investigated. They may be effective when the preparers are unusually experienced in the type of action being examined and require only reminders.

Overlays

These methodologies rely upon a set of maps of a project area's environmental characteristics (physical, social, ecological, aesthetic). These maps are overlaid to produce a composite characterization of the regional environment. Impacts are identified by noting the congruence of inherently antagonistic environmental characteristics within the project boundaries. The Geographic Information System (GIS) is a modern development of this method.

Checklists

These methodologies present a specific list of environmental parameters to be investigated for possible impacts, *or* a list of agency activities known to have caused environmental concern. They may have considerable value when many repetitive actions are carried out under similar circumstances. They do not, in themselves, establish a direct cause-effect link, but merely suggest lines of examination. They may or may not include guidelines about how parameter data are to be measured and interpreted.

Matrices

The matrix methodologies incorporate both a list of project activities *and* a checklist of potentially affected environmental characteristics. In a way, the matrix presents both alternatives from the checklist approach (i.e., both attributes and activities) to be considered simultaneously. The two lists are then related in a matrix which identifies cause-and-effect relationships between specific activities and impacts. Matrix methodologies may either specify which actions affect which environmental characteristics or simply list the range of possible actions and characteristics in an open matrix to be completed by the analyst.

Networks

These methodologies work from a list of project activities to establish cause-condition-effect relationships. They are an attempt to recognize that a series of impacts may be triggered by a project action. Their approaches generally define a set of possible networks and allow the user to identify impacts by selecting and tracing out the appropriate project actions.

Combination computer-aided

These methodologies use a combination of matrices, networks, analytical models, and a computer-aided systematic approach to (1) identify activities associated with implementing major federal programs, (2) identify potential environmental impacts at different user levels, (3) provide guidance for abatement and mitigation techniques, (4) provide analytical models to establish cause-effect relationships to quantitatively determine potential environmental impacts, and (5) provide a methodology and a procedure to utilize this comprehensive information in responding to requirements of EIS preparation.

6.3   Review Criteria

To serve the purpose of NEPA, an environmental impact assessment must effectively deal with four key problems: (1) impact identification,

(2) impact measurement, (3) impact interpretation, and (4) impact communication to information users.

Experience with impact assessments to date has shown that a set of evaluation criteria can be defined for each of these four key problems. These review criteria can be used for analyzing a methodology and determining its weaknesses and strengths. The criteria follow.

Impact identification

Comprehensiveness. A full range of direct and indirect impacts should be addressed, including ecological, physical-chemical pollution, social-cultural, aesthetic, resource supplies, induced growth, regional economy, employment, induced population or wealth redistributions, and induced energy or land-use patterns.

Specificity. A methodology should identify specific parameters (subcategories of impact types), such as detailed parameters under the major environmental categories of air, water, ecology, etc., to be examined.

Isolating project impacts. Methods to identify project impacts, as distinct from future environmental changes produced by other causes, should be required and suggested.

Timing and duration. Methods to identify the timing (short-term operational versus long-term operational phases) and duration of impacts should be required. (Data sources should also be listed for impact measurement and interpretation.)

Data sources. Identification of the data sources used to identify impacts should be required. (Data sources should also be listed for impact measurement and interpretation.)

Impact measurement

Explicit indicators. Specific measurable indicators to be used for quantifying impacts upon parameters should be suggested.

Magnitude. A methodology should require and provide for the measurement of impact magnitude, as distinct from impact significance.

Objectivity. Objective rather than subjective impact measurements should be emphasized. Professional judgments should be identified as such, although they may be the only criteria available in many cases.

Impact interpretation

Significance. Explicit assessment of the significance of measured impacts on a local, regional, and national scale should be required.

Explicit criteria. A statement of the criteria and assumptions employed to determine impact significance should be required.

Uncertainty. An assessment of the uncertainty or degree of confidence in impact significance should be required.

Risk. Identification of any impacts having low probability but high damage or loss potential should be required.

Alternatives comparison. A specific method for comparing alternatives, including the no-action alternative, should be provided.

Aggregation. A methodology may provide a mechanism for aggregating impacts into a net total or composite estimate. If aggregation is included, specific weighting criteria or processes to be used should be identified. The appropriate degree of aggregation is a hotly debated issue on which no judgment can be made at this time.

Public involvement. A methodology should require and suggest a mechanism for public involvement in the interpretation of impact significance.

Impact communication

Affected parties. A mechanism for linking impacts to the specific affected geographical areas or social groups should be required and suggested.

Setting description. A methodology should require that the project setting be described to aid statement users in developing an adequate overall perspective.

Summary format. A format for presenting, in summary form, the results of the analysis should be provided.

Key issues. A format for highlighting key issues and impacts identified in the analysis should be provided.

NEPA compliance. Guidelines for summarizing results in terms of the specific points required by NEPA and subsequent CEQ regulations should be provided.

In addition to the above "content" criteria, methodological tools should be evaluated in terms of their resource requirements, replicability, and flexibility. The following considerations, used in arriving at the generalized ratings for these characteristics (shown in Table 6.1), may be useful when considering the appropriateness of other tools. Table 6.1 provides a framework for methodology evaluation.

TABLE 6.1 Methodology Evaluation

TABLE 6.1 Methodology Evaluation

Criteria	Adkins (C)	Dee (1972) (C)	Dee (1973) (C-M)	Univ. of Georgia (C)	Jain/ Urban (CO)	Jain (1974) (M)	Krauskopf (O)	Leopold (M)	Little (C)
Comprehensiveness									
Specificity									
Isolate project impact									
Timing and duration									
Data sources known									
Explicit indicators									
Magnitude provided									
Objective measurement									
Significance scaled									
Criteria explicit									
Uncertainties made known									
Risks identified									
Alternatives compared									
Impacts aggregated									
Public involvement seen									
Affected groups visible									
Setting described									
Format for summary									
Key issues highlighted									
Match NEPA regulations									
Resource requirements									
Reliability									
Flexibility									

(Column header: Methodology and type)

METHODOLOGY TYPE KEY: A = Ad Hoc; C = Checklist; CO = Combination, computer-aided; M = Matrix; NW = Network; O = Overlay

EVALUATION SYMBOLS FOR USE IN SCORING: S = substantial compliance, low resource needs, or few reliability-flexibility limitations; P = partial compliance, moderate resource needs, or moderate limitations on reliability or flexibility; N = minimal or no compliance, high resource needs, or major limitations on reliability or flexibility; — = evaluation not attempted

NOTE: Methodologies listed are described in Sec. 6.5 of this chapter.

McHarg (O)	Moore (M)	Central N.Y. Reg. Planning Board (M)	Smith (C)	Sorenson (NW)	Stover (C)	Bureau of Reclam. (C)	USACOE (C)	Walton (C)	Western Systems (A)

Resource requirements

Data requirements. Does the methodology require data that are presently available at reasonable acquisition or retrieval cost?

Personnel requirements. What special skills are required? How many persons will be needed to implement the methodology? Do you have them available?

Time. How much time is required to learn to use and/or apply the methodology?

Costs. How do costs using a methodology compare to costs of using other tools?

Technologies. Are any specific technologies (e.g., use of a particular computer software) required to use a methodology?

Reliability

Replicability. Can the results be repeated given the same or similar conditions?

Ambiguity. What is the relative degree of ambiguity in the methodology? Does it measure what it says is measured?

Analyst bias. To what degree will different impact analysts using the methodology tend to produce widely different results? How much of the "methodology" is really subjective?

Flexibility

Scale flexibility. How applicable is the methodology to projects of widely different scale?

Range. For how broad a range of project or impact types is the methodology useful in its present form?

Adaptability. How readily can the methodology be modified to fit project situations other than those for which it was designed?

Comparison of methodologies. Methodologies may be rated for their degree of compliance with the 20 content criteria discussed above. Three rating characteristics on one possible rating scale are suggested as follows:

S = Substantial compliance, low resource needs, or few replicability-flexibility limitations

P = Partial compliance, moderate resource needs, or major limitations

N = No compliance or minimal compliance, high resource needs, or major limitations

These ratings may be applied to various methodologies in order to choose one best suited for a particular application. Table 6.1, a summary of methodology evaluation, can be completed as a practical exercise for the methodologies discussed herein or for other emerging methodologies.

Cumulative impact analysis

For some time, evaluators of environmental effects have realized that the most significant environmental effects may result not from the direct effects of a particular action but, rather, from the cumulative effects of multiple actions over time. Historically, federal agencies have addressed the direct and indirect effects of a proposed action on the environment in their analyses. This is, of course, the one that they propose to put into action. What has regularly been overlooked is the effect of the proposed action taken in the context of many other actions, proposed and real, of many other entities. Cumulative impact assessment, however, has been given less attention due to limitations in structured methodologies and procedures, as well as difficulties in defining the appropriate geographic (spatial) and time (temporal) boundaries for the impact analysis (Canter and Clark, 1997).

The CEQ defines cumulative effect as "the impact on the environment which results from the incremental impact of the action when added to other past, present, and reasonable foreseeable future actions regardless of what agency (Federal or non-Federal) or person undertakes such other actions" (40 CFR 1508.7). Actions by businesses and other nongovernmental groups are also relevant in many cases. Although the CEQ has defined cumulative impact, additional guidance on cumulative impact assessment has been lacking, thus prompting additional questions and concerns by the analyst. As a result, federal agencies have independently developed procedures and methods to analyze the cumulative effects of their actions on the environment. In order to address these issues, the CEQ developed the handbook *Considering Cumulative Effects under the National Environmental Policy Act*. This document presents a framework for addressing cumulative effects in either an environmental assessment or an environmental impact statement. The handbook provides practical methods for addressing the cumulative effects on specific resources, ecosystems, and human communities of all related activities, not just the proposed project or alternatives that initiated the assessment process. The methods described hereafter for developing cumulative impact analysis have been adapted from the CEQ handbook.

The CEQ-defined process for analyzing cumulative effects is very similar to the traditional components of an environmental impact assessment: (1) scoping, (2) describing the affected environment, and (3) determining the environmental consequences. Additionally, it should be noted that it is important to incorporate cumulative impact analysis in developing alternatives for an EA or EIS, as well as in determining appropriate mitigation efforts. A summary of the steps for cumulative effects analysis can be found in Table 6.2.

In many ways, scoping is the key to analyzing cumulative effects; it provides the best opportunity for identifying important cumulative effects issues, setting appropriate boundaries for analysis, and identifying relevant past, present, and future actions. Describing the affected environment sets the baseline and thresholds of environmental change that are important for analyzing cumulative effects. Recently developed indicators of ecological integrity and landscape condition can be used as benchmarks of accumulated change over time. In addition, remote sensing and GIS technologies provide improved means for displaying and analyzing historical change in indicators of the condition of resources, ecosystems, and human communities. Determining the cumulative environmental consequences of an action requires delineating the cause-and-effect relationships among the multiple actions and the resources, ecosystems, and human communities of concern. The significance of cumulative effects depends on how they compare with the environmental baseline and relevant resource thresholds.

Selection of which actions to include and which aspects of them to evaluate is the greatest challenge here. There are no fixed standards as to which are relevant in any one case, and the choice of which to include or exclude is of utmost importance. A special application of scoping is indicated here. We note that in the case of *Fritiofsen v. Alexander,* 772 F.2d 1225 (5th Cir. 1985), the court, ruling against a decision by the Galveston, Tex., district of the Corps of Engineers, said that reasonably foreseeable actions, not solely permits already in hand, must be the basis of the analysis of cumulative actions. The action here was the granting of a wetland fill permit on Galveston Island, and the Corps had originally evaluated the cumulative effect of granting all permits that had been filed. The proper focus, said the court, was that of all likely actions, present and future, given that development was continuing and that many more applications would likely be received.

Successfully analyzing cumulative effects will depend on the appropriate application of individual methods, techniques, and tools to the environmental impact assessment of concern. The unique requirements of cumulative effects analysis must be addressed by developing

TABLE 6.2 Steps in Cumulative Effects Analysis (CEA) to Be Addressed in Each Component of Environmental Impact Assessment

EIA components	CEA steps
Scoping	1. Identify the significant cumulative effects issues associated with the proposed action and define the assessment goals. 2. Establish the geographic scope for the analysis. 3. Establish the time frame for the analysis. 4. Identify other actions affecting the resources, ecosystems, and human communities of concern. 5. Identify those organizations, agencies, and businesses whose actions will be incorporated
Describing the affected environment	6. Characterize the resources, ecosystems, and human communities identified in scoping in terms of their response to change and capacity to withstand stresses. 7. Characterize the stresses affecting these resources, ecosystems, and human communities and their relation to regulatory thresholds. 8. Define a baseline condition for the resources, ecosystems, and human communities.
Determining the environmental consequences	9. Identify the important cause-and-effect relationships between human activities and resources, ecosystems, and human communities. 10. Determine the magnitude and significance of cumulative effects. 11. Modify or add alternatives to avoid, minimize, or mitigate significant cumulative effects. 12. Monitor the cumulative effects of the selected alternative and adapt management.

SOURCE: CEQ, 1997.

an appropriate conceptual model. To do this, a combination of methods can be used, including questionnaires, interviews, and panels; matrices; networks and system diagrams; modeling; trends analysis; and overlay maps and GIS. General principles for cumulative effects analysis are presented in Table 6.3.

For a more complete description of cumulative effects analysis, refer to CEQ's *Considering Cumulative Effects under the National Environmental Policy Act.*

6.4 Methodology Descriptions

Nineteen methodologies or tools listed in Table 6.1 are examined here in detail. The brief description given for each methodology discusses some or all of the following points:

Methodology type

General approach used

Range of actions or project types for which the methodology may be applicable

Comprehensiveness of the methodology in terms of the range of impacts addressed

Resources required (data, labor, time, etc.)

Limitations of the methodology (replicability, ambiguity, flexibility)

Key ideas or particularly useful concepts

Other major strengths and weaknesses as identified by the review criteria

Because of the brevity and subjectivity of these characterizations, they should not be considered fully adequate critiques of the tools examined. They may instead serve as a useful introduction to the range of techniques available. Many other methodologies, beyond those discussed here, are available for use by different agencies. The list of methodologies discussed here should not be considered exhaustive because of the dynamic nature of this subject area.

6.5 Methodology Review

Interim Report: Social, Economic, and Environmental Factors in Highway Decision Making (Adkins and Dock, 1971; Checklist). This methodology is a checklist which uses a +5 to −5 rating system for evaluating impacts. The approach was developed to deal specifically with the evaluation of highway route alternatives. Because the bulk of the parameters used

TABLE 6.3 Principles of Cumulative Effects Analysis

1. **Cumulative effects are caused by the aggregate of past, present, and reasonably foreseeable future actions.**
   The effects of a proposed action on a given resource, ecosystem, and human community include the present and future effects added to the effects that have taken place in the past. Such cumulative effects must also be added to effects (past, present, and future) caused by all other actions that affect the same resource.

2. **Cumulative effects are the total effect, including both direct and indirect effects, on a given resource, ecosystem, and human community of all actions taken, no matter who (federal, nonfederal, or private) has taken the actions.**
   Individual effects from disparate activities may add up or interact to cause additional effects not apparent when looking at the individual effects one at a time. The additional effects contributed by actions unrelated to the proposed action must be included in the analysis of cumulative effects.

3. **Cumulative effects need to be analyzed in terms of the specific resource, ecosystem, and human community being affected.**
   Environmental effects are often evaluated from the perspective of the proposed action. Analyzing cumulative effects requires focusing on the resource, ecosystem, and human community that may be affected and developing an adequate understanding of how the resources are susceptible to effects.

4. **It is not practical to analyze the cumulative effects of an action on the universe; the list of environmental effects must focus on those that are truly meaningful.**
   For cumulative effects analysis to help the decision maker and inform interested parties, it must be limited through scoping to effects that can be evaluated meaningfully. The boundaries for evaluating cumulative effects should be expanded to the point at which the resource is no longer affected significantly or the effects are no longer of interest to the affected parties.

5. **Cumulative effects on a given resource, ecosystem, and human community are rarely aligned with political or administrative boundaries.**
   Resources typically are demarcated according to agency responsibilities, county lines, grazing allotments, or other administrative boundaries. Because natural and sociocultural resources are not usually so aligned, each political entity actually manages only a piece of the affected resource or ecosystem. Cumulative effects analysis on natural systems must use natural ecological boundaries and analysis of human communities must use actual sociocultural boundaries to ensure including all effects.

6. **Cumulative effects may result from the accumulation of similar effects or the synergistic interaction of different effects.**
   Repeated actions may cause effects to build up through simple addition (more and more of the same type of effect), and the same or different actions may produce effects that interact to produce cumulative effects greater than the sum of the effects.

7. **Cumulative effects may last for many years beyond the life of the action that caused the effects.**
   Some actions cause damage lasting for longer than the life of the action itself (e.g., acid mine drainage, radioactive waste contamination, species extinctions). Cumulative effects analysis needs to apply the best science and forecasting techniques to assess potential catastrophic consequences in the future.

8. **Each affected resource, ecosystem, and human community must be analyzed in terms of its capacity to accommodate additional effects, based on its own time and space parameters.**
   Analysts tend to think in terms of how the resource, ecosystem, and human community will be modified given the action's development needs. The most effective cumulative effects analysis focuses on what is needed to ensure long-term sustainability of the resource.

SOURCE: CEQ, 1997.

relates directly to highway transportation, the approach may not be readily adaptable to other project types.

The parameters are broken down into categories of transportational, environmental, sociological, and economic impacts. Environmental parameters are generally deficient in ecological considerations. Social parameters emphasize community facilities and services.

Route alternatives are scored +5 to −5 in comparison with the present state of the project area, not the expected future state without the project.

Since the approach uses only subjective relative estimations of impacts, the data, labor, and cost requirements are very flexible. Reliance upon subjective ratings without guidelines for such ratings reduces the replicability of analysis and generally limits the valid use of the approach to a case-by-case comparison of alternatives only.

The detailed listing of social and, to a lesser extent, economic parameters may be helpful for identifying and cataloging impacts for other types of projects. An interesting feature of possible value to other analyses using relative rating systems is the practice of summarizing the number and the magnitude of plus and minus ratings for each impact category. The number of pluses and minuses may be a more reliable indicator for alternative comparison, since it is less subject to the arbitrariness of subject weighting. These summaries are additive, and thus implicitly weight all impacts equally.

Environmental Evaluation System for Water Resources Planning (Dee et al., 1972; Checklist). This methodology is a checklist procedure emphasizing quantitative impact assessment. While it was designed for water-resource projects, most parameters used are also appropriate for other types of projects. Seventy-eight specific environmental parameters are defined within the four categories of ecology, environmental pollution, aesthetics, and human interest. The approach does not deal with economic or secondary impacts, and social impacts are partially covered within the human interest category.

Impacts are measured via specific indicators and formulas defined for each parameter. Parameter measurements are converted to a common base of "environmental quality units" through specified graphs or value functions. Impacts can be aggregated by using a set of preassigned weights.

Resource requirements are rather high, particularly data requirements. These requirements may restrict the use of the approach to major project assessments.

The approach emphasizes explicit procedures for impact measurement and evaluation and should therefore produce highly replicable results. Both spatial and temporal aspects of impacts are noted and explicitly weighted in the assessment. Public participation, uncertainty, and risk concepts are not dealt with. An important idea or approach is the highlighting of key impacts via a "red flag" system.

Planning Methodology for Water Quality Management: Environmental System (Dee et al., 1973: Checklist/matrix). This unique methodology of impact assessment defies ready classification, since it contains elements of checklist, matrix, and network approaches. Areas of possible impacts are defined by a hierarchical system of 4 categories (ecology, physical-chemical, aesthetic, social), 19 components, and 64 parameters. An interaction matrix is presented to indicate which activities associated with water-quality treatment projects generally affect which parameters. The range of parameters used is comprehensive, excluding only economic variables.

Impact measurement incorporates two important elements. A set of "ranges" is specified for each parameter to express impact magnitude on a scale from 0 to 1. The ranges assigned to each parameter within a component are then combined by means of an "environmental assessment tree" into a summary environmental impact score for that component. The significance of impacts for each component is quantified by a set of assigned weights. A net impact can be obtained for any alternative by multiplying each component score by its weight factor and summing across components.

The key features of the methodology are its comprehensiveness, its explicitness in defining procedures for impact identification and scoring, and its flexibility in allowing use of best available data. Sections of the report explain the several uses of the methodology in an overall planning effort and discuss means of public participation. While the data, time, and cost requirements of the methodology when used for impact assessment are moderate, a small amount of training would be required to familiarize users with the techniques.

The methodology possesses only minor ambiguities and should be highly replicable. Because the environmental assessment "trees" are developed specifically for water-treatment facilities, the methodology cannot be readily adapted to other types of projects without reconstructing the "trees," although the parameters could be useful as a simple checklist.

One potentially significant obstacle to use of this approach is the difficulty of explaining the procedures to the public. Regardless of the validity of the "trees," they are devices developed by highly specialized multivariant-analysis techniques, and public acceptance of conclusions reached by their use may be low.

Optimum Pathway Matrix Analysis Approach to the Environmental Decision-Making Process: Test Case: Relative Impact of Proposed Highway Alternatives (University of Georgia, 1971; Checklist). This methodology incorporates a checklist of 56 environmental components. Measurable indicators are specified for each component. The actual values of alternative plan impacts on a component are normalized and expressed as a decimal of the largest impact (on that one component). These normalized values are multiplied by a subjectively determined weighting factor. This factor is the sum of 1 times a weight for "initial" effects plus 10 times a weight for "long-term" effects.

The methodology was developed to evaluate highway project alternatives, and the components listed are not suitable for other types of projects. The wide range of impact types analyzed includes land use, social, aesthetics, and economic.

The potential lower replicability of the analysis produced by using subjectively determined weighting factors is compensated for by conducting the analysis over a series of iterations and incorporating stochastic error variation in both actual measurements and weights. This procedure provides a basis for testing the significance of differences in total impact scores between alternatives.

The procedures for normalizing or scaling measured impacts to obtain commensurability and testing of significant differences between alternatives are notable features of potential value to other impact analyses and methodologies. These ideas may be useful whenever several project alternatives can be identified and compared. This methodology may place rather high resource demands, because computerization is necessary to generate random errors and make the large number of repetitive calculations.

Environmental Impact Assessment Study for Army Military Programs (Jain et al., 1973) and *Computer-Aided Environmental Impact Analysis for Construction Activities: User Manual* (Urban et al., 1975; Combination computer-aided). This is a computer-aided assessment system employing the matrix approach to identify potential environmental impacts. The system relates Army activities from nine functional areas to attributes contained in eleven technical areas of specialty describing the environment. The nine functional areas are construction, research and development, real estate acquisition or outleases of land, mission change, procurement, training, administration and support, industrial activities, and operation and maintenance.

Three levels of attributes are identified: detailed level, review level, and controversial attributes. Ramification remarks regarding potential impacts are presented along with mitigation procedures for minimizing adverse impacts. Potential impacts are identified on a need-to-consider scale, using A, B, and C as indicators, instead of a numerical system.

Given the appropriate input information for a particular program, the computer-aided system developed will provide relevant environmental information to allow the user to respond to the requirements of

CEQ guidelines. In addition, analytical models are being developed to quantitatively assess the environmental impacts. One early such model, the Economic Impact Forecast system, was put into operation in 1975 and used by the army for several years.

Significant features of this methodology are (1) it is cost-effective, (2) it provides analytical models for cause-effect relationships, (3) it is a comprehensive methodology, (4) the output matrix is modified, based upon site-specific input, to produce a project-specific input matrix, (5) it provides information regarding environmental laws and regulations, and (6) it includes information about abatement and mitigation techniques.

This methodology was designed for Army military programs. Its applicability to programs of other agencies is limited and would thus require some systematic modifications. Problems associated with effective community participation and evaluation of trade-offs between short-term areas of environmental resources and long-term productivity are not adequately addressed.

Handbook for Environmental Impact Analysis (Jain et al., 1974; Matrix). Employing an open-cell matrix approach, this handbook presents recommended procedures for use by Army personnel in the preparation and processing of environmental impact assessments and statements. The procedures outline an eight-step algorithm in which details of the proposed actions and associated alternatives are identified and evaluated for environmental effects in both the biophysical and socioeconomic realms. Briefly, the procedural steps are outlined as follows:

1. Identify the need for an EA or an EIS.
2. Establish details of the proposed action.
3. Examine environmental attributes, impact analysis worksheets, and summary sheets.
4. Evaluate impacts, using attribute descriptor package.
5. Summarize impacts on summary sheet.
6. Examine alternatives.
7. Address the eight points of the CEQ guidelines.
8. Process final document.

The handbook provides examples of representative Army actions that might have a significant environmental impact (step 1) and guidance on the identification of Army activities (steps 2 and 4) for Army functional areas.

Environmental attributes (steps 3 and 4) are identified and characterized. After evaluating the effect of the proposed action and the alternatives (step 6) on the interdisciplinary attributes, and summarizing the effects (step 5), it is recommended that the assessment be documented in the format suggested by the CEQ guidelines (step 7). Each of the eight points in the CEQ guidelines is discussed in detail, and Army-related examples are presented. (The guidelines were superseded in 1979 by the NEPA regulations, though much of the discussion is still relevant.) In addition, the handbook gives information regarding processing of assessments and statements (step 8).

Because the methodology is designed for Army military programs, its applicability to programs of other agencies is limited and would require systematic modifications. In addition, this methodology does not provide the depth and comprehensiveness of environmental information made available by the computer-aided study previously discussed.

Evaluation of Environmental Impact Through a Computer Modeling Process, Environmental Impact Analysis: Philosophy and Methods (Krauskopf and Bunde, 1972; Overlay). This methodology employs an overlay technique via computer mapping. Data on a large number of environmental characteristics are collected and stored in the computer on a grid system of 1-km-square cells. Highway route alternatives can be evaluated by the computer (by noting the impacts on intersected cells), or new alternatives may be generated via a program identifying the route of least impact.

The environmental characteristics used are rather comprehensive, particularly regarding land use and physiographic characteristics. Although the methodology was developed and applied to a highway setting, it is adaptable (with relatively small changes in characteristics) to other project types with geographically well-defined and concentrated impacts. Because the approach requires considerable amounts of data about the project region, it may be impractical for the analysis of programs of broad geographical scope. The labor skills, money, and computer technology requirements of the approach may limit its application to major projects or to situations where a statewide computer database exists (e.g., New York, Minnesota, Iowa). The 1-km resolution would be considered unacceptable today. The Geographic Information System, a successor to the overlay, may use 20- to 100-m resolution for similar purposes.

Impact importance is estimated through the specification of subjective weights. Because the approach is computerized, the effects of several alternative weighting schemes can be readily analyzed.

The methodology is attractive from several viewpoints. It allows a demonstration of which weighted characteristics are central to a particular alternative route; it presents a readily understandable graphic representation of impacts and alternatives; it easily handles several subjective weighting systems; its incremental costs of considering or generating additional alternatives are low; and it fits well with developing regional and statewide databank systems.

The mechanics of the approach (how impacts are measured and combined) may not be readily apparent from the reference cited. Considerable training beyond the information available in this reference may be required prior to using the approach.

A Procedure for Evaluating Environmental Impact (Leopold et al., 1971; Matrix). This is an open-cell matrix approach identifying 100 project activities and 88 environmental characteristics or conditions. For each action involved in a project, the analyst evaluates the impact on every environmental characteristic in terms of impact magnitude and significance. These evaluations are subjectively determined by the analyst. Ecological and physical-chemical impacts are treated comprehensively; social and indirect impacts are discussed in part; and economic and secondary impacts are not addressed.

Because the assessments are subjective, resource requirements of the approach are very flexible. The approach was not developed in reference to any specific type of project and was very widely applied in the 1970s, usually with some local alterations.

Guidelines for use of the approach are minimal, and several important ambiguities are likely in the definition and separation of impacts. The reliance upon subjective judgment, again, without guidelines, reduces the replicability of the approach.

The approach is chiefly valuable as a means of identifying project impacts and as a display format for communicating results of an analysis.

Transportation and Environment: Synthesis for Action: Impact of National Environmental Policy Act of 1969 on the Department of Transportation (DOT, 1969; Checklist). This approach is basically an overview discussion of the kinds of impacts that may be expected to occur from highway projects, and the measurement techniques that may be available to handle some of them. A comprehensive list of impact types and the stages of project development at which each may occur is presented. As broad categories, the impact types identified are useful for other projects as well as highways.

The approach suggests the separate consideration of an impact's amount, effect (public response), and value. Some suggestions are offered for measuring the amount of impact within each of seven general categories: noise, air quality, water quality, soil erosion, ecological, economic, and sociopolitical impacts.

Five possible approaches to handling impact significance are presented. Three of these are "passive" (requiring no agency action), such as "reliance upon the emergence of controversy." The other two involve the use of crude subjective weighting scales. No specific suggestions are made for the aggregation of impacts either within or between categories.

In general, the reference cited is a useful discussion of some of the important issues of impact analysis, particularly as they apply to transportation projects; however, it does not present a complete analytical technique.

A Comprehensive Highway Route-Selection Method, and Design with Nature (McHarg, 1968 and 1969; Overlay).   This approach employs transparencies of environmental characteristics overlaid on a regional base map. Eleven to sixteen environmental and land-use characteristics are mapped. The maps represent three levels of the characteristics, based upon "compatibility with the highway." While these references do not indicate how this compatibility is to be determined, available documentation is cited.

This approach is basically an earlier, noncomputerized version of the ideas presented in Krauskopf (1972). Its basic value is a method for screening alternative project sites or routes. Within this particular use, it is applicable to a variety of project types. Limitations of the approach include its inability to quantify and identify possible impacts and its implicit weighting of all characteristics mapped.

Resource requirements of this approach are somewhat less demanding in terms of data than those of the Krauskopf approach, because information is not directly quantified, but rather is categorized into three levels. However, high degrees of skill and training are required to prepare the map overlays.

The approach seems most useful as a "first-cut method" of identifying and sifting out alternative project sites prior to preparing a detailed impact analysis. Historically, McHarg was the primary popularizer of the concept of "compatibility" in planning major development projects. His background led him to a visual rather than a mathematical representation of "incompatible" elements, but most or all of his elements correspond to environmental problems, such as noise, soil loss, and ecological disturbance.

A Methodology for Evaluating Manufacturing Environmental Impact Statements for Delaware's Coastal Zone (Moore et al., 1973; Matrix).   This approach was not designed for impact analysis, although its principles could be adapted for such use. Employing a network approach, it links a list of manufacturing-related activities to potential environmental alterations, major environmental effects, and, finally, human uses affected. The primary strength of the set of linked matrices is their utility for displaying cause-condition-effect networks and tracing out secondary impact chains.

Such networks are useful primarily for identifying impacts. The issues of impact magnitude and significance are addressed only in terms of high, moderate, low, or negligible damage. As a result of these subjective evaluations, the approach would have low replicability as an assessment technique. For such a use, guidelines would likely be needed to define the evaluation categories.

The approach incorporates indicators especially tailored to manufacturing facilities in a coastal zone, although most indicators would also be pertinent to other types of projects. It would perhaps be valuable as a visual summary of an impact analysis for communication to the public.

Environmental Resources Management (Central N.Y. Reg. Planning Board, 1972; Matrix). This methodology employs a matrix approach to assess in simple terms the major and minor, direct and indirect impacts of certain water-related construction activities. It is designed primarily to measure only the physical impacts of water-resource projects in a watershed and is based upon an identification of the specific, small-scale component activities that are included in a project of any size. Restricted to physical impacts for nine types of watershed areas (e.g., wetlands) and fourteen types of activities (e.g., tree removal), the procedure indicates four possible levels of impact-receptor interaction (major direct through minor indirect).

Low to moderate resources, in terms of time, money, and personnel, are required for this methodology, due principally to its simple method for quantification (major versus minor impact). However, the procedure is severely limited in its ability to compare different projects or the magnitude of different impacts.

Since there is no spatial or temporal differentiation, the full range of impacts cannot be readily assessed. Impact uncertainty and high-damage/low-probability impacts are not considered. Since only two levels of impact magnitude are identified, and the importance of the impacts is not assessed, moderate replicability results. The lack of objective evaluation criteria may produce fairly ambiguous results. NEPA requirements for impact assessments are not directly met by this procedure.

This methodology may be less valuable for actual assessment of the quantitative impacts of a potential project than for the "capability rating system," which determines recommended development policies on the basis of existing land characteristics. Thus, guidelines for desirable and undesirable activities, with respect to the nine types of watershed areas, are used to map a region in terms of the optimum land-use plan. The actual mapping procedure is not described; therefore, that aspect of the impact assessment methodology cannot be evaluated here.

Quantifying the Environmental Impact of Transportation Systems (W. L. Smith, nd; Checklist). This approach, as developed for highway route selection, is a checklist system based upon the concepts of probability and supply and demand. The approach attempts to identify the alternative with least social cost to environmental resources and maximum social benefit to system resources. Environmental resources elements are listed as agriculture, wildlife conservation, interference noise, physical features, and replacement. System resources elements are listed as aesthetics, cost, mode interface, and travel desired. Categories are defined for each element and used to classify zones of the project area. Numerical probabilities of supply and demand are then assigned to each zone for each element. These are multiplied to produce a "probability of least social cost" (or maximum social benefit). These "least social cost" probabilities are then multiplied across the elements to produce a total for the route alternative under examination.

The approach is tailored and perhaps limited to project situations requiring comparison of siting alternatives. While the range of environmental factors examined is limited, it presumably could be expanded to more adequately cover ecological, pollution, and social considerations.

Since procedures for determining supply and demand probabilities are not described, it is difficult to anticipate the amounts of data, labor, and money required to use the approach. The primary limitations of this methodology are the difficulties inherent in assigning probabilities, particularly demand probabilities, and the implicitly equal weightings assigned each element when multiplying to yield an aggregate score for an alternative.

A Framework for Identification and Control of Resource Degradation and Conflict in the Multiple Use of the Coastal Zone (Sorenson, 1970) and *Procedures for Regional Clearinghouse Review of Environmental Impact Statements— Phase Two* (Sorenson and Pepper, 1973; Network). These two publications present a network approach usable for environmental impact analysis. The approach is not a full methodology but rather a guide to identifying impacts. Several potential uses of the California coastal zone are examined through networks relating uses to causal factors (project activities), to first-order condition changes, to second- and third-order condition changes, and, finally, to effects. A major strength of the approach is its ability to identify the pathways by which both primary and secondary environmental impacts are produced.

The second reference also includes data types relevant to each identified resource degradation element, although no specific measurable indicators are suggested. In this reference, some general criteria suggested for identifying projects of regional significance are based upon project size and types of impacts generated, particularly land-use impacts.

Because the preparation of the required detailed networks is a major undertaking, the approach is presently limited to some commercial, residential, and transportation uses of the California coastal zone for which networks have been prepared. An agency wishing to use the approach in other circumstances might develop the appropriate reference networks for subsequent environmental impact assessment. This is one of many examples of a special-purpose tool constructed for a repetitively applied function. Such a tool may be excellent for its original purpose while only mediocre for generalized use.

Environmental Impact Assessment: A Procedure (Stover, 1972; Checklist). This methodology is a checklist procedure for a general quantitative evaluation of environmental impacts from development activities. The type and range of these activities is not specified but is believed to be comprehensive. The 50 impact parameters are sufficient to include nearly all possible effects and thereby allow much flexibility. Subparameters indicate specific impacts, but there is no indication of how the individual measures are aggregated into a single parameter value. While spatial differences in impacts are not indicated, both initial and future impacts are included and explicitly compared.

The moderate to heavy resource requirement, especially in terms of an interdisciplinary personnel team, increases as more subparameters are included and require additional expertise in specific areas. However, the actual measurements are not based on specific criteria and are only partially quantitative, having seven possible values ranging form an extremely beneficial impact to an extremely detrimental one. Therefore, there may be room for ambiguous and subjective results with only moderate replicability.

The assumption that impact areas are implicitly of equal importance allows aggregation of the results and project comparisons, but at the expense of realism. A specific methodology is mentioned for choosing the optimum alternatives in terms of the proportional significance of an impact vis-à-vis other potential alternatives. There is no explicit mention of either public involvement in the process or environmental risks.

The impact assessment procedure is presented as only one step in a total evaluation scheme, which includes concepts of dynamic ecological stability and other ideas. An actual description of the entire process is not indicated, however.

Guidelines for Implementing Principles and Standards for Multiobjective Planning of Water Resources (Bureau of Reclamation, 1972; Checklist). This approach is an attempt to coordinate features of the Water Resources Council's Proposed Principles and Standards for Planning Water and Related Land Resources with requirements of NEPA. It develops a checklist of environmental components and categories organized in the same manner as the council guidelines. The categories of potential impacts deal comprehensively with biological, physical, cultural, and historical resources, and pollution factors, but do not treat social or economic impacts. Impacts are measured in quantitative terms wherever possible, and also rated subjectively on "quality" and "human influence" bases. In addition, uniqueness and irreversibility considerations are included where appropriate. Several suggestions for summary tables and bar graphs are offered as communications aids.

The approach is general enough to be widely applicable to various types of projects, although its impact categories are perhaps better tailored to rural than urban environments. While no specific data or other resources are required to conduct an analysis, an interdisciplinary project team is specified to assign the subjective weightings. Since quality, human influence, uniqueness, and irreversibilities are all subjectively rated by general considerations, results produced by the approach may be highly variable. Significant ambiguities include a generally inadequate explanation of how human influence impacts are to be rated and interpreted.

Key ideas incorporated in the approach include explicit identification of the "without project" environment as distinct from present conditions, and a uniqueness rating system for evaluating quality and human influence (worst known, average, best known). The methodology is unique among those examined because it does not label impacts as environmental benefits or costs, but only as impacts to be valued by others. The approach also argues against the aggregation of impacts.

Matrix Analysis of Alternatives for Water Resource Development (USACOE, 1972; Checklist). Despite the title, this methodology can be considered to be a checklist under the definitions used here. Although a display matrix is used to summarize and compare the impacts of project alternatives, impacts are not linked to specific project actions. The approach was developed to deal specifically with reservoir construction projects but could be readily adapted to other project types.

Potential impacts are identified within three broad objectives: environmental quality, human life quality, and economics. For each impact type identified, a series of factors is described to show possible measurable indicators. Impact magnitude is not measured in physical units but by a relative impact system. This system assigns the future state of an environmental characteristic without the project a score of zero; it then assigns the project alternative possessing the greatest impact on that characteristic a score of +5 (for positive impact) or −5 (for negative impact). The raw scores thus obtained are multiplied by weights determined subjectively by the impact analysis team.

Like the Georgia approach (University of Georgia, 1971), this methodology tests for the significance of differences between alternatives by introducing stochastic error factors and conducting repeated runs. The statistical manipulations are different from those used in the Georgia approach, however, and are considered by Corps writers to be more valid.

Resource requirements of this methodology are variable. Since specific level types of data are not required, data needs are quite flexible. The consideration of error, however, requires specific skills and computer facilities.

Major limitations of the approach, aside from the required computerization, are the lack of clear guidelines about exactly how to measure impacts and the lack of guidance about how the future "no project" state is to be defined in the analysis. Without careful description of the assumptions made, replicability of analyses using this approach may be low, since only relative measures are used. Since all measurements are relative, it may be difficult to deal with impacts that are not clearly definable as gains or losses.

The key ideas of wider interest incorporated in this approach include reliance upon relative, rather than absolute, impact measurement; statistical tests of significance with error introduction; and specific use of the "no project" condition as a baseline for impact evaluation.

A Manual for Conducting Environmental Impact Studies (Walton and Lewis, 1971; Checklist). This methodology is a checklist, unique in its almost total reliance upon social impact categories and strong public participation. The approach was developed for evaluating highway alternatives and identifies different impact analysis procedures for the conceptual, corridor, and design states of highway planning. All impacts are measured either by their dollar value or by a weighted function of the number of persons affected. (The weights used are to be determined subjectively by the study team.) The basis for most measurements is a personal interview with a representative of each facility or service affected.

Resource requirements for such a technique are highly sensitive to project scale. The extensive interviewing required may make the approach impractical for many medium-sized or large projects, because agencies preparing impact statements seldom have the necessary labor or money to contract for such extensive interviewing.

Analyses produced by the approach may have very low replicability. This results from the lack of specific data used and the criticality of the decision regarding boundaries of the analysis, since many impacts are measured in numbers of people affected. There is also no means of systematically accounting for the extent to which these people are affected.

The key ideas of broader interest put forth by the approach are the use of only social impacts, without direct consideration of physical impacts (e.g., pollution, ecology changes); the heavy dependence upon public involvement and specific suggestions about how the public may be involved; and the recognition of the need for different analyses of different project development stages. If vigorously applied, the intensive incorporation of public sentiment may achieve a desirable endpoint in spite of the lack of technical rigor.

Environmental Guidelines (Western Systems, 1971; Ad hoc). The environmental guidelines are intended primarily as a planning tool for siting power generation and power transmission facilities. However, they address many of the concerns of environmental impact analysis and have been used to prepare impact statements. Viewed as an impact assessment methodology, the approach is an ad hoc procedure, suggesting general areas and types of impacts but not listing specific parameters to examine.

The approach considers a range of pollution, ecological, economic (business economics), and social impacts; however, it does not address secondary impacts, such as induced growth or energy use patterns. The format of the approach is an outline of considerations important to the selection of sites for each of several types of facilities (e.g., thermal generating plants, transmission lines, hydroelectrical and pumped storage, and substations). An additional section offers suggestions for a public information program.

Since the approach does not suggest specific means of measuring or evaluating impacts, no particular types of data or resources are required. The application of this approach is limited to the siting of electric power facilities, with little carryover to other project types.

6.6 Future Directions

This chapter has provided guidance for choosing an environmental impact assessment methodology, a description of six general categories of methodologies, criteria for reviewing a given methodology to determine its weaknesses and strengths, a description of selected methodologies, and a reference listing of other methodologies, with a notation of the general category in which each of these methodologies can be classified. As mentioned previously in this chapter, depending upon the specific needs of the user and the type of project being undertaken, one particular methodology may be more useful than another. While it is possible to select one of the methodologies mentioned here for use by an agency to solve its specific needs for environmental impact analysis, no one methodology can effectively and economically be utilized for major agency programs. An agency, using the information and systems developed under existing methodologies, should investigate the feasibility of developing procedures and systems to address its specific needs for environmental impact analysis. In the long run, this can provide substantial cost savings and allow the agency to prepare meaningful and comprehensive environmental analyses.

Several new methodologies have been introduced since an earlier version of this chapter was first prepared in the mid-1970s as a research report. The information presented here is designed to acquaint the reader with general types of methodologies and provide illustrative examples of some available methodologies. Any written text captures only a small window in time and cannot be considered to cover comprehensively all existing methodologies for impact analysis. Other approaches termed "multiattribute utility theory," "systems diagrams," and "simulation modeling" may be viewed as other ways of grouping the basic methodologies described here. The Further Readings section provides information about these approaches and methodologies.

It is important to note that the CEQ regulations emphasize using an analytic rather than an encyclopedic approach to impact analysis. This approach is expected to cut down the unnecessary bulk of environmental documents and should make the documents more useful to the decision makers. Consequently, in evaluating an impact analysis methodology, one should consider the extent to which the methodology provides analytic information as one of the important criteria for its usefulness. New methodologies are expected to include more analytic techniques than in the past.

6.7   Discussion and Study Questions

1   Take another look at the local project you studied in Discussion Question 4 following Chapter 5. What would be the advantages and disadvantages of each of the six methodologies (presented in Sec. 6.2) if they were to be applied to assess the environmental impact of this project? Which of the techniques would you recommend? Why?

2   Briefly review recent EISs developed by three different federal agencies. Is the type of assessment methodology stated? May it be inferred from the content and coverage of the document?

3   Assume you are charged with the responsibility of producing, for a government agency, a handbook to be used as guidance in preparing their EISs. How would you set about deciding which assessment methodology (or combination) would best be used as a basis for this handbook?

4   Identify a major federal agency with offices in your area. Obtain the EIS preparation guidelines for that agency or your own agency, and review them to determine which assessment methodology (or combination) is used within the agency.

6.8 Further Readings

Brouwer, Floor. *Integrated Environmental Modeling: Design and Tools*. Boston: Kluwer Academic Publishers, 1987.

Costanza, Robert, and Matthias Ruth. "Using Dynamic Modeling to Scope Environmental Problems and Build Consensus." *Environmental Management*, 22:183–195, 1998.

Morgan, R. K. *Environmental Impact Assessment: A Methodological Perspective*. Boston: Kluwer Academic Publishers, 1998.

Rossini, Frederick A., and Alan L. Porter, eds. *Integrated Impact Assessment*. Boulder, Colo.,: Westview Press, 1983.

Generalized Approach for Environmental Assessment

Most federal agencies are large organizations with diversified activities and programs. To assess the environmental impact of implementing agency programs, most agencies have developed systematic procedures and agency-specific guidelines for preparing environmental documentation. A generalized approach for environmental assessment system development for an agency is shown in Fig. 7.1. Figure 7.2 provides a generalized flowchart for integration of the NEPA requirements into the agency planning process.

7.1 Agency Activities

In utilizing this generalized approach, the first thing one has to do is to become familiar with and categorize agency activities and actions such that these activities could be related to potential environmental impacts. When categorizing agency activities, one has to intimately understand the various functions, programs, and operations of the agency and its components. The agency activities may be categorized into a hierarchical structure as shown below:

Functional area

  Program

    Subprogram

      Basic activities

To provide the reader with an example of how this is typically accomplished, the following paragraphs describe a case study for U.S. Army military programs.

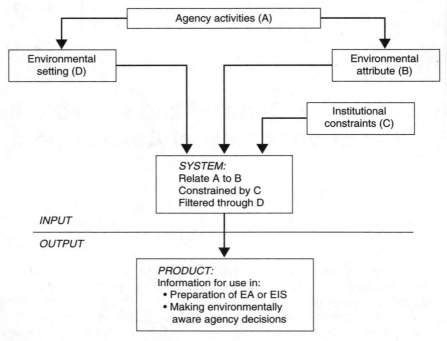

Figure 7.1   Generalized approach for performing environmental assessment.

Case study

In developing a methodology for relating Army activities to potential environmental impacts, it was necessary to develop a scheme for categorizing and classifying all Army activities in a systematic way (Jain et al., 1973). To develop a classification system, consideration was given to

1. Classification, based on the Fiscal Code, as documented in Army regulations
2. Classification of Army activities by installation
3. Classification based on the Army environmental impact guidelines

It was recognized that, individually, each of the above approaches created unique problems regarding the scope and amount of detail required. For example, if only existing installations were inventoried, the system would have been inflexible and would not have been capable of incorporating potential impacts in areas other than those specifically identified in the database. New installations would then have to be totally assessed and entered as a specific addition to the database. Also, in order to assess impacts at a specified installation,

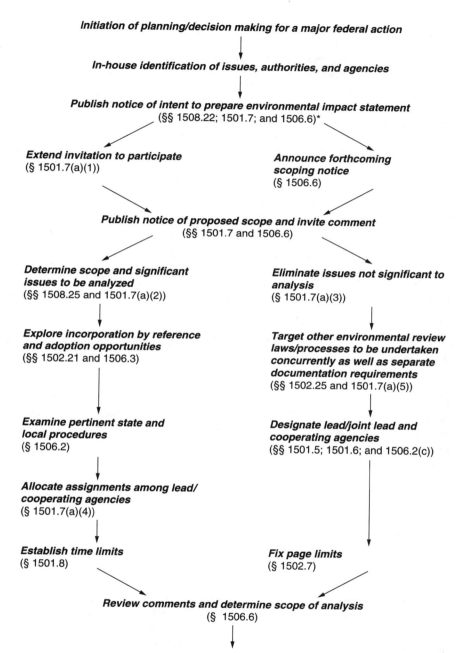

Figure 7.2 Building NEPA considerations into decision making.
* Section numbers refer to the CEQ regulations, 40CFR1500 (see Appendix D).

it would be necessary to review the baseline data for that specific site. Such information is usually not available in sufficient detail or appropriate format. Hence, specific installation review for development of basic activities associated with implementing Army programs was not possible.

Therefore, after active consultation with the potential users and careful review of agency guidance, a classification scheme was developed which synthesized the above approaches. This scheme generated the nine Army functional areas shown below:

1. Construction projects
2. Operation, maintenance, and repair
3. Training—basic to large-scale maneuvers
4. Mission changes which increase or decrease the number or type of personnel at the installation or change the activities of the people
5. Real estate acquisition or outleases or disposal of land
6. Procurement
7. Industrial plants
8. Research, development, test, and evaluation
9. Administration and support

These functional areas were defined to encompass all Army activities. For each functional area, basic activities were identified. In most cases, the activities identified were at such a level of detail that it was necessary to relate them to the functional area through a hierarchy of activities. Therefore, for most functional areas, a hierarchy of Army activities was established as follows:

Functional area

Program

Subprogram

Aggregate activities

Detailed activities

Due to variations in the nature of the functional areas, some of the hierarchical levels were omitted in some functional areas. Further details concerning how the activities for Army military programs were developed are described in the research report that formed the basis for developing a comprehensive environmental impact assessment system for application to Army military programs (Jain et al., 1973).

In addition to categorizing agency activities, it is necessary to develop a list of representative major actions and programs of an agency that might have a significant environmental impact, or whose impact, if implemented, might be considered controversial. If the agency is active in development programs, past experience alone may serve as a basis for selection of activities known to have caused problems in the past. Such experience is surely the best possible basis for a selection.

7.2 Environmental Attributes*

In order to relate agency activities to potential environmental impacts, it is desirable to categorize the elements of the environment into subsets. It should be recognized that the environment is a continuum and that there is interaction between the various environmental parameters. A minor impact on an environmental parameter could have more serious and far-reaching secondary or indirect impacts on other parameters of the environment. For example, removal of vegetation may cause excessive soil erosion, which may cause excessive sediments in the receiving stream. This, in turn, will reduce the amount of sunlight that can penetrate the water, thus reducing the dissolved oxygen in the water. Dissolved oxygen plays an important role in the biological economy of water. Reduction of dissolved oxygen will adversely affect aquatic life and water quality of the stream.

The environmental attributes can be categorized into different subsets, depending upon the level of detailed information required and the particular needs of the agency. For example, it might be desirable to develop three different types and levels of attributes. These could be

1. Detailed attribute

2. Review level attribute

3. Controversial attribute

The detailed attribute may be used to describe the conditions of the environment; any changes in the attribute would indicate changes in the environment. A review level attribute may provide an overview of the nature of the potential impacts. As such, this kind of parameter could be useful for summarizing the potential environmental impacts and providing synoptic information for personnel at the management level. Controversial attributes may be those parameters which, when affected by the agency's activity, are likely to produce an adverse public reaction or controversy.

---

*May also be referred to as environmental parameters, or environmental elements.

It is not sufficient to develop a list of environmental attributes. It is also necessary to give substance and meaning to these parameters by providing scientific information, such as definition of the attribute, how human activity might affect this attribute, how this attribute can be measured, and how this relates to other environmental attributes. Information of this type for 49 selected attributes is included in Appendix B. These 49 attributes would correspond roughly to review level attributes—clearly the environmental parameters for all specific proposals cannot be fully described using only this set of attributes. Using these as examples, the reviewer should determine the actual attributes for the project at hand, which may mean adding to or deleting from this list.

Description.

The following paragraphs give further information and delineate the items included in characterizing the attributes.

Definition of the attribute.   This item defines the environmental attribute. The definition also explains how the attribute relates to the environment.

Activities that affect the attribute.   This item contains *examples* of human activities and suggests what type of activity affects the subject attribute.

Source of effects.   This item provides a brief discussion of some of the potential ways human activities will cause an impact on the subject environmental attribute.

It should be noted that these descriptions are intended to give the reader an overview of each attribute in the context of its role in impact analysis. None of the descriptions should be considered complete, as many of the individual subject areas themselves form the basis for complete texts. It is anticipated that familiarity with these 49 attributes can serve to expedite communication between disciplines. This communication problem can be overcome when the participants attain some understanding of each other's terminology, problems, and difficulties in achieving solutions to those problems.

Variables to be measured.   This item discusses the real world variables that are to be measured to indicate environmental impact. If necessary, the relationship of the measurement to the attribute is also discussed.

How variables are measured.   This is one of the most important items in the attribute description. To the greatest extent possible, the methods for measuring impact on the variables are presented here. This includes information on sources of data that can be used to assist in measuring impact, primarily secondary data sources. References to additional technical materials that are required to adequately measure changes in the variables may be included. The types of skills that may be required in measuring impact on the variables are also discussed. For example, no special skill is required for collecting census data from published reports, but for measuring sound levels, detailed technical capabilities and equipment may be required. The need for these capabilities is identified in this item. Special instruments for measuring impact, to the extent that they are required, are also identified.

Evaluation and interpretation of data. When the data regarding the impact have been collected, an additional step is required to determine whether the impact on the subject attribute is favorable or unfavorable. In addition, the evaluation of the severity of impact is also discussed. For some attributes, the method for converting the changes in the variable into another indicator of impact is presented. This permits comparison to other environmental attributes. Other attributes are not as easily evaluated, and evaluation of the impact may require considerable professional expertise.

Special conditions. This item discusses the special measurement problems or difficulties that may be encountered in determining the impact on the subject attribute. These special conditions stem from poor availability of secondary data. If necessary, this item discusses the type and necessity of special measurement techniques. Examples of the special conditions would be the necessity for survey data regarding community values to provide baseline data for some of the impacts in the human environment category. Another example would be the need for extremely complicated measurement instruments, which may require special expertise.

Geographical and temporal limitations. Discussed here are the potential problems that might arise because of different geographical or time locations of impacts on the attribute. For example, many of the land attributes will have varying impacts, depending on the geographical location of the subject activity.

Mitigation of impact. Each environmental attribute has the potential for being affected by human activities. However, it is also possible for the activities to be modified in such a way as to reduce the impact on the attributes. In this section, the methods for reducing impacts are discussed.

Secondary effects.   Impacts on other aspects or attributes of the environment may result in a secondary or an indirect manner. For example, an aircraft runway modification project may alter flight patterns, directly changing the sound levels in adjacent areas. These could lead to a shift in land-use development, followed by a variety of biophysical and socioeconomic effects.

Other comments.   This item is reserved for information that does not fall within any of the other items relating to the environmental attributes.

Procedure for using the attribute descriptor package

The evaluation of environmental impact on an attribute-by-attribute basis involves a straightforward review of each attribute description, keeping in mind the activity that may cause the impact. As the attribute is reviewed, the data collected, and the impacts identified, entries should be made in an environmental attribute list to indicate the potential impact of the human activity on the environment. A procedure for using the attribute descriptor package in the preparation of an EA or EIS is given in Appendix C.

7.3   Institutional Constraints

Implementation of a project, action, policy statement, or regulation is subject to institutional constraints, such as emission standards for air quality control, effluent standards for wastewater discharge, and noise pressure levels for acceptable land uses. These institutional constraints could include federal, state, regional, or local environmental regulations, standards, or guidelines; as such, these could place severe constraints on the implementation of projects or actions. It is, therefore, important to carefully consider these institutional constraints in the environmental impact analysis process.

Since there are vast numbers of environmental regulations, and there is also the overlapping of agency jurisdictions, it is not always possible to obtain information regarding institutional constraints easily and expeditiously. To help solve this problem, many environmental legislative data systems have been developed. An overview of environmental laws and regulations and the various regulatory and legislative data systems is provided in Chapter 2. If you identify nonregulatory constraints which may drive decision making, such as public opinion or internal agency goals, do not be afraid to capture these in the same manner as is described here for regulatory constraints.

7.4 Environmental Setting

Depending upon the environmental setting (or environmental baseline) at a location where the project or action is to be implemented, the relative importance or even the *existence* of an impact would vary. Consequently, when utilizing a generalized EA system, provisions need to be made for incorporating the site-specific environmental setting or baseline. In a systematic procedure, environmental baseline information serves as a quasi-filtering mechanism, eliminating consideration of impacts unrelated to the specific site. Some types of impacts are, in fact, *generated* by the setting alone. An example would be proposing to locate a new building on a site that happens to be the critical habitat for an endangered species.

7.5 System

The following refers to the input sources identified in Fig. 7.1. After developing a set of typical agency activities A, applying appropriate environmental parameters B, identifying relevant institutional constraints C, and applying information on environmental baseline characteristics D, a system needs to be developed to relate A to B, using C as a constraint and D as a filtering mechanism. This "system" could be just as simple as an organized thought process or a manual storage and manipulation system, or as elaborate as a computer-aided system. Benefits and rationale for utilizing a computer-aided system for such an analysis are discussed in Section 7.7.

7.6 Output

Output from such a system should be structured to provide information necessary for preparing an EA or EIS and for making environmentally compatible management decisions. This output could include

1. An impact matrix relating activities to potential environmental impacts

2. Abatement and mitigation techniques

3. Analytical cause-and-effect relationships providing quantitative information for some environmental areas

4. Institutional constraints which must be considered

7.7 Rationale for a Computer-Based System

As discussed previously, one of the options for systematizing the generalized approach for environmental assessment analysis is to use a computer-aided system. When one discusses utilization of computer-aided systems for environmental impact analysis, many questions arise, such as

1. Can a meaningful computer-aided system be developed which is practical, useful, and cost-effective, but does not provide mechanical solutions to important environmental impact analysis problems?

2. Can any systematic procedure, computer-aided or otherwise, be developed for environmental impact analysis?

Before establishing a need for a computer-aided system, it might be well to look at some of the general problems associated with preparing an EA or EIS. After discussions with agency personnel charged with preparing these documents, the following problems have been identified:

1. The cost of preparing an EA or EIS is (often) extremely high.

2. The interdisciplinary expertise required by NEPA to prepare an EA or EIS is not always available within the staffs of agencies.

3. Even with availability of interdisciplinary expertise, it is not always possible to determine secondary and cumulative impacts which would result from implementation of a given action. This means that additional fundamental research is needed to identify, in a meaningful way, the secondary and cumulative impacts of an action.

4. A vast amount of environmental information is scattered in various publications, reports, standards, and technical manuals. It is neither convenient nor economically feasible to scan all these information sources to make environmentally compatible decisions or to prepare an environmental impact assessment. It may not be economically feasible, for example, to obtain the necessary environmental regulatory information for preparing a comprehensive EA or EIS. For this reason alone, an efficient and cost-effective system for storing and accessing data is needed. This requirement leads, almost inevitably, to a computer-aided system.

5. For some environmental impact analysis problems, it is necessary to develop cause-and-effect analytical models. It would not be possible to operate these analytical models economically without the aid of computer systems.

To address the above-cited problems, a computer-aided system may be the answer. A computer-aided system does not imply a mechanical system which would solve complex environmental problems mechanistically, but rather a system that would provide a tool to allow the user to address these problems in a comprehensive and systematic manner. One such system, called the Environmental Impact Computer System, was developed at the U.S. Army Construction Engineering Research Laboratory (Jain et al., 1973; Lee et al., 1974; Urban et al., 1974).

Geographic Information Systems as a tool for environmental assessment

Geographic Information Systems (GIS) are becoming a standard tool for use in environmental assessment and analysis due to the heightened complexity and volume of information gathered. In recent years, an increased demand for the efficient storage, analysis, and display of environmental data has led to use of computers and the development of sophisticated information systems, including GIS. GIS enables users to display and compare spatial data from a geographic location for a particular set of objectives. The combination of GIS with associated data sources, such as remote sensing imagery, is becoming common in environmental monitoring and assessment. The ability to manage voluminous sets of data from different origins, formats, and scales allows analysts to approach environmental studies in different manners (Silveira et al., 1996).

GISs developed in the late 1960s and by the mid-1970s were already being used for EIA. The overlay technique, discussed in Chapter 6, was computerized in the early 1970s and first used for siting power lines and roads. Improvements in GIS enabled its use for environmental assessment and analysis. These applications, however, have yet to make full use of current GIS capabilities. (Haklay et al., 1998).

Using GIS as an environmental modeling tool allows modelers to incorporate database capabilities, data visualization, and analytical tools in a single integrated environment. Recent surveys, however, have shown that while GIS is widely used as a tool in environmental assessment, its use is largely limited to basic GIS functions such as map production, overlay, and buffering (Haklay et al., 1998). This utilization does not take full advantage of the spatial analysis and modeling capabilities of GIS. Future applications of GIS in environmental assessment must evolve from the simple storage and display of data to include more sophisticated data analysis and modeling capabilities. An example might be evaluation of the compatibility of a proposed activity with the soils and vegetation at several possible project sites. While simple overlays may show the intersection of several elements, advanced GIS programs are able to evaluate and rank suitability for many factors simultaneously. The development of intelligent GIS (IGIS) to support spatial analysis decisions will play a large role in environmental research in the near future (Silveira et al., 1996).

Current GISs manage data through four processes. *Encoding* is the process of creating digital abstractions of the real world, *storage* is the ability to effectively handle these data, *analysis* is the correlation of spatial data to variables, and finally, the results are shown through a *display process*. For modelers to take full advantage of GIS in complex modeling capabilities, the integration of the two systems must be tightly coupled (Karimi et al., 1996). Limitations in current GIS make tight coupling with other systems difficult; for GISs to provide a simple environment for modeling activities, it must be improved.

Although the use of GIS in EIA provides many benefits, there are several factors that may limit its applicability. Many of these limitations are related to economics. A substantial amount of time and cost are required for compiling the necessary data, establishing a GIS, and analyzing the system's output. Adding to the cost, specialized personnel will be required for the operation and maintenance of a GIS. When using GIS in preparation for EIA, the personnel would need to be technically knowledgeable not only about the system, but also in the environmental issues it would address. The economic concerns may be particularly relevant in using GIS for EIA preparation because, oftentimes, EIAs are conducted by private consultants operating in a highly cost-competitive market (Haklay et al., 1998).

In addition to economic limitations, there are other concerns with using GIS, or other computer aids, for EIA. The lack of data, the cost of such data, and their level of accuracy often reduce the applicability of GIS for low-cost, small-scale projects. Additionally, as with many highly technical systems, there is the danger of "tunnel vision." It is easy for the user to assume that all factors and considerations have been accounted for within the system. Consequently, users may overlook other factors that are essential to the local environment and not covered by the system. Similarly, as with the use of expert systems, there is the danger that the user will view the system as a "black box." The system takes inputs and generates outputs; the reasoning process has been hidden away within the system, and the internal process may be unknown and its potential shortcomings not considered. Furthermore, individual judgments and values have been internalized within the system. The knowledge bases contain "facts" (actual data or sometimes estimates) gathered by various specialists. Choices concerning what information should be included within these knowledge bases are based upon the judgments of individuals. These choices will reflect individual values as well as more objective criteria related to the specialization of the experts involved. The use of computer systems does not allow these choices to be openly scrutinized by the user and/or other peers; the information is stored away within the computer. Further, some data sets may contain sensitive spatial data whose release is not allowed, such as the location of archaeological sites. These data are necessary to prepare the analysis, but should not be visible to observers without a need to know. Overall, the increased use of technology to process large amounts of data is establishing a barrier between the user and the process impact identification. The danger is that users will unquestioningly take expert system results and act on them without understanding the process and considering the outputs more carefully (Morgan, 1998).

In summary, although the potential of GIS for EIA analyses is understood, few actual applications of GIS have made full use of its analytical capabilities. Only a small number of agencies and consultants possess the full complement of skills and resources to perform analyses at this higher level. Broader use of this higher-level approach will require improvements within GIS as well as the development of a higher level of personnel expertise and significant reduction in the time and cost required to do so. These problems can be expected to be an especially significant constraint on the regular use of advanced GIS techniques for EIA, considering the stringent time and cost constraints under which EIAs need to be completed. With improvements in these limiting factors, however, much of the EIA process could potentially be largely automated through advances such as use of universal local or regional databases available to all users, and standardized analytical tools developed specifically for this purpose. In time, the GIS may be the best ally of the environmental impact profession.

7.8 Discussion and Study Questions

1 Select a proposed (or hypothetical) project which will be (or might be) sited in your area. An airport, landfill, highway bypass, and prison are good examples. Using a life-cycle approach, develop an outline which includes planning, land acquisition, construction, operation, and decommissioning. Develop additional levels of detail to the degree necessary to adequately describe the project to an interdisciplinary group which would evaluate possible environmental impacts of the project.

2 Create an interdisciplinary team and, utilizing the approach outlined in this chapter and the project description and related activities described in question 1, above, develop a draft environmental impact statement. If time does not allow preparation of full text, a detailed outline will illustrate most of the principles.

3 Obtain examples of environmental analyses that used GIS. How was the GIS used—mapping, analysis, or both?

7.9   Further Readings

Bosselmann, Peter, and Kenneth H. Craik. *Perceptual Simulations of Environments.* Berkeley, Calif.: Institute of Urban and Regional Development, University of California, 1985.

Environmental Restoration Risk Assessment Program. "Guide for Developing Conceptual Models for Ecological Risk Assessments." Document No. ES/ER/TM-186. May 1996.

Gardner, Julia E. "Decision Making for Sustainable Development: Selected Approaches to Environmental Assessment and Management." *Environmental Impact Assessment Review,* 9:337–366, 1989.

Goodchild, M. F., et al., eds. *GIS and Environmental Modeling: Progress and Research Issues.* Fort Collins, Colo.: GIS World Books, 1996.

Hyman, Eric L., and Bruce Stiffel. *Combining Facts and Values in Environmental Impact Assessment: Theories and Techniques.* Boulder, Colo.: Westview Press, 1988.

Procedure for Reviewing Environmental Impact Statements

It may be said that an agency's work is just beginning, rather than completed, when an EIS has been prepared. In fact, EISs are intended to be reviewed at many different levels within the proponent agency, as well as by other federal and state agencies with jurisdiction by law or special expertise with respect to environmental impacts. It is normal that formal findings of one federal agency may be reviewed at higher levels. This is also the case for EISs. Reviews of these documents are also made by conservation, environmental, and other public interest groups and by concerned members of the community, especially those who might be affected by the implementation of the project or the action.

Finally, and perhaps most importantly, EIS findings are reviewed by decision makers, who must consider the results of the NEPA process along with economic and technical considerations prior to the implementation of an agency decision. In view of the involvement of persons at various levels and organizations in the review of EIS documents, and the number of such documents that may be encountered, it is reasonable for an agency to develop specific procedures for reviewing EISs in an efficient and objective manner. This chapter discusses procedures that may be utilized to accomplish these goals.

A review procedure can be used by both the reviewer and the preparer of an EIS document for ascertaining the completeness, accuracy, and validity of the document. However, it should be kept in mind that as new requirements for the EIS documents are levied, and as environmental concerns include new areas, such as energy and resource conservation, any review procedure would also require updating to meet the new demands.

In general, a review procedure should allow the reviewer to (1) ascertain the completeness of the EIS document, (2) assess the validity and accuracy of the information presented, and (3) become familiar with the project very quickly and ask substantive questions to determine whether any part of the document needs additional work and/or strengthening. The concerns of the many different persons at different levels are quite variable. A single technique or procedure may not meet all needs. Processes which were designed for EIS *preparation* may not be ideal for use in a review mode. Therefore one "procedure" may consist of several very different steps with widely varying characteristics.

8.1 Types of EIS Review

Who needs or wishes to review an EIS? Is this an occasional requirement or a daily routine? Do the reviewers have special expertise? Is the reviewer also a decision maker? There are a wide variety of individuals, groups, and/or agencies who may be involved in the review process. Each review may be conducted for a different purpose, at a different location, and from a different perspective by the reviewer(s). The following are typical of the review situations that may occur.

Internal review

In order that EIS documents meet the test of scrutiny by other agencies and the public while fulfilling NEPA and CEQ requirements, it is essential that a sound system of intraagency review be established and followed. Pending or threatened litigation, potentially costly delays, presentation of a poor public image, and the likelihood of embarrassing internal and external squabbles can be minimized or (in most cases) avoided if systematic steps are taken to ensure that all NEPA-related environmental documents are reviewed for administrative (or legal) compliance, objectivity, writing style, and technical content. If inadequacies are uncovered in a rigorous internal review process, these problems may be solved prior to the public release of the document.

Interagency review

Following the preparation of a draft EIS and before completion of a final EIS, the proponent agency is required to obtain the comments of any federal agency which has jurisdiction by law or possesses "special expertise" with respect to any environmental impact involved or which is authorized to develop and enforce environmental standards. These comments are required to be solicited *in addition to* other specific statutory obligations requiring counsel or coordination with other federal or state agencies (such as that resulting from legislation such as the Fish and Wildlife Coordination Act, the National Historic Preservation Act of 1966, the Endangered Species Act of 1973, and other environmental review laws and executive orders).

Beyond the statutory reviews, the agency must request the comments of (1) appropriate state and local agencies which are authorized to develop and enforce environmental standards, (2) American Indian (Native American) tribes when potentially affected, and (3) any other agency which has requested that it receive statements on actions of the kind proposed. A system of state and area clearinghouses of the Office of Management and Budget (OMB) provides a means for obtaining state and local review, and this mechanism may be used through mutual agreement of the lead agency and the clearinghouse. As noted in Chapter 4, however, a delay in receipt caused by slow redistribution by the clearinghouse may result in serious consequences to the proponent agency. Critical reviewers should receive documents directly.

EPA review

Each draft EIS and final EIS, together with comments received and responses made (in the case of a final EIS) must be filed with the EPA as specified in Sec.1506.9 of the CEQ regulations. Five copies, accompanied by a letter of transmittal prepared by the agency filing the EIS (or usually the lead agency if more than one is involved, are sent to the EPA at the appropriate address specified on its web site. The EPA, in turn, delivers one copy to the CEQ, thereby satisfying the NEPA requirement of availability to the president. The EPA follows a formal review procedure in evaluating the statements and publishing the results of its review in the *Federal Register*; summaries of its findings are also published on its web site.

For draft statements, the EPA considers two categories: environmental impact of the action and adequacy of the statement. Under environmental impact, the statement may be classified as lack of objections (LO), environmental concerns (EC), environmental objections (EO), or environmentally unsatisfactory (EU). Under adequacy of the impact statement, the document may be rated as Category 1 (adequate), Category 2 (insufficient information), or Category 3 (inadequate). A summary explanation of these classifications is presented in Fig. 8.1. For each draft EIS which was rated EO, EU, or Category 3, the EPA must initiate a formal consultation process with the lead agency. These consultations will continue at increasing levels of management until the EPA's concerns are resolved or until it is determined that further negotiations are "pointless."

Summary of EPA Rating Definitions

- EPA's rating system was developed as a means to summarize EPA's level of concern with a proposed action.
- The ratings are a combination of alphabetical categories that signify EPA's evaluation of the environmental impacts of the proposal and numerical categories that signify an evaluation of the adequacy of the EIS.

Environmental Impact of the Action

LO (Lack of Objections)
The EPA review has not identified any potential environmental impacts requiring substantive changes to the proposal. The review may have disclosed opportunities for application of mitigation measures that could be accomplished with no more than minor changes to the proposal.

EC (Environmental Concerns)
The EPA review has identified environmental impacts that should be avoided in order to fully protect the environment. Corrective measures may require changes to the preferred alternative or application of mitigation measures that can reduce the environmental impact. EPA would like to work with the lead agency to reduce these impacts.

EO (Environmental Objections)
The EPA review has identified significant environmental impacts that must be avoided in order to provide adequate protection for the environment. Corrective measures may require substantial changes to the preferred alternative or consideration of some other project alternative (including the no action alternative or a new alternative). EPA intends to work with the lead agency to reduce these impacts.

EU (Environmentally Unsatisfactory)
The EPA review has identified adverse environmental impacts that are of sufficient magnitude that they are unsatisfactory from the standpoint of public health or welfare or environmental quality. EPA intends to work with the lead agency to reduce these impacts. If the potentially unsatisfactory impacts are not corrected at the final EIS stage, this proposal will be recommended for referral to the CEQ.

Adequacy of the Impact Statement

Category 1 (Adequate)
EPA believes the draft EIS adequately sets forth the environmental impact(s) of the preferred alternative and those of the alternatives reasonably available to the project or action. No further analysis or data collection is necessary, but the reviewer may suggest the addition of clarifying language or information.

Category 2 (Insufficient Information)
The draft EIS does not contain sufficient information for EPA to fully assess environmental impacts that should be avoided in order to fully protect the environment, or the EPA reviewer has identified new reasonably available alternatives that are within the spectrum of alternatives analyzed in the draft EIS, which could reduce the environmental impacts of the action. The identified additional information, data, analyses, or discussion should be included in the final EIS.

Category 3 (Inadequate)
EPA does not believe that the draft EIS adequately assesses potentially significant environmental impacts of the action, or the EPA reviewer has identified new, reasonably available alternatives that are outside of the spectrum of alternatives analyzed in the draft EIS, which should be analyzed in order to reduce the potentially significant environmental impacts. EPA believes that the identified additional information, data, analysis, or discussions are of such a magnitude that they should have full public review at a draft stage. EPA does not believe that the draft EIS is adequate for the purposes of the NEPA and/or CAA Section 309 review, and thus should be formally revised and made available for public comment in a supplemental or revised draft EIS. On the basis of the potential significant impacts involved, this proposal could be a candidate for referral to the CEQ.

Figure 8.1   Draft EIS classification from EPA review.

It is the EPA's policy to conduct detailed reviews of those final EISs which the EPA found to have significant issues at the draft stage. Although a rating system is not used, the EPA will conduct a detailed review for those draft EISs rated EO, EU, or Category 3, and will report its actions in the *Federal Register* and on the EPA's Office of Enforcement and Compliance Assurance web site.

Throughout the EPA review process, a high degree of coordination between the EPA and the lead agency is encouraged. There is normally no legal or procedural reason why agency and EPA personnel may not simply discuss potential problem areas. The willingness of the agency to initiate a discussion often has a positive effect on the tone of the comments submitted. Normally, before a low rating is given, an attempt is made first to obtain a revision of the discussions in the statement or the specifications of the proposed action, whichever one is believed by the EPA reviewers to fail to meet the necessary standards. It must be noted that even a severely negative EPA rating does not, in itself, constitute rejection of the proposed action, unless it is associated with a finding that a legal standard, such as a waste discharge, will be violated by the action. A low rating may be, and often is, cited by plaintiffs in subsequent legal action as evidence of inadequate evaluation under NEPA. Thus while the EPA review is not, in law, an *approval* process per se, it is still a vital step in the successful implementation of the agency's proposal.

Public review

In addition to federal, state, and local agency review, the lead agency must also request comments from the public, "affirmatively soliciting comments from those persons or organizations who may be interested or affected." Usually, this is accomplished by publishing newspaper notices regarding the availability of the draft statements, by holding public hearings, and by maintaining lists of interested conservation groups and individuals and providing them with project information and copies of the draft statement.

The review given an EIS document at this level is typically less formal than those previously described. In addition, the reviewer is likely to be biased toward or against the proposed action (or some phase of it), and the review may be conducted with the objective of identifying those aspects of the document which support that bias. Experience has shown that even those persons with strong feelings about a proposal may be largely or partially accommodated through keeping them informed at all stages. The courtesy shown in providing timely information thus substitutes, at least partially, for making those changes in the project which fully answer the objections.

Review for decision making

Ultimately, the EIS document is reviewed again at the agency level. However, this time it accompanies the proposal through existing review processes so that agency officials use the statement in making decisions. Specifically, the EIS is utilized in preparing the Record of Decision, which includes (1) a statement of the proposed decision, (2) an identification and discussion of alternatives considered, and (3) a discussion of mitigations associated with the project.

8.2   General Considerations in EIS Review

Even with the wide variation in reviewers and objectives described above, it becomes apparent that there are at least three common areas of concern among the different types of review. These specific areas of concern may be identified as follows: (1) administrative compliance, (2) general document overview, and (3) technical content. These areas are discussed below.

Administrative compliance review

This aspect of review seeks to determine the adequacy of the EIS document with respect to the law, the NEPA regulations (40 C.F.R. 1500-1508), and specific agency EIS preparation and processing requirements. The basic philosophy of NEPA and the specific requirements of NEPA Section 102(2)(C) should serve as a primary basis for evaluation. Current CEQ regulations provide guidance regarding format, length, general content outline, and other details which must be included (see Chapter 4 and Appendix D). Some types of proposals do not lend themselves to the exact format suggested in the NEPA regulations. To the greatest degree possible, however, all major points must be included. Finally, specific agency requirements may form the basis for further review comparisons. As previously mentioned, scrupulous attention must be paid to the completion of all statutory and regulatory publication, distribution, and processing steps. When a step has been omitted or modified, this may become an easy target for complaint or litigation.

General document review

The second aspect of EIS review is concerned with clearness, completeness, and correctness. Clearness refers to the utilization of visual aids, the use of language and organization (including arrangement and presentation of data), utilization of headings, and consistency in physical layout. Completeness refers to the inclusion and coverage of all reasonable alternatives, incorporation of all necessary supporting data and information, and the limitation of that information to only what is relevant to the project being analyzed. Correctness refers to ascertaining the validity of the EIS document content.

Specific concerns include reflection of current information, use of acceptable analysis techniques and adequate references, and presentation without bias. A common complaint which is made in this respect refers to "conclusory" statements. These are areas where what may be termed "advertising claims" are stated as fact without supporting evidence, or are tied to possibly unrelated scientific results. Claims of economic benefits are among the most common problem areas. Exaggeration of real but limited benefits may also fall into this category. The proponent and preparer should probably believe that the proposed action is *capable* of being carried out without undue environmental damage, but the evidence presented in the EIS should provide adequate, verifiable information which will serve to allow the reviewer to reach the same conclusion.

Technical review

Evaluating an EIS for technical content is perhaps the most difficult aspect of review; however, it is also probably the most important. Many of the concerns in technical review are the same as those voiced in general document review, only now these aspects are more subtle, often almost hidden in discussions of complex processes and interrelationships. Just as no one person can possess the expertise in all technical specialty areas necessary for the preparation of an EIS, it is doubtful that any one individual can accurately determine the technical adequacy in *all* categories of a completed EIS document. The technical review is thus usually the *sum* of several reviews by specialists.

8.3 EIS Review Procedures

Each of the various groups and persons has a purpose and need for systematic, structured review procedures for utilization in EIS evaluation. Of primary importance are those held by decision makers who will act upon the statement contents to approve or disapprove a proposed action. Of secondary importance is the viewpoint held by those within an agency who check to determine whether the administrative and legislative requirements for environmental statements have been met. Furthermore, they must determine whether the statement contents are complete and accurate prior to release for extramural review by other governmental agencies and private interests. Members of the first group (the decision makers) are the primary ones addressed in the preparation of statements according to the provisions of NEPA. The act states that environmental measures are to be incorporated into the decision-making process. This means that this group of factors must be considered along with the other parameters normally used in formulating a decision. Ideally, if a review procedure for EISs is developed for the decision maker, this should not hinder the correctness or accuracy of the statement, but should enhance its value by giving both direction and additional guidance to the authors of the statement.

The two theoretically possible approaches for review procedures were first outlined by Warner et al. (1974). The first approach calls for the decision maker or reviewer (1) to examine the problem, (2) to develop an independent analysis of the problem situation, and (3) to compare the results with the document being reviewed. This could be both time-consuming and expensive in terms of project delay and labor. It assumes either a small, simplistic document or a large, skilled available staff to assist in the review. The second alternative is to utilize a set of predetermined evaluation criteria by which the completeness and accuracy of a statement can be tested. This approach can be utilized with a minimum of labor in a short period of time. The primary disadvantage is the uncertainty associated with the complete identification of all inadequacies of the statement. It is possible that by expanding the number of criteria used in evaluating a statement, fewer impacts will be missed, since more topic coverage is that it be required. Another danger in the use of criteria by the decision maker is that it will tend to increase the possibility that pertinent impacts in some situations may be missed, especially when the proposed action does not fit preestablished criteria.

It appears that these two categories (or a combination of the two) do indeed encompass the alternatives that can be utilized to evaluate environmental impact statements in a structured, systematic manner. The development of either an independent analysis or predetermined evaluation criteria can follow the general methodologies utilized to identify and assess impacts from proposed government actions prior to their inclusion in an environmental statement. These methodologies include checklists, matrices, networks, overlay techniques, and combination computer-aided techniques. (Descriptions, uses, and procedures for evaluating the various types of methodologies used in the preparation of EISs are presented in Chapter 6. Again, each type exhibits varying advantages and disadvantages when applied to different problem situations and conditions.)

8.4   Approaches to Systematic EIS Review

Following the implementation of NEPA, a proliferation of methodologies were developed by which environmental impacts stemming from governmental actions can be identified and assessed. This is the major focus of Chapter 6. Documentation of these methodologies and other NEPA-related literature has been concerned primarily with the measures that can be utilized to identify, assess, and compare impacts *prior* to their incorporation in the initial statement. Most of the few pieces of literature which have addressed the problem of reviewing and evaluating EISs have attempted to increase the evaluator's depth of understanding in the subject matter associated with the problem area. The evaluator, familiar with the document, then presumably is better prepared to examine the statement and either agree or disagree with its contents as developed by the authors. It can be concluded that there appear to be only a few examples available by which an evaluator can compare statements or determine the "worth" of a statement. This section presents various approaches to systematic EIS review and suggests other examples related to the review procedure classifications previously identified.

Independent analysis

In the ideal situation, in order to conduct an independent analysis, the reviewer should have complete familiarity and knowledge of the proposed projects and alternatives. Utilizing this information, a "mini-EIS" is then developed and the resultant analysis compared with the document being reviewed. If a particular EIS methodology was utilized in the analysis, this "perfect" reviewer would repeat the analysis, utilizing either the same or a different methodology, and compare results. Obviously, the majority of reviews and reviewers outside the proponent agency would not have this degree of familiarity with the project and its associated alternatives and impacts. In the real world, most reviewers are short of time and can call upon only a small support staff—or none at all! At second best, the project purpose and discussion of alternatives and description of the affected environment must be sufficiently detailed in the EIS for the reviewer to evaluate the environmental consequences of the proposal.

During this independent analysis, the reviewer can utilize a checklist which can be developed from the outline of EIS content shown in Fig. 4.2. Other summaries can be developed utilizing general document review and technical review considerations. After the review has been completed, summaries can be reported utilizing the form of the example suggested in Fig. 8.2. The responsible official and/or decision maker may then utilize these summaries in determining (1) changes or modifications needed in the EIS, (2) decisions to release the document for public and interagency review, or (3) decisions to proceed with, modify, or halt the project and/or alternatives.

Administrative Compliance Summary

Review Factor	Meets Standards		Remarks
	YES	NO	
Interdisciplinary Preparation			
EIS			
Format			
Page Limits			
General Content			
Cover Sheet			
Summary			
Table of Contents			
Purpose & Need Clear			
Alternatives Examined			
Affected Environment			
Environmental Consequences			
List of Preparers			
Distribution List			
Index			
Appendix			
Original Studies			
Data Support EIS			
Not Overly Lengthy			

Recommendation:	Concur: _____ Nonconcur: _____
Approve _____ Disapprove _____	
	Signature: _____
	(Responsible Official)
Signature: _____	
Date: _____ Title: _____	Title: _____ Date: _____

Figure 8.2 Sample Administrative Compliance Summary form.

Predetermined evaluation criteria

Evaluation criteria for use by reviewers could take many forms. The form could range from a short, concise statement answering certain questions concerning the proposed activity to a weighted checklist which portrays numerical values for different criteria which can be compared to index values. The contents of this analysis could be attached to the EIS and utilized in the decision-making process. Majority and minority opinion of the reviewers could also be included as another decision parameter to be considered by the responsible official.

Wide variation in missions and programs may exist between agencies and even within one agency. This increases the difficulty in developing a *single* set of criteria that can be utilized to evaluate all federally related projects. The more specialized the agency activities, the more detailed the criteria that can be utilized, whereas the more variable the projects that can be encountered, the greater the generalization of the criteria. Generalized criteria, if properly selected, still have the capability of directing the statement review so that it is an effective tool for decision makers.

A review procedure first suggested in 1975 (Jain et al., 1975) makes use of evaluation criteria whereby the level of significance of construction projects could be determined. After determining this level, specific review criteria are applied to the corresponding level. In applying this procedure, the characteristics must be known and combined with a set of screening questions (shown in Table 8.1). These questions broadly categorize construction projects by their characteristics according to the extent of potential impacts. The response rating of these questions is recorded along with the response score. Example response ratings are shown in Table 8.2 and may be used to guide the determination of the appropriate response rating and associated score. The scores may then be summed for the project to provide a total score. The score provides a rationale to categorize construction project impacts into three major levels (I, II, and III). Next, the detailed EIS review criteria are used to review the document.

Project screening questions. The 12 project screening questions in Table 8.1 were developed (Jain et al., 1975) to categorize potential project impacts according to project characteristics, and are slightly modified here for the present purpose. The questions cover a broad range of major environmental impacts associated with the construction projects. These questions are answered either by "yes" or "no," or by "high," "medium," or "low." Determination of an answer is based upon response rating criteria.

Response rating criteria. Specific numeric and qualitative criteria were developed to determine the answer to each project screening question. Such criteria prescribe what is meant by a "high," "medium," or "low" (or "yes" or "no") rating for a particular question.

Example rating criteria presented in Table 8.2 for each screening question were developed by use of informed professional judgment and were meant to apply to construction projects. Suggested response rating

TABLE 8.1   Screening Questions

No.	Questions	Rating	Score
1.	What is the approximate cost of the construction project?	High Medium Low	10 5 0
2.	How large is the area affected by the construction or development activity?	High Medium Low	10 5 0
3.	Will there be a large, industrial type of project under construction?	Yes No	10 0
4.	Will there be a large, water-related construction activity?	Yes No	10 0
5.	Will there be a significant waste discharge or generation or hazardous waste?	Yes No	10 0
6.	Will there be a significant disposal of solid waste (quantity and composition) on land as a result of construction and operation of the project?	Yes No	10 0
7.	Will there be significant emissions (quantity and quality) to the air as a result of construction and operation of the project?	Yes No	10 0
8.	How large is the affected population?	High Low None	10 5 0
9.	Will the project affect any unique resources (geological, historical, archaeological, cultural, or endangered or threatened species)?	Yes No	10 0
10.	Will the construction be on a floodplain?	Yes No	10 0
11.	Will the construction and operation be incompatible with adjoining land use in terms of aesthetics, noise, odor, or general acceptance?	Yes No	10 0
12.	Can the existing community infrastructure handle the new demands placed upon it during construction and operation of the project (roads/utilities/health services/vocational education/other services)?	No Yes	10 0

criteria shown in Table 8.2 would have to be modified to apply to other types of projects and as experience in their use shows shortcomings.

Project screening criteria. Each response rating from Table 8.2 is assigned a point value of 10, 5, or 0. For each "yes," a project gets a score of 10; for each "no," the score is 0; for "high," "medium," or "low" ratings, scores assigned are 10, 5, and 0, respectively. Possible total scores for all combinations of various construction projects range from 0 to 120. Within this range, the following three levels of projects are defined:

Level I:	Small-impact projects	scores 0–60
Level II:	Medium-impact projects	scores 60–100
Level III:	High-impact projects	scores >100

Remember, however, that there is no "magic" in the number 100, 120, or any other number at all! This entire system is merely an *example,* and an entirely different one may be constructed which is based on any set of values across any range.

Review criteria. Review criteria are employed to assess the completeness and accuracy of the impact statement. The review level is established by the score of the project screening exercise. These levels (or other appropriate ranges) may be used to discriminate between projects that require detailed versus less detailed review. The potentially high-impact project should be given the most thorough review, while the others should be given a less intensive review, particularly in the technical area. Administrative compliance may be evaluated on criteria developed from CEQ regulations (see Fig. 4.2) and general document review criteria as suggested in Table 8.3.

Ad hoc review

A third form of review is summarized for the many persons who may find themselves in the position of occasionally, or even on a one-time basis, needing to review an EIS but not desiring to employ the detailed, structured approaches suggested above. For those reviewers, the following sequence of activities is suggested. It is equally applicable to persons with technical background and to those whose capabilities are entirely administrative.

To perform an ad hoc review,

1. Familiarize yourself with the CEQ-prescribed outline and content (Fig. 4.2) and the agency's format and outline, if available. This will provide you with an idea of the general sequence and format to be expected as you examine the body of the EIS.

2. Read the summary. This will provide an overview of the project, its alternatives, and the anticipated environmental consequences. Does it lack a summary?

TABLE 8.2   Example Response Rating Criteria

No.	Criteria	Rating
1.(a)	The construction is less than or equal to $10 million.	Low
1.(b)	The construction cost is >$10 million but <$100 million.	Medium
1.(c)	The construction cost is >$100 million.	High
2.(a)	The area affected by construction is ≤10 acres.	Low
2.(b)	The area affected by construction is >10 and <50 acres.	Medium
2.(c)	The area affected by construction is >50 acres.	High
3.(a)	An industrial-type project costing more than $10 million is involved	Yes
3.(b)	Otherwise.*	No
4.(a)	The large water-related construction project consists of one or more of the following: A dam A dredging operation of 5 miles or longer; disposal of dredged spoils A bank encroachment that reduces the channel width by 5 percent Filling of a marsh, slough, or wetland >5 acres Continuous filling of 20 or more acres of riverine or estuarine marshes A bridge across a major river (span: 400 feet)	Yes
4.(b)	Otherwise.	No
5.(a)(1)	At least one of the following waste materials may be discharged into the natural streams: Asbestos PCB Heavy metals Pesticides Petroleum products Cyanides Solvents Radioactive substances Other hazardous materials or waste (specify)	Yes
5.(a)(2)	Rock slides and soil erosion into streams may occur because No underpinning is specified for unstable landforms. No sluice boxes, retention boxes, retention basins are specified for excavation and filling.	Yes
5.(b)	Otherwise.	No
6.(a)(1)	At least one of the following solid wastes may be disposed of on land: Asbestos PCB Heavy metals Pesticides Cyanides Radioactive substances Any designated hazardous waste	Yes
6.(a)(2)	The solid waste generated is greater than 2 pounds per capita per day.	Yes
6.(b)	Otherwise.	No

TABLE 8.2 Example Response Rating Criteria (Continued)

No.	Criteria	Rating
7.(a)(1)	If there are to be Concrete aggregate plants—EIS does not specify dust control devices.	Yes
7.(a)(2)	Hauling operations—EIS does not specify use of dust control measures.	Yes
7.(a)(3)	Road grading or land clearing—EIS does not specify water or chemical dust control.	Yes
7.(a)(4)	Open burning—EIS does not specify disposal of debris.	Yes
7.(a)(5)	Unpaved roads—EIS does not specify paved roads on construction sites.	Yes
7.(a)(6)	Asphalt plants—EIS does not specify proper dust control devices.	Yes
7.(b)	Otherwise.	No
8.(a)	Fewer than 20 persons are displaced by the project.	Low
8.(b)	From 20 to 50 persons are displaced by the project.	Medium
8.(c)	More than 50 persons are displaced by the project.	High
9.(a)(1)	A rich mineral deposit is located on the construction site.	Yes
9.(a)(2)	A historical site or building is located at or near the construction site.	Yes
9.(a)(3)	A known or potential archaeological site is located near the construction project.	Yes
9.(a)(4)	A state or federally listed endangered species is found in the project area, or habitat is found on the site.	Yes
9.(b)	Otherwise.	No
10.(a)	The construction project is on a 100-year floodplain.	Yes
10.(b)	Otherwise.	No
11.(a)(1)	No visual screening is specified in the EIS for the construction site.	Yes
11.(a)(2)	No progressive reclamation of quarry and/or disposal sites is proposed.	Yes
11.(a)(3)	No permissible noise level specifications are stated for vibrators, pumps, compressors, piledrivers, saws, and paving breakers.	Yes
11.(b)	Otherwise.	No
12.(a)	The projected demand for community services exceeds existing or planned capacity. These services include Water supply Wastewater treatment and disposal Electric generation Transportation Educational and vocational facilities Cultural and recreational facilities Health-care facilities Welfare services Safety services: fire, flood, etc.	Yes
12.(b)	Otherwise.	No

* "Otherwise" implies that none of the previously mentioned situations are applicable to the project.

TABLE 8.3   General Document Review Criteria

Area of concern	Criteria
A. Readability	1. Write clearly. 2. Remove all ambiguities. 3. Avoid use of technical jargon; all technical terms should be clearly explained.
B. Flavor and focus	1. Do not slant or misinterpret findings. 2. Avoid use of value-imparting adjectives or phrases. 3. Avoid confusion or mixup among economic, environmental, and ecological impacts and productivity. 4. Avoid unsubstantiated generalities. 5. Avoid conflicting statements.
C. Presentation	1. Use consistent format. 2. Use tables, maps, and diagrams to best advantage. 3. Avoid mistakes in spelling, grammar, and punctuation.
D. Quantification	1. Use well-defined, acceptable qualitative terms. 2. Quantify factors, effects, uses, and activities that are readily amenable to quantification.
E. Data	1. Identify all sources. 2. Use up-to-date data. 3. Use field data collection programs as necessary. 4. Use technically approved data collection procedures. 5. Give reasons for use of unofficial data.
F. Methods and procedure	1. Use quantitative estimation procedures, techniques, and models for arrival at the best estimates. 2. Identify and describe all procedures and models used. 3. Identify sources of all judgments. 4. Use procedures and models acceptable by professional standards.
G. Interpretation of findings	1. Consider and discuss all impact areas before any are dismissed as not applicable. 2. Give thorough treatment to all controversial issues, and discuss the implications of all results. 3. Consider the implications for each area of a range of outcomes having significant uncertainty. 4. Analyze each alternative in detail and give reasons for not selecting it. 5. Scrutinize and justify all interpretations, procedures, and findings that must stand up under expert professional scrutiny.

3. Examine the table of contents to determine the location of various parts of the EIS. Depending on your familiarity with the project and/or the affected environment, you may wish to go directly to a specific section of the document.

4. Study the content of the EIS. Look for those items specifically identified in Table 8.3.

5. Is there any area or topic on which you *do* have specialized knowledge or technical expertise? Is the discussion of these points reasonable? Are there obvious errors of fact or confused application of basic principles in these areas?

6. Focus next on issues and concerns regarding administrative, general document, and technical review concerns previously identified.

7. Evaluate the EIS on the basis of your review, using as *examples* those topics where you possess specialized knowledge.

8.5 Summary

In order to assist the many different reviewers and the decision makers in the NEPA process, this chapter has presented a discussion of procedures for reviewing and evaluating EIS documents. These procedures focus on three areas of concern:

1. Administrative review

2. General document review

3. Technical review

This chapter has described two types of approaches to developing a systematic, structured review procedure. By using such procedures, the reviewer can become familiar with the project very quickly and ask substantive questions to determine whether any part of the EIS document needs additional work or strengthening.

If review procedures are developed and are acknowledged during the preparation process, EIS contents will not only contain the information necessary to satisfy CEQ requirements but will also reflect the evidence in the statement at hand. The statements should therefore become more analytic rather than encyclopedic, in line with the CEQ regulations.

8.6 Discussion and Study Questions

1 Obtain a draft EIS and a final EIS from any federal agency. Conduct an ad hoc review of each document, and prepare a classification based on the EPA

review criteria. Then, and not before, determine the actual evaluation assigned to the document by the EPA through its own review. It will be given in the *Federal Register.* Do you agree with the EPA's classification? Discuss the differences you find. Were you more severe than the EPA? Do you think some problems were overlooked? Were you more lenient?

2   Review several final EISs which include the comments received. Examine the agency comments (they are usually placed at the beginning of the comment section). Do you feel that the content of the comments furthers the letter and spirit of NEPA?

3   Examine again the same final EISs which you used for question 2, above. Look for comments from the general public and environmental groups. Is there evidence that the public was *informed* when the comments were made, or are they simply expressions of opposition (or support)? How would you proceed to increase constructive participation on the part of the unorganized public?

4   Obtain a draft and the following final EIS, for any project. Compare the two. In what ways does the final differ from the draft? Are there *any* changes? Examine the comments (included with the final) which were made on the draft. Do any changes in the final appear to have resulted from these comments? Was the EIS *improved* by these changes? Were new alternatives added? Were *any* changes made in the proposed action, or were changes merely in the way the effects were described? Are these changes for the better, in an environmental context?

8.7   Further Readings

In the considerations involved when reviewing an EIS there appears to be no substitute for getting your own hands on actual examples of an EA or EIS. Thus, we recommend no specific additional readings beyond those involved in pursuing the Discussion and Study Questions above. Review as many NEPA documents, long and short, good and bad, as possible within the available time. It is only through becoming familiar with actual examples of NEPA documentation that your understanding is advanced. Many NEPA documents are now available on the web.

International Perspectives on Environmental Assessment

The international community is increasingly concerned about environmental issues. This is reflected in the increase of international environmental organizations, the investment nations are making to protect the environment, and the fact that environmental issues are taking center stage during meetings between world leaders.

There is a general consensus that national and international security has an important environmental dimension. How nations use natural resources to foster economic development often determines what kinds of societies are likely to emerge. When long-term viability of the environment is ignored, economic development is not likely to be sustained long; eastern Europe provides a vivid example. Such policies also provide an indication of governmental attitudes toward other social issues internally and toward international responsibilities. The cost of cleanup of past environmental degradation can become a significant proportion of a nation's GNP, and thus exceed its ability to undertake the cleanup effort, as is the case in eastern Europe at this time. Regional environmental degradation could severely affect the health of its population and its economic base to a point that national and international security could be perceptibly affected for years to come.

Would a process for environmental impact assessment (EIA) process have helped to eliminate some of the environmental problems in eastern Europe? Should donor countries insist on a formal EIA process before providing financial aid? Should other countries replicate the formalized EIS process of the United States? These are all important questions for the international community to address.

Whether all provisions of NEPA are applied to U.S. projects in other countries and whether other nations use the EIS process as in the United States are not, in and of themselves, vital. The conduct of environmental assessment of projects undertaken by the U.S. agencies in other countries, cooperating with host nations in their environmental assessment activities, and in all cases assessing the environmental consequences on the global commons are prudent courses of action.

9.1 International Implications of NEPA

The overseas actions of many federal agencies, such as the Agency for International Development (AID) and military bases operated by the Department of Defense, have the potential to create major environmental impacts in foreign countries. NEPA contains no unequivocal language on the extent to which it was intended to apply to overseas federal actions, and court cases have not provided a definitive answer to this ambiguity. As a result, the extraterritorial application of NEPA continues to be debated.

There is concern that applying the full procedural content of NEPA to overseas actions could interfere with United States' foreign policy and national security objectives, that such activities may be viewed by some nations as interference in their sovereign rights, that delays from preparing EISs and possible litigation could hamper the United States' ability to compete internationally, and that on-site assessments could, in many cases, be difficult to carry out.

Concerns about the authority of Congress to require the application of NEPA overseas center around the issue of national sovereignty. The United States, or any other nation, in traditional international law does not have the right to extend its own laws extraterritorially except under certain conditions in which the conduct of other nations affects its well-being in a material way. The bases of extraterritorial jurisdiction include such conduct as that which affects national security or a nation's citizenry (Goldfarb, 1991).

NEPA includes certain references which are clearly domestic in scope, such as "the nation" and "Americans." There are also references to "man" and "his environment" without reference to specific locality. These nondomestic references can support, although they do not clearly specify, an extraterritorial interpretation. Section 102(2)(F) of NEPA refers explicitly to international activities and directs federal agencies to support any program which enhances international cooperation in recognizing the global and long-term character of environmental problems. A direction to "support" programs which enhance global environmental protection may indicate that NEPA was *conceived* as having international scope, but it does not, by itself, constitute a clear requirement for preparing rigorous environmental documentation, such as an EIS, for overseas actions.

One consideration for the application of NEPA extraterritorially is the reasonableness of the application. An extraterritorial application of domestic law is reasonable if it respects the sovereignty of other nations, does not generate conflict, and balances the interests of the countries affected. The cases addressing this issue demonstrate that the courts are inclined to exclude NEPA from situations in which the statute may conflict with foreign policy objectives or infringe on the sovereignty of other nations. The courts are also inclined to rule against NEPA application when the interests of the United States are minimal. Although several cases have addressed the issue of applicability overseas, there has been no conclusive determination.

NEPA was ruled *to apply* in the following cases, due to the absence of foreign policy conflict and/or the presence of strong United States interest (Goldfarb, 1991):

Nuclear testing on a United States trust territory (*Eneweitak v. Laird*)

The construction of a highway in Panama and Colombia which (it was alleged) could provide a route to infect United States livestock with disease (*Sierra Club v. Adams*)

A proposed program to spray pesticides in 20 developing countries (*Environmental Defense Fund v. USAID*).

Due to limited United States interest and potential foreign policy conflicts, the courts have ruled *against* the application of NEPA in these cases:

The licensing of private corporations by the Nuclear Regulatory Commission to sell nuclear reactor components to western Germany (Babcock & Wilcox hearing) and to the Philippines (*Natural Resources Defense Council, Inc. v. Nuclear Regulatory Commission*).

Movement of chemical munitions by the U.S. Army across West Germany (*Greenpeace USA v. Stone*). The U.S. Army had prepared necessary environmental documentation (EISs) for the construction of a chemical incinerator at Johnston Island and for the operation of the incinerator, and had prepared an EIA (under Executive Order 12114) for the transportation of the munitions from West Germany across the global commons to Johnston Island. The movement of the munitions *within* the territorial boundaries of West Germany, performed by German authorities with U.S. oversight, was not covered in the documentation. The court was not persuaded that transporting munitions within West Germany pursuant to an agreement between heads of state warranted preparation of NEPA documentation.

The draft regulations issued by the CEQ in 1978 included a statement that for any action affecting the U.S. environment, the global commons, or Antarctica, full environmental assessment would be required, whereas actions affecting only another national environment would require an assessment of reduced scope. However, this provision was withdrawn in response to the protest of various agencies, particularly the State Department, which maintained that foreign policy considerations must have priority over environmental assessment (Goldfarb, 1991).

President Carter's Executive Order 12114 was intended to resolve the stalemate. The order limited application of NEPA to those actions which would

1. Affect a country not involved in the action

2. Affect the global commons

3. Expose a country to toxic or radioactive emissions

4. Affect resources of global concern

The executive order excludes activities of concern to the State Department such as military and intelligence activities, arms transfers, export licenses, votes in international organizations, and emergency relief actions. EIS requirements may also be modified in consideration of potential adverse impacts on foreign relations, other nations' sovereignty, diplomatic factors, international commercial competition, national security, difficulty of obtaining information, and inability of the agency to affect the decision. Critics of Executive Order 12114 maintain that it is not enforceable, and that the many listed exemptions create loopholes for most actions.

9.2   Future NEPA Trends

In light of the limited scope of Executive Order 12114, several proposed bills in 1989 to 1991 demonstrated congressional interest in affirming the applicability of NEPA abroad. SB 1089 proposed to close the exempted activities loophole of the executive order by limiting exempted activities to those which are necessary "to protect the national security of the United States." HR 1113 would have amended NEPA to require agencies to "work vigorously to develop and implement policies, plans and actions designed to support national and international efforts to enhance the quality of the global environment" where the existing language states that agencies must only "lend appropriate support to initiatives." Both bills would have required EISs to include assessment of the effect of federal actions on the global commons and on extraterritorial actions. Neither bill was enacted into law.

Goldfarb (1991) argues that basing the main objection to applying NEPA overseas on the question of sovereignty of other nations is unjustified. International law has always maintained that sovereignty is limited by the responsibility to avoid causing harm to other nations, and that nations may voluntarily restrict their sovereignty by entering into agreements or treaties. In recent years, increasing concern about the global environment, and recognition that it is not possible to limit environmental impacts to specific geographical areas, has led to international agreements which limit sovereignty. The existence of these treaties, as well as the existence and activities of various international organizations concerned with environmental issues, demonstrates the recognition in the international community that voluntary limitations on sovereignty are important to protecting the global environment and a reasonable expectation for the concerns of our age. The challenge is: Is it possible to find a middle ground that responds to the many legitimate and real concerns on all sides of the issue?

The 1992 Earth Summit that attracted over 100 heads of state to Rio de Janeiro served to highlight renewed interest in assessing long-term environmental consequences of human activities. In the United States the NEPA reviews have provided a meaningful mechanism for incorporating environmental considerations in major governmental undertakings. The Earth Summit deliberations, though not completely embraced by environmentalists or the business community, generated interest in adopting similar processes in other lands.

9.3 Environmental Impact Assessment in Other Countries

The application of EIA in other countries has been inspired by the example of NEPA in the United States and the 1972 Stockholm United Nations Conference on the Human Environment, and in developing countries by various multilateral and bilateral assistance organizations which promote EIA. Despite wide-ranging interest, the comprehensive application of EIA as it exists in the United States is not widely duplicated in other countries. Rather, it is represented in a variety of legislative, institutional, and procedural manifestations, which reflect the variety of resources, institutions, and unique interests of the nations.

Industrialized nations have carried the implementation of EIA to the greatest extent, and highly developed systems are found in Canada and the Netherlands. The developing countries of the Asian and Pacific regions have achieved a partial implementation of EIA; many countries have federal agencies responsible for the environment, national environmental policies, and requirements at the legislative level. Latin American countries have been able to accomplish somewhat less in EIA development, and EIA in African countries is very limited. A study undertaken by Sammy in 1982 indicates that the percentage of countries with legislation requiring EIA for some projects is 66 percent in the southeast Asian and Pacific region, 57 percent in Latin America, and 41 percent for Africa and the Middle East (Kennedy, 1988). The World Resources Institute (1998) has developed an extensive directory of impact assessment guidelines from other countries.

9.4   EIA in Developing Countries

The EIA process now found in the Philippines, Korea, and Brazil exemplifies general trends in developing countries. Analysis of the EIA process in these countries was conducted by Lim (1985) and is summarized here.

A presidential decree of 1977 established a national environmental policy and a requirement for EIA in the Philippines. A previous decree had established the environmental agency. Guidelines specify the projects to be included in environmentally critical areas. The environmental agency is made up of heads of various agencies, which undermines its legal authority. Responsibilities are divided among six agencies, and accountability by the participants is low. Public hearings are not mandatory. Between 1978 and 1983, an average of only eight EISs were filed each year, while several hundred new projects were registered. EIA is not welcomed by many participants. EIA in the Philippines performs an agency adjustment function.

EIA in Korea was legislated by 1980 revisions to the Environmental Conservation Law, which also created the Office of Environment under the Ministry of Health and Social Welfare. The 5-year plan for 1982–1987 was the first to state environmental conservation as an official national goal and EIA as a tool to achieve it. The EIA system is centralized. The legal authority of the Office of Environment is limited by its status as a subministry. Public participation is lacking, and procedural rules are not clearly defined. EIA is required only for large projects; on average only seven have been prepared annually, and they have resulted in minimal modification of plans. In Korea EIA provides an environmental remediation function.

Brazil's Special Environmental Agency was established in 1974. The National Environmental Policy Law requiring EIA and establishing the National Environmental Council was passed in 1981.

Those projects requiring EIA are not delineated, and the roles of various agencies are unclear. The rule-making body has limited legal authority. Only a small number of projects have been evaluated (averaging 11 yearly), but several have been modified as a result of the assessments. The role of Brazil's EIA process would be perfunctory except for this last fact.

9.5 Limitations to EIA Effectiveness in Developing Countries

Limited technical abilities, such as lack of data-gathering capability, lack of scientific understanding, and lack of expert staff hamper EIA in developing countries. But perhaps more importantly, an institutional and legislative framework which can promote assessment and make use of the results is lacking. The establishment of government offices that are responsible for environmental concerns, as well as offices that are responsible for EIA, are recent accomplishments even for industrialized countries. A legal framework to ensure cooperation between agencies is also necessary. Effective EIA requires a political context which recognizes value in environmental protection and can allow public review of governmental activities. In addition, economic resources to commit to the EIA process are not always available in developing counties.

Observers find several general tendencies in the application of EIA in developing countries which limit its effectiveness. Assessments are undertaken too late in the planning to contribute to decision making and are used instead to confirm that the environmental consequences of the project are acceptable. The environmental management plans discussed in the EIS documents are often not carried out, and there is no mechanism for monitoring compliance. The studies which have been completed are relative only to projects, as opposed to policies or programs. Few studies have evaluated projects for social or economic consequences. Many countries limit projects that are subject to EIA such that projects which may have significant environmental impact are excluded from the EIA requirement. A final observation is that external review of the process, essential to limiting abuse and mismanagement, is often lacking.

Horberry suggests that EIA often functions as a "device for promoting a realignment of relationships among domestic institutions" (Horberry, 1985, p. 205), as a way to enhance the power of the environmental agency, or to change the operating routine of other agencies, rather than as a tool to consider environmental issues early in the planning and decision-making process.

9.6   EIA in Asia and the Pacific

Lohani (1986) classifies the situation of EIA within various Asian and Pacific countries in four categories:

1. Countries with specific legislation for EIA include Australia, Japan, and the Philippines.

2. Countries with general legislation on environmental protection that empowers a government agency to require EIA for particular projects, but no specific EIA legislation, include Iran, Malaysia, Hong Kong, New Zealand, Republic of Korea, Thailand, Trust Territory of the Pacific Islands.

3. Countries with no formal requirements for EIA but with informal procedures to incorporate environmental consideration into the planning of specific projects include Bangladesh, Indonesia, India, Pakistan, Sri Lanka, Papua New Guinea.

4. Countries lacking any formal requirements for EIA include Afghanistan, Cook Island, Nepal, Fiji, Tuvalu.

Various regional groups have been formed to facilitate the sharing of information about environmental protection between neighboring countries. These include the Association of Southeast Asian Nations Environment Program; the South Asian Cooperative Environment Program; the Mekong Committee of ESCAP; and the South Pacific Regional Environment Program.

9.7   EIA in Latin America

The impetus for the development of legislation, scientific resources, and community interest in EIA in Latin America has come from external aid organizations, including the United Nations Environment Programme and the Pan-American Health Organization, which have sponsored development projects. Although this influence has been extensive, more comprehensive EIA development is limited by the nature of the region's governments.

Uruguay and Peru have no legal requirements for EIA. Argentina does not require EIA, although voluntary EIA studies are promoted. Colombia, Venezuela, Mexico, and Brazil have environmental policy laws which include some provisions for EIA studies. In Brazil, EIA is required for potentially polluting industrial plants within the Rio de Janeiro and São Paulo regions.

9.8   EIA in Canada

Canadian requirements for EIA were established in 1973 at the cabinet level as the Environmental Assessment and Review Process. The Federal Environmental Assessment Review Office oversees the EIA system. Consultation by the public is extensive, being called for at various stages of the assessment, and ending in a series of public meetings. The review is conducted by an independent panel appointed by the Minister of the Environment, and the public has access to all panel information. In addition to the federal EIA process, each province has its own program, usually mandated with legislation.

9.9   EIA in Europe

The Commission of the European Communities (CEC) Directive on environmental impact assessment, 85/337/EEC, came into force in 1988. It took 20 drafts and more than 15 years to finalize (Wood, 1988). As with the framers of NEPA in the United States, the CEC felt that ". . . effects on the environment should be taken into account at the earliest possible stage in all the technical planning and decision making processes" (Wood, 1988). It was updated and amended in 1997.

The commission decided that the EIA system should promote, among other things, two sets of objectives (Wood, 1988):

- To avoid distortion of competition and misallocation of resources by harmonizing environmental controls

- To ensure that a common environmental policy is applied throughout the EEC

The directive contains 14 articles and four annexes. Listed in Annex I are the types of projects for which EISs should normally be prepared. These include: large oil refineries and storage facilities, large power stations and major electric transmission lines, toxic or radioactive waste disposal sites, integrated steelworks, large-diameter pipelines, integrated chemical plants, and major airports, ports and canals (CEC, 1985). Annex II contains a much longer list of types of projects than does Annex I. This list includes projects which "...shall be made subject to an assessment where member states consider that their characteristics so require" (CEC, 1985). For the purpose of preparing an EIA, projects listed under Annex I are considered mandatory while projects listed under Annex II are discretionary. Annex III contains selection criteria to help determine if Annex II projects must be assessed. The type of information to be included in an EIA is outlined in Annex IV, as shown in Table 9.1.

To focus on some of these crucial environmental issues, in 1990 the European community created a European Environmental Agency (EEA) to develop common environmental policies for the region. The EEA is headquartered in Copenhagen, Denmark, and provides a wide variety of information and services throughout the EEU. In the words

TABLE 9.1 Annex IV of 1985 CEC Environmental Directive (as amended in 1997)

Information Referred to in Article 5(1)

1. Description of the project*, including in particular
   - A description of the physical characteristics of the whole project and the land-use requirements during the construction and operational phases.
   - A description of the main characteristics of the production processes, for instance, nature and quantity of the materials used.
   - An estimate, by type and quantity, of expected residues and emissions (water, air and soil pollution, noise, vibration, light, heat, radiation, etc.) resulting from the operation of the proposed project.
2. An outline of the main alternatives studied by the developer and an indication of the main reasons for his or her choice, taking into account the environmental effects.
3. A description of the aspects of the environment likely to be significantly affected by the proposed project, including, in particular, population, fauna, flora, soil, water, air, climatic factors, material assets, including the architectural and archaeological heritage, landscape, and the interrelationship among the above factors.
4. A description of the likely significant effects of the proposed project on the environment resulting from
   - The existence of the project
   - The use of natural resources
   - The emission of pollutants, the creation of nuisances and the elimination of waste, and the description by the developer of the forecasting methods used to assess the effects on the environment
5. A description of the measures envisaged to prevent, reduce, and, where possible, offset any significant adverse effects on the environment.
6. A non-technical summary of the information provided under the above headings.
7. An indication of any difficulties (technical deficiencies or lack of know-how) encountered by the developer in compiling the required information.

*This description should cover the direct effects and any indirect, secondary, cumulative, short-, medium-, and long-term permanent and temporary, positive and negative effects of the project.

of the executive director of the EEA, " The European Environment Agency (EEA) is a European Community institution with the aim of serving the Community and the Member States with information to support policy making for environmental protection put in the perspective of sustainable development" (Jimenez-Beltran, 2001). The EEA has thus become a facilitator rather than a regulator, with this stated goal: "The EEA aims to support sustainable development and to help achieve significant and measurable improvement in Europe's environment through the provision of timely, targeted, relevant and reliable information to policy making agents and the public" (EEA, 2001).

Wathern (1988) points out that the directive, despite over 10 years of debate and deliberation, is limited and simply formalizes some of the provisions already in place in member states. The various member states find ways to comply with the directive's formal requirements while ensuring that the substance of the directive does not conflict with their domestic policies. Some European countries have been requiring EA for major projects long before the CEC directive was finalized. Two prominent examples are France and the Netherlands.

France passed a Nature Protection Act in 1976 which requires EIA for all major public and private projects. However, Monbailliu (1984) asserts that the effectiveness of EIA in France is hampered by several limitations in the process. Social and economic impacts are excluded, as well as all developments costing under 60 million francs. Public participation is very limited, and only published EISs may be discussed. In 1990, the Agency for the Environment and Energy Management was formed, and administers a national budget for environmental projects of about $600 million per year.

Starting in the mid-1970s, the Netherlands has been working to develop comprehensive environmental program, which culminated in 1989 with the passage of the National Environmental Policy Plan. This system represents the state-of-the-art in EIA procedures in Europe. It is implemented by a single, integrated law that applies to legislation, plans, and projects at the national, provincial, and municipal levels. A positive list is used that specifies the type of projects to be assessed. Public involvement and independent review are provided for. A biennial "report card" is prepared, showing accomplishments and shortfalls.

9.10   International Aid Organizations

Various international organizations promote EIA in developing countries by making it part of the funding process and by recommending its adoption by recipient countries.

The World Health Organization (WHO) actively promotes the development of EIA procedures in its member states, with special emphasis on health and safety impacts. It presents courses and seminars, assists member states directly in establishing and improving their EIA procedures, and commissions research. WHO is particularly interested in improving the state of knowledge about health impacts and developing methodologies to enhance health impact assessment.

EIA assistance programs are maintained by various agencies of the United Nations, including UNESCO, FAO, ESCAP, the Development Programme, and the Environment Programme. The regular programs provide policy review, technical advice, and information management to their members. Field programs provide direct technical assistance and project funding.

The World Bank and the Asian Development Bank both attempt comprehensive environmental evaluation of projects. Others which attempt less extensive evaluation include the Inter-American Development

Bank, the Organization of American States, the United Nations Development Programme, the European Commission, and the European Investment Bank.

Bilateral aid agencies may require EIA in the recipient country as a prerequisite to receiving funds, although the donor may provide financial and technical assistance to the recipient in meeting this requirement. USAID is unique in that it must fully comply with NEPA. Other countries which incorporate some EIA into aid programs include Canada, Germany, the Netherlands, Sweden, and Norway.

Horberry (1985) points out that the political needs of an aid organization are reflected in its handling of and commitment to EIA. Multilateral banks are primarily concerned about maintaining a reputation for creditworthiness and secondarily with increasing disbursement of funds; EIA requirements which increase project costs and difficulty are not welcomed by the project staff and are at odds with their main objective. Bilateral agencies are primarily concerned with foreign policy objectives and secondarily with fund disbursement; their environmental commitment is an extension of the domestic pressure toward environmental responsibility. Recipient countries are interested in maximizing their funding and minimizing the restrictions placed on the funds, and tend to resist EIA requirements because many developing countries have not yet completely realized the long-term economic costs of environmental neglect and the importance of sustainable economic development.

The attention given to EIA by these agencies may increase the attention given to environmental policy in the recipient countries, and help legitimatize it and attract political support. Horberry (1985) feels that it is unclear if aid can actually improve the ability of government agencies to carry out EIA. Aid program EIA assistance needs to be appropriate for the recipient country, and it should be designed to help develop host nation capabilities to undertake analyses internally with minimal outside assistance.

9.11   Discussion and Study Questions

1 Do you think the existence of an EIA process would have minimized the severe environmental problems now facing eastern Europe? If so, what are the impediments in implementing such a process?

2 How should the various interests, such as foreign policy objectives, sovereignty, resource requirements, and fiscal viability, be balanced in developing an EIA process applicable to extraterritorial projects?

3 Should the formalized EIS process be replicated in other industrialized countries? Prepare a discussion paper for your response.

9.12   Further Readings

Gilpin, Alan. *Environmental Impact Assessment (EIA): Cutting Edge for the Twenty-First Century.* New York: Cambridge University Press, 1995.

Goldfarb, Joan R. "Extraterritorial Compliance with NEPA Amid the Current Wave of Environmental Alarm." *Boston College Environmental Affairs Law Review,* 18:543–603, 1991.

Lohani, B. N. "Status of Environmental Impact Assessment in the Asian and Pacific Region." *Environmental Impact Assessment Worldletter,* Nov./Dec. 1986.

von Moltke, Konrad. "Impact Assessment in the United States and Europe." In *Perspectives on Environmental Impact Assessment.* Boston: Reidel, 1984, pp. 25–34.

Wood, Christopher. "The European Directive on Environmental Impact Assessment: Implementation at Last?" *The Environmentalist,* 8(3):177–186, 1988.

Wood, Christopher. "The Genesis and Implementation of Environmental Impact Assessment in Europe." In *The Role of Environmental Impact Assessment in the Planning Process.* London: Mansell, 1988, pp. 88–102.

Economic and Social Impact Assessment

The consideration of the consequences of proposed actions on the social and economic aspects of human life is, at one and the same time, very easy and extremely difficult. One of the easier aspects is that of identifying the concerns of the public. If the public—any public—expresses a concern—any concern—then it may be established that a valid concern exists. The twist on this is that it need not be further "proven." At least with respect to the existence of a social concern, the expression of a problem may be equated with its presence. The converse need not be true, however. Valid problems may exist which are not necessarily perceived by the public or voiced by any group. The problem here, in the context of the National Environmental Policy Act (NEPA), is that of determining which concerns, and to what degree, might be valid foci for inclusion within an environmental assessment or impact statement.

Since a very large percentage of government proposals have the stated purpose of *deliberately altering* some aspect of human life, very many of these proposals contain elements of social and/or economic change. Is each of them to be examined under NEPA? How do we determine *which* aspects of *which* actions must be so assessed? The application of NEPA to concerns about social and economic consequences of government actions was originally unclear, was developed in almost an accidental manner, and remains equivocal. Interestingly, many observers of the development of the field of social impact assessment (SIA) trace the beginnings of the preparation of formal SIAs to

NEPA and the implementing CEQ guidelines and regulations (Friesema and Culhane, 1985; Rickson et al., 1990; Burdge et al., 1990).

A formal or informal SIA may, of course, be prepared without there existing any NEPA requirement whatsoever. In Australia, the requirements for the regional economic impact assessment, very similar in intent to NEPA, incorporate the concept of SIA in many cases (McDonald, 1990), although other roots are placed in the 1950s (Craig, 1990). Thus, in the United States, the SIA may exist either outside or within the NEPA context. It may take very different forms in different settings. The discussions later in this chapter will emphasize the inclusion of social considerations within the NEPA-driven environmental assessment process.

Similarly, economic analyses are extremely old. Since economic consequences are normally measured in terms of the local currency, it seems reasonable that quantification is desirable. Econometric models date back to at least the nineteenth century. Totally outside the NEPA context, estimates of the economic benefits associated with a particular proposed action have been used as a selling point in the legislative arena for more than a century. This is particularly true for "public works" projects, those massive development efforts which were conceived and promoted specifically for the purpose of bringing economic benefits to a locale or region. These projects proliferated during and after the great depression, up through the 1970s. For a variety of reasons, such proposals have become rare in the 1980s and 1990s. It may even be proposed that the elaborate propositions made for economic benefits in these project plans actually laid the basis for the inclusion of economic impact analyses within the NEPA context. The many international development projects, focused on economic development within underdeveloped countries, are based on the premise that they will derive economic benefits to the population of that country. In practice, these benefits may come at great social cost, an interesting point of tension between the social and economic spheres.

10.1 Socioeconomic Assessment within NEPA

Just what *is* the place of examination of social and economic considerations within NEPA? First, it is clear that when NEPA was originally debated, the focus of Congress itself was directed toward the requirement to prepare environmental assessments and impact statements. The often-quoted words of Sec. 102(2)(C), which begin "Include in every recommendation or report on proposals for legislation and other major Federal actions significantly affecting the quality of the human environment, a detailed statement...," were not originally interpreted to include aspects of the social, cultural, and economic environment. Examination of Sec. 101 of the act, however, finds one clear reference to considerations which do not relate to the physical or biological environment. The wording of Sec. 101(b)(2), "[to] assure for all Americans safe, healthful, productive and esthetically and culturally pleasing surroundings," could certainly be interpreted to incorporate many aspects of the social and economic environment. From the beginning, some federal government agencies, notably the Department of the Army, prepared guidelines—which had no regulatory status, even within the Army—that did include aspects of the social and economic environment. Even where it was allowed, there was no guarantee that employees or contractors would incorporate these factors. Other agencies attempted, up through the 1980s, to dismiss or minimize social and economic consequences when preparing environmental assessments and statements.

The general consensus, if there may be said to be one in this turbulent arena, is that the examination of economic and social considerations is generally undertaken, under NEPA, only within rather specific limits. What are these limits? First, it is acknowledged that NEPA is *primarily* an act for purposes of examining consequences of government actions on the biophysical environment. If there are no potentially *significant* consequences to the biological or physical environment, the requirement to prepare an EIS is not triggered. Thus, in the absence of potentially significant (usually interpreted to mean adverse) effects to the biophysical environment, socioeconomic consequences alone, even if potentially significant, will not serve to trigger the requirement to prepare an EIS. If, however, there are sufficient potential effects on the biological and physical environment to require that an EIS be prepared, a full examination of social and economic effects is required. Congress has frequently, however, directed the preparation of an EIS through riders on appropriations legislation, even when—or especially when—socioeconomic issues are highly debated.

Thus, the guidelines of some agencies specifically omit examination of socioeconomic factors in the environmental assessment phase of the EIS process. This is the stage at which an agency examines an action to determine if the potential consequences are severe enough to require preparation of an EIS (see Chapter 4). This omission is ostensibly designed to avoid prejudicing the decision about requiring an EIS. It may, however, place the agency at a disadvantage in understanding the relative overall importance of the issues involved if it delays this examination. Other agencies require or suggest full development of socioeconomic issues from the beginning of the process. The authors generally concur with this approach, since an understanding of these concerns will often assist in the scoping of the project.

10.2 Economic Impact Analysis

Economic impact analysis is a component of environmental impact analysis that is frequently misunderstood. The relevance of economics as an element of the environment is difficult to rationalize, particularly when economics has been set forth as an equal and opposite factor to be traded off against the environment. However, just as the ambient environmental setting within which a project is to take place determines the effect that project will have on the environment, so the economic setting within which a project is to take place will affect the environment. This is based on the fact that the environment, in its broadest sense, covers all of the factors that affect the quality of a person's life. This quality is determined by all the factors contributing to health and welfare, for both the short term and the long term. A general list of factors that describe the environment in this context includes both ambient biophysical conditions such as air, ecology, water, land, and noise and the existing social, political, and economic structure of a community. The economic conditions per se might be affected just as is air or water.

Certainly, today lesser-developed countries and regions often state themselves to be willing to trade environmental (ecological) quality for a beneficial change in their economic condition. Likewise, the fairness of displaced environmental degradation, such as the intercontinental shipment of hazardous waste or the international effect of acid rain, is being considered widely in national and international environmental debates. Knowledge and understanding of the economic consequences of an action (positive and negative) can no longer be separated from the environmental impact analysis.

Economic impact analysis would normally consider effects on both economic *structure* (e.g., the mix of economic activities such as forestry, agriculture, industry, commerce) and economic *conditions* (e.g., income, employment levels, inflation rate). Measurement of effects on both the economic structure and conditions is appropriate. As a result, consequences of projects such as changes in employment, income, and wealth for a community are used to describe the economic aspects of environmental impact. These factors, however, should be weighed with environmental (i.e., biophysical) gains and losses. In this analysis, it is useful to divide economic factors into two categories, the first relating to a description of the economic structure, and the second to a description of economic conditions.

Structure:

  mix of: employment by industry

           public versus private sector income

  mix of: economic activity by industry and commercial sector

           income distribution

           wealth distribution

Conditions:

           income per capita

           employment level

           changes in wealth

           levels of production by sector

The relationship of economic impact to environmental impact has its basis in the fact that changes in economic conditions lead to direct or indirect effects on the environment. Increases (or decreases) in income, production, or output lead to changes in effluents from production and consumption of goods and services. Changes in the quantity and nature of these effluents affect the environment. International development projects provide a model here.

Direct observation of economic structure and conditions is difficult, although generally easier than many other environmental attributes. Economic effects have been modeled formally for many years. Because of this, a model of the economic system is usually used to estimate and project resulting effects. Models are constructed so that changes resulting from project activity can be traced through to the effect on the economic variables of structure and conditions. Further, currency is naturally quantified, and many data on such factors as income, tax collections, public expenditures, and investment are already collected by various state and federal agencies for other purposes.

Project activity is the force (exogenous) that drives the economic model, as shown in Fig. 10.1. The model estimates impacts on economic conditions and/or structure. The changes in economic conditions are translated, usually through another model, into impacts on other environmental attributes.

10.3 Economic Models

In the schema in Fig. 10.1, the economic model plays an important role in estimating and projecting the effects of a project. There are several types of models that might be employed in this framework to help in

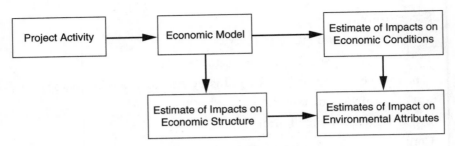

Figure 10.1   The relationship between project activity and impacts.

estimating the effects of project activities on the environment. Two of these models, the *input-output* model and the *economic base* model, are the most commonly employed, and are discussed below. The even more common *cost-benefit* (or benefit-cost) analysis is normally used primarily for project justification and support. It has occasionally been used in the environmental impact assessment context (Hundloe et al., 1990), where problems arise which are similar to those of the application of the other models.

Input-output model

The study of economics and its relationship to environmental quality has most frequently been approached by analyzing environmental considerations separately from economic considerations. Individual environmental factors such as air, water, and solid waste have also been treated separately from one another. As Ayres and Kneese (1970) noted, "the partial equilibrium approach is both theoretically and empirically convenient, but ignores the possibility of important tradeoffs between the various forms in which materials may be discharged back to the environment." Recent attempts at model development have recognized the limited value of this partial perspective. Isard analyzes the economic and ecologic linkages based on a linear flow model. The Isard model requires a detailed matrix of ecologic resource flows to describe all of the interrelated processes that take place within the ecosystem (Isard, 1972). Cumberland (1971) developed a model that adds rows and columns to the traditional input-output table to identify environmental benefits and costs associated with economic activity and to distribute these costs by sectors. Leontief's general equilibrium model is an extension of his fundamental economic input-output formulation, in which the model assumes one additional sector in the basic input-output table (Leontief, 1970). Pollution generated by the economy is consumed, at a cost, by an antipollution industry, represented by this additional sector.

An important modification of Leontief's approach was developed by Laurent and Hite (1971). This model is composed of an interindustry matrix, a local use matrix, an export matrix, and an ecological matrix. For each economic sector, it shows the physical environmental change in terms of natural resources consumed and pollutant emission rates per dollar of output. These effects are computed by deriving the Leontief inverse of the interindustry matrix and multiplying the environmental matrices by that inverse. This section discusses an extension of this approach to environmental impact analysis.

A regional analysis model based on a standard input-output table may be expanded to incorporate industrial land use and natural resource requirements as well as pollutant waste characteristics of industry into the table (Davis et al., 1974). The regional model may be viewed as a standard interindustry input-output matrix that has been supplemented with land use, natural resource, and emission sectors. It is expressed as follows:

$$A = RP (I - A)^{-1}$$

where R = resource matrix specifying land and other resource requirements of each sector.

P = pollution matrix specifying the nonmarketed by-products of each sector.

A = input-output table including resource and pollution sectors.

In applying this model, the Leontief inverse $(I - A)^{-1}$ is calculated and the land-use, natural resource, and pollutant matrix is multiplied by the inverse. This calculation provides an estimate of the impact of a proposed project on the land-use changes, natural resources, and waste-generation characteristics of the region.

The data on comparative land-use and natural resource inputs and waste emission characteristics may be organized in matrix form as shown in Table 10.1, where land-use and natural resource requirements are estimated for each Standard Industrial Classification (SIC) characterizing economic activity. Specific information must be collected to derive environmental coefficients for water and land input requirements and air, water, and other pollutant output emissions. This type of analysis is particularly applicable when the subject of the agency decision making relates to competing development proposals.

Applying the model to analyze the impact of specific project activity (adding new employees within differing economic sectors) can produce output illustrated by the example in Table 10.2. In this table, the effect of adding activities equivalent to 600 employees in two different economic sectors is compared. The major advantage of this model is that

TABLE 10.1   Land Use, Natural Resource Inputs, and Pollution Emissions by Sector

SIC_1...compared to...SIC_2
Natural resource inputs
Total land area, ft^2/employee
Floor space, ft^2/employee
Parking area, ft^2/employee
Building site area, ft^2/employee
Domestic water, gal/\$ output
Cooling water, gal/\$ output
Process water, gal/\$ output
Total water, gal/\$ output
.
.
.
n
Pollutant emissions
Particulates, lb/\$ output
Sulfur dioxide, lb/\$ output
Water discharge, gal/\$ output
5-day BOD, lb/\$ output
Solid waste, yd^3/\$ output
.
.
.
m

it produces the detailed information necessary to analyze the effect of project activity on the environment, both in terms of the structure of the economy and in terms of the secondary effects of changed economic activity on the environment. The main disadvantage is that it is relatively expensive to operate because, for reliability, some primary data collection is frequently necessary. This is because new activities may not correspond exactly to existing already-characterized activities in their need for space and type and amount of pollutants generated.

Economic base models

Another approach to modeling the economic elements of environmental impact analysis is represented by the army's Economic Impact Forecast System, developed for use in assessing the effects of military projects (Robinson et al., 1984). This model is based on the principle that the total effect of an injection of new money into an economy can be estimated by determining how much of the money remains in the economy and is respent, and how much is removed from circulation.

The model's principal objective is to answer the question, "What would happen to the local economy if certain activities affecting the economy were to take place?" To answer this question, the nature of

TABLE 10.2 Total Economic and Environmental Impacts Generated by Adding 600 New Employees—An Example

	Cotton finishing plant (Sector 2261)	Fabricated structural steel plant (Sector 3441)
Economic factors		
Value added by industry ($)	7,982,000	8,761,000
Employment opportunity (pn)	2,046	2,118
Land use and natural resources		
Domestic water, gal	291	317
Cooling water, gal	4,771	8,235
Process water, gal	15,023	11,979
Total water intake, gal	16,938	17,665
Land area, ft^2	14,300,350	14,728,435
Floor space, ft^2	1,073,721	1,173,006
Parking area, ft^2	1,291,594	1,622,903
Building site, ft^2	754,078	879,064
Waste emissions		
Particulates, lb	2,710,845	4,166,001
Hydrocarbons, lb	1,205,817	1,328,205
Sulfur dioxide, lb	147,225	164,735
Gaseous fluoride, lb	0	0
Hydrogen sulfide, lb	15,997	16,976
CO$_2$, lb	87,382	104,641
Aldehydes, lb	3,481	3,861
NO$_2$, lb	54,887	61,561
Discharge, gal	12,031	9,453
5-day BOD, lb	1,395,944	1,023,066
Suspended solids, lb	930,809	592,683
Solid waste, yd^3	53,231	56,835

the local economy must be characterized and the type and magnitude of the reactions presented.

The three basic participants in the local economy are local government, households, and business. Local households purchase some goods and services from local business, receive wages and profits from the sale of their productive services, and pay taxes and consume services provided by local government. Local businesses sell goods and services, purchase inputs, pay taxes, and also receive services from local government. Local government purchases goods and services from business; purchases inputs such as labor from households; collects taxes; and provides public goods and services such as police protection, fire protection, and libraries. Thus, it can be seen that there is significant interrelationship among all the various elements of the economy. So far, the concepts involved do not differ from those of input-output analysis.

The effect of even one household in the economy obtaining additional money can be traced using this type of model. In the army model, the flow of this money, as it works its way through the various sectors, is traced. Part of the money received would be put into the household's savings, and the rest would be used to finance purchases. Some of the products that are purchased would be purchased locally; others would be purchased from other regions. Purchases that are made from other regions require dollars to flow out of the local economy, while money received by local business would be used locally to hire labor, purchase products, pay taxes, and become profits. The wages received by local labor would, in turn, be partly saved and partly spent. Some of the products purchased from labor income would have been produced in the subject region; others would have been produced elsewhere. Thus, the cycle repeats until the original injection is completely dissipated. The calculations are, however, expressed in *annualized* terms (i.e., what happens in the first year only). We note that many locally developed models extend the cycle for 5 or even 10 years, yielding a much larger multiplier.

The general idea is that money injected into a local economy would be partly retained (and respent within the area) and partly "dissipated," or spent for goods and services available only from other regions. The total effect of the initial injection depends upon many factors, but the sum total will normally be greater than the initial injection; that is, the initial injection will have a multiple effect upon the local economy. This concept is called the *multiplier effect* and is extremely important to the assessment of impacts.

Any change in injections into the economy will consequently lead to a multiple change in income. The model described above system assumes that, in the short run, the variable most likely to change is exports. As a result, exports are considered basic to economic growth. Other activities in a region are nonbasic in the sense that they do not result in any money inflows, at least not under the assumptions made about the short-run model.

If the relationships postulated in the multiplier analysis are constant, the multiplier can be written as:

$$\frac{1}{1-S} = \frac{1}{1 - \text{nonbasic income/total income}}$$

$$= \frac{1}{\text{basic income/total income}} = \frac{\text{total income}}{\text{basic income}}$$

where S = the proportion of total income attributable to nonbasic economic activity

An estimate of the proportion of total income of the region, based upon export sales or basic industry sales, is necessary to use this multiplier. Fortunately, there are many techniques that can be used for an indirect estimation at low cost. The central assumption of indirect techniques is that there is a fixed relationship between the export industries in a region and the other local businesses. Perhaps the most widely used method to isolate export industries is the *location quotient technique* (Miernyk, 1968).

Location quotients are based upon a comparison of regional employment with national employment. Because the United States is basically self-sufficient, if a region has a greater percentage of its employment in a particular industry than does the nation, it is assumed to be specialized in the production of that commodity. Producing in excess of its own requirements, such a region must export that commodity to other regions. A hypothetical example of the calculation of location quotients is given in Table 10.3.

Next to each industry grouping is the percentage of the total national employment that an industry contains. In the next column is given the total employment in the region for each industry, and the percentage of total regional employment that industry contains is calculated in column 3. The location quotient is derived by dividing column 3 by column 1. A location quotient greater than 1 indicates that the subject region exports that commodity to other regions. Location quotients less than 1 imply that the good is not produced locally in sufficient quantities to satisfy local needs and hence must be imported.

Given that basic industries have been identified, how is employment in that industry allocated to exports? In column 5, the location quotient minus 1 is divided by the original location quotient. This provides an estimate of the percentage of employment in the industry that is involved in export activity. Multiplying column 5 by column 2 provides the estimate of the number of export employees for each industry. The multiplier is simply the ratio of export employment to total regional employment. In this example, the multiplier would be 5, indicating that a $1 increase in export demand would cause regional income to change by $5.

The multiplier concept is the basis for the development of this model. The details of the model take this general concept and use it to convert project activity (usually in dollars) into changes in business and economic activity. The strength of this approach is that results can be obtained relatively quickly and inexpensively. The major weakness is that the results are presented primarily in terms of changes in economic conditions, and changes in terms of structure or secondary effects on the environment from the changed economic activity are not dealt with in this approach. This means it is difficult to convert the estimates directly into such environmental impacts as pollution.

The concept of multipliers is subject to considerable variation in actual practice. Within the army model, for example, actual multipliers calculated vary from about 2 to about 4. This is in the range of economic

TABLE 10.3 Location Quotients for a Hypothetical Region

Industry or sector	Percentage of national employment	Regional employment	Percentage of regional employment	Location quotient (LQ)	LQ − 1/LQ	No. of export employees
Services	.40	400	.40	1.00	—	
Durable goods manufactured	.20	75	.075	.375	—	
Nondurable manufactured	.10	25	.025	.25	—	
Trade	.30	500	.50	1.667	.40	200
Total		1000				

$$\text{Multiplier} = \frac{\text{total employment}}{\text{basic employment}} = \frac{1000}{200} = 5$$

base model multipliers, *when considered on the basis of annual effects.* The emphasis is important. Some input-output models have a much longer time frame of operation, counting circulation of money input until *all* effects leave the region, possibly 3 or 4 years or more. The multipliers used in this case may be 8 to 10 or greater. Neither approach is right or wrong, but the differences must be realized when comparisons are made among different estimates.

10.4 Future Direction for Economic Impact Analysis

Both of the models discussed in this section were operational and have been applied hundreds of times in specific impact analyses. Thus, they represent applied approaches to dealing with the economic aspects of environmental impact analysis. The first approach can be used to develop detailed estimates of changes in structure and the secondary impacts in the local economy, while the second approach provides a broad estimate of the effect on economic conditions in a community where changes have been introduced by project activity. For the larger purpose for which it was originally developed the army model had a, larger benefit. Since the data are derived from surveys prepared by the Department of Commerce, any counties (or equivalent) in the United States can be assembled into an economic region appropriate to an action within a few minutes. Thus many alternatives can be examined rapidly, and at almost insignificant additional cost. The time and cost of developing the data required to prepare fully adequate inputs for input-output models often militates against full study of other than the preferred alternative. While the economic base model has drawbacks, its ease of use and uniformity of data mean that it is more widely applied than any other in actual practice. Data from 1991 suggest that it may have been applied more than 1000 times by agencies within the Department of Defense within 1 year. This may indicate that early application at planning stages of the project is being routinely performed with the model.

10.5 Social Impact Assessment

Of particular importance is the consideration of qualitative effects, which are not easily captured by conventional methods. It is in this area, an area called social impact assessment, that specific considerations of the effects on people and their relationships is considered. The origins of social impact assessment are difficult to define because almost all historical literature and scientific inquiry has at its base an inquiry into the conditions of humans. Finsterbusch and Prendergast

(1989) explained the social impact assessment aspects of the work of Condorcet on the development of canal systems in France in 1775–1776. Birth and death records were collected and potential changes estimated from the proposed project.

Work on social impact assessment, identifying issues, and making recommendations for mitigation and compensation, has a fairly long history in the United States. Some notable early concerns were for the massive government projects at Los Alamos and Oak Ridge National Laboratories during World War II. In the quasi-governmental and non-governmental area, the work on boom towns associated with energy developments in the western United States is an example of early social impact assessment (Gilmore and Duff, 1975). These projects shared a common factor in that it was well realized that massive changes were to be caused by government action, and it was acknowledged that this growth should be planned and coordinated to the extent reasonable.

Today, social impact assessment is recognized as important because it represents a method to capture the effects of programs and projects on the quality of life. The parameters range from health and education to recreation and community cohesion. It has also been viewed, and correctly so, as very difficult to conduct because the measurement of social impacts, which are of necessity qualitative, is not easy. Once they have been measured, there are no solid objective standards against which the changes can be compared to say if they are "good" or "bad." Of course, some can obviously be evaluated—live births per capita should probably be higher rather than lower. But, are bowling alleys per capita an indicator of recreational quality? The more the better?

In this context, checklists have been developed to document social impact parameters, and it is frequently left to the reader to decide if the impact or the sum of the impacts is good or bad. Further complicating the situation is the fact that different checklists are used from one study to another because issues of social importance vary significantly from one project that is analyzed to another. An example of a social impact analysis list is shown in Table 10.4. This list was originally prepared in 1984 for the Office of Nuclear Waste Isolation (ONWI-505) (Stacey, 1985).

Historically, the most widely used method for social impact assessment was the case study. This approach relies extensively on the creativity of the person conducting the study to find the critical factors to be analyzed. In addition, the data collected, which provide the historical, current, and projected future, tend to be qualitative and anecdotal. Involvement of people in the community is practiced in case

TABLE 10.4   A Social Effects Checklist Tailored to Siting and Construction of Radioactive Waste Depository

Social impact	Effect issues

People must relocate because the project will take their land.	1. Potential effects related to relocation:
	   a. Disruptions will occur to familial and friendship patterns problems.
	   b. Individual adjustment problems.
	   c. Psychological ties to property will be destroyed.
	   d. Inadequate compensation for relocating (real or perceived).

Substantial numbers of people will inmigrate because of project-related employment (either operation or construction or both). The construction and operating phases may have different effects.	1. The differences in social composition between old and new residents will create social problems.
	   a. Increase in social deviance
	     (1) Crime
	     (2) Alcoholism
	     (3) Drug abuse
	     (4) Child abuse
	     (5) Divorce
	     (6) Mental illness
	   b. Disruptions to current way of life
	     (1) Norms challenged
	     (2) Values challenged
	     (3) Local customs ignored
	     (4) Loss of sense of community
	   c. Increase in social conflict
	     (1) Confrontations between new and long-term residents for power, status, and group position.
	     (2) Delivery of services may be perceived as being inequitable.
	
	2. The sudden change in economic activity will create localized inflation because the supply of goods and services will be less than the demand.
	   a. Buying power of some old residents will be decreased.
	   b. Housing costs will increase.

People perceive health and safety risks from the presence of radioactive material.	1. Potential effects related to fear of radioactivity:
	   a. Marketability of farm products will fall.
	   b. Tourists will avoid the area.
	   c. Property will lose its value.
	   d. Current residents will outmigrate.
	   e. Stress and other psychological disturbances will occur.

studies through interviews and meetings. In the end, the case study is useful as a description of a situation, but it is difficult to appeal to relative or absolute measures for criteria to assess the extent and desirability of impacts (Stacey, 1985).

More problematic is the fact that the approaches that have been used do not conveniently lead to a general social impact method like the specific approaches documented and required for assessing effects on biophysical parameters such as air quality, land use, and water quality. The literature on social impact assessment is extensive, as evidenced by the more than 60 examples reviewed by Stacey in 1985. The main parameter of an ideal social impact assessment model emerged from a consideration of needs, usefulness, and value to the purposes at hand, and is presented in Table 10.5 (Carley, 1981).

This is a sketch of an ideal social impact assessment program which is described in steps which could be redefined to be expressed as tasks. The method(s) to be used, which underlie social impact assessment, range from trend analysis to scenarios. These methods are all aimed at obtaining a view of the future with respect to social parameters. Some methods are very objective and analytical, and others are subjective and qualitative. People react differently to the method that is being employed. The more analytical and abstract the method, the more argumentative and defensive are people in the community being analyzed. The qualitative and opinion-based approaches have the strong advantage of involving people from the affected area directly in the analysis. This improves communications, understanding, and involvement. These factors are critical to the success of social impact assessment, a fact which significantly distinguishes this aspect of environmental impact assessment from the other traditional dimensions.

TABLE 10.5   Schema for Social Impact Assessment

1. Establish a baseline
    a. Identify key issues
    b. Identify data sources
2. Forecast changes
3. Evaluate changes
4. Identify how to respond
    a. Weigh available mitigation
    b. Weigh need for compensation
5. Evaluate how to respond
    a. Recommend mitigation
    b. Recommend compensation
6. Monitor
    a. Evaluate effectiveness
    b. Make adjustments

10.6   Social Impact Analysis Methods

Another important factor is that, for social impact assessment, there is seldom a definitive answer or forecast. There are lots of "if this, then thats" and significant uncertainty and risk. When methods and approaches are used to derive a definitive answer which disguises the uncertainty and risk, people in the affected community realize it and tend to be argumentative and contentious. The methods that expose uncertainty and risk, although difficult to apply to decision making, can be highly useful in clarifying the situation for the community affected. Decisions are more difficult to make but are made with better consideration of risk and uncertainty.

Some of the principal methods that have been used in social impact assessment are examined here.

Trend analysis

This method is based upon extrapolation of past developments and changes into the future. It is simple to do, and the techniques can be as ordinary as visual interpretation of directions (from a graph or chart) or as complicated as multiple regression techniques based on statistics and mathematical modeling. This method is very useful as a "first cut" at possible future outcomes. The main weakness of trend analysis is that usually the models are simple relationships that include time and, as a result, may not be particularly accurate or compelling. The "behavioral" content of trend analysis models tends to be very weak. As a result, the models may not capture the true underlying forces that are likely to be the reasons for change.

Content analysis

This method is useful and popular because it relies on the analysis of secondary sources (newspapers, journals, magazines) for expressions of opinion, judgment, and expectations. One weakness of this method is that ideas about unexpressed or unexplained issues would not be analyzed (for example, the problems of aged or retired people when prices inflate dramatically in a community). Another weakness is that it remains an indirect indicator of social concerns. It is really an evaluation of the *newsworthiness* of an issue, and is dependent totally upon the perspicacity of the reporter and editor, much less upon the feelings of members of the general public, and still less upon objective analysis of the probable change.

Case study

The case study is the most popular approach utilized so far for social impact assessment. It has the advantage of flexibility, which permits the assessment to be tailored to the specific issues important to the situation. The main disadvantage of the case study approach is that the future views are not produced systematically and are generally not reproducible. It is also an approach which usually provides an "external" view of the social issues and thus can be less compelling than an approach which admits more community participation. Studies may be highly controversial and not be accepted by many of the groups included in the study.

Delphi

This method involves the assembly of judgment and opinion by a group of people. This is done by a survey (mail or in a group setting), and the results are rapidly communicated to the participants. Following the communication of results, the participants are asked to revise their opinions based on what they have learned from the other participants. This process is reiterated to arrive at a consensus (Chapter 5). The Delphi technique is also widely used to crystallize the opinions of small groups, especially in the public involvement phases of the NEPA process. It is, perhaps, not as well suited to acquiring the full spectrum of public thought at any one time. The main shortcoming of this method is that the "outlying" opinions and ideas tend to be submerged in the mass judgment. This means that the uncertainty and risk are masked by this process, and important, but less prevalent, concepts may be lost from view. While the median level of concern is noted, we may not know how spread out the full range was. It is the outliers who often polarize opinion, and an understanding of the full range is important when developing project plans.

Participant observation

This method relies on the observation of patterns of behavior in the affected community. These patterns are then used to extrapolate relationships for the future. The methods used are similar to those used by the cultural anthropologists who live with and study primitive cultures. This is tantamount to an individualized form of a case study. As it relies on specific observable and recordable behavioral relationships, it can deal only with those relationships and those individuals and institutions for which observations have been or can be made. As it is a data-driven method, it has statistical appeal. The problem is that usually historical data are not available for critical behavioral relationships and these data must be re-created. The second problem is that many important behavioral relationships are not easily recorded in a quantitative fashion. For example, community cohesion is not easily measured (if it can be measured at all). Nevertheless, it is viewed as an important measure of community characteristics which is extremely susceptible to change when a project or program is introduced. Finally, here, the Heisenberg principle* acts critically (as with other parts of the social impact assessment process) to result in the population being measured having its behavior affected by the act of measurement.

Similarity

This is a catchall category which represents the collection of attempts to use the results of what has happened with respect to one project at one site to infer what will happen for another project at another site (or at another time for the same site). It is a very weak form of extrapolation for application to social impact variables, but it is still widely used, or at least attempted. Personal experience forms the basis for a personal "similarity analysis" in many cases. This may explain why "experienced" observers are frequently wrong in their analyses! These products are usually intensely personal, reflecting the ideas and experiences of the preparer. In practice, this approach is not usually very successful because

It is not always clear what has happened in the original situation. Data may be inadequate, or understanding of the changes *and the reasons behind them* not clear.

No two sites are ever really the same in terms of population, geography, and the proposed project. Furthermore, they are usually displaced in time as well. The demonstrated values and tastes of people evolve over time.

The knowledge of the principal issues and relationships for the new site may be insufficient, or the preparer's background may be inadequate for this location and situation.

The behavior of the population will differ from one site to another. The behavior patterns of people at a site in the western United States (in the Rocky Mountains) will not be transferable for forecasting to a similar program in eastern Kentucky or Tennessee, although the project (e.g., coal mining) might be similar.

Dynamic simulation

Systems dynamics modeling has been successfully used to illustrate complex behavioral relationships and their evolution over time. This

---

*From Werner Heisenberg (1901–1976), who voiced the "uncertainty principle," under which a scientist was more confident of the overall effect of a phenomenon rather than the exact value of any one part at one specific time.

method is robust enough to capture and analyze the range of quantitative and qualitative relationships necessary for a sound social impact assessment. It has two main drawbacks. It is very expensive to construct and calibrate such a model. It is most instructive for the model builder and practitioner, but involvement of the community in the construction and operation of the model is probably not feasible. The need to use a complex mathematical and statistical model renders it a "black box" in which people (particularly people in the affected community) do not have confidence.

Inference from theory

Theoretical constructs of behavior in different project situations can be used to infer newly developing changes. This method can be very useful for constructing hypotheses about change but is not good for conveying possible effects to people in the community. The "boomtown" phenomenon of energy development in the western United States has, from time to time, been used as a theoretical model for rapid development for large energy projects. For the Portsmouth Nuclear Enrichment Facility expansion, boomtown models were used by the local population as a model of what could happen. In this situation, the theoretical model did not apply at all, as there were buffers in effects in terms of the extent of migration of the workforce and its permanent versus transitory character that were outside the theoretical boomtown model. As a result, false expectations about effects (both good and bad) were raised by inferring effects from the (misapplied) theory (Battelle, 1979).

Surveys

Among all the methods that have been applied to social impact assessment, surveys must be ranked the most popular. They are easy to design and, relatively easy to administer, and the results can be organized and displayed to reflect a summary of the surveyed population. The results are often useful in scoping and in planning public involvement activities (see Chapter 11). The unfortunate aspect is that the surveys are of very little value in forecasting the future. Surveys are a description of a situation and might even find historical information. The major need, which is generally lacking, is the conduct of longitudinal surveys over a significant time period with the same population group and concerning the same project so that time-related behavioral patterns can be identified. This is especially difficult for a project that involves a significant transitory workforce. These people are very difficult to trace and resurvey at a later time.

Scenarios

This is a little-applied method which has significant potential value. The techniques have been developed and are readily applicable to social impact assessment (Stacey, 1990). Current techniques have made the approach simple and easily accessed by the affected population. In addition, realistic perspectives on uncertainty, the reality of the current situation, and the potential future situation with and without the program are possible. The use of the tool is also a direct form of communication and can be combined with the development of specific mitigation and compensation actions. These modeling approaches do not rely on statistical data or quantitative information exclusively; thus the qualitative uncertainty and risk can be included in the method. The fundamental analytical technique used is cross-impact analysis, which has now been in use in a variety of very practical applications for over 20 years.

There are many new and developing methods and tools that are useful for social impact assessment. The needs of this type of assessment place a premium on methods that are flexible, easy to access, and easy to understand, and promote communication and understanding. The main needs are to be able to produce long-term forecasts, to reflect clearly the uncertainty and risk, and to have enough experience in the application of methods to actual projects to gain confidence in the results and bring understanding and value to the affected community.

10.7   Assembling the Socioeconomic Impact Assessment

In recent years, impact assessment has taken on a new and important direction. Decision makers at all levels, as well as community members, have developed an increasing awareness of the need for estimating the effects of large projects on communities (Verity, 1977). The energy-related "boomtown" development in the west, as related to coal mining, in which small towns have increased 100-fold in size in a very short time period, is one example of a source of socioeconomic impacts (USDI, 1975). The purely social consequences were discussed above in Section 10.5. While not originally receiving a great deal of emphasis in the context of environmental impact assessments and statements, the origin of the concept that socioeconomic assessment would be useful may be attributed directly to the requirements under NEPA.

Estimation and analysis of these impacts have direct and immediate application in planning for change and growth that might occur as a result of a large project; such estimation and analysis is being done in

TABLE 10.6   Example of List of Socioeconomic Attributes—Two Levels

Demographic and population effects	Physical environmental quality effects
Age	Particulates (air)
Sex	Odor (air)
Race/ethnicity	Suspended solids (water)
Education completed	Thermal (water)
Occupations	Communication (noise)
Household composition	Social behavior effects (noise, etc.)
Government fiscal effects	Public health status effects
Tax rates	Number/type of facilities
Tax burden	Number/type of personnel by skill level
Expenditures	Occupancy patterns
Revenues	Cost of health care
Debt	Special services (elderly, low income)
Educational effects	Quality of drinking water supplies
Enrollment	Family status effects
Facilities	Marital status
Teacher supply/qualifications	Family size
Student-teacher ratio	Marriage
Achievement (graduates/dropouts)	Divorce
Finance	Composition
Housing status effects	Public safety effects
Enumerations	Fire protection
Ownership/rental patterns	Police protection
Characteristics by type, age, size	Ambulance service
Cost/rent	Rescue service
Construction starts	Recreational opportunity effects
Availability ratios by type	Type of facilities
Labor force effects	Ownership
Employment	Participation
Labor force participation	Distribution/accessibility
Employment distribution (by sector)	Cultural alternative effects
Employment opportunities	Historical/prehistoric sites
Economic status effects	Unique human settlements
Regional economic stability	Local government (functions-responsiveness,
Income	access to) effects
Income distribution	Planning
Energy expenditures	Regulation, standard setting
Industrial sector effects	Protection of welfare
	Education
	Administration
	Enforcement

support of studies of such large projects (ANL, 1978; TVA, 1976). The categories of effects that may be covered in socioeconomic impact analysis include those shown in Table 10.6. Some or all of these factors are of interest to planners, developers, businesspeople, and public officials who must deliver public services.

All of these people need to know the potential effects on the community or region of large construction projects to enable them to plan for potential changes in temporary and permanent employment in an area.

Changes in employment and in locally produced and consumed goods and services are the cornerstones of information needed to estimate impacts. The added people and activities will require augmented public and private services that will cost more money to deliver. Increased income to the population and resultant increases in assessed value of property will, in turn, generate additional public revenues. Before the community can deliver the services demanded, careful planning by responsible community entities is required. A detailed projection of the expected effects of a project on expanding the labor force should be made as a first step in this planning process. Figure 10.2 shows a simplified schematic of the flow of effects that can be expected from expanded local employment opportunities (Battelle, 1979).

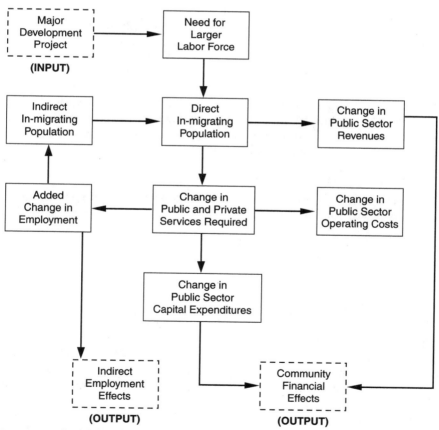

Figure 10.2 A simplified schematic of the flow of effects from requirements for an expanded local labor force.

There are two major consequences that result from both the increase in population and the resulting changes in the demand for public and private services and in the revenues collected to finance these services. One of these is the increase in employment in the public and private sectors providing these services. The other major effect is the increase in the cost of providing these services. These costs potentially include both capital and operating costs. The overall impacts are ultimately reflected as financial effects on local political subdivisions and as indirect employment effects.

Social impacts may be regarded as impingements upon community social conditions and processes. Socioeconomic impacts, then, are community impacts which are social and economic in character. A socioeconomic impact analysis is focused on tracing these effects. It begins with changes in the labor force and ends with expected impacts on a variety of factors, including the financing of local public services, private enterprise, and indirect employment opportunities.

Any analysis involves a number of key assumptions. However, it is possible that as a study progresses, new evidence and/or data can result in the desirability of changes in these assumptions. To be responsive to this need, computer-based models are used to estimate the effects. With models, it is possible to replicate results and to rapidly reiterate the analysis using new and/or differing assumptions.

Given the assumptions, both the quality and the planning usefulness of socioeconomic impact analysis are dependent upon a number of key factors. Four of the most important are

1. Introduction of a time dimension
2. Characterization of the labor force and estimation of percentage of "mover" workforce
3. Estimation of indirect workforce
4. Estimation of revenues and expenditures

Time dimension

To be able to compare the effects of a given project with other projects, the planner needs to know the timing of the effects. Impacts on parameters should, therefore, be estimated on an annual basis. This does not seem difficult until it is attempted. Information on a baseline for each parameter must be forecast for each year; then, estimates of each parameter's change resulting from the project must be prepared. To do this, requirements for the construction labor, materials, permanent labor, and operating inputs must be prepared on an annual basis. With this information, the annual effects on parameters may be estimated and the concomitant requirements for public and private sector services and associated expenditures and revenues may be established on an annual basis.

In the process of forecasting baseline and changes, error will occur. Adjustments for error should be accommodated as soon as the value of the adjustment (improved accuracy in impact assessment) exceeds the cost of making the adjustment (collecting new data and rerunning the model). One example of the adjustment process is shown in Fig. 10.3. The actual value for population size does not equal the forecast at $t = 1$ and $t = 3$. Obviously, the benefits of such forecasts must be compared with the possible costs.

Characterization of labor force and estimation of "movers"

The labor force (both permanent and temporary) is the key to assessing the need for various public and private services. The workforce should include both direct and indirect employment. The labor force must be characterized for each year of the analysis (or forecast). To do this, four steps are essential:

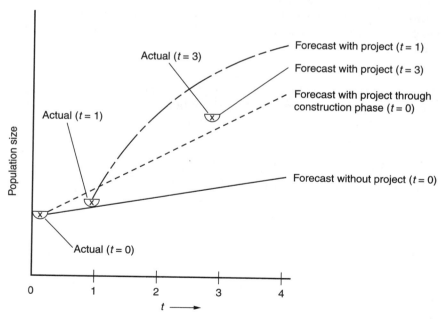

Figure 10.3 Relationship between forecast baseline (population size) and accrued impacts over time.

1. Identify labor force requirements
2. Determine labor force availability
3. Estimate "mover" labor force requirements
4. Define the composition of the workforce

Identify labor force requirements. The purpose here is to estimate the project workforce requirements for each year of the construction period. The specific number of workers by occupation for construction stages such as the following should be sought:

- Site work
- Underground and utilities
- Structural
- Equipment installation
- Finishes

It is necessary to identify the time frame for each construction stage and the number of workers required, by occupation or craft. Estimated salary/wage schedules also should be obtained for each occupation. In addition, similar information should also be sought for the permanent workforce. The completion of this work provides an estimate of the labor demand.

Determine labor force availability. At the same time the workforce requirements are being estimated, the labor force availability for relevant worker classification in the affected region should be estimated. The key sources of data for this include

- The U.S. Census Bureau
- Labor union officials in the affected region
- State and regional employment records
- Review of other similar experiences

Availability should be estimated for each occupation category or craft, to include

- Employment/unemployment status
- Wage/salary currently earned
- Distance from site (i.e., 1, 2, 3 hours average commuting time)

Estimates for availability should be generated for appropriate time intervals for the analysis. This constitutes an estimate of the supply of labor.

Estimate "mover" labor force requirements. Once the demand and supply for each time period have been determined, the "moving" requirements of the estimated required labor force must be identified. The movers will be workers who fill the gap (if one exists) between the demand and the supply of workers for each time period. The moving labor force may be identified by the following parameters:

- *Commuters*—that part of the required labor force able to travel to and from the construction site for work on a daily basis.

- *Permanent*—that part of the labor force that takes up permanent residence in or near the site. Such workers typically bring their families to their new residences.

- *Relocated*—that part of the labor force that takes up residence near the site for fewer than 5 years (more typically, just for the period of required employment). Such workers may or may not bring their families with them for all or a portion of their time in residence.

- *Travelers*—that part of the labor force that resides near the site during the normal workweek (Monday–Friday) and returns to their permanent places of residence on weekends and holidays. Such workers typically do not bring their families with them.

Define the composition and requirements of the workforce. The permanent and moving labor force should then be characterized so that the demands for public and private sector needs can be identified. The labor force would be characterized for the following factors:

- Family size
- Spouse's employment
- Number of school-age children
- Wages and salaries
- Housing requirements

Estimation of indirect workforce

In order for planners to prepare for changes in required services, a careful estimate of the indirect workforce is essential. This workforce may represent as much as 50 percent of the construction and/or permanent workforce. When family members (spouses and children) are also counted, the indirect workforce becomes significant in its demand for services and in its ability to generate revenues. Most work in this area so far has been conducted on the basis of observations of past experiences and the adoption of an appropriate "ratio" of indirect to direct employment. However, this is a method that may result in substantial error. Therefore, new approaches are required to make better estimates of this workforce.

These workers can be estimated using a regional input-output model. If carefully constructed and adjusted by sector to the local economy, such a model will provide highly defensible forecasts of indirect employment requirements. These forecasts serve two purposes. First, they allow estimates of the demand for public services and the additional revenue generated by the added employees. Second, the changes in output by sector of the economy will enable business planners to anticipate changes in the demands for private sector goods and services.

Estimation of revenues and expenditures

The estimated changes in the demand for public sector goods and services constitute the "real" effects of the project. The financial effects are of equal (or, to some community members, perhaps greater) importance. The real effects must be converted into fiscal effects.

As the temporary and permanent workforce moves into an area, community services experience expanded demand. At the same time, the new community members generate new revenues to finance the provision of these services. Thus, the new residents represent both a benefit to the community (more business and more tax revenues) and a burden (additional services that have to be provided) (University of Tennessee, 1973).

The estimation of revenues and costs is critical to the analysis of socioeconomic impact, because it is here that the results of supply-demand analysis, projections of budgets, and revenue and cost forecasts are brought together. The ability of affected entities to finance the delivery of additional services is essential for planning purposes. A preliminary examination of effects would show which years are likely to be the most heavily affected with respect to capital and operating costs.

The average per capita costs of services should be identified and multiplied by the expected change in population to set the costs of the services. There are many data available on the costs of various types of services provided by local government jurisdiction. National data may be compared with current local experience to arrive at an estimate of requisite service costs.

The budgetary process and revenue sources combine to determine the amount of funding available to finance the delivery of services.

General revenues are usually appropriated for use through the budgetary process. The availability of state-collected revenues to counties and towns (intergovernmental) is subject to rules and some uncertainty. All of these factors affect the lead times involved in estimating the availability of funds to finance additional services. In addition, mileage rates, assessed value, and limits on bonded indebtedness all affect the ability of a community to generate revenues to finance services.

The results of such analysis would be a report on each key entity (school district, township, city, town, county) for each affected service. The format should be such that the user can anticipate potential financing problems and begin to take steps to alleviate possible difficulties.

10.8 Problem Areas

For most projects for which you and your agency will be preparing NEPA documentation, many of the details of the socioeconomic impact assessment procedure described above will seem overdetailed and unnecessary. It is true that they were developed specifically for the very large project, a construction phase which might involve tens of thousands of persons and a permanent workforce of more than 1000. Each of the procedures does apply, however, to the smaller project as well. In practice, in the 30-plus years of NEPA examination of environmental consequences of government programs, some guidelines have emerged which may help you to assess your proposal in realistic terms. The "problem" areas which follow are a selection of those where errors have frequently been seen in the preparation of social and economic portions of NEPA documents.

Time phasing of effects

As discussed in Section 10.7, a firm knowledge of the time dimension is important. Suppose the project is stated to have a construction cost of $25,000,000. The "logical" approach is to enter this value in an economic model to determine the effect on the local region. This may lead to gross overestimation of positive effects! Why? Because an economic base model, such as the army's, reports its results on an *annual* basis. Thus, a model input of $25,000,000 implies a stimulus of $25,000,000 *per year!* In fact, your project is likely to take 3 or more years to complete, so the stimulus in any 1 year is much less than the eventual total. When using an input-output model which has no "expiration date," however, the *total* positive effect over a multiyear period may often require the input of the entire value. Know the time frame implicit in the dollar figures used. This also applies to employment figures. Not every new employee stated in the proposal will be hired at one time. Start-up may be spread over several years. Overall, misunderstanding of the time-phased aspects of project proposals has led to severe overstatement of many of the benefits of the project. While it is natural to be a proponent for your agency's idea, avoid making this severe misstatement in the assessment.

Selection of the economic region

What is the region in which economic (and social) effects will be felt? How do we define it? As suggested above, the use of commuting time is a common basis. But how do you relate commuting time to distance? This varies widely from one part of the country to another. In the northwest, 1.5 hours may mean only 30 miles, while in the southwest it may mean 100. One acceptable method is to inquire about commuting patterns of employees already in the area. This may not be possible. Many of the largest development projects are intentionally sited in remote areas with few present employees of the type proposed. The best advice we can provide suggests matching the size of the region to the magnitude of the action. If the region is a major urban area, a change of 100 or 200 persons probably would not impact rural counties beyond the center of the region. If, however, these 200 new employees will be located in a low-population county far outside a major urban area, then an aggregation of five or six surrounding counties is logical even for an action of this magnitude. When the numbers are in the thousands, then multicounty regions are the norm in economic analysis.

What are the consequences of having too large a region? In general, the effects of your action will show larger *total* dollar effects, but they will represent a smaller *proportion* of the normal annual income of the region. Conversely, selecting a very small (i.e., low-population) region for a major action will result in calculations of smaller absolute dollar value, but it will represent a greater proportion of the annual norm. There are no absolute standards for this selection. In some cases, one is asked to estimate the economic consequences to a much smaller area than a county. In the northeast, the town (township) is the local unit of association for many people. Analysis of effects on units of this size is difficult for economic base models for several reasons. First, data are not usually available on that basis to calculate local multipliers. Second, even if data were available, few areas as small as a town will have export employment values which are meaningful. The model would thus show *lower,* rather than higher, effects when the smaller area is considered, the reverse of the effect sought by the local proponents. A site-specific input-output model would be usable if developed, however.

Employment and unemployment

Many government proposals, especially for development projects, have as one of their goals to increase local employment. Usually, this is stated, or at least implied, to result in decrease in *un*employment. In practice, however, the two values are only very loosely related. There are several reasons for this. A newly created job, for example, is very likely to be filled by a person already employed, but seeking to improve his or her status. It may also be filled by a person not in the workforce—a woman with proper skills, but staying home to raise children. The availability of the new job provides the stimulus to move into the workforce. She was never counted, however, as unemployed, so no change in that figure is seen. Much secondary, or induced, employment is incremental. An example is the salesperson who sells refrigerators. When more people are hired at the new factory, the store sells more appliances, and she receives more commission. Her income has increased, counted as a fraction of a person-year of new employment, but nobody has been hired. There is more employment without a change in unemployment.

This elasticity of employment also applies to decreases. That same salesperson could lose a measurable percentage of her income without being unemployed. If your agency will be decreasing employees at a site, one must remember that loss of a job does not equal being unemployed, at least not in the official measurements. Many persons may transfer; some will be eligible for retirement; some will find other positions, even in very bad times; and only a minority are likely to file for unemployment benefits. Avoid making direct relationships between the effects of your proposed action and the unemployment rate in the area. There is seldom a one-to-one relationship.

10.9 Environmental Justice

Negative environmental impacts tend to fall more heavily upon the minority members of society. Studies have shown that chemical manufacturing plants, hazardous waste landfills, highways, and other developments with negative environmental consequences are more likely to be located in low-income and minority communities. In order to combat this trend and move toward the pursuit of equal justice and protection for all people under environmental statutes and regulations, the concept of environmental justice was developed.

What is environmental justice?

Environmental justice is a term that refers to the federal government's obligation to ensure that ethnic minority or low-income sectors of the population are not disproportionately affected by adverse environmental impacts or hazards. Specifically, the term refers to impacts that might be caused by programs, policies, or actions of the federal government. The underlying tenet of environmental justice is that agencies must take proactive measures to ensure that communities with concentrations of minority or low-income people will not be exposed to adverse environmental burdens or hazards at a rate greater than the population at large (Institute of Medicine, 1999).

The environmental justice movement traces back across the twentieth century, and in particular is interwoven with the United States civil rights movement of the 1960s and 1970s and the concerns about "environmental racism" brought forth in the 1980s (EPA, 1992a; Newton, 1996, Bullard, 1990). Environmental justice became a widely recognized national issue in 1982 when approximately 500 demonstrators gathered in Warren County, North Carolina, to protest the siting of a polychlorinated biphenyl (PCB) landfill in a predominately African-American and low-income community. This protest led to a 1983 investigation by the U.S. General Accounting Office (GAO), which found that three of the four major hazardous waste landfills in the south were located in minority (predominantly black) and low-income communities. In addition, two other major environmental conferences were held in the 1990s, further increasing awareness of environmental justice: The First National People-of-Color Environmental Leadership Summit and The University of Michigan School of Natural Resources Conference on Race and the Incidence of Environmental Hazards. NEPA, signed into law in 1970, made an early attempt to establish a national policy to "stimulate the health and welfare of man," and acknowledged the responsibility of the federal government to "assure for all Americans safe, healthful, productive, and aesthetically and culturally pleasing surroundings" (NEPA, 1970). Environmental justice, or "environmental equity," issues of the 1980s centered on the exposure of minority or low-income populations to environmental toxins, such as those in contaminated landfills, and occupational health issues, such as uranium mining (EPA, 1992b). However, the issue of environmental justice came into its own over the last two decades of the twentieth century, culminating in specific federal directions promulgated on an agency-by-agency basis. While acknowledging that there may be a correlation between "dirty" or "dangerous" activities and areas inhabited by minority/low-income peoples, many observers see this as due to the operation of basic price-demand economics, and reject the premise that the areas are selected because of the minority population. They observe that, in many cases, the activity was present before the minorities chose to live there.

Executive Order 12898, "Federal Actions to Address Environmental Justice in Minority Populations and Low-Income Populations," was signed by the President on February 11, 1994. The order and an accompanying Presidential memorandum (Clinton, 1994) direct each federal agency to identify and address "disproportionately high and adverse human health or environmental effects of its programs, policies, and activities on minority populations and low-income populations." Further, the order directs each federal agency to develop a strategy to

- Promote enforcement of all health and environmental statutes in areas with minority populations and low-income populations
- Ensure greater public participation
- Improve research and data collection relating to the health and environment of minority populations and low-income populations
- Identify differential patterns of consumption of natural resources among minority populations and low-income populations

Additionally, it is to include, where appropriate, a timetable for undertaking identified revisions to prior programs, policies, and processes, and consideration of economic and social implications of the revisions.

In relation to the environmental analyses performed under NEPA, an environmental justice discussion should address *adverse* impacts that are *significant* within the meaning of NEPA, and that are *disproportionately high* within minority or low-income populations. If impacts of a given proposed program, policy, or action are not adverse, or if adverse impacts are not significant, the NEPA review is not required to discuss environmental justice issues. Similarly, if impacts are both adverse and significant, but do not disproportionately affect minority or low-income populations, the NEPA review need not discuss environmental justice issues (CEQ, 1998).

How are environmental justice issues determined?

Both the CEQ and the EPA have prepared written guidance to help federal agencies determine when and how to consider environmental justice issues. Each agency is responsible for promulgating its own process in this regard.

The EPA (EPA, 1998) defines environmental justice as

> The fair treatment and meaningful involvement of all people regardless of race, color, national origin, or income with respect to the development,

implementation, and enforcement of environmental laws, regulations, and policies. Fair treatment means that no group of people, including racial, ethnic, or socioeconomic group should bear a disproportionate share of the negative environmental consequences resulting from industrial, municipal, and commercial operations or the execution of federal, state, local, and tribal programs and policies.

One way to determine if impacts are adverse, and disproportionately affect minority or low-income populations, is through the NEPA process. The Executive Office and the CEQ, which oversees the NEPA process, suggest this approach. It is especially pertinent for assessing the potential for environmental justice issues related to federal actions, but it can also be useful for looking at federal programs or policies. In 1998 the Council (CEQ, 1998) issued guidance for including environmental justice considerations through the NEPA process:

> Agencies should consider the composition of the affected area, to determine whether minority populations, low-income populations, or Indian tribes are present in the area affected by the proposed action, and if so whether there may be disproportionately high and adverse human health or environmental effects on minority populations, low-income populations, or Indian tribes.

What is an affected population?

Environmental justice seeks to identify significantly adverse impacts that would disproportionately affect minority or low-income populations. In order to do this, the reviewer must have some replicable way to determine what the affected population is, and whether the affected population is predominantly minority or low-income.

The terms *minority* and *low-income* are subjective. Because there may be differences in interpreting these terms, the CEQ (CEQ 1998) has defined these terms as follows:

- *Minority.*   Individual(s) who are members of the following population groups: American Indian or Alaskan Native; Asian or Pacific Islander; Black, not of Hispanic origin; or Hispanic.

- *Low-income.*   Low-income populations in an affected area should be identified with the annual statistical poverty thresholds from the Bureau of Census's Current Population Reports, Series P-60 on Income and Poverty.

Factors that the reviewer might consider in determining the affected population, and whether it is predominantly minority or low-income, could include

- Demographic information
- Geographic information
- Economic information
- Indigenous uses of resources
- Other localized sensitive issues

In accordance with CEQ guidance, the reviewer may wish to start with standard sources of demographic information, such as U.S. census data. Depending on how long it has been since the information was compiled and the accuracy of the data for a given region, there may be shortcomings in using only census data. Other sources of information may be state, tribal, or local economic development reports; universities; or private researchers. As a final point, one value of the public participation process required by NEPA and Executive Order 12898 is its use in fleshing out demographic information, or providing other information of interest in determining whether minority or low-income populations would be affected by a proposed federal action.

Minority or low-income populations are determined on a *comparative basis* to the population at large in a given "area of influence" or affected area. In some reviews, the agency may choose to use a set area, such as looking at the resident population within a 50-mile radius of the site for a proposed action. However, this method may overlook transient populations, such as nomadic indigenous population or migrant workers. Additionally, localized or indigenous use of natural resources (such as hunting or collecting certain plants for ceremonial use) may bring a specific minority or low-income population into contact with an effect that might be otherwise localized, or even overlooked, if only the dominant population within the area is considered (Hayes et al., 2000; Institute of Medicine, 1999).

There are various ways of calculating the percentage of minority or low-income populations within a given area. Minorities may be counted in comparison to national ethnic norms, or in comparison to state or local areas. The reviewer may want to consider appropriate alternative approaches to avoid dismissing a localized population where an ethnicity is in the majority locally or regionally, such as on the Navajo Indian Reservation (which is approximately the size of the state of West Virginia), but in the minority on a state or national basis. In some cases it may be more useful to consider the local population as a minority compared to the nation, even though the local population may be an ethnic majority within the local area.

Similarly, economic indicators and baselines vary considerably across the country, and must be examined in comparison to local or regional standards as well as national standards. For example, the dollar level that may legitimately be considered "low income" in San Francisco or another city with a high cost of living may not be seen in the same light in a prosperous, but rural, area with a lower cost of living. That is, a low income in a city might be similar to a moderate income on a national level, but would be low in comparison to the local norm. Or, on the other hand, if a local area has a uniform, but low, income level, it might be difficult to identify a comparatively low-income population unless a larger area is examined. Again, in some cases it may be more useful to compare incomes against regional or national norms rather than local norms.

Reviewers interested in environmental justice issues may want to look at comparisons against more than one set of statistics to ensure that a relevant measure of local and regional income has been considered to establish what comprises a "low-income population" or a "minority population" within a given impact area. This is especially important in NEPA reviews, which are open to public scrutiny, so that readers will know that a range of aspects have been considered. In any case, the agency's reasons or rationale for choosing one method over another should be adequately explained.

What is a disproportionate effect?

Environmental justice considerations are directed to those cases where an adverse impact is "disproportionately" directed at a minority or low-income population. In order to determine disproportionality, the environmental justice reviewer must first determine the effect on the total population in order to determine if the effect on low-income or minority populations is proportionate or disproportionate (CEQ, 1998). Under a NEPA review, an agency will identify and determine the extent of direct, indirect, or cumulative effects to human health and the environment at large. The NEPA process is a forecasting tool, however, and its determinations of effect, no matter how well-intentioned, may in fact prove inaccurate over time. NEPA monitoring, environmental monitoring, or public health studies may provide additional information that is more accurate in determining the actual effect of a program, policy, or action over time albeit after-the-fact (Institute of Medicine, 1999).

One hallmark of a disproportionate, adverse environmental effect is that a local population may be exposed to different impact pathways—that is, the people in a localized area may use soils, plants, and animals in a different way than the public at large, and therefore may be exposed to adverse health impacts in a different way than the majority population (Hayes et al., 2000; Institute of

Medicine, 1999). For example, in many parts of the country, Native American populations favor wild game and fish as a substantial part of their diet. If a proposed action, policy, or program might result in an adverse impact to wildlife, whether through habitat loss or exposure of wildlife to toxicity, the local people may be brought into greater contact with an adverse impact due to their higher consumption of wild game or fish (Fresquez et al., 1998a). Accordingly, an impact that might well be considered of no consequence to a regional or urbanized population may be adverse to this localized segment, and would disproportionately affect the local segment. Again, to use the same example, because of the reliance on game as part of the diet, the local population may be more susceptible to long-term, cumulative impacts of several smaller actions that would become significant in the aggregate. As a hypothetical example, a federal action from one agency might be to construct a new power plant in a wildlife habitat area that would result in a concentration of wild game in the remaining habitat. A second, unrelated federal action from a different agency might be to build a new highway, which might both further fragment game habitat by interrupting migration corridors and introduce toxicity (e.g., paving oils and road salts) into the habitat. While neither of these might be significant separately, taken together they could have a dramatic effect on the availability, health, and suitability of game as a food source in a local area. The adverse impact on local populations dependent on game as a food source would then derive as a secondary impact.

Another type of consideration when dealing with localized minority populations is that there might be localized uses of soils, plants, and animals that are not readily apparent to a reviewer not familiar with local customs (Fresquez et al., 1998a; Hayes et al., 2000). The reviewer may have to modify the typical assessment scenarios and incorporate additional pathway assumptions related to traditional or ceremonial activities (Fresquez et al., 1998b). For example, Native American tribes make use of certain plants for food or ceremonial purposes, and Hispanic *curanderas* use a wide variety of herbal tonics and medicines. A plant that does not have an apparent economic value to the majority population may be of extreme importance to a local indigenous population. The plant may be used in such a way that unique exposure paths are encountered; for example, an herb may be placed on a fire or boiled and the smoke or steam inhaled, or the plant extracts or ash rubbed directly on the skin. Since many of these ceremonies or cures are held in confidence, the local populations may not choose to divulge the use, and hence the exposure pathway, to the reviewer (Institute of Medicine, 1999).

Public involvement

As is the case with the remainder of the NEPA process, effective environmental justice depends upon and benefits from strong public involvement (Clinton, 1994). Both the CEQ guidance and the guidance published by the EPA provide for several steps to identify and address environmental justice issues (CEQ, 1998; EPA, 1998).

The NEPA process requires that public input be solicited at specific points in the review of a new proposal. Agencies can take proactive measures to ensure that indigenous, minority, tribal, ethnic, or low-income people are adequately heard. These measures can be simple, such as holding public meetings in neighborhood centers that serve minority, tribal, or low-income populations; having appropriate written materials and translators available for non-English-speaking people (such as fact sheets prepared in Spanish); and establishing a meeting format that is amicable to the culture and education level of the affected people.

Through the NEPA process, the agency can also ensure that the public is made aware of, and agrees with, mitigation measures designed to lessen adverse impacts to public health and welfare (Institute of Medicine, 1999). This is especially important when the agency is uncertain about exposure pathways, such as in the case of traditional tribal uses of the environment, or the acceptability of specific mitigation measures to a given population segment.

Other requirements of environmental justice

Executive Order 12898 directs federal agencies to improve research, data collection, and analysis in order to better capture information on environmental justice issues. Broadly, this requirement has three facets (Institute of Medicine, 1999):

1. Research to improve the science baseline
2. Research among the affected populations
3. Communication of research results in a meaningful way

The health baseline of minority or low-income populations is not well understood. Since poverty tends to result in both a poor state of nutritional health compared to the general population and a barrier to receiving adequate health care, and since some diseases occur in distinctive patterns among certain minorities, the health baseline of the low-income or minority populations may be quite different from the baseline of either the general population or the majority population (Institute of Medicine, 1999). Therefore the reaction of a minority or low-income population to an environmental stressor may be different from the effect upon the population at large. Agencies are required by the Executive Order to collect, maintain, and analyze data comparing environmental and health risks to different segments of the population. As agencies develop research strategies to gather information to supplement the demographic and health baselines, they may consult the affected population to gather additional information to determine the optimum way to proceed. They are also expected to share the information with the affected populations, in a way that can be readily understood (Clinton, 1994).

The EPA guidance suggests that certain populations may be at high risk from environmental hazards or exposed to substantial environmental hazards due to geographic factors that isolate them from other surrounding communities or that tend to allow pollutants to accumulate in the environment surrounding the population. Population age, population density, literacy rates, and the stability of a neighborhood may also play an important role in the health baseline of the affected population (EPA, 1998). Older or younger populations may be more susceptible to environmental risk, either because of the amount of time they are exposed to a potential toxin or because of the stage of development of the body's immune system. Individuals with a lower education level may have difficulty understanding complex technical documents, or be unaware of or unable to identify an environmental risk at an appropriate time.

Summary of CEQ guidance

The CEQ issued guidance for addressing environmental justice under the National Environmental Policy Act in 1998. The CEQ guidance elaborates on how agencies may take environmental justice into consideration during specific phases of the NEPA process. A summary of the recommendations is presented below.

Scoping. Considerations should be made during the scoping process to determine whether disproportionate impacts on minority communities may occur. In determining whether minority communities may be affected, it is necessary to consider both residents and people who use the affected area.

Establishing the affected environment. The preparer must take into account all aspects, including physical, social, cultural, and health, of the potential impacts resulting from the proposed action on the community. It is important to consider that the impacts within minority populations, low-income populations, or Indian tribes may be different from impacts on the general population due to a community's distinct cultural practices.

Environmental assessment. As defined earlier, an EA examines the intensity of a project's environmental consequences and their significance, and determines whether an EIS is necessary. The interests and concerns of potentially affected minority communities should be taken into consideration when determining the intensity of environmental consequences.

Analysis. Minority communities that may suffer disproportionate and adverse effects from the proposed action should be encouraged to participate in the development of alternatives, and in the identification of the environmentally preferred alternative in the Record of Decision (ROD). Involving members of the community in the development of alternatives may lead to the identification of alternatives that have fewer adverse impacts on minorities and reduced environmental effects.

Mitigation. If the preparer finds that the proposed action will have a disproportionately high and adverse effect on a minority community or any impact to tribal, cultural, or natural resources, then measures should be taken to mitigate these effects. Mitigation efforts should be developed in consultation with affected community members and groups, and should provide for ongoing participation and coordination after the measures are implemented.

10.10   Conclusions

Although socioeconomic impact analysis has been improved in the recent past, considerable effort is still required to improve the methods. The key areas are

- The inclusion of time in the analysis
- Better estimates of the labor force and its composition
- Better estimates of indirect employment
- More detailed fiscal impact analysis

These improvements are being incorporated in current work in the field. They result from the need to make the results of the analysis useful to the community as inputs to the planning processes. The recommended improvements lead directly to greater utility of socioeconomic impact analyses as planning tools.

10.11   Discussion and Study Questions

1 Discuss why you believe—or do not believe—that examination of the economic consequences of a proposed action deserves a place within the context of an EA or EIS prepared under NEPA. Is there a difference between effects on

individuals and effects on governmental entities in this respect? Do you have different opinions about positive (i.e., stimulating) effects than about negative consequences? Is the inclusion of one type more logical than for the other?

2 In a similar manner, does examination of social consequences belong in a NEPA document? Are *all* effects on individuals part of the social environment? When may such effects be safely omitted? May they *ever* be omitted? Discuss the circumstances.

3 Select any U.S. government agency for use as an example, or use the agency for which you work. Examine its NEPA regulations for mention of social and economic consequences. Are they handled equivalently to discussions of air and water quality and other elements of the biophysical environment? Do they have special rules? Is this area given any treatment at all?

4 Discuss why it may be logical for an agency to suggest that social and economic issues not be covered in NEPA documentation. Do you believe it is permissible for an agency to do so if it wishes? What overall purposes are furthered or hindered if social and economic effects are omitted from coverage?

5 Consider which of these principles of social and economic impact are appropriately applied when the action under consideration will result in negative changes in employment and income, rather than the increases discussed throughout Chapter 10.

6 Review a recent EIS prepared for an action proposed in an area with a large minority or low-income population. Are environmental justice issues discussed? How was the effected population determined? Were possible mitigations identified?

10.12   Further Readings

Coughlin, Steven S. "Environmental Justice: The Role of Epidemiology in Protecting Unempowered Communities from Environmental Hazards." *The Science of the Total Environment*, 184:67–76, 1996.

Environmental Impact Assessment Review, volume 10, number 1, 1990. New York: Elsevier Science Publishing Co. The issue is devoted to discussions of various aspects of social and economic consideration within the environmental assessment context. Among the individual articles of particular relevance are:

Brown, A. Lex. "Environmental Impact Assessment in a Development Context," 135–143.

Burdge, Rabel, and Robert A. Robertson. "Social Impact Assessment and the Public Involvement Process," pp. 81–90.

Craig, Donna. "Social Impact Assessment: Political Oriented Approaches and Applications," pp. 37–54.

Hundloe, Tor, Geoffrey T. McDonald, John Ware, and Leanne Wilks. "Cost-Benefit Analysis and Environmental Impact Assessment," pp. 55–68.

McDonald, Geoffrey T. "Regional Economic and Social Impact Assessment," pp. 25–36.

Wildman, Paul, "Methodological and Social Policy Issues in Social Impact Assessment," pp. 69–79.

Gilmore, J. S., and M. K. Duff. *Boom Town Growth Management: A Case Study of Rock Springs–Green River, Wyoming.* Boulder, Colo.: Westview Press, 1975.

Public Participation

What is meant by the term *public participation?* For that matter, what do the words mean separately? What is (or who are) the public? What constitutes participation? Where did the term originate? What is its contemporary meaning?

11.1 Beginnings

Virtually all government-sponsored activities have the potential to affect some aspect of the life or environment of the area within which they are to take place. Normally, this is openly stated as the basis of need for the proposed action (i.e., that something needs to be changed). Generally, public agencies are charged with the responsibility of acting on behalf of the constituency they serve or represent. Actions that require environmental impact assessment and statements are usually extensive and are likely to affect the community and the environment in a variety of ways, and these effects may be perceived as "good," "bad," or "of no consequence." This perception is, however, personal to the extreme. One person's "beautiful" proposal is someone else's "disaster waiting to happen." However, the need, or at least the desirability, for the project to be shaped in response to the requirements of the local community establishes the necessity for effective public participation. Without such participation, the project may take on a direction or emphasis that (although ostensibly directed toward public benefit) is counterproductive to the community's needs.

We have heard many times the epithet "Taxation without representation," usually with the implication that it is unfair, unjust, uncalled for, not desirable, and generally *not* in the best interests of the subject population. Similarly, public sector activities which are stated to be in our interest but which evolve without our inputs to guide direction, quality, and quantity also seem equally misguided. The value of public participation at many stages in the NEPA process is widely acknowledged (Stein-Hudson, 1988; Ketcham, 1988; O'Brien, 1988).

11.2 Early American Experiences in Public Participation

The role of public participation in seventeenth- and eighteenth-century decision making was examined in *The Puritan Oligarchy—The Founding of American Civilization* (Wertenbaker, 1947). Wertenbaker points out that there was a clear conflict between the Jeffersonian concept of a participatory democracy and the reality of the church society in Massachusetts. From its inception, the Massachusetts Bible State exemplified the government of the many by the few, represented in the comparatively small body of church members. All significant decisions were made by a still smaller body of powerful men who represented, alternately, the church and the political government.

The theoretical political base of the United States and most other democratic governments accepts, as one of its central tenets, the Jeffersonian concept of participatory democracy. This concept establishes the need for political figures to seek the consent of the governed when making decisions affecting the welfare of the state and its citizens. This theory finds classic expression in the town meeting and assumes the educability of the citizen public, the predominance of reason, the availability of full information, and free access to the decision-making process, with the end product being understanding, consensus, harmony, and sound decisions. Can we ever meet this ideal?

James Madison recognized the basic incongruity of this concept and wrote in the *Tenth Federalist Paper:*

> Those who hold and those who are without property have ever formed distinct interest in society. Those who are creditors, and those who are debtors, fall under a like discrimination. A landed interest, a manufacturing interest, a mercantile interest, a moneyed interest, with many lesser interests, grow up of necessity in civilized nations, and divide them into different classes, actuated by different sentiments and views. The regulation of these various and interfering interests forms the principal task of modern legislation and involves the spirit of party and faction in the necessary and ordinary operations of government.

Problems associated with this "principal task of modern legislation" to respond equally to various "publics" have been rearticulated many times since Madison's attempt. Many of the problems revolve around the question of citizen involvement in governmental decision making and have resulted in great difficulty identifying and defining pragmatic approaches to operationalize American government.

The eighteenth and nineteenth centuries were dominated by the frontier. Settlers in those centuries perceived the American continent both as a savage wilderness which should be conquered and as the new world, full of inexhaustible resources of every kind. So, basically, the destiny of humans appeared at the time to be to tame the wilderness and exploit its resources. In the nineteenth century, conservationists were philosophers and not activists. For example, Henry David Thoreau quietly and eloquently recorded in his journal his conviction that preservation is a worthwhile goal and that wilderness is justified by the inspiration that people can draw from it. Persons like Thoreau were out of the mainstream of the commercial and political life of the nation, and had only a few sympathizers. Their perspective had little impact on development policies. For them, preservation of natural amenities was an aesthetic, ethical, and moral issue. It appears that their philosophical ideas had little practical influence on the real problem, but what their writings did provide were philosophical foundations for the next generation of conservationists.

These philosophical concepts proved to be insufficient, in themselves, to persuade a majority of the public. For example, in 1910, in the period of recovery from the earthquake and fire of 1906, the city of San Francisco proposed to create a water supply reservoir in the spectacular Hetch-Hetchy Valley in Yosemite National Park. The question was whether a human-made impoundment should be built within a national park. Other sites were available, but the Hetch-Hetchy site was the least costly. We must remember, in retrospect, that this was a time when even most residents within California had never contemplated actually *visiting* Yosemite. The trip was lengthy and visitor accommodations within the park were too costly for the ordinary working-class person. The park was known almost entirely through black-and-white photographs.

John Muir, founder of the Sierra Club and a strong proponent of wilderness, argued that the reservoir would be inconsistent with the national park concept. Also, he argued that it would consume a magnificent scenic area and would offer no recreational benefits. Muir's philosophical and ethical arguments proved to be insufficient when put against the economics-based arguments of the proponents. In 1913, the Hetch-Hetchy reservoir was approved by the Congress (CEQ, 1973).

In the early 1950s, environmentalists and conservationists, in addition to arguing for preservation as a philosophical concept, utilized engineering and hydrologic studies to support their views. The case in point was the Echo Park Dam in western Colorado. As a result of the arguments set forth by the conservationists, and as a result of public participation and involvement, this particular project was dropped from the development plans.

11.3 Alternative Terminology

There are several other closely related terms which may be used more or less interchangeably with "public participation." Community involvement, public input, public involvement, community participation, and community relations are but a few of the terms which have been used in various contexts. Nor is NEPA the only legislation where public participation plays an important role. In the implementation of CERCLA, the EPA Superfund program mandates an extensive set of activities termed *community relations*. The stated objectives are to (1) *give the public the opportunity to comment on and provide input to technical decisions*, (2) *inform the public of planned or ongoing actions*, and (3) *focus on and resolve conflict* (EPA, 1988). These terms do have other aspects in common as well. While they all seek to further the provision of timely information to the public, they differ from traditional public information or public affairs activities in that they seek to operate in a *two-way flow of ideas*.

11.4 Public Involvement Requirements within NEPA

The term *public involvement* was introduced into the NEPA context with the publication of the NEPA regulations in 1978 (40 CFR 1500-1508). Clauses which implicitly or explicitly require notification to or consultation with some publics appear in several places in these regulations. In Part 1501, for example, in the section dealing with preparation of assessments, at a time prior to the determination that an EIS is required, the following sentence appears: *"The agency shall involve environmental agencies, applicants, and the public, to the extent practicable, in preparing assessments..."* [40 CFR 1501.4(b)]. As another example, in Part 1503, where the process of inviting comment on a draft environmental impact statement is described, the regulation states that *"After preparing a draft environmental impact statement and before preparing a final environmental impact statement the agency shall:...Request comments from the public, affirmatively soliciting comments from those persons or organizations who may be interested or affected"* [40 CFR 1503.1(a)(4)].

These are not, however, the instances normally considered to represent the most difficult aspects of public participation, although problems associated with bringing the public into the EIA process will be examined later. Section 1506 of the regulations, devoted to "Other Requirements," provides an extensive set of requirements at 40 CFR 1506.6 entitled "Public Involvement" (Appendix D). In this section, it is made clear that by use of the term *public,* the CEQ intends that all publics be included. The introductory words say, for example *"Agencies shall: (a) Make diligent efforts to involve the public in preparing and implementing their NEPA procedures. (b) Provide public notice or NEPA-related hearings, public meetings, and the availability of environmental documents so as to inform those persons and agencies who may be interested or affected."* The intent is very clear.

The term *public involvement* in close to its present meaning was apparently first used systematically within the U.S. Army Corps of Engineers at some time in the late 1960s or early 1970s. By the mid-1970s, the term was well established in water resources planning procedures. A pamphlet (Hanschy, 1976) was published outlining the procedures recommended to be used, and formal training courses were taught several times each year for Corps of Engineers planners. These courses emphasized use of various group dynamics activities under the guidance of a strong facilitator. The "players" in these groups were assumed to be representatives of the various "publics" interested in one or more aspects of a proposed water resources project.

It must be noted that the Corps, at this stage of the development of NEPA, or prior to its passage, had no specific requirement to involve the public. It chose to do so, however, as a means to its own ends. The emphasis during these meetings was primarily that of gaining a consensus which could then be used to represent the opinion(s) of the group assembled at that time. These assembled opinions, pro and con, were then taken into the planning process so that the widest practicable range of public opinion might be shown to have been considered. After receiving intense criticism from legislators, environmental advocacy groups, and the public for not incorporating a wider range of concepts and values into its water resources planning during the 1950s and 1960s, the processes implemented were an attempt to answer the critics. The Corps, in a sense, had "anticipated" the spirit of the forthcoming NEPA regulations, although its usage was not strictly in the NEPA context.

11.5   What Is a Public?

It was earlier asked what was really meant by the terms *public* and *participation.* This is far more than a rhetorical question. Successful implementation of the public participation aspects of any proposed project or action demands a closer discrimination within commonly used terms. When we read in the newspapers about public opinion, just what image do we create? What *is* the public? A more correct way to phrase this might be to ask, Who are the publics? It is a fact of life that the image of a large, cohesive, like-thinking public is obsolete, if, in fact, it ever existed. In the management of every proposed action, we must deal with many different publics, each with its own special interests and peculiarities.

When we propose a new project or action which is significant enough that an EIS (or even an EA) is required, we may automatically assume that there are several significant publics who feel they may be affected by the outcome. Some, such as environmental activist groups, are *always,* practically by definition, interested in the proposed action and its outcome. Similarly, elected and appointed government officials at every level form another public, one which must be handled with extreme care. Property owners, outdoor recreationists, farmers and ranchers, real estate developers, retirees, and officials of state and federal government agencies are other examples of publics which may be involved in the action in one way or another. As you may see, some, such as the Chamber of Commerce, are easily identifiable—with a listing in the telephone directory—while others may have no formal organization and be hard to define and locate.

In the development of a contact list for conducting community activities in connection with Superfund activities (EPA, 1988, Chapter 3), the EPA has prepared a set of recommendations which organize contact activities by target group. The groups suggested for targeting are

   State agency staff

   Local agency staff and elected officials

   Citizens' groups organized because of the proposed action

   Residents and individuals not affiliated with a group

   Local business organizations

   Local civic and neighborhood associations

   Local chapters of public interest groups

The point is also made (EPA, 1988) that a *variety* of persons within the designated category be included in discussions. The risk of being accidentally (or intentionally) misled about the position(s) of the group as a whole is much lower when many persons from several groups within the identified category are included in the discussions.

Collectively these different publics may be in favor of, be opposed to, or have no strong feeling about the *technical* consequences of the proposed action. Personally, however, each of the groups *potentially* affected by the action, no matter how obliquely, will believe that it deserves extensive information about the action, including the reasons behind it and the economic justification used for implementing it—topics which probably go far beyond the intent of NEPA disclosures. It is the *feeling*, on the part of these persons, of being left out, or of the proponent agency "putting something over on us," that has led to the institutionalization of public involvement. This feeling of alienation may be more important, in the final evaluation, than the presence or absence of measurable effects. Remember that the Corps of Engineers, as mentioned above, developed and used public involvement procedures and processes many years before the NEPA regulations required such actions. Their purpose at that time was the building of a local consensus which could agree, at a minimum, that nothing was being kept from them, and that they had had a fair chance to have their ideas heard and incorporated into the decision-making process. It is just this sort of benefit which may be derived from sound public participation activities today.

11.6   What Is Participation?

If we may agree that there are many publics present in association with each issue, what types of activities constitute participation? In the early days of the NEPA regulations, and for some agencies until the present, public notices of the availability of the draft EIS and the holding of a public hearing on the matter [following exactly the form given in 40 CFR 1503.1(a)(4)] constitute the sole participation activities. Frankly, for many agencies and in many regions, this minimal notification appears to be adequate to meet statutory requirements. For smaller, less controversial actions, the adequacy of these processes has not been severely tested. In such instances, it would appear that the degree of *secondary* publicity associated with a proposed action may serve the agency's purpose in making all interested parties knowledgeable about the process. An active, interested press and active local officials frequently serve to substitute for possible inadequacies on the part of the proponents of the action.

It is clear that merely providing public notice is a minimalist level of "participation." It seems likely that belief that this is an adequate procedure probably derives from reliance on the (now largely outmoded) concept of "federal supremacy." The normal procedure was, in many agencies, one which could best be described as "Get it in place quick, before they have a chance to collect the opposing forces." Requirements for even minimal public participation make this style of project implementation very difficult, which was certainly one of the purposes. No full-scale definition has ever been attempted, however, of all activities which could be considered acceptable for public participation.

11.7 Contemporary Experience in Public Participation

The actions of government administrative and management agencies frequently seem, to most citizens, remote decisions by a faceless bureaucracy. To the extent that such decisions affect their lives and environment, this isolation places citizens in a restrictive position. Either they must approach environmental management agencies to request assistance in dealing with a problem, or they may demand solutions to a problem through the judicial process. In both cases, the citizen is responding to an administrative decision which has already been made.

Rather than only respond to decisions, there is a need to involve more citizens in the decision-making process itself. This approach increases citizens' presence in the administrative agencies. It also reduces the need for antagonistic and legalistic behavior by the citizens (Sax, 1970; Stein-Hudson, 1988). Postdecision citizen protest and legal battles are increasingly seen by government agencies as expensive and time-consuming alternatives to involving citizen groups in the planning process from the beginning (Ketcham, 1988).

Examples of participation

Since *participation* is, in itself, a generalized, all-encompassing term which lacks specificity, it is probably best to define it in terms of examples. The actions specifically named within Section 1506.6 include (direct) notice (presumably by mail), publication in local newspapers, publication in newsletters, direct mailings, posting of notice on and off the site of the proposed action, and the holding of hearings or public meetings. While these avenues are explicit, the implicit charge is to make use of any appropriate means of communication. The Notice of Intent (NOI) (40 CFR 1508.22), announcing that a major action is planned for which an EIS will be prepared, is an example of a required announcement which is a part of the public involvement program for a project or action. The Notice of Availability, published in the *Federal Register* by the EPA following submission of the draft EIS, is another form of required public involvement. Together, these two notices serve to inform the public (1) that a major action is contemplated and (2) that an environmental analysis has been prepared and is available for review and comment. These two steps, which constitute minimum implementation of public participation, may be all that is reasonable to apply for some actions, especially smaller, relatively noncontroversial ones. As noted above, these are the only conscious steps which many agencies find necessary. If actions have the potential to be controversial, then a much more complex public participation plan is recommended.

Informing the public

Every major agency has a public affairs (or press) office, which should be knowledgeable about spreading information at the local level. Normally, the promulgation of project information will be left almost entirely to the existing public affairs office. Your task may be limited to providing information to the public affairs office. There is, however, a trap in this approach. The conventional methods used by many less activist and less innovative public information offices may not be suited to NEPA needs. How does the public affairs office operate? Is their function limited to mailings of one- or two-page press releases? Do they ever use follow-up contacts? Are their contacts limited to the business community? Do they have sources and contacts within interest groups? How do they deal with confrontational interest groups? Many public affairs officers, to their credit, have been able to initiate two-way communication with such groups. Many more have avoided the issue, and may not even know who represents such organizations locally. NEPA support may require education (or reeducation) of an organization's public affairs personnel. Do make use of their skills, but don't rely entirely on their existing knowledge.

The first step in informing the public is to identify who or what the publics are (EPA, 1988). If, as suggested above, no up-to-date lists of points of contact are known, then the construction of a relevant mailing list is a priority action. In some cases, this may be as simple as keeping a list of the names and addresses of persons calling, writing, or visiting as a result of articles appearing in the local press. Since these stories are likely, in many cases, to be oversimplified beyond recognition, highly alarmist, inaccurate in the facts presented, or all three, an important function of early contact and mailings may be to (gently) correct the concept of exactly what is proposed to be done in your action.

Some publics may be located in the local area through solicitation of national or state-level parent organizations. While the local Audubon Society may only rarely have a telephone directory listing, the National Audubon Society will be more than pleased to provide a local point of contact. Such similar national organizations as the Sierra Club, Isaak Walton League, and others may be identified in the same manner. Essentially all national interest groups, and most local groups, have a web page with contact information. Many agencies may believe that by locating potential opponents of the action, they are "asking for trouble," and avoid such efforts. The converse of this is that, even if activist organizations oppose your preferred alternative, they are gratified that *you* took the effort to locate *them*. It is this outreach which benefits the agency in the long run—even your opponents are left feeling that you are being fair with them. This is one major value to good public participation.

Once at least some publics are identified, the techniques of involvement appropriate to each public may be selected. Where there is some internal organization, as with a membership group, mailings to the officers may suffice. It may be more effective to provide the officers with flyers or newsletters for each member, and request that they be passed out at a future meeting. It is also acceptable to request an opportunity to speak at a meeting, if the meeting format of that organization is appropriate. The actual presentation may be required, by your agency's rules, to be made by the public affairs personnel. It is then your task to prepare the presenter well. One caveat exists in association with this approach. Do not avoid those groups anticipated to be unfriendly, such as the Sierra Club, and stick to the "safe" organizations, such as the Chamber of Commerce. Since all such public involvement activities will be listed (or summarized) in the draft EIS, the appearance of favoritism could be used against your agency.

If there is really no extant organizational structure upon which to build your plan, newspaper "advertising" has been used to build a list of interested persons. Whether in the form of a press release, resulting in a "news" article, or a (purchased) display advertisement, a summary of the proposed action and alternatives may be accompanied by a request to call or write if the reader is interested in providing comments or wants more information. One word of advice—the address or phone number should be a local one wherever possible. Toll-free telephone numbers may be established for this purpose as well, a practice which is much less complicated to implement than in the past. Or, a web site may be established.

More detailed participation programs may involve one or more mailings, a series of public meetings which are preceded by press notices, a series of meetings with governmental bodies and officials, telephone surveys, liaison with major interest groups, and solicitation of ideas and comments from within the agency itself. The dividing line between "public relations" and public participation is often hazy, and may disappear entirely. Broadly speaking, there are two purposes to public participation. First, you must get across to the various publics a reasonably clear picture of the action which you propose to take, and what alternatives exist. Second, there is the explicit obligation in the regulations to solicit the ideas held by the *public* on the effects and consequences of the proposed action. One type of activity may be better suited to the first purpose, another to the second.

11.8   Getting Input from the Public

Following identification of the publics, the next problem is retrieving meaningful expression of public ideas and concerns. It is of little value to prepare a list of those groups and persons opposed to your preferred action versus those in favor. We aren't dealing with a popularity contest—or not *only* a popularity contest. The charge in the NEPA regulations [40 CFR 1506.6(d)] to *"Solicit appropriate information from the public"* suggests that other important purposes are served by this reverse flow of information. The entire concept of scoping (40 CFR 1501.7) is that the focus of the EIS (or EA) should be on those (relevant) issues which the public(s) deem to be significant. This concept is in direct response to the "encyclopedic" approach to EIS preparation which flourished briefly in the mid-1970s. In this older approach, anything that might possibly be affected was examined and described. A draft EIS and its appendices (under this approach) might be 5000 to 10,000 pages long, and require a shipping carton for each "copy" mailed. The conduct of scoping is an extremely important and specialized form of public participation. It is treated separately in Chapter 4.

11.9   Commenting on the Draft EIS—A Special Case of Public Participation

Commenting on the draft (or final) EIS is another specialized form of public participation. It is mandated in several places in the NEPA regulations (Sections 1502.19, 1506.6), but is the subject of a separate section in itself (1503). Here, among others, the specific injunction is seen [1503.1(a)(4)] to *"Request comments from the public, affirmatively soliciting comments from those persons or organizations who may be interested or affected."* We note that the term *interested* is shown as more important than, or at least equal in importance to, *affected*. The meaning of this wording is clear. The *technical* (i.e., the agency's) opinion as to who is affected is not a determining factor in soliciting comment. Everyone who *believes* him- or herself to be affected, as well as those who are merely interested, are to be solicited. Some writers have gone so far as to suggest that the scoping process may be the most important aspect of public participation, in that it may serve to shape the final action itself (Ketcham, 1988; O'Brien, 1988).

If only a handful of comments are received, it may not be necessary to establish a formal scheme for classification. In most cases, however, the comments received following distribution of a draft EIS will be numerous enough that development of a systematic classification structure is recommended. For example, responses may commonly be grouped into at least three sets: (1) those which express support or opposition, with little or no explanation of the reasons for the position, (2) those which ask for more information, or raise questions about the completeness and accuracy of supporting data included within the draft EIS, and (3) those which propose different alternatives, or modifications and combinations of those alternatives already included. The responses in area 2 may further be subdivided as to the focus of the question (i.e., which aspect of the environment is flagged?). Social consequences, employment, wildlife, health effects, public safety, noise, drinking (or irrigation) water availability, and similar headings are among those regularly arising following public examination of the proposed action. In very many of these cases, the issue will have been anticipated—or flagged as important during the scoping process—and will have been examined in the document to some degree. Case Study 1 presents an approach to the analysis of thousands of responses.

11.10   Case Study 1

EPA and the Corps of Engineers propose to redefine navigable waters

Following passage of the Clean Water Act of 1974, the U.S. Army Corps of Engineers was required, in Section 404 of the act, to develop a significantly broader definition of the term *navigable waters*. The background was this: Section 404 of the act severely regulates the conditions under which fill may be placed in designated wetlands. The Corps was given jurisdiction for administration of the fill permits when they were "...in association with navigable waters...." (The EPA was given jurisdiction over other wetlands within Section 404. There have been several changes in jurisdiction since 1974.)

In May 1974, the Corps published a Notice of Intent to define navigable waters in such a manner that headwaters of streams which were navigable in fact were redefined as navigable. The definition of "navigable in fact" was also proposed to be expanded to specifically include recreational watercraft, including canoes, craft which were specifically excluded under earlier definitions. This would have had the effect of expanding, possibly by as much as 10 times, the streamside acreage considered to be a "wetland" and falling under the Corps' jurisdiction. (The 1991–1992 controversy about wetlands definitions had similar overtones, but was based on somewhat different criteria.) An extremely speculative story, which was widely disseminated by the national news services, stated that "The Corps wanted jurisdiction over all farm ponds...." This view of the proposed action was widely reprinted in local newspapers nationwide, and in bulletins and newsletters sponsored by farmers' organizations. As a result, the Chief of Engineers was asked to hold daily news conferences and provide almost daily briefings to several congressional committees. More than 6000 letters were received within a month, almost all expressing outrage at the "proposal." Many were transmitted from members of Congress and the White House, where they had originally been directed.

How does one assimilate the concepts expressed in thousands of letters? Clearly, the first task is to group, or organize, the ideas in some manner. Following planning meetings and review of the first several hundred letters, it was decided to create a "tally sheet" which attempted to identify the following characteristics of each letter:

State of origin: City or town; size of community if determinable.

Group membership: Farmer vs. nonfarmer; member or officer of identifiable public organization; if an organization member, are the ideas expressed personal feelings or the result of a formal resolution or petition.

Nature of comment: Exactly which aspects of the proposal did the commentor find unpalatable? Economic freedom? Government interference? Antibusiness character of the action (as understood by the writer)?

Level of understanding evidenced: Did the writer show that the actual proposal had been examined? A summary of the proposal? A news story *about* the proposal? A news story or newsletter article based on the misconceptions spread about the proposed action?

Through use of this tally sheet, several "letter readers" were able to rapidly examine each item of correspondence, fill out the tally form to show the above characteristics, and contribute the form to the statistical summary desk. First two, then four, and finally six persons were assigned to examining the flood of mail. Daily and cumulative summaries were prepared which provided to the Chief of Engineers and, thus, to the Congress, information on just who was commenting and what their ideas were.

More difficult to assimilate are those comments, stated to be in response to the draft EIS, which raise issues about environmental (and other) effects which are definitely, at least in the opinion of your agency's professional staff, *not* expected to occur. Who is correct? Again, this is an area in which it is prudent to be open-minded. Admit that outside ideas *may* be worth examination after all. It is best to attempt to provide a niche in which these "unqualified" commentors may be addressed. An added discussion that explains why these issues were rejected, and may serve to show that they really were evaluated in the first place—or even after the fact—will serve to accommodate these commentors. It is the dismissal of the idea *without explanation or consideration* which leads to feelings of mistrust and rejection on the part of the publics. Frankly, it may be more important to reject an idea thoughtfully and gracefully than to attempt to find a way to accept it.

Still more perplexing is the situation where the responses are not to the action which is actually proposed. How, you ask, could this even happen? In fact, it is not uncommon at all to encounter this in the case of controversial actions. The opponents of the action conduct their own public involvement campaign, describing the action in their own (usually alarmist) terms. National organizations may even publish *their* version of a summary of the draft EIS in their newsletter or magazine. If they then give *your* address for responses and angry letters, you may accumulate scores—or even thousands—of responses which address this flawed definition of the proposed action. How should one respond to this type of letter? There is a tendency to ignore them. If you receive only one or two such responses, then we suggest the proper response may be to note their contents and state that the issue(s) addressed are not relevant to the action. When they make up a large proportion of the comments, however, such dismissal is not prudent, and could even provide the basis for a legal challenge, perhaps on the grounds that X percent of the comments received were ignored. The next case study example discusses what can happen when there is substantial misinformation promulgated in association with a proposed action, and how the resulting dilemma was approached.

11.11   Case Study Example 2

Public response to the biological defense research program EIS

In 1987, the Secretary of Defense was sued on the grounds that the ongoing research programs designed to develop detectors for biological weapons and provide better diagnosis and treatment for personnel infected with diseases with potential for use as biological weapons had never been examined programmatically under NEPA. It was agreed between plaintiff and the Department of Defense (DOD) that these research programs, administered for DOD by the Army, would be the subject of a programmatic EIS. It was the contention of many activist groups that biological *weapons* were actually being developed and tested within this research program. Further, many of these groups alleged that these weapons were being tested outdoors, in proximity to civilian populations, and presented incredibly high risks to the surrounding citizens. These latter allegations received wide publicity, and a *majority* of the responses received following review of the draft EIS requested cessation of weapons development and outdoor tests of disease-producing organisms. Many respondents then proceeded to build arguments showing why and how the development of these weapons was unnecessary, in violation of international treaties, and otherwise highly questionable.

Since none of these actions was, in fact, ongoing *or* proposed, a reasoned response was extremely difficult. The respondents were so convinced that there was a hidden agenda that simple denial was believed to be counterproductive. In the responses to questions and allegations of this nature, a classification of types of question was prepared exactly similar to those for other issues. The (mistaken) contentions were approached *as though they were credible*, since, to the writer, they were. The issue is, again, one of fostering the perception of respect for an unpopular idea. It would have been an error in public relations to simply designate questions 196 through 315 (for example) as being "Based on misconceptions, therefore not answerable." The purpose of this exercise became the *education* of those with misconceptions, not *confrontation* with them. Certainly, in preparing the responses, feel free to state your position frequently (i.e., that the action(s) in contention were not proposed and are not part of the agency's action plan), but also state why these alleged actions are not needed or will not be done. Responses to this set of questions became much more complex than response to "real" (i.e., scientific) issues. In fact, this draft EIS received very few comments on those biological or medical issues which were based on calculations or studies included in the EIS (U.S. Army, 1989).

11.12   Application of Public Input

As strange as it may seem, we use input from the public to modify our plans and programs—not necessarily drastically, but observably. This is clearly the intent of the NEPA regulations, and is taken as gospel by the public interest organizations. At a minimum, and this minimum is specified in the act, the comments must be taken into consideration. Moreover, they should be taken *visibly* into consideration if a project with any degree of controversy is to succeed. What do we mean by "visibly"? Just that the commentor must recognize that some change in the description of the action or of the alternatives, or the incorporation of a new alternative, or the examination of environmental effects, or all of these, follows logically and directly from comments made by the commentor (O'Brien, 1988). Note that this does not mean that the project must be *drastically* altered, or that the agency's original purpose must be forgone—although this may happen—but just that each responsible commentor feels that his or her comment should result in some modification of the document. This is *not* unreasonable.

No one agency has a corner on technical expertise, or on professionalism. It is a mistake for anyone to acquire a strong ego commitment to the exact wording in a draft EIS and therefore be unwilling to modify it in any way. It was just this recalcitrance to accept ideas from sources other than within the agency which resulted in the promulgation of the NEPA public participation requirements in the first place.

"Not invented here" is not an acceptable reason for an agency to disparage any concept presented to it by the public. Now, these publics may be strongly prejudiced and word their ideas in highly florid terminology, thus alienating professional scientists within the agency (or its supporting contractors). They regularly bring up wild ideas which are (from the agency point of view) totally unassociated with the "real" action. The assessment specialist's responsibility is to "mine" these responses to identify underlying concepts which *may* represent responsible variations of scientific opinion.

11.13 Participation in Developing Regulations

We must remember that all major federal agencies have prepared their own NEPA regulations, which may expand greatly on the general wording in 40 CFR 1500-1508. Examination of the wording of 40 CFR 1506.6(a) finds the term "diligent efforts" used in describing required activities "...to involve the public in preparing and implementing (their) NEPA procedures" (i.e., preparing agency NEPA regulations). This injunction is separate from the requirements in 1506.6(b–f), which provide, among others, for "...public notice of NEPA-related hearings, public meetings, and the availability of environmental documents...." The CEQ clearly wished to be able to assure the public that the NEPA regulations, at least, of federal agencies were an open book. To a large degree, this has succeeded. All of the largest agencies have complied, though some have shown little imagination— their "regulation" amounting to little more than a cover letter attached to the CEQ NEPA regulations.

11.14 An Effective Public Participation Program

Effective public participation is characterized by the community acting with full information, equal access to decision-making institutions, and implementing its jointly articulated objectives. Based on this definition, several important objectives should be achieved to attain effective public participation.

First, there must be as much information as possible made available to the public. There often is considerable misinformation about the nature of most proposed projects, even when they do not involve withholding of information. This lack of communication precludes effective citizen participation in many cases.

An agency often allows its image as a public-spirited service institution to be maligned because organizations and individuals construe the agency's failure to provide adequate information as cavalier or inconsiderate. If instead an active program of public information and public participation were undertaken, not only would there be more useful public input, and therefore a better project, but there would probably be less criticism of the agency.

Second, community members, general public as well as officeholders, must have access to the decision process. Allowing or encouraging community involvement in problem identification and discussion, without influence on the ultimate decision, is not an answer to the problem—rather, it becomes a charade.

Third, for community participation to be effective, the input provided by citizenry should result in a course of action consistent with their desires and with the needs of their fellow community members. The agency must have the power to act on behalf of the citizens, and the decision must reflect the joint objectives of the agency and the community.

In the simplest form, the elements of an effective participation system are

1. Information exchange
2. Access to decision making
3. Implementation powers

Various types of communication exchange provide for the elements of an effective program. For a communication technique to function as a public participation tool, it must allow for citizens to become involved in decision making. This definition means that techniques that allow only one-way communication, such as newspaper articles, are not very useful. Newspaper articles may, however, be one prerequisite communication step in a public participation program that includes other forms of interaction with a well-informed public. A wide range of techniques contain some or all of the characteristics necessary for a public participation program.

Figure 11.1 presents a list of selected techniques for public participation and communication. This list may be used as an aid in determining which techniques are best suited to particular planning programs. It must be recognized that a comprehensive and operational community participation program would be composed of a variety of these communication techniques. A comprehensive handbook designed for Superfund community relations by the EPA discusses the relative effectiveness of different techniques in different situations (EPA, 1988).

Recommendations are made in Fig. 11.1 for the best application of 21 common public participation techniques, and are of four types: (1)

COMMUNICATION TECHNIQUES

The techniques and formats given below vary greatly in time, cost, and efficacy, as well as in level of acceptance in different settings. Some may be required by law or regulation; for some the format may be set by the agency.

Communication Techniques	Effective Public Contact	Impacts Decision Makers	Attendee Sophistication	Time & Effort Required	Useful to Receive Inputs	Flexibility to New Issues	Size of Group	Decision Makers	General Public	Mass Media	Interest Groups	Regulatory Agencies	Educate and Inform	Identify Issues/Problems	Solicit Original Ideas	Respond to Agency Proposals	Resolve Conflicts
	CHARACTERISTICS							**TARGET PUBLIC(S)**					**OBJECTIVES**				
Public Hearings	2	2	2	2	1	1	L	+	+	+	+	+	1	2	0	2	0
Traditional Public Meeting	3	2	2	2	2	2	L	+	+	+	+	+	2	2	1	2	1
"Open" Public Meeting	4	3	3	5	5	4	L	+		+	+	+	5	3	3	2	3
Speak to Interest Groups	2	2	1	2	2	1	L	+	+		+		2	1	1	1	1
Form Advisory Group	4	5	5	5	5	4	S	+	+		+	+	2	4	4	3	5
Exercises with Role Playing	5	1	4	5	3	3	M	+	+		+		1	4	2	2	3
Hold Seminars and Workshops	4	2	3	3	3	4	M		+	+		+	3	2	2	2	2
Conduct Delphi Group Sessions	5	3	5	5	5	5	S	+	+		+	+	2	4	4	2	4
Meet with Local or Regional Officials	3	4	2	4	4	3	S	+		+		+	3	4	3	4	3
Add Public Members to Project Planning Group	4	3	4	3	4	3	S	+	+		+	+	2	4	3	2	4
"Walk-in" Resource Center	4	1	1	2	1	2	P		+	+			3	2	1	1	1

						Technique	Group Size									
1	1	1	3	0	1	Distribute Project Brochures/Pamphlets	G	+	+			2	0	0	0	0
2	2	1	5	0	1	Prepare/Distribute Videotapes	G		+	+		2	2	0	0	0
3	2	3	4	3	2	Establish Project Website for Information	G		+	+	+	4	2	1	3	2
4	3	3	2	4	3	Provide Q&A and FAQ's via website	P	+	+	+	+	4	3	4	2	3
4	1	1	3	4	3	Create Toll-Free Hotline for Q&A	P	+		+	+	2	3	3	2	1
2	1	1	2	0	1	Prepare Direct Mailings	P	+	+	+	+	2	1	1	1	1
4	2	1	3	3	3	Use Local Radio/TV Call-In Programs	G	+	+	+		4	3	3	1	1
2	1	2	1	0	1	Issue Press Releases	G		+	+		3	0	0	1	0
3	4	2	5	4	4	Conduct Surveys of Public Opinion	G	+	+	+	+	1	4	3	1	3
5	4	2	4	3	4	Reply individually (by mail) to Inquiries	P	+	+	+	+	2	4	2	1	4

Techniques are rated on a scale of 0 through 5, where 0 means "normally of little or no value" and 5 means "rather effective if well planned and executed."

Group size column used on this scale:
P = Personal, one person
S = Small (3–10 persons)
M = Medium (10–25)
L = Large (25–100+)
G = General public access

The Plus sign (+) in a cell indicates that the technique is of some value in communicating with the public noted. Not all + scores are equal, and situational variation is the norm.

The techniques in the center column are rated on a scale from 0–5 for suitability in meeting objectives. 0 means "Of No Value." 5 means "Can be Effective if Well Done."

Figure 11.1 Comparison of public participation techniques.

Explanation of terms used in Fig. 11.1

Characteristics refers to various aspects of the application of the technique as described under each of the columns.

Effective Public Contact describes the level of interaction with affected publics.

Impacts Decision Makers indicates the degree to which the technique is likely to affect relevant officials.

Attendee Sophistication refers to the level of education and experience needed for effective understanding.

Time and Effort Required gives the relative difficulty for the agency personnel to carry out this technique.

Useful to Receive Inputs notes whether or not the technique is useful to receive public inputs on the issue.

Flexibility to New Issues shows the ease with which the technique may alter direction to reflect changes in content.

Target Publics refers to the type of group which the agency may wish to target through use of the technique.

Decision Makers are the appropriate elected and appointed officials as well as business leaders and NGO officials.

General Public is representative of those members of the public not identified with any special interest group.

Mass Media refers to reporters and editors in the traditional press and broadcast media.

Interest Groups is a term used here to identify all permanent and temporary groups formed to relate to the issue.

Regulatory Agencies range from the EPA to state and regional agencies charged with enforcing applicable rules.

Objectives includes many of the likely reasons why an agency may wish to use a particular participation technique.

Educate and Inform is the basic dissemination of relevant project information about alternatives and effects.

Identify Issues and Problems suggests that more detailed elements will be addressed, and known issues discussed.

Solicit Original Ideas represents the basic receptivity to receiving constructive suggestions and new alternatives.

Respond to Agency Proposals refers to presentation of revised concepts and data following changes in the project.

Resolve Conflicts identifies applicability to resolution of ongoing areas of difference and disagreement.

the effectiveness of a communication technique for different goals, (2) the size of the group with which the technique is best applied, (3) the sectors of the public to which a technique may be targeted, and (4) those objectives which may be accomplished through use of the participation technique. This figure is loosely based on a table from Isard (1972), as modified and presented in Jain et al. (1981). In general, the techniques which are the most effective are also the most time-consuming and difficult for the proponent agency. Note also that the "traditional" public hearing is seen to be of low overall value in achieving most goals, but is relatively easy to implement. Few procedures are presented as being of high value for more than one or two purposes; some serve only a single purpose at best, and may be of only fair value for that one! If one is selecting a possible procedure to use, it is also useful to note those areas where the process has little or no value, and avoid attempting to use it for the wrong purpose. While not directly observable in Fig. 11.1, some techniques work well with small groups and fail when used with larger audiences.

In a further examination of Fig. 11.1, several generalizations may be made. One of these is that many of the traditional "public affairs" processes are considered relatively ineffective for most purposes. Since the mid-1980s, many factors have altered the traditional relationships between government agencies and the public. Where, once, an agency simply announced that it planned to do some action and publicized that proposal, now it is likely to be accompanied by a great deal of "public support" activity. While the NEPA developments associated with formal public participation requirements may have formed some of this expectation, much of the desire to achieve public acceptance appears to go beyond the regulatory requirement.

As noted above, the Corps of Engineers, in the mid-1960s, developed a program of public involvement to achieve acceptance (especially) of localized urban projects. This was many years before the process became associated with NEPA, or the several other programs, such as RCRA and CERCLA, which now contain community involvement requirements. Why was the Corps impelled to create such a program, anyway? There was no specific requirement to do so, but it was seen as a way to obtain a more general "level of satisfaction" on the part of the affected citizens. This leads, further, to the statement of the ultimate purpose of all public involvement activities, namely, that all parties feel their position has been heard and understood. This "feeling" may be the most important possible outcome of the entire process. It is essentially an interpersonal reaction between a citizen and a representative of the proponent agency. The importance of developing this level of understanding cannot be overstated. This does not mean that the parties will always agree with the outcome. In many cases agreement with the decision may be impossible, and is not necessarily the major goal. A consensus does not have to be reached on the *outcome*, just on the *fairness* of the process.

It is worth noting that this is one of the important ways in which U.S. practice within NEPA differs significantly from the otherwise similar environmental assessment processes found in Western Europe. In most countries in Europe, it is required that all parties agree before the environmental documentation is approved. Under NEPA, all that is statutorily required is that the consequences be made public (as well as made known to the decision maker). So long as relevant regulatory compliance is present, it is legal for an agency to proceed with a decision even though many of the affected parties disagree with the final decision. This said, fewer and fewer actions are being taken where significant contrary opinion and opposition exists following comments on the final EIS. Agencies are usually still sensitive to opinion, especially as expressed in the press and through pressure on Congress, and may well be reluctant to execute an unpopular action even when it is otherwise legal to do so. In many cases, this may be through fear of losing appropriations for this or future programs.

11.15 Benefits from an Effective Public Participation Program

The catalog of reasons why decision making should not be made in a public forum but should reside in a central locus is extensive. Centralized decision making leads to more rapid, cost-effective, decisive decisions, permitting effective and efficient leadership. Most bureaucracies, including the military, are built on this decision-making mode. Congress seems to act on issues lethargically, appearing to be inefficient and ineffective in comparison with the executive branch. However, this slow action has benefits: It provides an opportunity for diverse views to be accommodated. This perspective on the value of public participation suggests that decisions made on behalf of the public by centralized agencies can be substantially enhanced by providing channels for public input.

There is a greater likelihood that more viable or innovative alternatives to a project will be identified by opening up the process to the public. Community members are well aware of their own resources, limitations (most often), and problems. The diverse perspectives of the community's citizens provide input that could otherwise be obtained only through extensive fieldwork by the agency sponsoring the project. There is, further, the possibility that there might be a closer integration of planning and development with existing area planning efforts in which major input has already been made by the public. A community may be expected to react unfavorably when previous input to other pertinent plans is summarily disregarded by agency planners and decision makers.

One executive branch agency, the U.S. Forest Service (Ketcham, 1988), worded its reasons for belief in the effectiveness of public participation, especially in the scoping process, because it

> builds agency credibility and public support
>
> provides an excellent opportunity for dispute resolution, even before documents are prepared and decisions made
>
> substantially reduces the number of subsequent appeals and law suits.

If the proponent of a major action feels that these are among the benefits of the process, then there seem to be few reasons to oppose its full implementation.

Active public participation may also ensure that the final product, which the community has helped to develop, will be successfully implemented. Implementation is much more likely where the community has taken an active concern in planning problems and has played an important role in generating and evaluating alternative solutions. An important spinoff from a positive program of public involvement is a positive public attitude not only toward the proposed project, but toward the agency as well.

11.16 Response to Public Participation Format Variations

The different methods of effecting various public participation activities are discussed in Section 11.14. It was noted that the public hearing is not very useful in conveying information to the public, although the fact that it is a "traditional" format makes it "comfortable" for the agency holding the hearing. This familiarity works in several ways, however. Just as the agency and its personnel know the processes and procedures, and little internal education or planning is needed, so is it familiar and comfortable to "professional objectors" who oppose the action. This may have some unintended results.

Consider the typical hearing. The presiding officer calls for a description of the proposal, including its alternatives, from a staff officer. The proponent then describes the benefits which he or she believes will accrue, and may also discuss how the known adverse effects are proposed to be managed. Some statements of support are made by persons and groups who favor the proposal. Then it is the turn for the opponents. Their statements of opposition and the reasons why they believe the action should *not* take place may be lengthy and noisy, and often take the form of a pep rally. In fact, this "rallying of the troops" aspect of a typical hearing, no matter how carefully the hearing is managed by the agency holding it, is often a high point in the week for the opposing forces.

What happens if the hearing is dropped in favor of one or more of the other techniques? We can say from experience that this departure from the "norm" is often viewed with dismay by the opponents. Their podium has been denied them, and the pep rally element so often a part of public hearings disappears. The organized opposition feels that their chance to generate support for their position has been unfairly taken away. In the authors' experience, this degree of resentment may be severe where the opposition groups have been planning on a major confrontation and find it "defused." They may go so far as to hold an alternative meeting at the same time (or at another time) where they may bring their message to their supporters.

The issue here appears to be one of mistaking the purpose of holding a public meeting. In NEPA terms, the purpose is clearly to provide information to the public, and to solicit inputs from them. As one may see from Fig. 11.1, we believe the "hearing" format to be relatively useless in soliciting input and in resolving apparent (or real) conflict. The less formal public meeting is somewhat better in meeting both objectives, and the open meeting is superior to both in many respects. Remember, however, that it may take weeks to create a good "open meeting" product, with many display tables and several teams of professionals to answer subject-specific inquiries, and require 100 to 200 hours of personnel time to execute each iteration. It is thus much more costly than the traditional "send three people and a note-taker to the high school for 2 hours" approach. A good planner will start early in the process to convince decision makers *within* the agency that the additional effort is worthwhile.

In defense of the more formal hearing, it may be noted that the creation of a transcript, usually similar to that prepared by a court reporter, may later provide needed confirmation that certain actions were taken or that a particular item of information was provided to those attending. For both proponents and opponents, the existence of a record of this testimony may be desirable for reference in later stages of the EIS processes, including its use in court.

11.17 Public Participation and the Internet Revolution

In the second half of the 1990s, the rise of Internet capability and the proliferation of web sites devoted to different purposes has brought sweeping change to many aspects of public participation and community involvement. There cannot be a significant project proposed by any agency in any western country (and many in the third world) which has not produced one or more web sites devoted to providing information about the project. This may be promotional information from the proponent or derogatory information from opposition persons or groups. The problems involved in providing more accurate information about a proposed action are both lessened by the ease of creation of an agency web site for this purpose, and increased by the dissemination of disinformation or negative commentary from opposing groups.

The whole practice of public involvement has been changed by this form of almost instantaneous information dissemination. In the twenty-first century, an agency is likely to receive comments on its draft EIS not only from the community, but from persons and groups in dozens of other states, and even from other countries. In almost every case, this level of nationwide or worldwide awareness has been made possible through the Internet and the World Wide Web. Is this good or bad? In one sense, the goal, required by regulation, of making the public aware of the consequences of a proposed action, and of providing an opportunity for them to provide input, is furthered by this development. In another sense, it has engendered a sort of competitive setting, in which groups, including the proponent agency, compete to provide more and more information and more and more replies to issues which arise. One clear outcome has been the need for the proponent to be ready to respond to quite sophisticated replies and queries, which may have been prepared by nationally known scientists brought into the controversy through e-mail solicitation and web postings. Some of these persons and groups appear to be "specializing" in opposing certain types of proposals based on their acknowledged area of expertise. It may be very difficult, indeed, to refute the calculations (or allegations) of a well-known scientist or former official who is providing an expert opinion.

11.18 Internet Capability to Support Public Participation

At one time, the concept of public involvement and public participation was not very well known. A large part of the effort of the environmental professional was often devoted to educating the publics and the agency on the requirements and responsibilities. To a great degree, both of these objectives may be said to have been well met. During the 1990s, especially, there was hardly a major action in which massive attention to the publics and their points of view was not one of the original planned activities. How has this come to pass? There are several contributing factors, and, again, the World Wide Web and Internet availability have played a major role.

A cursory search of the Web in early 2001 found that there were *at least* 11 major web sites devoted to dissemination of information about public involvement, and *at least* 42,000 web pages which related either to the process itself or to a specific action where public involvement was playing a role. When the search was shifted to "public participation," eight major sites and more than 75,000 web pages were found. It is worth noting that these sites included many major U.S. government agencies, where the information was focused on telling users what the public involvement responsibilities of the agency are and how to participate. Guidance of many types was also available for use by the agency on how to establish a public participation program. Of course, most of those tens of thousands of web pages relate to a specific action or proposal. These were by no means restricted to the United States, or even to North America. Tens of thousands of these pages related to actions and issues in Europe, Asia, South America, and Africa. Virtually every country which has an environmental assessment regulation of any type also must treat the challenge of public involvement in the decision-making process. In many of these instances, it is apparently *not* an element of law or regulation, but rather a self-generating process, where affected publics, often with support of an international *nongovernmental organization* (NGO), create an atmosphere of wishing to be heard. We note that, in many cases, negative focus is often taken on major development projects financed by the World Bank. In turn, this may have resulted in more assessment of proposals by the Bank and greater attempts to resolve localized social and economic issues which are associated with the massive projects. In this sense, the negative publicity may have been effectively applied from the point of view of the opponents and NGOs involved.

11.19   Discussion and Study Questions

1   Examine the many and varied terms used within the concept of public participation. What are the differences among participation, involvement, information, coordination, and input? What are the similarities? Why do you think the CEQ regulations use the term *involvement* for most instances where such actions are required?

2   In case study example 2 above, discuss what your response would have been to the flood of public comments which raised issues derived from actions not proposed to be taken, and which were outside NEPA. What would your agency have done?

3   Using your community as an example, how many "publics" are you able to identify? Define them and say how you have grouped persons and organizations into categories. Exchange these lists with a study partner, and discuss the way(s) in which the two treatments differ.

4   Taking the list of groups developed in question 3 above, propose two substantially different major actions in or near the community, making them different in character. Examples might be airport development, a water supply reservoir, urban redevelopment, an interstate highway connector, and a major manufacturing complex which required public financing. How might the definition of the various groups previously prepared differ between two different proposed projects?

5   Taking the same two projects defined for question 4 above, try to assign the different sets of publics into those who are likely to favor, oppose, or be neutral about the project. Discuss what this means about the definition of and accessing the opinions of "the public."

6 From the point of view of the agency, summarize what is gained and what is lost when there is an extensive public participation program in association with a proposal as opposed to a minimal program. In your opinion, are the benefits always worth the cost? Usually worth the cost? Seldom worth it?

11.20 Further Readings

Brown, Jennifer, ed. *Environmental Threats: Perception, Analysis, and Management.* London; New York: Belhaven Press, 1989.

Coenen, Frans, Dave Huitema, and Laurence O'Toole, Jr., eds. *Participation and the Quality of Environmental Decision Making.* Boston: Kluwer Academic, 1998.

Goldenberg, Sheldon, and J. S. Frideres. "Measuring the Effects of Public Participation Programs." *Environmental Impact Assessment Review,* 6:273–281, 1986.

Energy and Environmental Assessment

Critics of the environmental protection movement frequently blame "environmentalists," at least in part, for the energy crisis. While many of these claims are unfounded, it should be recognized that many interrelationships indeed exist. The production and consumption of energy both inevitably result in environmental consequences, and environmental protection measures also have effects on energy production and use patterns. For example, the shift away from coal following the 1970 Clean Air Act Amendments and what is effectively a moratorium on nuclear power plant construction have both resulted in increased oil and gas demand. Other examples originally cited included decreased gasoline mileage due to emission control requirements for new automobiles and delays in Alaskan pipeline construction and offshore drilling efforts.

U.S. energy problems reached crisis proportion with the OPEC oil boycott in October 1973, although many other factors contributed to the dilemma. This situation brought about an almost overnight recognition by the overall American public that energy sources are indeed finite and valuable. Furthermore, the situation pointed out that many of these resources are in short supply, and that significant progress is essential in areas of conservation and development of domestic supplies in order to meet projected demand requirements. Questions about the energy-environment relationship continue to be raised, and energy consideration in environmental assessment takes on an important role. Energy demands again came to the forefront in 2001.

Although the term *energy* is not specifically used, an easily inferred basis for the inclusion of energy considerations in environmental assessment may be found in several sections of Title 1 of the National Environmental Policy Act (NEPA). Recognizing that energy and fuels constitute a resource, perhaps the most obvious reference is made in Section 102(2)(C), where it is required that a detailed statement be made for federal actions on "...any irreversible and irretrievable commitments of resources which would be involved in the proposed action should it be implemented." Indirect implications are also included in Section 101(b)(6), where it is stated that the federal government has the continuing responsibility to "...enhance the quality of renewable resources and approach the maximum attainable recycling of depletable resources."

These sections thus suggest at least four areas where energy considerations become a part of environmental impact analysis. These areas are (1) commitments of energy as a resource, (2) environmental effects of fuel resource development, (3) energy costs of pollution control, and (4) energy aspects of materials recycling. The following sections examine these areas through relating energy considerations to the analysis of environmental impact.

12.1 Energy as a Resource

Energy resources include all basic fuel supplies that are utilized for heating, electrical production, transportation, and other forms of energy requirements. These resources may take the form of fossil fuels (oil, coal, gas, etc.), radioactive materials used in nuclear power plants, or miscellaneous fuels, such as wood, industrial wastes, municipal solid waste, or other combustible materials. Solar, hydroelectric, and wind resources or other energy sources currently in a developmental state may be important in particular projects.

When a proposed project consumes energy (and this is almost inevitable), this consumption should be considered as a primary or direct impact on resource consumption. Actions requiring consumption of energy can be categorized into (1) residential, (2) commercial, (3) industrial, and (4) transportation activities.

Residential activities include space heating, water heating, cooking, clothes drying, refrigeration, and air conditioning associated with the operation of housing facilities. Also included is the operation of energy-intensive appliances such as hair dryers, power tools, and the like. Most of these are not used for long periods of time, and so are less important, overall, than the first-level functions above.

Commercial activities include space heating, water heating, cooking, grain drying, refrigeration, air conditioning, feedstock heating, and other energy-consuming aspects of facility operation. Facilities that consume particularly significant amounts of energy include bakeries, laundries, and hospital services.

Industrial activities which inherently require large amounts of fuel resources include power plants, boiler and heating plants, and cold storage and air conditioning plants. Other industrial operations that require process steam, electric dryers, electrolytic processes, direct heat, or feedstock may have a heavy impact upon fuel resources.

Transportation activities involving the movement of equipment, materials, or persons require the consumption of fuel resources. The modes of transportation include aircraft, railroad, automobile, bus, truck, pipeline, and watercraft.

The most important variables to be considered in determining impacts on fuel resources are the rate of fuel consumption for the particular activity being considered and the useful energy output derived from the fuel being consumed. Various units may be utilized in describing consumption rates: Miles per gallon, cubic feet per minute, and tons per day are commonly used in describing the consumption of gasoline, natural gas, and coal, respectively. Similarly, the energy output of various fuel- and energy-consuming equipment and facilities may be described in many different units. Horsepower, kilowatthours, and tons of cooling are a few examples.

A common unit of heat, the British thermal unit, or Btu, may be applied to most cases involving fuel or energy consumption. A Btu is the quantity of heat required to raise the temperature of 1 pound of water 1 Fahrenheit degree. In the evaluation of transportation systems, for example, alternatives may be compared on a Btu per passenger-mile or a Btu per ton-mile basis.

Other variables of concern include the availability (short and long term) of fuel alternatives, cost factors involved, and transportation distribution and storage system features required for each alternative.

Data on the consumption of fuel resources may be applied to almost any environmental impact analysis, but the depth and degree to which such data are required depend upon the nature of the project under consideration. For an analysis of existing facilities or operations, sufficient information should be available from existing records and reference sources. Where alternative fuels or transportation systems are under consideration, additional background information may be necessary to evaluate not only efficiencies, but also cost-effectiveness and long-term reliability.

Because of the complexities in the nature of the variables discussed above, most are measured by engineers or energy economists, although the results may be applied by most individuals with technical training.

Once the heat contents of fuels are known, comparisons may be made on the basis of the heat content of each required to achieve a given performance. An energy ratio can be established as the tool for comparison, defined as the number of Btu's of one fuel equivalent to one Btu of another fuel supplying the same amount of useful heat.

$$\text{ER} = \frac{\text{amount of fuel No. 1 used} \times \text{heat content of fuel No. 1}}{\text{amount of fuel No. 2 used} \times \text{heat content of fuel No. 2}}$$

where ER = energy ratio

Determination of energy ratios requires careful testing in laboratory or field comparisons, but yields usually reliable results when conducted under impartial and competently supervised conditions. These ratios have been determined in various tests and are summarized in such publications as the *Gas Engineer's Handbook*.

The consumption of fuels for a particular use may be determined from procurement and operational records. Measurements may be made using conventional meters, gauges, and other devices.

The fuel resource data can be used in an environmental impact analysis for the benefit of planners and decision makers for either (1) evaluating the alternatives where either fuel consumption or fuel-consuming equipment or facilities are involved or (2) determining baseline fuel and energy consumption. This analysis includes the evaluation of irreversible and irretrievable commitments of resources resulting from the action, the short-term/long-term trade-offs, and the identification of areas for potential conservation and mitigation of unnecessary waste. The analysis would include evaluation of efficiency, availability, cost of fuel and support facilities (transportation, distribution, storage, etc.), and projected changes in these values that might occur in the future.

Conversion of fossil and nuclear fuels into usable energy can lead to both direct and secondary effects on the biophysical and socioeconomic environment. Some of the effects that may occur are listed in Table 12.1. These impacts would also be considered in the analysis.

If a project results in significant additional demands for waste of fuels already in short supply, public controversy may be expected to follow. Natural gas supplies, presently limited or unavailable in some areas, should be considered with special emphasis. Electric consumption, in most cases, bears directly upon fuel resources, the effects of which should be included in the analysis.

Concern for fuel resources typically peaks during summer (when air conditioning loads are high) and winter (when demand for heating fuels, especially fuel oil, is high). Thus, projects in northern climates would be expected to have the greatest concern for heating fuels, while in the south, the emphasis would be on projects with heavy cooling

TABLE 12.1   Environmental Effects Related to Energy Consumption

Environmental area	Environmental problems
Air	Pollutant emissions
	Carbon monoxide
	Sulfur oxides
	Hydrocarbons
	Nitrogen oxides
	Lead
	Mercury
	Other toxic compounds
	Smoke
	Smog
	Greenhouse gases
Water	Oil spills
	Brines
	Acid mine drainage
	Heat discharges
Land	Land disturbance
	Aesthetic blight
	Loss of habitat
	Subsidence
Solid waste	Leachates
	Radioactive waste
	Storage/disposal of waste

requirements in the summer, although exceptions to this may occur due to localized demands or geographical or climatic effects (e.g., *el niño* and *la niña* cycles). Proximity to natural supplies also plays an important role in fuel selection, since transportation may affect the availability and economic desirability of certain fuels.

Mitigation of impacts directly and indirectly attributable to energy and fuel resources falls into two categories. The first pertains to mitigation by alternate fuel selection and is based on a number of complex variables—availability, cost, environmental effects, and pollution control requirements, to name a few. Other factors to be considered in the selection are the short-term/long-term effects of a particular choice, and the irreversible and irretrievable commitment of resources associated with the selection. The second category of mitigation is associated with the conservation of energy, regardless of the type or types of fuel being consumed. These mitigations, however, bring up other environmental questions, as shown in the following sections.

Of the four categories of energy consumption (residential, commercial, industrial, and transportation), changes in transportation will have the most direct effect on individual populations. Transportation-related goods and services within the United States account for about one-tenth of the nation's gross domestic product, and the economy relies heavily on the low-cost, highly flexible movement of goods and services.

The U.S. transportation system is about 95 percent petroleum dependent and is the only sector of the economy that consumes significantly more petroleum today than it did in 1973. In 1997, oil demand driven by transportation uses, along with declining domestic production, gave rise to the highest levels of oil imports ever (CEQ, 1997). Over the 1990–1996 period, highway passenger-miles increased about 20 percent, while air passenger-miles grew about 24 percent; travel by other public transit stayed about the same, and rail travel declined slightly. Many factors have contributed to the increase in passenger miles traveled, including increases in U.S. population, the number of people in the labor force, and the number of people commuting to work (CEQ, 1997).

Americans are generally traveling more miles annually in their vehicles. In 1990, the average passenger car traveled 10,280 miles during the course of the year; by 1997, average vehicle-miles for passenger cars had increased to 11,575 miles. This can be partially attributed to changes in the labor force and income, as well as increases in the size of households and the number of vehicles per household. An increased number of households and vehicles leads to more trips for shopping, recreation, and taking care of children. Private vehicle trips soared as metropolitan areas expanded and low-density suburbs spread into rural areas, offering more mobility and direct connections between destinations (CEQ, 1997).

The costs of mobility, however, are not paid directly by the individuals and businesses who are the beneficiaries. Transportation has a significant impact on environmental quality in a wide variety of ways, including air quality, land use and development, habitats and open space, and energy use. The form and shape that cities and suburban areas take in the next several decades will affect future mobility and air quality. Certain land-use and transportation strategies can lead to a reduction in vehicle trips and vehicle-miles traveled by allowing a shift to other modes of travel, especially in congested urban and suburban areas. Such strategies can make it easier for people to walk, bicycle, or use transit service (rail or bus), instead of relying primarily on automobiles for mobility. To gain a better understanding of the benefits of transportation and land-use strategies in reducing vehicle use and related emissions, the California Air Resources Board funded a research study entitled "Transportation-Related Land Use Strategies to Minimize Motor Vehicle Emissions: An Indirect Source Research Study." The study recommended a set of transportation-related land-use strategies that are designed to assist communities in achieving improved environmental quality. These strategies are presented below:

- *Strong downtowns:* A strong commercial and cultural center (not solely offices generating only workday traffic) can become a focal point for a regional transit system and can facilitate pedestrian travel.

- *Concentrated activity centers:* Combining higher-density development into concentrated areas increases the opportunity for providing and using more efficient transit service and also facilitates pedestrian travel.

- *Mixed use development:* Locating different types of compatible land uses in close proximity to one another or within a single building can result in higher levels of walking, as compared to segregated single-use projects.

- *Redevelopment and densification:* Encouraging the redevelopment and reuse of vacant or underutilized property within developed areas also supports the use of transit systems.

- *Increased density near transit stations and corridors:* Intensifying land uses within ¼- to ½-mile walking distance of existing or planned high-capacity transit stations and corridors encourages higher levels of transit use.

- *Pedestrian/bicycle facilities:* Providing good pedestrian accessibility supports the other strategies and can reduce vehicle travel. This strategy includes adequate and direct sidewalks and paths, protection from fast vehicular traffic, pedestrian-activated traffic signals, traffic calming features, and other amenities.

- *Interconnected travel networks:* Ensuring direct routes for vehicles, pedestrians, and bicycles can result in slower vehicle speeds while maintaining travel times that are comparable to current street patterns.

- *Strategic parking facilities:* Parking availability should be adjusted to reflect increased rates of transit use, walking, and bicycling that result from implementing the strategies listed above. Ideally, the amount and cost of parking should vary according to the type and location of land use.

Implementation of these strategies could have significant long-term environmental benefits. The air quality improvements that may result from implementing these strategies depend on a number of factors, including whether a community is urban, suburban, or exurban. For example, they could help reduce air emissions from mobile sources, which, to date, are attributable almost entirely to technological advances and to regulatory requirements. Other environmental elements that may be positively affected include noise levels, fuel consumption, aesthetics, and environmental health.

12.2 Fuel Alternatives and Development of Supplies—Environmental Considerations

Not all fuel alternatives produce the same effects on the environment either directly or indirectly. The CEQ undertook a study of the environmental impact of electric power alternatives and concluded that such a comparative discussion is useful for discussion purposes and provides a basis for further analysis. CEQ recognized the difficulty in making comparisons of very different systems and stressed that regional differences, emission control variability, and other factors should be considered in each individual case. Next, each fuel is examined for specific environmental effects as presented in Tables 12.2 through 12.5.

Coal: Although coal is our most abundant fossil fuel resource, its use in the production of electrical energy is judged the most environmentally damaging of alternatives. Table 12.2 details some of the problems associated with the use of coal.

Oil: Environmental effects of oil are different in both character and magnitude from those of coal, as may be seen in Table 12.3.

TABLE 12.2   Environmental Effects of Use of Coal

Operation	Major environmental effects
Surface mining	Land disturbance Acid mine drainage Silt production Solid waste Habitat disruption Aesthetic impacts
Underground mining	Acid drainage Land subsidence Occupational health and safety Solid waste
Processing	Solid waste stockpiles Wastewater
Transportation	Land use Accidents Fuel utilization
Conversion	Air pollution Sulfur oxides Nitrogen oxides Particulates Greenhouse gases Carbon dioxide Solid wastes Thermal discharge
Transmission lines	Land use Aesthetics

TABLE 12.3   **Environmental Effects of Use of Oil**

Operation	Major environmental effects
Extraction	Land use (drilling)
	Spillage
	Brine disposal
	Blowouts
Transportation	Land use (pipelines)
	Leakage and rupture (pipelines)
	Spills
Refining	Air pollution
	Water pollution
Conversion	Air pollution
	Sulfur oxides
	Nitrogen oxides
	Hydrocarbons
	Greenhouse gases
	Carbon dioxide
	Thermal discharge
Transmission lines	Land use
	Aesthetics

TABLE 12.4   **Environmental Effects of Use of Natural Gas**

Operation	Major environmental effects
Extraction	Land use (drilling)
	Brine disposal
Transportation	Land use (pipelines)
Processing	Air pollution (minor)
Conversion	Air pollution (relatively minor)
	Carbon monoxide
	Nitrogen oxides
	Greenhouse gases
	Carbon dioxide
	Methane
	Thermal discharge
Transmission lines	Land use
	Aesthetics
	Safety hazards

Gas: Natural gas is significantly more desirable from a pollution production standpoint, although not problem-free, as may be seen from Table 12.4.

Nuclear fission: A different set of environmental effects results from the nuclear fission process, as indicated in Table 12.5. The accident potential in conversion and disposal represents a highly controversial issue in evaluating nuclear fission utilization, although the short-term effects of operation are notably less polluting.

TABLE 12.5   Environmental Effects of Use of Nuclear Fission

Operation	Major environmental effects
Mining	Land use (not extensive)
Milling (separation)	Radioactive wastes Air Water Solid waste
Enrichment	Minor release of radioactive material
Conversion	Thermal discharge Release of radionuclides (minor) Accident potential
Transmission lines	Land use Aesthetics
Reprocessing	Radioactive air emissions
Radioactive waste disposal	Accident potential (handling, storage)

Fuel selection must be made on the basis of many factors in addition to environmental consequences. The cost, availability, and facilities and equipment requirements also must be considered, as well as the political acceptability. Both short- and long-term aspects would be included in the life-cycle analysis of a proposed system, and decision makers should consider all aspects in making fuel selections.

To mitigate or reduce the adverse environmental effects of energy and fuel utilization, various procedures have been initiated, some of which have been controversial, and/or have not been effective, and/or have resulted in further limitations in fuel supplies. For example, supplies of some fuels have, effectively, been restricted or limited through such actions as strip mine regulations and limitations on oil drilling and exploration, particularly in offshore coastal waters and in the Alaskan wildlife refuges.

The conservation of energy may be accomplished through (1) voluntary means, such as cutbacks in heating and lighting use, (2) economic incentives, such as taxation, or (3) legislative means, such as mandatory speed limits. Conservation will undoubtedly continue to play a key role as the nation moves toward energy self-sufficiency.

Conservation measures may vary greatly with project type and magnitude. Such measures can be applied to new construction, in the form of additional insulation and design, incorporating energy conservation features related to color, orientation, shape, lighting, etc. Conservation of energy can be applied to existing facilities, in the form of added insulation and programs to reduce loads on heating, cooling, and other utility consumption. Likewise, in the operation and maintenance of equipment, steps may be taken to reduce fuel consumption further by increasing efficiencies through proper equipment maintenance, reducing transportation requirements, and scheduling replacement of old equipment with newer, highly efficient models.

Special efforts toward energy conservation should be pointed out in an environmental impact statement because, generally, adverse impacts on the biophysical environment tend to become reduced with decreased energy production and consumption. However, some question arises as to the socioeconomic effects of a substantially lowered growth rate of energy consumption.

12.3   Energy Costs of Pollution Control

Energy requirements for the operation of pollution control systems are an area of conflict that probably will continue to be present as long as pollution regulations are in effect.

Generally speaking, the energy requirements for various aspects of pollution control will vary with type of process, quantities involved, and degree of treatment or removal. Energy required to meet pollution control regulations at stationary sources in 1977 amounted to about 2 percent of total U.S. energy consumption (Serth, 1977). This requirement may have increased by as much as 50 percent during the 1990s. Since generation of energy produces many adverse environmental effects, any increased consumption of energy to control pollution may reduce the net pollution control benefits.

If reduction of net environmental degradation is the main goal, two strategies are suggested. First, marginal benefits from stronger pollution control requirements should be compared with marginal costs, including environmental consequences of increased energy consumption. Second, research and technology development efforts should be focused on high energy consumptive industrial categories and pollution control processes. Industrial categories include primary metals, chemicals, paper and paper products, and petroleum and coal products. Pollution control processes include municipal wastewater treatment and control of sulfur oxides from industrial and utility boilers (Serth, 1977).

12.4   Energy Aspects of Recycling Materials

The recycling of materials such as paper, metals, and glass has long been known to reduce environmental problems such as solid waste and litter, while at the same time conserving supplies and preserving resources. In addition, a renewed look at recycling has come about as a result of the energy aspect of materials manufacture.

Some indication of the potential for energy conservation may be determined by examination of the energy requirements for various

TABLE 12.6   Fuel Consumption by End-Use Sector 1987, 1990, 1993, 1996, and 1999, in Quadrillion Btu's

Sector	1987	1990	1993	1996	1999
Residential and commercial	28.49	29.48	31.12	33.67	34.17
Industrial	29.68	32.15	33.30	35.71	36.50
Transportation	21.46	22.54	22.89	24.52	25.92
By fuel					
Coal	2.83	2.92	2.64	2.56	2.36
Natural gas	14.27	15.72	17.45	19.02	18.25
Petroleum	31.61	32.30	32.79	35.03	36.74
Electricity	8.37	9.24	9.74	10.56	11.12

SOURCE: Energy Information Administration/Annual Energy Review, 1999.

sectors as indicated in Table 12.6, and the distribution of energy consumption in the manufacturing sector shown in Table 12.7. The primary products industries (food, paper, chemical, petroleum, stone, clay and glass, and primary metals) in 1971 accounted for over 83 percent of the energy consumed by manufacturing (FEA, 1974), a proportion which was almost unchanged in 1994 (see Table 12.7). In the face of fluctuating energy prices and great uncertainty surrounding the promise of future supplies, these industries are forced to examine programs to improve their energy efficiencies. One approach that is advocated is the greater use of recycled materials.

Recycling and recovery of materials from waste streams depends, in the practical world of business, primarily on economics. Depletion allowances, capital gains treatments, transportation costs, and other factors have had the effect of inhibiting a greater movement toward recovery efforts. Increases in energy costs along with increases in material costs and shortages in many materials can stimulate recycling through the creation of new markets and increasing the demand for certain recycled products. Explicit governmental policies to use recycled products can also provide a portion of the necessary impetus for the recycling industry.

Recycle of specific materials

The "recyclability" of different basic materials differs greatly. In the following discussion, many of the energy, economic, and environmental considerations associated with the recycling potential of several basic materials are examined. Remember, however, that development of new technology which enables recycling, actions which artificially limit supplies of virgin materials, and legislation which allows or prohibits use of specific manufacturing processes may alter this picture almost overnight.

TABLE 12.7   Manufacturing Total First Use of Energy for All Purposes, 1994, in Trillion Btu's

SIC code	Major group	Total consumption	Net electricity*	Fuel oil**	Natural gas	Coal and coke
Total		21663	2656	648	6835	2554
20	Food and Kindred Products	1193	198	49	631	165
21	Tobacco Products	W†	3	1	W†	W†
22	Textile Mill Products	310	111	24	117	40
23	Apparel and Other Textile Products	W†	26	1	25	W†
24	Lumber and Wood Products	491	68	27	48	W†
25	Furniture and Fixtures	69	22	1	24	3
26	Paper and Allied Products	2665	223	182	575	307
27	Printing and Publishing	112	59	2	48	0
28	Chemicals and Allied Products	5328	520	124	2569	304
29	Petroleum and Coal Products	6339	121	93	811	W†
30	Rubber and Miscellaneous Plastics Products	287	149	14	110	5
31	Leather and Leather Products	W†	3	2	W†	0
32	Stone, Clay, and Glass Products	944	123	30	432	282
33	Primary Metal Industries	2462	493	56	811	1346
34	Fabricated Metal Products	367	115	4	220	W†
35	Industrial Machinery and Equipment	246	109	4	111	11
36	Electronic and Other Electric Equipment	243	113	5	88	W†
37	Transportation Equipment	363	132	18	157	30
38	Instruments and Related Products	107	46	5	29	0
39	Miscellaneous Manufacturing Industries	W†	19	2	19	1

*Net electricity is obtained by aggregating purchases, transfers in, and generating from noncombustible renewable resources minus quantities sold and transferred out.
**Includes distillate and residual.
†W = Withheld to avoid disclosure of data for individual establishments.
SOURCE: Energy Information Administration/Annual Energy Review, 1999.

Glass.   Glass can be recycled back into glass furnaces, but difficulties in the glassmaking operation present problems which make recycling unattractive in many cases. First, glass "formulas" include not only silica but limestone, soda ash, and, in many cases, coloring agents that are blended, melted, and refined in precise operations. Reclaimed glass necessarily results in the blending of formulas and the inclusion of many foreign substances, the end products of which are highly unpredictable. As a consequence, recycled glass is considered usable for only a limited range of products, which offsets much of any cost saving.

Some glass products are manufactured with about 25 percent cullet (waste glass) as a component. The use of cullet reduces energy consumption in two ways: (1) The heat required to melt cullet may be 33 to 50 percent less than that required to produce glass from the virgin raw materials, and (2) the use of cullet requires the addition of fewer additives, thus saving the energy required to mine the inorganic chemicals usually added. These energy savings from the use of cullet are partially offset, however, by the energy required to collect, beneficiate, and transport waste glass (Renard, 1982).

The separation of glass from other wastes poses a second problem to glass recycling. This process may vary from simple hand classification, accomplished during time of collection, to complex automated separation operations employing air classification, dense media separation, or froth flotation. Color separation must also be accomplished and may be done at time of collection or via automated optical systems. So-called source separation, where glass of different colors is separated at each household, is a feature of many U.S. community recycling programs. The separation may take place in each home, for curbside pickup, or may be accomplished at the time of drop-off at neighborhood centers. In Europe, especially Germany, this is accomplished through placement of large metal bins in densely populated neighborhoods. Three bins are provided, one for each glass color, green, brown, and white (clear), and each station serves several thousand residents. The cullet obtained through this separation is much more likely to be useful than mixed materials containing different colors.

Utilization of returnable bottles and containers assures that the effective use of a given container will be greatly increased, thereby decreasing the necessity for more containers and the waste produced as each container is emptied. Discouragement of "throwaway" containers promotes not only less waste production, but less energy expenditure for manufacturing as well. When the total energy consumption involved in collecting, returning, washing, and refilling glass bottles is compared to that required in delivering the same volume of beverage to the consumer in a throwaway container, a significant energy savings is apparent. One study has indicated that "…a complete conversion to returnable bottles would reduce the demand for energy in the beverage (beer and soft drink) industry by 55 percent, without raising the price of soft drinks to the consumer" (Hannon, 1972). Unfortunately, bottling companies see mandated recycling, especially through use of deposit containers, as an unmitigated horror. They lobbied successfully against deposits and returnable containers in many states in the 1980s. Reclaimed glass may be used for secondary products other than glass containers, such as for aggregate in road construction, manufacture of insulating materials, or brick production.

Tires and rubber products. Rubber is a natural forest product resource that is critical to military and civilian transportation and to the production of mechanical rubber goods. Natural rubber is used primarily for tire production, and approximately 70 percent of the world's production comes from Southeast Asia (IRSG, 2000). Disposal of tires and other rubber products represents a potential loss in several ways. Disposal represents a problem from an economic standpoint of collection, shipment, storage, and ultimate disposal. Disposal of rubber goods presents several environmental questions, as the long-term effects of slowly disintegrating tires and rubber products have not been determined. Large piles of discarded tires have caught fire, causing air and water pollution effects. Recognizing these problems, many states now prohibit the disposal of whole tires in municipal landfills.

Recycle and reuse potential for scrap tires and rubber products include (1) direct reuse as artificial reef construction, (2) reprocessing for retreaded tires or other rubber products, (3) alternate use such as in road surfacing, or (4) use as a fuel in boilers. All these represent a possible resource enhancement or savings, and some are directly or indirectly related to energy savings as well. Only a minority of discarded tires are reused in any way, however.

Paper. Recycled paper can be manufactured relatively easily, with end products competitive in quality to those made from virgin materials. Major difficulties arise from the economics of collection and transportation of waste paper products to centers for reprocessing.

Shredded wastepaper and other forms of wastepaper products may be utilized as packaging material or as mulches for erosion control, or may form a portion of compost material for soil enrichment. When solid waste is utilized for incineration and heat recovery, the paper and cardboard content provides much of the energy content which is converted to heat.

Estimates of energy savings that can be realized due to recycling of paper products vary greatly. Most studies indicate that energy savings of 7 to 57 percent are possible for paper products such as newsprint, printing paper, packaging paper, and tissue paper. On the other hand, paperboard products require *more* energy (40 to 150 percent more) when manufactured from recycled material (OTA, 1989).

Metals. High costs of metals and metal products have resulted in extensive programs to reclaim stainless steel, precious metals, lead, and copper in particular. Significant amounts of steel, aluminum, and zinc are also recycled, but not to the extent that could, or perhaps should, be returned for reuse. As with other waste materials, metal recycling reduces the quantity of solid waste to be disposed of, reduces the consumption of natural resources, and further reduces the energy requirements for the production of manufactured products.

Steel. Studies done in the 1970s indicate that about 75 percent of the energy required to produce raw steel from ore is saved in the production of steel from scrap metal. When the mining, beneficiation, and transportation processes are also considered, the energy savings drops to 47 to 59 percent. The production of finished steel from scrap reduces energy consumption by about 45 percent overall (Renard, 1982).

Aluminum. Recycling aluminum has a natural economic impetus because of the high energy costs associated with producing primary aluminum. Manufacturers have voluntarily established recycling centers for aluminum soft drink cans since the 1980s. In some areas, up to 50 percent of all cans sold are recycled. This is an exceptional success in view of the general failure of many container recycling efforts. The recovery of aluminum from scrap saves up to 90 to 95 percent of the energy required to produce the same product from alumina (Renard, 1982).

Plastics. Making products from recycled plastic can save considerable energy. The use of recycled resin in plastics manufacture can reduce energy consumption by 92 to 98 percent of the energy required to produce virgin resins (OTA, 1989). Some of these energy savings will be reduced when energy required to collect and transport the used containers is included. Lack of collection is the major factor limiting plastics recycling (OTA, 1989). One effort started in the 1990s was the uniform marking of plastic containers so that their resin classification may be easily determined, and delivery back to processing of a more uniform batch of cullet is possible. This increases usefulness to the manufacturer, and will probably increase the price manufacturers are willing to pay for the used material.

Oil wastes. Waste oil and petroleum products originate from crankcase and lubrication wastes generated during the normal maintenance of motorized vehicles and machinery. Waste oils may be used directly without reprocessing as road oils for dust control, or may be mixed with virgin fuel oil for use in boilers for heating or electrical power generation. Emissions of heavy metals and other related environmental problems should be carefully evaluated before burning or otherwise recycling waste oil.

The process of refining waste oil to produce lubrication oils or fuel oils is technologically possible and currently is being practiced in many areas. Difficulties in removing impurities of lead, dirt, metals, oxidation products, and water, along with environmental standards and product specifications, have hampered the widespread practice of recycling in the past. However, the improvement of recycling technology, coupled with economic incentives, may result in a resurgence in recycling of petroleum products in the near future.

Waste oil and its impurities possess potential threats to the environment, whether the waste oil is indiscriminately dumped on land or into water courses or burned. Even the refining process may produce acid sludges and contaminated clays that must be disposed of in a manner that is environmentally safe.

General solid waste.   Municipal solid waste has been termed by some an "urban ore" with a great potential for materials and energy recovery. Currently, a great variety of approaches are being investigated and demonstrated to tap this potential resource. Typical content includes the following:

Paper	Lead
Glass	Textiles
Ferrous metals	Rubber
Aluminum	Plastics
Tin	Food, animal, plant, and other wastes
Copper	Miscellaneous materials

In addition to materials recovery and the potential savings represented, many solid wastes may be incinerated with significant energy recovery. The energy value produced in 1990 through energy generation from 128 waste-to-energy plants in North America has been estimated at approximately equivalent to 27 million barrels of oil per year (Kiser, 1990).

12.5   Discussion and Study Questions

1   Do the pollution control laws in your state encourage or discourage industrial expansion? What would be the consequences of relaxing their requirements? Of tightening them?

2   Consider the electricity you are currently utilizing for lighting, heating/cooling, etc. Tracing back through transmission, conversion, and extraction/

transportation of the original fuel source, what are the environmental effects resulting from its consumption, and where are they occurring?

3 Compare the potential positive and negative environmental effects of the following actions:

   a A statewide ban on nonreturnable beverage containers.

   b A regulation requiring all federal (or state) agencies to utilize *only* recycled paper.

   c A requirement that all gasoline sold in your state contain at least a minimum amount of alcohol distilled from grain or similar product.

   d A requirement for state-owned and all commercial fleet vehicles to convert to alternative fuels (liquid petroleum gas, natural gas, ethanol, etc.).

   e A proposal to allow homeowners to tap into the municipal water distribution system to use potable water as a heat sink for residential heat exchanger units.

4 Does your community currently have a recycling program?
   a If so, find out how it is structured and financed. Is it successful? What are the measures of success? What environmental trade-offs are associated with the program? Identify any problems it is experiencing and suggest ways in which it could be further improved.

   b If not, outline a program you believe would be successful. Anticipate any problems that would be encountered and suggest ways to overcome them.

12.6   Further Readings

Chadwick, M. J., N. H. Highton, and N. Lindman, eds. *Environmental Impacts of Coal Mining and Utilization.* New York: Pergamon Press, 1987.

DOE. "Alternatives to Traditional Transportation Fuels: An Overview." Publication No. DOE/EIA-0585/O, June 1994.

DOE. "Wind Energy: Incentives in Selected Countries." November 1998, by Louise Guey-Lee, DOE/EIA, in *1998 Renewable Energy Annual, 1998: Issues and Trends,* 1999, DOE/EIA.

The Environmental Developments Impacts of Production and Use of Energy: An Assessment Prepared by the United Nations Environment Programme, Essam E. El-Hinnawi, study director. Published for the United Nations Environment Programme by the Tycooly Press Ltd. and printed by Irish Printers Limited, Shannon, 1981.

Environmental Impacts of Renewable Energy. OECD Compass Project. Paris: Organisation for Economic Co-Operation and Development, 1988.

Fowler, John M. *Energy and the Environment.* New York: McGraw-Hill, 1975.

Gordon, Judith G. "Assessment of the Impact of Resource Recovery on the Environment." EPA-600/8-79-011, U.S. Environmental Protection Agency, Cincinnati, Ohio, 1979.

Hohmeyer, Olav, and Richard L. Ottinger, eds. *External Environmental Costs of Electric Power—Analysis and Internalization.* Proceedings of a German-American Workshop held at Ladanburg, FRG, October 23–25, 1990. New York: Springer-Verlag, 1991.

Odom, H. T., and E. C. Odom, *Energy Basis for Man and Nature,* 2d ed. New York: McGraw-Hill, 1981.

Porter, Richard, and Tim Roberts, eds. *Energy Savings by Wastes Recycling.* New York: Elsevier, 1985.

United Nations, *Environment and Energy.* New York: Pergamon Press, 1979.

Winteringham, F. Peter W. *Energy Use and the Environment.* Chelsea, Mich.: Lewis Publishers, 1992.

Contemporary Issues in Environmental Assessment

The range of issues which may need to be considered while preparing an environmental assessment is very large indeed. Some become relatively more important at one time than at another, while new problems arise constantly. It is for this reason, among others, that it is difficult to build into legislation or regulations a required set of items to be covered in every case. We present here seven contemporary issues which are currently important, and suggest ways in which consideration of these problems enter into an assessment. There are certainly many other problem areas which may be more important in certain instances, but each of these has some history of being relevant to national and international decision making. Issues examined are global warming, acid rain, deforestation, endangered species, biodiversity, cultural resources, and ecorisk.

13.1 Global Warming

Swedish chemist Svante Arrhenius (1896) coined the term *greenhouse effect* at the turn of the century. He postulated that if certain gases such as carbon dioxide were to be increased in the atmosphere due to combustion of fossil fuel, this would allow sunlight to penetrate, but retain outgoing infrared radiation, in a manner analogous to a greenhouse; this could cause appreciable global warming.

Scientific agreement

In the natural functioning of the earth's climate, atmospheric gases, most importantly water vapor and carbon dioxide, and less importantly methane, nitrous oxide, and ozone, trap solar heat reflected from the earth's surface and prevent it from escaping into space. Without this natural greenhouse effect, the earth would be 33°C cooler and could not support life as we know it.

As a result of industrial activity in the past century, however, atmospheric concentrations of these natural greenhouse gases, and other synthetic gases with similar effects, have increased. Combustion of fossil fuels and industrial use of synthetic gases in developed countries, and deforestation in developing countries, have released ever-increasing levels of greenhouse gases into the atmosphere. The scientific community agrees that carbon dioxide levels have risen 20 percent in the past century, and there is general agreement that the earth's global mean temperature has risen 0.3 to 0.6°C in the same time period (Abrahamson, 1989). If there is indeed a causal relationship between increased levels of greenhouse gases due to human activity, and warming on a global scale, the human community may be faced with changes in the earth's climate and resultant disruptions of our human and natural environments on an unprecedented scale.

Uncertainties

Yet there is no certainty about the challenge which may face us: that it will happen, when it may happen, and just how severe it might be. Although it is established that carbon dioxide emissions are expected to increase if no action is taken to limit them, we do not have scientific information adequate to predict confidently how the earth's climate will respond to this increase.

The Intergovernmental Panel on Climate Change (IPCC), a body of scientists and other experts from 30 countries, in 1990 produced a Scientific Assessment of Climate Change, which indicates the consensus that has been reached by the scientific community on the certainties and doubts surrounding this issue. The panel predicted that global mean temperature will increase by 0.2 to 0.5°C per decade to total 1°C higher by 2025 and 3°C higher by 2100. The panel acknowledged that our incomplete understanding of the impact of clouds, oceans, and polar ice sheets on earth's climate make this prediction very uncertain with regard to timing, magnitude, and regional changes (OTA, 1991). At the end of 1995, however, the IPCC released its Second Assessment Report. The report concludes that "the balance of evidence suggests that human activities are having a discernible influence on global climate."

At a 1985 conference sponsored by the United Nations Environmental Programme and the World Meteorological Organization, scientists concurred that a rise in mean global temperature between 1.5 and 4.5°C will accompany a 50 percent increase in carbon dioxide levels, with the midrange prediction being 3°C (Abrahamson, 1989).

Although many scientists accept the prediction that increased levels of greenhouse gases will cause global mean temperature rise, there are some dissenters. Kenneth Watt of the University of California at Davis maintains that greater cloud cover due to higher carbon dioxide levels may cause global cooling rather than warming. Reid Bryson at the University of Wisconsin maintains that dust and smoke, and not carbon dioxide levels, are the cause of climate change (Anderson and Leal, 1991).

Some of the uncertainty in predicting the extent of temperature rise is due to the role of the oceans, which may absorb excess heat and delay or offset higher temperatures. Vegetation may take up some portion of the carbon dioxide, and ice caps may melt at rates we cannot predict. Even small changes in cloud cover may affect global temperatures, and this mechanism is not well understood. Recent measurements show a lower concentration of carbon dioxide in the atmosphere than predicted, and there is speculation that increased growth of vegetation in some areas has caused some of the released carbon dioxide to be fixed as plant tissue through photosynthesis. Few facts can be stated with certainty.

Effects of global warming

Current climate models are inadequate to confidently predict the regional effects of a 3°C global mean temperature rise, although some tentative predictions can be made. Tropical areas may experience a smaller temperature rise with decreased rainfall in dryer regions and greater rainfall in moist regions; higher latitudes will experience the largest temperature increase; summer dryness will be more frequent in the middle latitudes of the Northern Hemisphere; and due to the expansion of ocean water and melting of polar ice, sea level may rise 20 to 140 cm (Abrahamson, 1989). Regional areas may experience changing patterns of temperature, storms, winds, and rainfall. Tropical hurricanes may become more frequent and severe. Flooding may be exacerbated in coastal areas.

Changes in weather patterns would directly affect agriculture, forestry, and natural ecosystems. Some agricultural areas and forest species may lose productivity, while others may benefit, shifting the current patterns of food and timber production. Such shifts may disrupt the equilibrium of present economies, both within national boundaries and among nations. Food supplies for some countries may be threatened, and the present network of trade relationships could be altered to favor some nations and disadvantage others. The effect of warming on recreation and tourism may also be mixed, shifting advantages and disadvantages among various geographic regions.

Sea level rise may account for the most extensive and expensive damage caused by global warming. A rise in global sea level of 1 meter would cause the loss of 5000 to 10,000 square miles of land in the United States, affecting more than 19,000 miles of coastland. The built structures and transportation, power, water, and drainage support systems of developed coastal areas may suffer severe damage (Titus, 1990).

Prevention strategies

The different greenhouse gases make varying contributions to the greenhouse effect, and are released by varying activities; thus, a policy mix will be needed to reduce the total atmospheric level of greenhouse gases. Carbon dioxide emissions account for 55 percent of the total warming effect of greenhouse gases, and "excess" releases are largely the result of fossil fuel consumption and biomass burning. Remember that carbon dioxide is the normal result of all plant and animal metabolism, however, and not all releases are from artificial sources. Because carbon dioxide is stored in biomass form in forests, the level of atmospheric carbon dioxide is also elevated by deforestation. Chlorofluorocarbons, synthetic chemicals used in air conditioning, refrigeration, insulating foams, aerosols, and solvents, contribute 24 percent. Methane, contributing 15 percent, is produced by anaerobic decay of organic matter in moist areas, such as in rice farming, and by ruminant animals. Nitrous oxide also results from fossil fuel consumption, particularly coal, and contributes 6 percent (OTA, 1991).

Earth Summit.   Because global warming is a global rather than a local phenomenon, it is necessary to incorporate international agreements in developing prevention strategies. At the Earth Summit in Rio de Janeiro in 1992, more than 150 governments signed the Framework Convention on Climate Change. Developed countries agreed to the "aim" of returning their greenhouse emissions to 1990 levels by the year 2000. Developing countries agreed to prepare inventories of emissions and strategies to mitigate climate change with financial support from the industrialized nations.

This largely voluntary effort, however, proved insufficient. By the end of 1997, emissions had increased in all but a few developed nations and prospects for meeting the year 2000 target were poor. In 1997, at the third Conference in Kyoto, Japan, more than 160 nations developed a Protocol to the convention. Under the Protocol, industrialized nations agreed to reduce their aggregate emissions of six greenhouse gases by at least 5 percent below 1990 levels in the period 2008–2012. Developing countries do not have a legally binding obligation to reduce greenhouse gas emissions under the Protocol. Programs such as emissions trading, joint implementation, and the Clean Development Mechanism are intended to provide flexibility to make these reductions both at home and abroad.

Methane (CH$_4$).   It is unlikely that U.S. methane production can be significantly reduced. Beef and dairy farming is responsible for most U.S. methane production; large reductions in the cattle population or dramatically improved animal waste management practices would be required to reduce methane emissions. Neither effort is believed likely to be well received or productive (OTA, 1991).

Nitrous oxide (N$_2$O).   Most U.S. nitrous oxide release is due to the use of nitrogen fertilizers in agriculture. Some reduction of emissions could be obtained with policies that discourage monocropping and heavy fertilizer use (OTA, 1991).

Chlorofluorocarbons (CFCs).   Limitation of CFC emissions promises to bring high returns from policy efforts. Substitute chemicals are already available for some CFC compounds, and others are under development. Technology exists for the recapture of CFCs from products currently in use, and may support a market for recycled gases. The chemicals remain in the atmosphere 75 years, exercising a powerful greenhouse effect and causing depletion of stratospheric ozone. Thus, limiting CFC emissions will provide substantial environmental benefit at a low level of economic hardship. Effective January 1, 1994, the EPA issued an Accelerated Phaseout Schedule for Class I Substances (including CFCs). This schedule limits the production of CFCs, in terms of percentage of baseline production allowed, to 25 percent in 1994 and 1995 and 0 percent for 1996 and beyond. The CFC phaseout not only will affect greenhouse gas concentrations, but will also have a direct effect on the protection of the stratospheric ozone layer, reduced health risks, and pollution prevention.

CFCs and the Montreal Protocol.   However, vigorous action is required; global emissions can be stabilized at present levels only with an 85 percent reduction in CFCs. The United States, Sweden, and Norway have already banned nonessential aerosol uses of CFCs. In 1987, an international agreement was reached in Montreal to address limiting CFC emissions worldwide. The Montreal Protocol has been ratified by over 100 nations and came into force as of 1990. It targeted a scheduled phaseout of the most damaging CFCs by 2000, employing a marketable permit system to raise CFC prices and encourage the use of substitute chemicals, the recovery of gases from used products, and reduction in overall use. It has been estimated that a 50 percent reduction of CFC levels in the United States will cost $0.6 billion, which is minimal (Hoeller, Dean, and Nicolaisen, 1991). Although the Montreal Protocol represents a landmark in international environmental cooperation and protection, there are already indications that more stringent targets are needed. The agreement was designed to extend leniency to the Soviet

Union, eastern Europe, and developing countries; these exceptions may need revision especially since the breakup of the Soviet Union.

Carbon dioxide. Controlling atmospheric carbon dioxide levels, which account for 55 percent of the greenhouse effect, is the primary focus of global warming prevention policy. Developing countries influence atmospheric carbon levels chiefly through deforestation, which removes from the global equation the carbon-storing function provided by forests. Industrial countries influence atmospheric carbon levels chiefly through the burning of fossil fuels, which releases carbon dioxide directly into the atmosphere. The contribution of carbon dioxide from third world fossil fuel consumption will undoubtedly increase dramatically in the future as the countries pursue development.

Of total carbon released worldwide, 6 billion tons are due to fossil fuel use, and 0.5 to 3.0 billion tons are due to deforestation (with accompanying burning of the plant material) (Hoeller, Dean, and Nicolaisen, 1991). It has been estimated that trees in active growth sequester carbon dioxide at a rate of 6 tons per hectare (Sedjo, 1990, cited by Hoeller, Dean, and Nicolaisen, 1991). The net sequestering of carbon slows at maturity, and the stored carbon is released when the trees decompose or are burned. To maintain continuity in carbon storage, forests must be regularly renewed.

Technological developments. The extent to which greenhouse gas emissions can be reduced by climate-friendly technologies will depend on how quickly and thoroughly these technologies penetrate the economy. The President's Council on Sustainable Development has suggested that the most significant barriers include

- High up-front cost of new technologies compared to the low cost of fossil energy
- Lack of awareness of the availability of climate-friendly technologies and their value for solving quality-of-life concerns
- Long time frame for natural turnover of capital stock
- Fiscal or regulatory policy disincentives that impede early retirement of carbon-intensive technologies or fail to encourage continuous improvement in technology and environmental performance
- Political uncertainty about future carbon control policy

Possible solutions to overcome these barriers are also presented:

- Fiscal policy should encourage the replacement of greenhouse gas-intensive technologies with those that are climate-friendly, and increase investment in innovation through performance-based incentives and other mechanisms.

- Statutory and regulatory authority should facilitate flexible and performance-based approaches that make it easier to install and employ climate-friendly technologies.

- Voluntary commitments should be used to learn how to reduce emissions and put these lessons into practice.

- Research and development efforts should help ensure that future emissions reductions can be met at low cost and in ways that contribute to sustainable development.

Carbon sequestration can be defined as the capture and secure storage of carbon that would otherwise be emitted to or remain in the atmosphere. The idea is to keep carbon emissions produced by human activities (anthropogenic) from reaching the atmosphere by capturing and diverting them to secure storage, or to remove carbon from the atmosphere by various means and store it. Carbon sequestration could be a major tool for reducing carbon emissions from fossil fuels. For example, Norway's state-owned petroleum company, Statoil, is currently sequestering the carbon dioxide content of the natural gas it is extracting from the Sleipner gas field off the coast of Norway back into an aquifer about 1000 meters below the seabed. Statoil has found this to be more economical than paying the $55/ton tax that would apply for emitting the carbon dioxide to the atmosphere (Allenby, 2000). Thus, there is proof of the concept; however, much work remains in order to understand the science and engineering aspects and realize the full potential of carbon sequestration options.

The U.S. DOE defines three requirements for the success of carbon sequestration technologies:

1. Be effective and cost-competitive
2. Provide stable, long-term storage
3. Be environmentally benign

Using present technology, estimates of sequestration costs are in the range of $100 to $300/ton of carbon emissions avoided. The President's Committee of Advisors on Science and Technology recommended increasing the U.S. DOE's research and development on carbon sequestration. The goal is to reduce the cost of carbon sequestration to $10 or less per net ton of carbon emissions avoided by 2015. Achieving this goal would save the United States trillions of dollars.

On April 12, 1999, the U.S. DOE issued *Carbon Sequestration— State of the Science.* The report defines six scientific/technical "focus areas" relevant to carbon sequestration. These focus areas are described below.

Separation and capture of CO_2.   The goal of CO_2 separation and capture is to isolate carbon from its many sources into a form suitable for transport and sequestration. The costs of separation and capture are generally estimated to make up about three-fourths of the total costs of ocean or geologic sequestration. Sources that appear to lend themselves best to separation and capture technologies include large point sources of CO_2. Dispersed sources of CO_2 emissions are especially challenging issues for applying cost-effective separation and capture methods.

The technology required to perform this function depends on the nature of the carbon source and carbon form(s) that are suitable for subsequent steps leading to sequestration. The most likely options currently available for CO_2 separation and capture include chemical and physical absorption, physical and chemical adsorption, low-temperature distillation, gas-separation membranes, mineralization and biomineralization, and vegetation.

Ocean sequestration.   The ocean represents a large potential sink for sequestration of anthropogenic CO_2 emissions. Currently, the ocean actively takes up one-third of our anthropogenic CO_2 emissions annually. On a time scale of 1000 years, about 90 percent of today's anthropogenic emissions of CO_2 will be transferred to the ocean. Ocean sequestration strategies attempt to speed up this process to reduce both peak atmospheric CO_2 concentrations and their rate of increase. Although the long-term effectiveness and potential side effects of using the oceans in this way are unknown, two methods of enhancing sequestration have been proposed: (1) the direct injection of a relatively pure CO_2 stream and (2) the enhancement of the net oceanic uptake from the atmosphere.

Technologies exist for direct injection of CO_2 at depth and for fertilization of the oceans with microalgal nutrients. However, we lack sufficient knowledge of the consequences of ocean sequestration on the biosphere and on the natural biogeochemical cycling. In addition, public perception of ocean sequestration will certainly be an issue for its broader acceptability. Much of the public, as well as ocean advocacy groups, believes that the oceans must remain as pristine as possible. Legal issues may also be complicated. With the exception of the coastal economic zones, the ocean is international in domain and is protected by international treaties or agreements. Ultimately, both scientific understanding and public acceptability will determine whether ocean sequestration of carbon is a viable option.

Carbon sequestration in terrestrial ecosystems (soils and vegetation). Enhancing the natural processes that remove CO_2 from the atmosphere may be one of the most cost-effective means of reducing atmospheric levels of CO_2. This program area is focused on integrating measures for the improvement of carbon uptake by terrestrial ecosystems, including farmland and forests, with fossil fuel production and use. This development has received much support by the public, and forestation and deforestation abatement efforts are already under way.

Sequestration of CO_2 in geological formations. CO_2 sequestration in geologic formations includes oil and gas reservoirs, unmineable coal seams, and deep saline reservoirs (ocean sequestration). One such process already in use is enhanced oil recovery. During this process, CO_2 gas is pumped into an oil or natural gas reservoir in order to push out the product. This process represents an opportunity to sequester carbon at low net cost due to the revenues from recovered oil and gas.

Advanced biological processes for sequestration. Advanced biological technologies will augment or improve natural biological processes for carbon sequestration from the atmosphere in terrestrial plants, aquatic photosynthetic species, and other microbial communities. Enhanced biological carbon fixation significantly increases carbon sequestration without incurring costs for separation, capture, and compression. Available technologies encompass the use of novel organisms, designed biological systems, and genetic improvements of metabolic networks in terrestrial and marine microbial, plant, and animal species.

Advanced chemical approaches to sequestration. Advanced chemistry shares significant common ground with separation and capture. Improved methods of separation, transport, and storage will benefit from research into advanced chemical techniques. The advanced chemical technologies designed for the future would work with technologies now being developed to economically convert recovered CO_2 to benign, inert, long-lived materials that can be geologically sequestered or that have commercial value. In addition, advanced chemical technologies can develop new catalysts needed to enhance geologic sequestration, develop new solvents and sorbents for gas separations, explore new formulations for fertilizers to enhance terrestrial or oceanic sequestration, and create membranes and thin films for advanced separations.

The policy dilemma: acting now or later

If most of the scientific community agrees that global warming will occur, but that its timing and extent cannot be accurately predicted, perhaps it is prudent to simply delay action until adequate information is available, and avoid committing large sums to address the possibility, only to find later that concern was unfounded. The energy practices which have released increasing quantities of greenhouse gases lie at the heart of our technology and cannot be altered readily or without cost.

However, the situation is not this simple. There is a significant time lag, on the order of decades to centuries, between emission of gases and climatic effects. The greenhouse gases, with the exception of methane, are long-lived in the atmosphere (50 to 200 years) and accumulate rather than decay. The climate does not respond immediately as the gases accumulate or as emission levels are reduced. Thus, if we delay, hoping to learn how best to proceed, the accumulating gases commit us to ever-increasing climatic effects into the future, which may not be fully felt for decades. Furthermore, the level of uncertainty about the impacts of global warming increases with the degree of warming. Although some limited and uncertain scenarios can be generated to predict the impacts of a 3°C temperature increase, temperatures higher than this exceed known conditions for the earth, and the potential impacts at higher temperature are thus totally unknown.

Addressing the prospect of global warming presents a fundamental choice for policymakers as well as for the persons charged with assessing the effects of these decisions: take action now, both to prevent climate change and to plan for adaption to change that cannot be prevented, or defer action until the issue is better understood or until climate changes actually occur. This is a choice based on weighing present costs against future benefits. Should we expend current resources and risk that they will be spent needlessly, or should we save current resources and risk encountering changes in the future that may be still more costly, and may exceed our adaptive ability?

In spite of a general agreement at the 1992 Earth Summit in Rio de Janeiro that something must be done to keep global warming under control, the controversy over who pays for this benefit remained. Should the industrialized countries that produce proportionately higher levels of CO_2 (than do developing countries) pay for controlling CO_2 in the developing countries? Should developing countries slow down their rate of industrialization and population growth in order to temper the CO_2 emission increase? The lines are easily drawn, and agreement elusive.

Three major industrialized countries, the United States, Japan, and Germany, with populations of 4.7, 2.3, and 1.5 percent, respectively, of the world, now emit 22.3, 4.8, and 2.9 percent, respectively, of the world's CO_2. The United States alone accounts for nearly one-fourth of the world's generations (*N.Y. Times,* 1992). Many economists feel that it would be much less expensive for industrialized countries to invest directly in reducing CO_2 emissions in the developing countries than to achieve comparable levels of reduction in industrialized economies. This would amount to the transfer of resources from the developed to the developing countries to address a problem affecting the global commons (i.e., the atmosphere). In spite of the economic justification, this course of action presents complex public policy problems. Few industrialized countries are eager to embrace such bold international initiatives for unquantifiable and unguaranteeable returns.

Delaying action. The cost of implementing prevention and adaptation strategies which in the end may be unnecessary, or may be inefficient or ineffective due to lack of information, is the justification for delaying action. If the impact of climate change is small or can be easily managed, then efforts to prevent and plan adaption to global warming will provide minimal benefits. If our current understanding of the issue is inadequate to ensure that policies conceived today will be effective in the future, then the cost of present action may not be justified. Perhaps effective and efficient adaption strategies can be designed only if and when climate changes have arrived. Perhaps prevention strategies incur costs without adequate assurance of future benefits.

Acting in the present. The benefit of avoiding or limiting unknowns (costs, damages, and environmental surprises) in the future is the justification for acting in the present. It may be preferable to pay known costs today rather than encounter unknown and far greater costs and unknown and far greater environmental damage in the future. It may be possible to limit future warming by actions taken today, but impossible to remediate warming in the future if it proceeds past some point of irreversibility. It may be possible to begin the decades-long process of policy design and implementation today and have policies in place in time to meet the situation, but impossible to put effective policies in place quickly enough once climate change has arrived.

The President's Council on Sustainable Development recommends an incentive-based early action program that includes broad participation; encourages learning, innovation, flexibility, and experimentation; grants formal credit for legitimate measures to protect the climate; ensures accountability; is compatible with other climate protection strategies and environmental goals; and is inspired by government leadership. The Council notes that an early action strategy must evolve over time in response to advances in scientific knowledge and technology. Improved understanding of the climate system and the sources and sinks of the various greenhouse gases will help determine how best to target appropriate incentives to protect the climate. As new and existing technologies are deployed more rapidly and as new

technologies are developed, improved cost-effective early action strategies may emerge.

Cost-benefit analysis

A cost-benefit analysis is thus implied in making this "now or later"
choice about global warming strategies. The calculation of environmental policy *costs,* while not an exact science, has been developed into
a useful evaluative tool. The calculation of policy *benefits* (the costs of
environmental damage which are avoided as a result of implemented
policy) has been included in policy analysis only in recent years, perhaps because of its substantial difficulty, and it is less well developed.
Despite the difficulty of valuing policy benefits, some observers (e.g.,
Pearce, 1989) maintain it is essential to include benefits in policy evaluation.

On the small scale, individual policies which are evaluated only for
cost effectiveness (by assuming a target and attempting to minimize
the cost of achieving that target) may assume a target that is inappropriate and thus waste resources on an ineffective policy. An appropriate policy target can be set at the point that costs equal or exceed
benefits, if the value of policy benefits is included in the equation.
Probability of occurrence for each event, if known, can also be included
in the analysis. Intangible costs and benefits can be arranged in a preference index and utilized in policy analysis.

Nordhaus (1990) completed one of the few cost-benefit analyses for
different levels of greenhouse gas reduction. This information is summarized in the following table:

Greenhouse gas reduction, %	Marginal cost per carbon ton, $	Global cost per year, $	Global benefits per year, $
11	8	2.9 billion	10.1 billion
25	40	30.7 billion	22.9 billion
50	120	191.0 billion	48.8 billion

Based on this type of information, strategies and policies can be developed to focus on resources needed to achieve certain reduced levels of
greenhouse gases.

Although it is important that benefits be considered in relation to
costs in policy evaluation, formal cost-benefit analysis is not appropriate in all situations. It is a decision-making tool to evaluate economic
efficiency; however, we also need to consider economic utility and equity.
Economic utility would depend upon preferences of individuals to
determine what constitutes a benefit; equity would require balancing

interests of "losers" and "gainers." In addition, many environmental amenities cannot be converted to monetary terms.

Adaptation strategies

Assuming that the greenhouse effect is real, that increasing atmospheric concentrations of greenhouse gases do indeed cause a rise in mean global temperature, some adaptive response to this warming is needed. It is clear that we cannot reasonably expect to stabilize greenhouse gases at current levels, but can only hope to limit their rate of increase. Faced with this prospect, policymakers have a challenging task of deciding whether adaption should begin only after climate changes have taken place, or if steps should be taken now to make future adaption more efficient and less costly.

Some expenditures to limit greenhouse gas emissions seem like a prudent course of action. Level of expenditures, distribution of expenditures among industrialized and developing countries, and market mechanisms used to implement these policies would require creative strategies and international community agreements on an unprecedented scale. Expenditures made at this time to limit greenhouse gas emissions can capture numerous other benefits regardless of the future extent of global warming or its adverse effects. For example, policies that reduce major greenhouse gases like carbon dioxide and CFC will improve energy efficiency, help develop alternate (non-fossil-fuel) energy sources, reduce air pollution, reduce ozone layer depletion, and provide incentives for developing efficient and less-polluting public and individual transportation systems and vehicles.

Environmental assessment implications

Global warming is an example of a particular type of problem which is extremely difficult to deal with in the context of an environmental assessment. First, unless the action being assessed is intended specifically to deal with the *issue* of global warming, very few actions will have a *significant* effect on the release of any of the greenhouse gases. Many actions, however, will have a *little* effect on them. Any action whose effect is to increase net vegetated land area may be said to have a minor positive effect; the converse is also, of course, true. Policy actions which increase the efficiency of energy generation, or rely on other than fossil fuels for power generation, may be said to have a positive effect. See Chapter 12 for a discussion of the relative position of different types of power generation on greenhouse gases. The authors' best advice is to remember to discuss the issue to the extent that it seems to be applicable, without either over- or understating the consequences (i.e., don't omit, but don't exaggerate).

13.2 Acid Rain

Because acid rain and global warming have common roots in the burning of fossil fuels, the two problems can appropriately be considered together in assessing environmental consequences of an action. As acid rain damages trees worldwide, it also contributes to global warming by reducing the carbon fixing function provided by forests. Preventing acid rain thus can assist in the control of global warming.

What causes acid rain?

Acid rain is produced when atmospheric sulfur dioxide (SO_2) and nitrogen oxides (NO_X) undergo transformations in the atmosphere to produce harmful compounds which then settle as dry fallout or are washed out by rain. The components are organic chemicals that are normally released by the oceans, volcanoes, lightning, and biological processes, and would not cause environmental damage at naturally occurring concentrations. However, sulfur dioxide emissions from the burning of fossil fuels, especially coal-fired power plants, and nitrogen oxide emissions from motor vehicles and secondarily from coal-fired power plants, make a significant contribution to the atmospheric levels of these chemicals.

The chemicals are easily carried long distances in the atmosphere; the use of tall smokestacks, originally intended to reduce local pollution, has had the effect of increasing their dispersion as well. During atmospheric dispersion, sulfur dioxide and nitrogen oxides interact with sunlight, moisture, ozone, and pollutants in complex chemical reactions to produce the compounds which may cause environmental damage.

In 1980, 81 percent of U.S. sulfur dioxide emissions were contributed by 31 eastern states: Ohio had the highest level of emission, followed by Pennsylvania, Indiana, Illinois, Missouri, Texas, Kentucky, Tennessee, and West Virginia. The main sources were coal-fired electric utilities and industrial boilers and smelters. The 31 eastern states also contributed two-thirds of U.S. nitrogen oxides, with highest emissions from Texas, California, Ohio, Pennsylvania, and Illinois. The primary sources for nitrogen oxides are automobiles and utilities (Webber, 1985).

Title IV of the Clean Air Act Amendments (CAAA) of 1990 called for a 10-million-ton reduction in annual emissions of SO_2 in the United States by the year 2010, which represents an approximately 40 percent reduction in anthropogenic emissions from 1980 levels. Implementation of Title IV is referred to as the Acid Rain Program (U.S. EPA, 1999). The overall goal of the Acid Rain Program is to achieve significant environmental and public health benefits through reductions in emissions of SO_2 and NO_x. To achieve this goal at the lowest cost to society, the program employs both traditional and innovative market-based approaches for controlling air pollution. In addition, the program encourages energy efficiency and pollution prevention. To achieve the reductions required by Title IV of the CA90, the law required a two-phase tightening of the restrictions placed on fossil fuel–fired power plants.

Phase I began in 1995 and affects 263 units at 110 mostly coal-burning electric utility plants located in 21 eastern and midwestern states. An additional 182 units joined Phase I of the program as substitution or compensating units, bringing the total of Phase I affected units to 445. Emissions data indicate that 1995 SO_2 emissions at these units nation-wide were reduced by almost 40 percent below their required level. Phase II began in 2000 and was focused on tightening the annual emissions limits imposed on large, higher-emitting plants; it also set restrictions on smaller, cleaner plants fired by coal, oil, and gas, encompassing over 2000 units in all (U.S. EPA, 1997).

Uncertainties

Debate over acid rain has been continuing, particularly in Great Britain and the United States, for the past decade. Those who maintain that environmental damage is caused by emissions of sulfur dioxide and nitrogen oxides are opposed by those who maintain that the causal link is not totally proven. One of the grounds for debate is the possibility that the formation of acid rain may depend more on the availability of oxidants such as ozone, rather than on the emission levels of sulfur dioxide and nitrogen oxides.

The two sides also disagree on the seriousness and irreversibility of observed damage, and on the value of attempting to control sulfur dioxide and nitrogen oxide emissions. The two links to be confirmed are therefore between emissions and acid rain and between acid rain and environmental damage. Policymakers are reluctant to act to limit emissions until these links are established. The results of the National Acid Precipitation Assessment Program 10-year study indicate that the causal link between forest damage and acid rain remains elusive, and that a reduction in sulfur dioxide levels may not cause a corresponding drop in acid rain levels (Anderson and Leal, 1991).

One of the many uncertainties about acid rain is why specific effects are seen in some areas and not in others. Forest damage is more extensive in Germany, whereas fish kills are more extensive in Norway. The form and level of acid deposition varies from region to region, as does the ability of the native ecosystem and soil to resist or buffer acid effect. Forestry management may also influence regional acid levels. Commercial conifer plantings are known to increase the acidity of runoff waters, unrelated to the effects of acid rain.

Damages due to acid rain

Important early reports of serious environmental damage attributed to acid rain (large fish kills in Sweden) were made at the U.N. Stockholm conference in 1972. Since that time, damage to rivers and lakes, forests and vegetation, buildings, and human health have been reported by many countries. Of European countries, Scandinavia and Germany have been most affected; the northeast United States and Canada have been most affected in North America. These highly affected areas are downwind of emission sources in Europe and, in the case of Canada, in the United States. Because acid rain can be thus exported from one country to another, the problem raises significant political difficulties between countries.

Lakes and rivers.   The alteration of lake and river chemistry caused by acid rain kills fish and other water species, damaging the aquatic ecosystem. The primary cause of fish death is likely to be aluminum, which is released from soils by acid fallout. Norway has lost fish from 13,000 km^2 of waters, with an additional 20,000 km^2 affected to some degree. Fourteen thousand lakes in Sweden are unable to support sensitive aquatic life, and another 2200 are nearly lifeless. Over 14,000 lakes in Canada are acidified, with one in seven suffering biological damage. In the United States, the Environmental Defense Fund has identified 1000 acidified lakes and 3000 marginally acidic lakes; the EPA has identified 552 strongly acidic lakes and 964 marginally acidic lakes (French, 1990).

The Adirondack Mountains in New York and the mid-Appalachian highlands contain many of the U.S. waters most sensitive to acidification. It has been documented that 180 lakes in the Adirondack Mountains have suffered loss of fish populations, acid rain being the suspected cause (Webber, 1985). Other sensitive areas include Florida, the upper midwest, and the high-elevation west.

The loss of fish occurs primarily in surface waters resting atop shallow soils that are not able to buffer, or counteract, acidity, most commonly in the northeast and mid-Atlantic regions. Acidification can be chronic or episodic. Lakes and streams suffering from chronic acidification have a constantly low capacity to buffer acids over a long period of time. A national surface water survey conducted in the mid-1980s found that more than 500 streams in the mid-Atlantic coastal plain and more than 1000 streams in the mid-Atlantic highlands are chronically acidic, primarily due to acidic deposition. In the New Jersey Pine Barrens area, more than 90 percent of streams are acidic, the highest rate in the nation. Many streams in that area have already experienced trout losses due to the high level of acidity. Hundreds of lakes in the Adirondacks have acidity levels unsuitable for the survival of sensitive fish species. Episodic acidification is the rapid increase in surface water acidity resulting from large surges of nitrate and/or sulfate, which typically occur during snowmelt or the heavy rains of early spring. Preventing these surges in winter and early spring is critical because fish and other aquatic organisms are in their vulnerable, early life stages. Temporary, episodic acidification can affect aquatic life significantly and has the potential to cause "fish kills" (U.S. EPA, 1999).

North American forests.   Acid deposition, combined with other pollutants and natural stress factors, can damage forest ecosystems. Damage could include increased death and decline of Northeastern red spruce at high elevations and decreased growth of red spruce in the southern Appalachians. In some cases, acid deposition is implicated in impairing a tree's winter hardening process, making it susceptible to winter injury. In other cases, acid deposition seems to impair tree health beginning with the roots. As acid rain moves through soils, it also can strip nutrients from the soil and increase the presence of aluminum ions, which are toxic to plants.

Long-term changes in the chemistry of some sensitive soils may have already occurred. In some regions, nitrogen deposition in forests can lead to nitrogen saturation, which occurs when the forest soil has taken up as much nitrogen as possible. Saturated, the soil can no longer retain nutrients, and they are leached away. Nitrogen saturation has been observed in a number of regions, including northeastern forests, the Colorado Front Range, and mountain ranges near Los Angeles. This phenomenon can create nutrient imbalances in the soils and roots of trees, leaving them more vulnerable to the effects of air pollutants such as ozone, climatic extremes such as drought and cold weather, and pest invasion.

European forests.   Damage to forests has been extensive and well documented in Europe. As of 1988, 35 percent of Europe's total forested area was showing signs of damage (French, 1990). The German *Waldsterben* problem has received widespread attention; 52 percent of forest trees are affected. Damage costs for German forests are estimated at $3 to $5 billion per year over the next 70 years (French, 1990). Such wide-scale damage to forests threatens the economies of affected countries through losses in timber production and tourism.

Visibility.   The pollutants associated with acid deposition also reduce visibility. Visibility impairment occurs when particles and gases in the atmosphere, including sulfates and nitrates, scatter and absorb light. Visibility tends to vary by season and geography because it also is affected by the angle of sunlight and humidity. High relative humidity heightens pollution's effect on visibility because particles, such as sulfates, accumulate water and grow to a size at which they scatter more light, creating haze.

Sulfate particles from SO_2 emissions account for more than 50 percent of the impaired visibility in the eastern United States, particularly in combination with high summertime humidity. In the west, nitrogen and carbon also impair visibility, and sulfur has been implicated as a major cause of visibility impairment in many of the Colorado River Plateau national parks, including the Grand Canyon, Canyonlands, and Bryce Canyon.

The Interagency Monitoring of Protected Visual Environments (IMPROVE) network monitors visibility primarily in the nation's national parks. Reductions in particulate sulfate, usually correlated to visibility improvements, have been measured at 13 eastern IMPROVE sites. It is still too soon to tell how much of these improvements can be attributed to the Acid Rain Program (U.S. EPA, 1999).

Buildings.   Sandstone, limestone, and marble structures are susceptible to acid rain damage, including erosion, crumbling, and discoloration. European countries are particularly affected by damage to structures of historical and touristic value. Damage has been recorded to structures and works of art in virtually every country, and is especially bad in Greece and Italy.

Human health.   The risks to human health from acid rain appear to be both direct and indirect. Both sulfur dioxide and nitrogen oxides contribute to respiratory diseases; some researchers estimate sulfur dioxide is responsible for 2 percent of the annual U.S. mortality (French, 1990). Indirectly, acid rain releases heavy metals from soils, which then can find their way into the food chain through water and fish.

Policy options

The use of lime to buffer acid conditions in lakes, rivers, and soils has shown some promise in temporarily improving conditions for plants and animals in affected areas. Sweden has been liming lakes experimentally since 1976, and has observed recolonization by fish and plankton populations (Park, 1987). However, this cannot be considered a permanent solution to acidification.

Because fossil fuel consumption releases the ingredients of acid rain as well as carbon dioxide, the major greenhouse gas, energy policies to reduce fossil fuel consumption will simultaneously limit both environmental problems. However, additional strategies specific to sulfur dioxide and nitrogen oxides are required in order to reduce the amount of these oxides released during the burning of fossil fuels.

Allowance trading. Allowance trading is the centerpiece of the EPA's Acid Rain Program, and allowances are the currency with which compliance with the SO_2 emissions requirements is achieved. An allowance authorizes a unit within a utility or industrial source to emit 1 ton of SO_2 during a given year or any year thereafter. At the end of each year, the unit must hold an amount of allowances at least equal to its annual emissions. However, regardless of how many allowances a unit holds, it is never entitled to exceed the limits set under Title I of the CAAA to protect public health. Allowances are fully marketable commodities. Once allocated, allowances may be bought, sold, traded, or banked for use in future years. Allowances may not be used for compliance prior to the calendar year for which they are allocated.

Through the market-based allowance trading system, utilities regulated under the program, rather than a governing agency, decide the most cost-effective way to use available resources to comply with the acid rain requirements of the CAAA. Utilities can reduce emissions by employing energy conservation measures, increasing reliance on renewable energy, reducing usage, employing pollution control technologies, switching to lower sulfur fuel, or developing other alternate strategies. Units that reduce their emissions below the number of allowances they hold may trade allowances with other units in their system, sell them to other utilities on the open market or through EPA auctions, or bank them to cover emissions in future years. Allowance trading provides incentives for energy conservation and technology innovation that can both lower the cost of compliance and yield pollution prevention benefits.

Clean coal technologies. Several promising technologies are under development. Most U.S. coal-fired plants continue to use low-sulfur coal rather than these technological solutions.

1. The sulfur content of fuels can be reduced *before burning* by these methods:
   a. Washing with water can remove 8 to 15 percent of the inorganic sulfur content, and it is inexpensive.
   b. Chemical cleaning can remove 95 percent of inorganic sulfur and 50 percent of organic sulfur.
   c. Coal can be converted to a gas or liquid fuel and sulfur removed in the process.
   d. Crude and gas oils can also be desulfurized.

All these methods but water washing add 10 to 25 percent to the cost of energy production, making them less than optimal choices (Park, 1987).

2. Fluidized bed combustion removes sulfur from fuel *at the time of combustion* by fixing it with lime. This new technology may be the most promising option.

3. Flue gas desulfurization removes sulfur gases from flues *after burning,* but before they are released. The dry approach recaptures the sulfur so it can be sold. The wet approach (known as "limestone scrubbing") removes 70 to 90 percent of sulfur for an additional cost of 8 to 18 percent, but produces sludge, which presents a disposal problem. Only 30 percent of U.S. plants are fitted with scrubbers, while the percentage of European coal-fired plants fitted with scrubbers ranges from 40 percent for Germany to 100 percent for the Netherlands (French, 1990).

Nitrogen oxides. Nitrogen oxides are produced during combustion of fossil fuels when air is introduced, and are not contributed by the fuels themselves. The strategy for nitrogen oxides is therefore to minimize the air present during combustion. In coal-fired plants, two-stage combustion, modified burner design, and flue gas recirculation are possible options. In motor vehicles, catalytic converters are most commonly used to reduce emissions.

International efforts

The year 1979 marked the beginning of initiatives in Europe to control acid rain for the benefit of all European countries. In that year the Long-Range Transport of Air Pollutants convention was signed by 35 countries. It was intended to be a statement of purpose, as opposed to a legally binding agreement, about the seriousness of acid rain and a commitment to cooperation in reducing emissions. The convention came into force in 1983.

Despite this consensus about the validity of acid rain as a problem and the necessity of reducing emissions, debate continued over how emissions should be reduced and who should bear the cost. By 1985, with damage evidence mounting from many countries, 21 had signed a legally binding protocol document committing themselves to reduce individual sulfur dioxide emissions 30 percent by 1993. Fourteen countries signed a declaration of intent to support the principle of emissions reductions. Both the United States and the United Kingdom, as in the past, continued to withhold their support for these efforts, asserting that the issue was too uncertain and required more research. In 1986, the United Kingdom agreed to a 14 percent reduction by 1996 and, finally, with Title IV of the CAAA of 1990, the United States agreed to a 10 million ton reduction in annual emissions of SO_2 by 2010, which represents an approximately 40 percent reduction in anthropogenic emissions from 1980 levels (U.S. EPA, 1995).

U.S. relations with Canada have also been strained in the past over the issue of acid rain, due to U.S. reluctance to control emissions which are affecting southeast Canada. It is estimated that 50 percent of acid deposition affecting Canada comes from the United States, while only 15 percent of U.S. acid deposition is produced in Canada. Canada is vulnerable to acid rain damage because a large area (the Canadian Shield) is already acidic. This same area is economically important for forests and recreational lakes. Canada's dependence on the timber industry makes preventing forest damage an urgent issue.

Acid rain and environmental assessment

Acid rain is another complex issue which may be examined in two ways. Unless the purpose of the action being assessed is the reduction of acid rain, it is an issue which should be discussed to the degree relevant. If, however, the reduction of acid rain *is* a focus of the action, then it becomes a national—or international—issue of the first magnitude. Again, as with global warming, many proposed actions may be seen to have some small aspect which is related to this question, especially if power generation or consumption is involved. The suggestion is the same. Do not fail to mention the relationship, but don't dwell upon it beyond the degree to which it is relevant to the action being assessed. It is easy to be drawn into a lengthy discussion of an element, such as acid rain, which is not closely related to your real action.

13.3 Deforestation

Deforestation is intimately linked to carbon dioxide release, global warming, acid rain, and extinction of plant and animal species. Deforestation not only contributes to the greenhouse effect, but also destroys the long-term ability of the land and forest resources to meet human needs, and inhibits the development of viable local economies. The extreme rate of extinction of plant and animal species due to loss of forest habitat, more than 100 per day, would result in the loss of one-fifth of species worldwide over a 15-year period (Postel, 1988). This represents an irreversible loss of resources to the rest of the world, as well as to the local peoples.

Forests are an important economic resource for some regions of the United States. Any reduction in timber harvesting, no matter how justified on environmental grounds or even overall economic grounds, is going to meet with strong opposition. Often, the very livelihood of a large proportion of the regional population depends upon the timber harvesting industry. The ultimate losses due to reduced aesthetic and tourism values may be just as serious.

Industrialized countries

In spite of these values, deforestation in industrialized countries is not usually a major economic problem. Impacts on forests due to acid rain, photochemical oxidants, overharvesting, and changed land use are not well understood. Most industrialized countries are able to make the necessary economic trade-offs and manage forest resources effectively. Regional problems to control timber harvesting to a renewable level are manageable, and some long-term policy options (as discussed later in this section) can be developed to balance economic and environmental requirements.

Developing countries

Problems of deforestation in developing countries are serious, and environmental and economic consequences significant. Of the world's annual carbon dioxide release due to deforestation, 40 percent is contributed by tropical America, 37 percent by tropical Asia, and 23 percent by tropical Africa. Five countries account for half of the total: Brazil, Colombia, Indonesia, the Ivory Coast, and Thailand (Postel, 1988). By the late 1980s, 45 countries around the equator that were practicing aggressive forest clearing destroyed 20 to 40 hectares every minute (Gradwohl and Greenberg, 1988). Clearing for crops, fuelwood, cattle ranching, and commercial timber harvest destroyed 39 million acres annually (Postel, 1988). Fortunately, the pace of deforestation seems to be slowing at the global level as well as in developing regions: The estimate of forest cover change in developing countries indicates an annual loss of 13.7 million hectares (Mha) between 1990 and 1995, compared with 15.5 Mha between 1980 and 1990 (Marcoux, 2000).

Tropical deforestation is driven by poverty, national development policies, and foreign debts. Much of this cleared land is unsuitable for the monocrop agricultural practices being adopted, and is barren within one to two crop seasons. Lands cleared for pasture may support livestock for only 5 to 10 years. Once forests are removed, rural people are unable to meet their pressing need for fuelwood, and soil erosion, floods, and drought become more severe. Forest regeneration on cleared lands is largely unsuccessful due to the lack of natural seed, predators which feed on the seeds and seedlings, and the hot, dry conditions of tropical pasture land compared to the forest environment.

Brazil, the site of 30 percent of the world's tropical forest area, alone contributes one-fifth of the total carbon dioxide emissions from deforestation. Although the annual release has been estimated at 336 million tons, 500 million tons of carbon dioxide were released in 1987 (Postel, 1988). Government programs are unfortunately responsible for most of this deforestation. Brazil's current problems have roots in the decision of the 1960s to provide overland access to Amazonia before there was adequate understanding of the resources available and how they could be developed in a sustainable manner.

Beginning in the 1960s, the Brazilian government undertook major road-building programs to open the Amazon, followed by subsidized settlement. In the 1970s subsidized programs for large-scale export projects in livestock, timber, and mining were initiated; 72 percent of the tropical forests altered up to 1980 were due to cattle-ranching efforts. Despite subsidies and tax incentives, the supported livestock projects have performed at only 16 percent of what was expected, because cattle ranching in this environment is intrinsically uneconomic (Mahar, 1990). The government also supports a policy that considers deforestation as evidence of land improvement and thus gives the tenant rights of possession, which can then be sold. In 1989, a program was initiated which will end subsidies for new livestock projects, and may support agroecological zoning for the country. Perhaps this is an indication of more appropriate government policies to come in the future. There are, however, powerful political and economic forces within the country which are opposed to change.

Policy options

As discussed earlier, forestry management in developed countries, for whom serious and ongoing deforestation is not a major problem, need only to focus on some long-term market-economy-based policies. Increasing total forested area and ending subsidies which support logging are two policy options. For example, the 13 million hectares of marginal U.S. cropland which have been set aside in the Conservation Reserve Program, if reforested, could absorb 65 million tons of carbon annually until the trees mature, reducing U.S. carbon emissions by 5 percent (OTA, 1991). Federal subsidies of below-cost timber sales in remote areas of national forests promote excessive timber cutting, cost billions in the early 1990s, and should be discontinued (Wirth and Heinz, 1991). Efforts to increase the productivity of forests and to plant and manage trees as a renewable biomass energy source are other possibilities for U.S. policy. Some regional issues related to economic impacts of reduced timber harvesting are important and would require creative, region-specific policy options.

Slowing deforestation in third world countries will require the financial and technical support of industrial nations to ease their international debt burden and to assist them in developing sustainable economies. Developing countries are encouraged by their debt burden to exploit forests for quick economic gain. "Debt-for-nature swaps" were devised by the World Wildlife Fund science director Thomas Lovejoy in 1984 as an innovative approach to this problem: A non-governmental organization (e.g., The Nature Conservancy) purchases a portion of the debt and then donates the debt instrument to the country's bank in exchange for environmentally appropriate action. Swaps of millions of dollars contribute much needed funds for environmental programs, but have little impact on national debts measured in tens or hundreds of billions of dollars.

A policy option for developed countries may be to require industry to make equal investment in reforestation whenever a carbon-emitting project is undertaken. A joint venture between Applied Energy Services, World Resources Institute, and CARE was planned to offset carbon emissions from a coal-fired power plant in Connecticut with forestation in Guatemala. Twelve thousand hectares of woodlot and 60,000 hectares of combined trees and crops would be planted, to be harvested on a sustainable basis. Large-scale forestation programs are faced by the difficulties of locating and financing the purchase of suitable land, and gaining cooperation from local governments. However, this project is relatively inexpensive ($16.3 million) because land would not be purchased and workers and families to benefit from the planting would not be paid (Flavin, 1990).

13.4   Endangered Species and NEPA

The extinction of an entire species, especially when the cause is human action (or inaction), is an event which evokes many emotions. Guilt is a common feeling, as are shame, sorrow, and regret. In the words of a widely quoted, but unattributable phrase, "Extinction is forever." In the United States, several approaches were made to dealing with this question, starting relatively early in the period of environmental awareness in the 1960s. The first act of Congress to use the term was the Endangered Species Preservation Act of 1966 (PL 89-669, Oct. 15, 1966). Its contents may best be termed a policy statement, which "encouraged" federal agencies to take precautions so as not to further erode habitats of species found to be in danger of extinction, and it covered only vertebrate animals. The Endangered Species Conservation Act of 1969 (PL 91-135, Dec. 5, 1969) added coverage of invertebrate animals and specifically prohibited interstate commerce in illegally taken species. Most importantly, it authorized the Secretary of the Interior to identify species threatened outside the United States, and to prohibit or limit import of these species or their products.

During this time, other proenvironment forces were extremely strong. The Marine Mammal Protection Act (PL 92-552, Oct. 21, 1972) focused on whales and other marine mammals, and provided protection for "depleted populations" in addition to species in danger of total extinction. In early 1973, an international convention, the Convention on International Trade in Endangered Species of Wild Fauna and Flora (CITES)—sometimes pronounced as a single word, as "sights" or "sy-tees"—developed a broad international agreement on restrictions for the import and export of endangered species and their products. Since this international treaty was stronger than existing U.S. law, new legislation was needed. The Endangered Species Act (ESA) (PL 93-205, Dec. 28, 1973) received extremely strong support from many environmental organizations, was passed by both houses with little organized opposition, and is the basis for the present regulations on threatened and endangered species. One title of the ESA implements CITES. It has been amended and reauthorized several times in the ensuing years.

CITES also lists endangered species in three categories: Appendix I lists the most threatened species for which no commercial trade is allowed and other trade only by permit; Appendix II lists species that may become threatened in the near future for which commercial trading is allowed only by permit; and Appendix III allows countries to list species unilaterally and require export permits. The United States supports CITES by prohibiting imports of species taken in violation of the convention. As its name would indicate, CITES is related to species preservation through diminishing the trade value of products related to endangered species. The Fish and Wildlife Service (FWS) is the U.S. permitting agency for CITES species.

As a result of the ESA, several species such as the bald eagle, peregrine falcon, brown pelican, and American alligator have been saved from extinction, and the grizzly bear, gray wolf, whooping crane, and California condor may be out of danger in the future. The ESA has also broadened our approach to conservation from a focus exclusively on game species to concern for all species. However, critics maintain that the FWS is not taking its responsibility seriously, because recovery plans have not been implemented for many species. Processing candidates is slow; species have become extinct while waiting for listing (CEQ, 1990). On the other hand, some feel FWS plans are too proscriptive.

Important aspects of the Endangered Species Act

Within our context (i.e., the relationships between NEPA and the ESA) only certain portions of the act are relevant. Some terminology becomes important, as do some procedural steps required by the rules implementing the act.

What is a listed species?   The term *listed species* appears regularly when examining NEPA documentation for actions where this issue is relevant. Simply put, *listed* means that the species appears on the list, or catalog, maintained by the Secretary of the Interior for species which appear to be in danger across all or a portion of their range. The term *endangered* is used if, in the opinion of the agency [Fish and Wildlife Service (FWS) for terrestrial species and National Marine Fisheries Service (NMFS) for marine species], the entire species is in danger of extinction. The term *threatened* is applied when populations are low enough that it appears likely that if no protection is offered, the species will become endangered (i.e., threatened with extinction). The provision for preendangerment listing is unique to the 1973 act, and provides an important management tool to wildlife agency biologists (see Table 13.1). Some distinction may be made for wide-ranging species, which may have different status in different portions of their range. Both categories are "listed," and the distinctions are relatively minor in the NEPA context.

TABLE 13.1   Summary of Listed Species

Listings and Recovery Plans as of January 31, 2001

Group	Endangered U.S.	Endangered Foreign	Threatened U.S.	Threatened Foreign	Total species	Total species with recovery plans
Mammals	63	251	9	17	340	47
Birds	78	175	15	6	274	76
Reptiles	14	64	22	15	115	30
Amphibians	10	8	8	1	27	11
Fishes	70	11	44	0	125	90
Clams	61	2	8	0	71	44
Snails	20	1	11	0	32	20
Insects	33	4	9	0	46	28
Arachnids	12	0	0	0	12	5
Crustaceans	18	0	3	0	21	12
Animal subtotal	379	516	129	39	1063	363
Flowering Plants	564	1	141	0	706	554
Conifers and Cycads	2	0	1	2	5	2
Ferns and Allies	24	0	2	0	26	26
Lichens	2	0	0	0	2	2
Plant subtotal	592	1	144	2	739	584
Grand total	971	517	273*	41	1802	947

Total U.S. endangered—971 (379 animals, 592 plants).
Total U.S. threatened—273 (129 animals, 144 plants).
Total U.S. species—1244 (508 animals, 736 plants).
*Nine U.S. species have dual status.
SOURCE: *U.S. Fish and Wildlife Service,* http://endangered.fws.gov/wildlife.html.

Between 1990 and 2001, almost every category of species has shown increases in listings. Most classes of animals have added about 50 percent in that decade, but the number of listed plant species has tripled. Why should this be? Have environmental changes been disproportionately harsh on plants? It seems likely that it is the case that attention was originally more focused on animal species, and examination of the status of the numerous, lesser known, plant species has been somewhat delayed in comparison.

In the 30-plus years of the listing of endangered species (since 1967), some species have been removed from this list. The reasons are hardly encouraging, as shown in Table 13.2. Three of the recovered species (the table includes species from all countries) are kangaroos (from Australia), another three are birds from Palau, and another is the gray whale. Thus, when considering conterminous U.S. species only, the score becomes four recoveries and seven extinctions, almost 2 to 1 in favor of extinction. This is not to say that the gains are without merit. The four species are the American alligator, the brown pelican, and the peregrine and Arctic falcons. Clearly, there are hundreds of species remaining which are in clear danger of disappearance.

What is critical habitat? Once a species has been listed as endangered, as described above, there may or may not be a concurrent designation of a particular land (or water) area as habitat critical to the continued survival of the species. Not every endangered species has been linked with a corresponding critical habitat. In fact, only a minority of endangered species have had such habitat identified. The reasons for this are many and are the subject of a lengthy controversy within several government agencies and between these agencies and outside environmental advocacy organizations. In the NEPA context, proposing activities within an area designated as critical habitat is controversial and delicate in the extreme. It should be noted here that critical habitat is not restricted to the area in which an endangered species is *now* found, but includes that area deemed necessary for "conservation" (i.e., recovery) of the species. This distinction is not well understood among many of the agencies with need for projects or actions near, but not within, the *present* actual distribution of such a species.

TABLE 13.2 Summary of Delisted Species

Reason for delisting	Number of species
Species considered recovered	11
Species considered extinct	7
Taxonomy revised; no longer considered a separate species	7
New information found on other populations	5
Original listing found to be in error	1
Total number of species delisted 1967–2000	31

Complying with Section 7 of the ESA

When most persons employed by those government agencies which may propose construction and development projects recall the Endangered Species Act, it is Section 7 of the act which they have in mind. Applying exclusively to federal government agencies, it directs them to ensure that the actions of the agency do not jeopardize any listed species or destroy or alter any designated critical habitat. Under the "consultation" provisions of this section, when a listed species (or critical habitat) is present in the area of a proposed project or action, it is mandatory to request a biological assessment* from the expert agency (FWS or NMFS). In practice, the proponent is asked to provide the expertise (and/or funding) to prepare this assessment. To avoid lengthy delay in project completion, the proponent usually prepares the assessment and delivers it, if the assessment shows any potential for conflict with the listed species, to the FWS for consultation as described below.

This assessment evaluates the likelihood that the proposed action may adversely affect the listed species. The proposing agency may conduct (or contract for) its own studies, and present them for evaluation. In practice, this is regularly done to assist in speeding the evaluation process. If the conclusion of the biological assessment is that the proposed action *is likely* to jeopardize the continued existence of the species (or will adversely affect a critical habitat), then "formal consultation" with the wildlife agency must be initiated. During this consultation, the ESA requires the best and most current data and procedures to be utilized in study of the situation. Consultation is discussed in the NEPA context below.

Complying with Section 9

When members of the general public recall the ESA, it is likely that they have one or more of the provisions of Section 9 in mind. This section prohibits, among other actions, (1) the import or export of endangered species and products made from them, (2) commerce (within the United States) in listed species or their products, and (3) possession of unlawfully acquired endangered species. Unlike the provisions of Section 7, these provisions apply to all persons within U.S. jurisdiction. The primary prohibition within Section 9 is against "taking" of endangered wildlife. Originally, the colloquial meaning of this term implied capture or killing of the animal. The act, however, defines *take*

---

*Note that this biological assessment is not an environmental assessment within the context of NEPA.

in broader terms: to "harass, harm, pursue, hunt, shoot, wound, kill, trap, capture, or collect, or to *attempt* to engage in any such conduct" [16 USC 1532(19), emphasis added]. While the provisions of this section are of broad general importance in commerce, government agencies normally are not involved in actions which violate them. One exception will be discussed below.

NEPA compliance with the ESA

Which has priority, NEPA or ESA? Many persons have seen conflicts between the provisions of the Environmental Policy Act and the provisions for protection of endangered species. As with so many complex situations, the actions required are intertwined. One step is taken under one act, to be followed by a step under the other. Both must be complied with fully. Neither has "priority" in the strict sense of the word. The process one must generally follow is outlined below.

Planning the project or action. When general project planning is initiated or, at the latest, when public scoping procedures are undertaken, it must be identified whether or not there is *any* possibility that an endangered species or critical habitat is present within the area affected (or influenced) by the action. This is a "must ask" question for every proposed activity which involves outdoor aspects. Don't be satisfied with a simplistic negative, such as "I have never heard of any in the area," especially when the person quoted is not an authority. Undertake to ask experts, including, at a minimum, representatives of the U.S. Fish and Wildlife Service and the (state) Natural Heritage Program. This is a form of public involvement. If listed species are present, then the input of the appropriate wildlife agency, usually the FWS, may not be avoided.

It is at this stage when internal versus external agency imperatives usually first arise. If the internal ego commitment among your agency planners and administrators is well developed, and they feel the proposed project or action is vital to agency interests, there is a strong tendency to proceed full speed ahead, trusting that a solution to the problem of this "insignificant" little animal may be found somewhere, later in the process. There have been a few cases where this has been true. At the very least, however, a lengthy, costly, and controversial battle will be joined. For many, if not most, agencies at this time, there is simply no desire to join in a heated conflict which pits them against another federal agency (the FWS), numerous environmental activist groups, a large segment of the public, and many members of Congress. Even if your agency eventually prevails, any victory may be pyrrhic.

Case study 1: The Tellico Dam and the snail darter. In the classic example of this struggle, the Tennessee Valley Authority had, in 1975, partially constructed, but not closed, the Tellico Dam. The history of the authorization and justification of the project was extremely controversial for many decades, due to economic and political disagreements (Wheeler and McDonald, 1986). During the construction phases, outside scientists determined that a minnow-sized fish, the snail darter, was a rare, distinct species, known at that time only from one small stream which would be destroyed in allowing the reservoir to fill. The snail darter was proposed for listing as an endangered species and was subsequently listed. More than $50 million had already been expended on the project. The FWS found that the project was likely to endanger the existence of the entire species, and moved, under the ESA, to stop construction. The Supreme Court upheld the application of Section 7 of the act, finding no grounds for an exception. In partial response, Congress did create a special committee (popularly called the "God squad") with authority to allow exemptions to Section 7 if there were extremely extenuating circumstances. In this case, the special committee voted not to exempt the Tellico project from the ESA. The fact that construction was more than 50 percent complete before the species was even recognized by science was considered adequate extenuation by many supporters in Congress, and the Tellico Dam was allowed to be completed through amending another appropriations bill. In a biological footnote, it may be noted that the snail darter was transplanted to several other locations, where it survives, and that other natural locations were found in later years. The species was not actually made extinct by the completion of the reservoir.

What is the "taking" of a species?

The popular concept of the taking of a species (i.e., killing or capturing) is only rarely the direct object of a proposed action which is the subject of NEPA documentation. In other circumstances, however, the issue of taking has often become much more complex, and may indirectly result from proposed federal agency actions. Broadly construed, the principle is that actions which affect habitat required for the continued existence of a species, even though not designated as critical habitat, may be called "taking" under Section 9 of the ESA. In a series of cases involving (largely) state fish and game agencies, several principles have been developed in case law under this section. One is that habitat modification must be shown to be harmful to the entire species, even though no individual deaths need be cited. While it is not clear that this was specifically envisioned by the drafters of the act, the term *harm,* when applied to habitat changes, must be examined with care when planning your action.

Preparing the environmental documentation

Assuming that the initial investigations show some possible interaction, but no irrevocable conflict, with respect to a listed species, preparation of the EA or EIS must include interaction with the FWS (or NMFS). While the provisions of Section 7 call only for the preparation of a biological assessment, and a request for consultation if necessary, there is normally no reason why the FWS may not be asked for guidance and advice early in the process. During the formulation of the alternatives, the possible interaction with the listed species (or critical habitat) must be evaluated before lengthy and costly commitments are made. In fact, it is specifically prohibited under the ESA to make "irreversible or irretrievable" commitment to any course of action which would preclude any alternative prior to receiving a biological opinion from the wildlife agency [16 U.S.C. 1536(d)].

It is normal, in the preparation of environmental documentation where listed species are involved, that only the characteristics of the preferred alternative are presented to the wildlife agency with a request for an opinion. There are, however, circumstances where this may not be possible. With the consent of the wildlife agency, a *range* of possible actions may be submitted with the biological assessment. This may occur when the proponent is able to identify several possible ways to complete agency goals, and needs to know whether the accommodations required by the biological opinion will make one option more practical or significantly less costly than another. In other words, instead of fixing on a course of action *before* requesting an opinion, the terms of the opinion, if they differ among the options, become one of the final steps in the decision-making process. This would appear to be totally in the spirit of NEPA, incorporating environmental considerations at all stages of the planning process.

Present areas of controversy

It is clear that many of the almost irresolvable conflicts between listed species and proposed federal agency actions are intimately related to economic factors. The $50-plus million which had already been sunk in the Tellico Dam weighed extremely heavily in the considerations which followed. Many persons who were less than committed to the principles of the act felt that the "wasted" dollars, alone, were more than enough justification to allow the project to be completed. While the fate of these species at the brink of extinction is of extreme importance to many environmentally aware persons, a very large number of supporters may be persuaded that costs and jobs are more important in the long run. The listing of the spotted owl in 1990, which effectively protected 8.4 million acres of old growth timber in the northwest, has been a subject of controversy since 1986. Vigorous protest was made by the logging industry, which claims that the economic effect of this restriction is unduly severe and would eliminate 131,000 jobs related to wood products; other estimates set the loss of jobs at 30,000. See Case study 2 for a summary of the northern spotted owl question. However, 60 to 90 percent of all old growth forests in the region has already been logged; the remainder will be gone in 30 years if present rates continue (Arrandale, 1991).

Case study 2: The northern spotted owl and its critical habitat.   In 1989, the Secretary of the Interior was sued by a coalition of environmental activist groups to designate the northern spotted owl an endangered species and, further, to designate as its critical habitat "old growth forests in the Pacific Northwest." The species was subsequently found to be endangered. The real controversy arose when it was realized that, if the entire area of "old growth" forest was determined to be a critical habitat, logging activity in the area would have to be severely curtailed. This, in turn, led to loss of employment for loggers and truckers, sawmill workers, and other persons whose jobs depended directly or indirectly on exploitation of such old growth timber. The exact numbers of jobs lost which were directly attributable to the spotted owl were a matter of extreme disagreement. Forest workers were inclined to say that all unemployment was due to this "insignificant bird," and put the number at 30,000 to 100,000. Others suggest that automation in the processing sector and the general economic downturn, especially in housing construction, accounted for almost all jobs that were lost, and put the number of jobs lost due to the spotted owl at no more than 3000. Whatever the numbers, almost all unemployed (and underemployed) forestry workers in northern California, Oregon, and Washington believed their social and economic trouble was caused by "outside agitators" who "had more concern for owls than for people." This is always extremely difficult to balance, and in 1992 the Special Committee ("God squad") eventually approved rules allowing some harvests to continue. Predictably, environmental activists protested that far too much logging was allowed, while timber interests protested that far too few areas were opened.

In still other actions in the northwest, petitions have been made to list five species of salmon, which would curtail fishing and require major changes to the operations of dams in the region. Electric utilities claim that rates would rise 33 percent and thus hamper industrial development (Arrandale, 1991). FWS recovery plans for the wolf included reintroducing animals to Yellowstone Park, where they would also help control herds of bison and elk. Plans for reintroduction were delayed from the early 1980s to 1995 due to protest from ranchers and sheep growers who maintain that wolves endanger their livestock on adjacent lands (Cohn, 1990; National Wildlife Federation, 1995; USFWS, 2000). An exactly similar set of concerns were raised in 1999 and 2000 over the reintroduction of the Mexican gray wolf into the Gila Wilderness in New Mexico (USFWS, 1999; Defenders of Wildlife, 2000). Conflicts have also arisen over listing the southwestern desert tortoise, the Florida panther, the Louisiana black bear, sea otters in California, and red squirrels in Arizona.

Listed species and NEPA documentation

First and foremost, the presence or absence of listed species must be absolutely, positively verified during early project planning. If present, the possible effects of the implementation of the action on the species or its habitat must be documented. Further, the cognizant agency (FWS or NMFS) must be consulted during the EIS process. This is not optional. It is required by both NEPA and the ESA. The omission or deferral of this step will, inevitably, lead to adverse consequences. The FWS and NMFS are extremely careful to examine proposed actions thoroughly when listed species are involved in any manner. The assessment which they must provide should become a part of a draft EIS, and a formal biological opinion must be prepared before the proposed action may be initiated. If an opinion is available prior to completion of the EIS, it should be included as an appendix. If the biological assessment leads to a "jeopardy opinion" (i.e., the FWS decides that the proposed action would jeopardize the continued survival of the listed species), this may be considered a fatal flaw in the proposal. Barring original Congressional action, the agency should always redesign either the project or the mitigation measures so that a violation of the Endangered Species Act does not result.

Failure to consider listed species or critical habitat may also lead to an assessment with a fatal flaw. It is not always adequate to rely on second- or third-party opinions in this respect. In the past, one frequently saw the phrase "not known to occur in the area" when dismissing concerns about listed species. Original surveys are now frequently required in cases where existing information is incomplete. Plan for these surveys, if necessary, and allow time and money to complete them.

13.5 Biodiversity

The term *biodiversity* suggests what it means without extensive interpretation: the diversity of biota. It became accepted in the last quarter
of the twentieth century that biodiversity is a good thing; that it is a
characteristic of healthy ecosystems; that its loss is, per se, a negative
characteristic; and that its maintenance and/or recovery are a goal
toward which planners and managers should strive. It thus represents
an idealized concept, a concept which has become an element in environmental assessments. Just what *is* biodiversity, and how does it
interact with some closely related topics, such as endangered species?

What is the problem?

Ecologists believe that the stability and vigor of life on earth depend
on biological variation (i.e., the presence of variety in types of ecosystems, species, and genes, as well as in their relative frequency).
Genetic diversity refers to variety in genes of individuals and populations of the same species. It may be considered proven that genetic
diversity is necessary for successful adaptation to changing conditions.
Species diversity refers to variety in the types of organisms which
inhabit an ecosystem. Theory suggests that the existence of some variety among these organisms is also desirable, also in terms of successful maintenance of the system in the face of changing conditions. This
aspect of diversity is somewhat less thoroughly proven than that for
genetic diversity. *Ecosystem* diversity refers to variety in ecosystem
types, such as grasslands, woodlands, and wetlands. An ecosystem is
an array of plant and animal communities and their physical setting
functioning as a unit through interdependent relationships.

The evolution and extinction of species is always occurring as a regular process of nature. The lifeforms existing today represent perhaps
2 percent of all species which have ever existed on earth. Gene muta-
tion occurs spontaneously, and this new genetic variation provides the
"raw material" which allows new species to develop through natural
selection. Lack of diversity in and within the genes of a species makes
it more vulnerable to extinction by limiting the number of revised
genetic combinations available to respond and adapt to change.

Approximately 1.7 million current species have been identified and
named. Estimates of the total number of species in existence ranges
from 3 million to 30 million, with 10 million most often suggested
(OTA, 1987). Two-thirds to three-quarters of the earth's species are
found in moist tropical forests, habitats which have been among the
least studied. La Amistad National Park in Costa Rica, for example,
has more bird species than the North American continent. It is
assumed that such diverse areas are "healthy" in terms of biodiver-

sity. Six-tenths of the world's species are insects, and one-sixth to one-thirtieth are plants (the exact number of either is, as yet, unknown). Other animal species make up the remainder (Rohlf, 1989).

By extension, healthy ecosystems may also be said to perform a variety of functions: soil building; erosion control; nutrient cycling; carbon storage; hydrological regulation including moderating streamflow, filtering water, and controlling flooding; waste disposal; pest management; maintenance of atmospheric quality; and regulation of climate. Changes in any component species or physical characteristics of an ecosystem affect the functioning of the ecosystem, causing repercussions and adjustments elsewhere in the system. As changes become more severe, they cause ecosystems to malfunction and eventually convert irreversibly to some other type of environment. The removal of forest trees in the tropics is an example; cleared forest land is degraded by soil erosion, loss of soil nutrients, heat, and drought, which lead to severe physical changes, and further loss of other, understory species. Because biodiversity is required for the functioning of ecosystems upon which human life depends, maintenance of biodiversity is essential for the support of those human populations tied to them. In many quarters, this loss assumes the character of a moral responsibility.

In addition to maintaining the balance of nature, which supports some human life activities indirectly, biodiversity also provides a storehouse of resources to directly meet human needs. Some products are harvested, such as fish and timber. The harvest of wild marine species totals $14 billion annually (CEQ, 1990). Wild plant species provide genetic material for plant breeders to develop desirable characteristics in domesticated crops, contributing 50 percent of the productivity increases and $1 billion annually to U.S. agriculture. Wild green tomatoes from Peru have contributed genes for increasing tomato pigmentation and soluble solids content worth nearly $5 million annually to the industry (OTA, 1987). It is worth noting that many third world countries are requesting "exploration fees" for the rights to survey wild populations for useful genes. Wild species also provide compounds from which pharmaceuticals are developed. Approximately 25 percent of prescriptions sold in the United States contain active plant components (OTA, 1987). Alkaloids from the rosy periwinkle flower used in the treatment of Hodgkin's disease and childhood leukemia are just one example. The use of derivatives from the bark of the Pacific yew to treat ovarian cancer caused great concern that the species would become extinct due to overgathering in 1991 and 1992. There are thus seen to be "undiscovered" values directly important to humans in these diverse, undisturbed ecosystems.

Functioning, biologically diverse ecosystems also provide some economic benefits through recreational hunting and fishing, other outdoor recreation, tourism, and the opportunity to view wildlife in its natural habitat. Wild habitats adjacent to agricultural areas provide food, cover, and breeding sites for crop pollinators and pest predators. For example, brambles adjacent to grape fields house alternate food sources for wasps, saving grape growers $40 to $60 per acre in reduced pesticide costs (OTA, 1987), Finally, many Americans feel that lifeforms have intrinsic value, unrelated to their immediate or direct relevance to human needs, and that we are ethically bound to preserve them.

Although extinction of species is a natural process, the current rate of extinction is far higher than the rate of evolution for new species. Because a complete inventory of the earth's species has not been made, it is not possible to accurately estimate the current worldwide rate of loss. Edward O. Wilson estimates the global loss rate at one species lost per hour (Wilson, 1991). Norman Myers predicted that with increasing pressure from population growth and development, the global rate by 2000 would be 100 species per day (Rohlf, 1989). During 30,000 years of the Pleistocene Era, including the last ice age, 50 mammal and 40 bird species were lost in North America. Since 1620, in the same area, over 500 species have been lost (Rohlf, 1989). The loss of species is only one part of the larger loss of biodiversity; species extinction is an indication of malfunctioning ecosystems, and we are also losing ecosystems at a rapid rate. Old growth forest and tallgrass prairies once defined the U.S. landscape, but as of 1990, 98 percent of both these ecosystems had been lost, and less than one-half of U.S. wetlands still remain (Arrandale, 1991).

In the past centuries, overexploitation of such marketable species as the bison and passenger pigeon was an obvious primary cause of species loss. These were exceptions. Most species loss has always been due to the modification or destruction of habitat. The loss has been by means of introduction of alien species, clearing of forests, draining of wetlands, livestock grazing, introduction of the monoculture of agricultural crops, construction of buildings and transportation systems, and, finally, through pollution. Species loss in developing countries is proceeding at an alarming rate due to all of these causes, with tropical deforestation being the most obvious. Poaching and illegal trade severely threaten specific species such as the African elephant and rhinoceros.

The primary policy tool to address loss of biodiversity in the United States is the Endangered Species Act (ESA), passed in 1973. See Chapter 2 regarding the ESA, and see Section 13.4 for a more extensive discussion of endangered species per se. The core of the ESA is the listing of species as endangered (in immediate danger of extinction) or threatened (likely to become endangered). The National Marine Fisheries Service (NMFS) is responsible for listing marine animals. Other animals and plants are listed by the Fish and Wildlife Service of the Department of the Interior. A total of 1244 U.S. species are now listed federally, with many candidates under evaluation. An additional 550 species are also listed by other countries, bringing the total listed species to about 1800 (see Table 13.1). It must be noted that the ESA does not directly address the issue of the biodiversity of a *habitat* or of an *ecosystem*. It addresses only single species and their defined (minimum) habitat requirements. As a primary tool, it is thus rather indirect. Ecological biodiversity per se is not addressed as a goal of the ESA.

Nongovernmental efforts

The Nature Conservancy (TNC), a private, nonprofit group in existence since 1951, has promotion of biodiversity as a major purpose for its existence. The group purchases and manages ecologically significant land tracts, and now owns the largest private sanctuary system in the world,With over 12 million acres in the U.S. (TNC 2001). TNC has initiated natural heritage programs with the states to list the species present in each; these are on the web page for the association for Biodiversity Information at http://www.abi.org. Many other programs involve a mixture of private and governmental effort. Here the governmental component is considered incidental to the *ecological* biodiversity issue. These "offsite" programs have proven rather effective, in certain instances, in saving and restoring species in immediate danger, and preserving genetic resources. Offsite methods, sometimes called "rescue" biology, include seed storage, in vitro culture, living collections, embryo transfer, botanic gardens, zoological gardens, field collections, captive breeding programs, seed banks, embryo banks, microbial culture collections, and tissue culture collections. The breeding of the California condor in captivity led to the 1992 release of young condors into their original habitat. Activities which take place outside the habitat in question are called offsite.

A botanical example of an offsite program is the National Plant Germplasm system run by the Agricultural Research Service of the U.S. Department of Agriculture, which has a collection representing 230,000 species for plant breeders and researchers. Similar institutes devoted to rice, wheat, and potatoes are located in the Philippines, Mexico, and Peru, respectively. These programs thus preserve some examples of commercially valued biodiversity in a sort of botanical "zoo." The varieties are available for genetic research even though they may, in the future, become extinct in their natural habitat.

International context

Considerable effort toward implementation of biodiversity goals has been made in other countries by TNC. Its international program has cooperating groups, private or public, in more than 40 countries outside the United States. As within the United States, TNC seeks to protect critical habitats through government action, private agreements with landowners, or purchase, as appropriate. Certain provisions of other U.S. legislation do go beyond the ESA, and mention biodiversity as a goal. The Lacey Act, amended in 1981, also supports the efforts of other nations in their conservation, and the Foreign Assistance Act of 1983 makes conservation of biodiversity one of several objectives of that U.S. assistance carried out by the Agency for International Development. The 1990 Neotropical Migratory Bird Conservation Program is intended to protect 200 songbird species by preventing habitat losses in the United States and Latin America. The 1986 North American Waterfowl Management Plan was signed by the United States and Canada to protect and restore 5.5 million acres of wetland habitat for waterfowl. All these acts recognize biodiversity as a goal.

The International Union for Conservation of Nature and Natural Resources (IUCN) is an important international organization within the United Nations structure representing governments, nongovernmental organizations, and research institutions. It strongly promotes the protection of biodiversity. Its Conservation Monitoring Centre prepares "Red Lists" and "Red Data Books" which list endangered species and provide information about their habitats and status. Biodiversity is also promoted by the United Nations' Man and the Biosphere Program.

Major unanswered questions

The biodiversity policy issues of current concern include the need to manage at the ecosystem level rather than the species level; the need for a national, integrated plan to address biodiversity; and the need for complete research data about species and ecosystems. It is clear that preservation on the ecosystem level is more effective and appropriate than scattered efforts to protect individual species, although some individual species also require focused attention. Of the 261 ecosystem types in the United States, only 104 are represented in the wilderness system, and most reserves are too small to provide enough diversity to maintain the species present. Wetland areas particularly require protection because they are habitat for 35 percent of U.S. endangered species.

A small preserve within an area of development is essentially an island, with small insular populations that lack the genetic flexibility to cope with change, and inbreeding promoting undesirable traits. Preserve areas should be buffered with zones of limited development, and linked to other ecosystems with undeveloped corridors. An integrated approach is required which coordinates the activities and programs of all types of landowners within a regional ecosystem. The Greater Yellowstone Coordinating Committee is an example of regional ecosystem management; the entire ecological region encompasses 19 million acres, including the park of 2.5 million acres surrounded by national forests, Bureau of Land Management holdings, FWS holdings, state holdings, and private holdings (CEQ, 1990). With a similar motivation for regional management, TNC started a campaign to preserve 150 large, basically intact areas by combining their own holdings with public lands.

The Office of Technology Assessment (OTA) reports that our current legislation to address biodiversity is ineffective because it is piecemeal, and recommends passage of a National Biological Diversity Conservation Act which would state the conservation of biodiversity as a national goal, define a national strategy, and eliminate duplication and conflict in existing programs (OTA, 1987). The Bureau of Land Management and the U.S. Forest Service manage 47 percent of the federal estate, 343 million acres, under a multiple use mandate. DOD manages 18.7 million acres of wildlands, and DOE manages 2 million undeveloped acres. There is often controversy over the land use of these departments, and their management objectives do not always coincide with diversity. The national parks have dual mandates for recreation and conservation, and game and timber management objectives favor selected species. OTA recommends that these agencies receive directives to manage in accordance with biodiversity conservation. OTA also finds that existing programs are hampered by limited funding. Congress has proposed amending NEPA to require application of new explicit biodiversity standards in preparing and reviewing future EISs, but has never passed such an amendment.

Addressing biodiversity is also hampered by lack of complete and accurate information. There is no comprehensive list of species or adequate information about their habitat and biology. There is no listing of communities or scientific scheme for naming and cataloging communities. There are gaps in our knowledge of the links between ecosystems and landscape processes, and of species interaction in ecosystems. Information about land management strategies to preserve diversity is lacking. Furthermore, the information which exists is scattered in various institutions. OTA recommends the establishment of a clearinghouse for biodiversity information, and increased funding for research and public education.

Environmental assessment and biodiversity

As presented above, it may be seen that biodiversity is a potentially controversial issue, though one which has no defined answer in most instances. What does this mean for preparers of environmental assessments? First, there is seldom likely to be a clear answer unless the site has previously been studied for decades. It is more usual that biodiversity becomes a suspected outcome whenever "natural" habitats are to be seriously disturbed. It is also an issue which will become a concomitant consequence to be presented whenever an endangered or threatened species is involved. It is *related* to endangered species, but is a separate issue in the minds of the scientific community. The authors believe that it is one which should be discussed when major land-disturbing actions are assessed, but one with which it will be very difficult to come to closure. Nobody has ever proven that any single action has resulted in an unacceptable loss of biodiversity, but each action may become part of a cumulative effect. Biodiversity thus becomes both controversial and cumulative, with no "solution" likely to be acceptable to many members of the scientific community.

During the 1990s, worldwide interest in biodiversity grew explosively. As one example, an Internet search in early 2001 for web pages related to the topic returned more than 50,000 pages with a United States focus, and more than 330,000 when the search was refocused worldwide. More than 100 organizations have taken an interest in the topic, and many were created specifically to further research and discussion about biodiversity and related issues. Some of these are the Association for Biodiversity Information, Arlington, Va., the Missouri Botanical Garden, St. Louis, Mo., and the World Resources Institute, Washington, D.C. Each of these has a web site devoted to the topic, with links to scores of other organizations with similar interests. This may be the best way to determine if your project site is currently of special interest to a biodiversity-related interest group. Again, this becomes a special form of scoping and/or public participation, one for which great sensitivity is recommended.

13.6   Cultural Resources

When NEPA was originally examined, a majority of the attention of government agencies was directed toward the requirements of Section 102, which contains the requirement to prepare environmental assessments and impact statements. The often-quoted words of Section 102(2)(C), which begin "Include in every recommendation or report on proposals for legislation and other major Federal actions significantly affecting the quality of the human environment, a detailed statement..." (see Appendix A) were not originally interpreted to include aspects of the social, cultural, and economic environment. Examination of Section 101 of the act, however, finds two separate references to considerations which relate to cultural resources. The wording of Section 101(b)(2), "[to] assure for all Americans safe, healthful, productive *and esthetically and culturally pleasing surroundings*" and of Section 101(b)(4) "[to] preserve important *historic, cultural and natural aspects of our national heritage...*" certainly show that Congress had not intended to ignore this type of consideration.

In fact, several years before NEPA, Congress had passed the National Historic Preservation Act of 1966 (NHPA) (Public Law 89-665; 16 USC 470 et seq.). In the declaration of policy in Section 2, the act provided, among others, that "It shall be the policy of the Federal Government, in cooperation with other nations and in partnership with the States, local governments, Indian tribes, and private organizations and individuals to (1) use measures, including financial and technical assistance, to foster conditions under which our modern society and our prehistoric and historic resources can exist in productive harmony and fulfill the social, economic, and other requirements of present and future generations; (2) provide leadership in the preservation of the prehistoric and historic resources of the United States and of the international community of nations" [(PL 89-665, Section 2 (1) and (2)].

Note that none of this wording specifically mentions the environmental impact assessment process—remember that it antedates NEPA—although its status as a relevant consideration cannot be questioned. The converse is also true. NHPA has its own assessment process. Many decisions which do not require NEPA consideration *will* require the application of one or more provisions of NHPA and its implementing regulations (36 CFR 800). Further, meeting the requirements of one does not, automatically, assure compliance with the other. While the NHPA is, and will likely remain, the basis of most cultural resource consideration and rule making, other legislation, such as the Native American Graves Protection and Repatriation Act of 1990, is also influencing the assessment of environmental consequences of many federal projects and programs.

What are cultural resources?

In the strict definition of the term *cultural,* it would appear that almost any element which relates to our culture, or any culture, would qualify for examination. In practice, the scope is somewhat more limited. First, the term has come to mean the sum of historic, archaeological, Native American, and other resources which antedate modern American culture (generally 1950, with some exceptions). The term *historic resource,* considered synonymous with *historic property,* is defined as "...any prehistoric or historic district, site, building, structure, or object included in, or *eligible for inclusion in,* the *National Register*; such term includes artifacts, records, and remains which are related to such a district, site, building, structure, or object" (ACHP, 1989). Thus a historic property does not refer strictly to a building, as is commonly believed, but has a much wider definition. The inclusion of the phrase "or eligible for" also has significance in practice, since it means that a property need not have been formally listed or nominated for inclusion in the *National Register of Historic Places* to be protected under the terms of the NHPA.

What is our responsibility under NHPA?

Broadly speaking, the holders of U.S. government property have two responsibilities under the NHPA. First, there is an explicit requirement to take into account the effects of overt actions, such as construction, remodeling, demolition and excess of the property, on those historic properties held. Note that, as discussed above, the "property" may be an archaeological site—even an *undiscovered* site—in addition to a building. Nor does it have to be "listed" (i.e., in the *National Register*) to be considered—just be eligible for listing. Second, there is also a requirement to survey these holdings to discover what may not be known already and to nominate eligible sites to the *National Register.* These two responsibilities derive from Sections 106 and 110, respectively, of the NHPA, and will be discussed in that relationship.

Section 106 responsibilities. The wording of Section 106 of the NHPA, as amended, requires the following (16 USC 470f):

> The head of any Federal agency having direct or indirect jurisdiction over a proposed Federal or federally assisted undertaking in any state and the head of any Federal department or independent agency having authority to license any undertaking shall, prior to the approval of the expenditure or any Federal funds on the undertaking or prior to the issuance of any license, as the case may be, take into account the effect of the undertaking on any district, site, building, structure or object that is included in or eligible for inclusion in the National Register. The head of any such Federal agency shall afford the Advisory Council on Historic Preservation established under Title II of this Act a reasonable opportunity to comment with regard to such undertaking.

How does one go about providing this "reasonable opportunity"? In fact, the 2001 regulations which implement Section 106 (36 CFR 800) are far more specific, especially with regard to minimizing potential for adverse effects on historic properties. As with endangered species, a consultation process is mandated. The main parties to this consultation are the proponent agency and the State Historic Preservation Officer (SHPO). The anticipated outcome of this consultation may be a

Memorandum of Agreement (MOA) in which the agency and the SHPO mutually agree on what steps will be taken to maintain, preserve, or mitigate adverse impact to the historic aspects affected. The MOA is then reviewed by the Advisory Council on Historic Preservation (ACHP), an independent agency founded under the NHPA for this purpose. If the agency and the SHPO agree upon the conditions of the MOA, the ACHP normally accepts the terms. It may, however, request changes or prepare an independent evaluation, which the agency is then obligated to consider. This process is outlined in Fig. 13.1. If informal consultation reveals no impacts to resources, an MOA may not be required.

Normally, the "undertaking" means a new proposal or project, such as construction. This may affect historic properties in four ways. First, a historic structure may be proposed to be remodeled or otherwise modified. Second, a register-eligible structure may be proposed to be demolished to make way for the new project. Third, the soil of the site—even under pavements and buildings—may contain historic or archaeological sites which are not visible. Fourth, the proposed construction site may be within a historic district or other area where the *character* of the district will be adversely affected by new construction, especially that of a design not consonant with the style of the district.

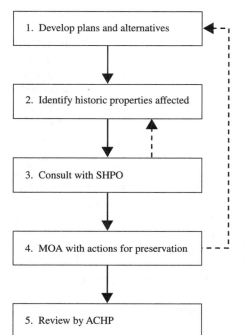

Figure 13.1 The Section 106 review process. (This figure is modified from Figure 1, p. 2 of *A Five-Minute Look at Section 106 Review,* ACHP, 1986 and 1989, and from Figure 1, p. 16 of *Section 106, Step-by-Step,* ACHP, 1986.)

All four possibilities must be considered in every proposed action. Many planners are aware of the first two problems; few consider the third and fourth, especially if no obvious, aboveground historic structures are present.

The steps shown are described as follows:

1. *Develop plans and alternatives:* This is the same as the first stage of the assessment process, and should meet all agency and NEPA criteria.

2. *Identify historic properties affected:* For each action under each alternative developed above, identify which, if any, historic properties may be affected, and to what degree. This is normally performed through a survey by professional architectural historians and/or archaeologists. Remember, as discussed above, the four possible ways in which a property may be affected.

3. *Consult with SHPO:* The identification of properties and the ways in which they may be affected is brought to the SHPO, along with your agency's plans to preserve or record the resources, if present. The SHPO may agree, may identify other cultural values not covered in your plans, may recommend other preservation measures, or may recommend against proceeding with the proposal at all. The process may end here if no significant effects are found after mitigation. The broken line from step 3 back to step 2 (Fig. 13.1) represents the possible identification of additional resources to be considered.

4. *Prepare memorandum of agreement:* Your agency and the SHPO prepare a mutually agreed upon statement of what accommodations to cultural resource values are required before your proposed action may be implemented. The broken line from step 4 back to step 1 (Fig. 13.1) represents the possible need to modify project actions or alternatives to comply with the provisions in this MOA.

5. *Review of proposal and MOA by ACHP:* The MOA and your project plans are reviewed by the ACHP. If agreement has been reached with the SHPO, few changes normally result. If there is no MOA, implying inability to reach agreement between your agency and the SHPO, and the potential for damage to a historic property is high, the ACHP may mediate or reach a finding through its own study. In this case, modification of plans and actions may also be required.

Section 110 responsibilities. The charge under Section 110 of the NHPA is much more open-ended and more general. Broadly speaking, it is a requirement to survey the agency's holdings in order to locate properties which may be eligible for listing on the register. Clearly, this means those buildings and structures which are standing, whether used or not. Less obviously, it means the survey of the land area for evidence of historic or prehistoric occupancy. These sites, *whether visible or still buried,* are also historic properties under the definition of the NHPA. Many large landholders have not completed this requirement. In practice, such studies, which must be performed by qualified professional personnel, may be rather costly. The normal response has been to conduct only the spot surveys required under Section 106 when a project is undertaken.

Native American cultural resources issues

The National Historic Preservation Act, important as it is, is not the only possible source of cultural resources conflict or need for consideration during the environmental assessment process. Many issues which are part of the vaguely defined aesthetic environment may be considered "cultural resources." Issues which relate to Native Americans (and other ethnic groups), especially those which have religious overtones, are especially important. Some of the issues which arise most frequently in the environmental assessment process are briefly discussed below.

The different tribes of American Indians, as well as Aleuts, Eskimos, and native Hawaiians, have an extremely different cultural background from that of the dominant western European culture of North America. Coming to a head in the 1990s, their concerns are being taken more and more seriously in government decision making, including environmental assessments. While this entire issue is very broad, most of the largest issues are related to the following questions.

Treaty rights to hunting and fishing. There are literally hundreds of treaties between the U.S. government and various tribes, many of which give almost unlimited rights to tribal members to hunt and fish on traditional lands. At the time they were written, the implications for this century could not have been foreseen. In general, no state, at that time, had ever set hunting or fishing seasons, required licenses, or set limits on numbers taken by any persons. Such "conservation" regulations came approximately at the turn of the last century. For decades, it was assumed that tribal rights were to be exercised only within the scope of such state regulations. In the 1980s and 1990s, court decisions have said that tribal rights could be exercised with only minimal control by state (or federal) fish and game departments. If the lands included within your department's proposed action are subject to such treaty rights, the issue may become serious. It is recommended that such problem areas be identified early in the planning process, and accommodation made to these rights, if present. Failure to include such issues in the assessment is a serious omission. These issues may be considered social issues or cultural resource considerations, as appropriate.

Sacred and traditional places. Many landscape features were held to be sacred or otherwise honored by Native Americans. Mountains and lakes are the most commonly identified. Since many native religions contain elements of secrecy as to the basis of their most revered beliefs, a full identification cannot be made by the agencies. Recent trends have allowed relatively full freedom of access to such sites for worship or other purposes, even if the land is managed by an agency and devoted to another purpose. It is also believed that many thousands of visits are made by tribal members without formal notification. Should the project or action which your agency is planning involve such a sacred site or traditional place, this must be determined early in the planning process or the scoping process, as discussed in Chapter 11. Generally the affected should be consulted. Because such questions are extremely sensitive, it may be necessary to locate a person with whom the Native Americans are willing to even discuss the question. The responsibility is still, however, that of the proponent.

Artifacts. In the past, many artifacts created by Native Americans were bought, traded, or taken from individual tribal members. Some were merely craft items which the individual had the "right" to sell or transfer. Many others were sacred tribal possessions, intended to be preserved for later generations. Many claims were made in the 1980s and 1990s for return of such objects, even when they had been in the possession of museums and collectors for decades. Increasingly, such claims are being honored in the courts when original ownership can be proven. Most government agencies are not involved to any great degree with artifacts—museums are an exception here—so this issue does not as often arise in a NEPA context. If the action proposed involves the collection or display of any such artifacts, their context must be discussed in the assessment. Artifacts associated with burials are a separate issue, and are discussed below.

Human remains and burial artifacts. The question of the disposition of the remains of Native Americans has become increasingly contentious, and led to the passage of the Native American Graves Protection and Repatriation Act of 1990 (NAGPRA). The provisions of this act state, in summary, that any remains of Native Americans and associated funerary objects must (1) be professionally curated in an accredited institution and (2) be made available, upon request, to the descendants of the person for return to a traditional tribal burial place. Even prior to passage of the act, many requests had been made to museums and other institutions for return of such materials. Some were honored; some were not. If there is any possibility that a burial may be encountered during the course of implementation of your agency's project, you must comply with the provisions of NAGPRA. The environmental documentation should specifically state how such remains will be handled, and what institution will assume responsibility. Neglect of these provisions may lead to extremely contentious public opposition to your proposals.

Human remains from postsettlement (i.e., European settler) burials are already covered by requirements in the laws of most states. The requirements are similar to those in NAGPRA, in fact, and usually involve notification of a state officer, removal of remains under supervision, an attempt to identify the remains, and reburial in a place agreed upon by the state and the descendants, if any, of the persons involved. The fact that Native American remains were usually exempted from these state laws was one of the reasons for passage of NAGPRA. Native Americans believe that the remains of their ancestors deserve proper respect and consideration.

The interest of the many organized Indian tribes in their cultural history has increased greatly since the 1980s. Thus, issues created by real or apparent conflicts are more likely to result in an organized response, frequently by a tribal attorney. Strict adherence to the regulatory requirements during the implementation of the project will need to be guaranteed. One way in which some larger agencies are handling this issue is through prenegotiated agreements on the manner in which events, such as inadvertent discovery of remains, will be managed. If such an agreement exists, its provisions must become a part of the discussion of cultural resources within the EIS.

Cultural resources and NEPA

The requirements of the NHPA, especially those of Section 106, are mandatory components of any environmental assessment. In practice, this means that it is the responsibility of the proponent of the action to positively verify, through original survey work if necessary, that no significant resources are present, *or* that they will not be affected by the proposed action, *or* that the mitigative measures are acceptable to the SHPO. Again, as with endangered species, it may not be acceptable to use vague wording such as "no historic resources are known from the site" if there has not been a qualified *survey* of the site. For construction projects of any kind, this has come to mean a site survey for every project *unless* there has previously been a complete, qualified survey of the area. This is where the requirements of Sections 106 and 110 meet. If the agency holding the property has long-term plans for development, it may be less expensive in the long run to conduct a full survey, thus meeting Section 110 requirements. The presence of a qualified survey under Section 110 will allow preparation of the environmental assessment and the Section 106 consultation to proceed as rapidly as possible.

The discussion of Native American issues, where present, is never to be omitted from the NEPA process. Native American tribes are specifically named, within the NEPA regulations, as bodies to be consulted when issues developed are relevant to them. The requirements of NAGPRA may not be the only ones which apply. This should be determined early in the scoping process.

In what ways may the consideration of cultural resources alter the plans which the agency has prepared? There may be no changes at all, especially if, in the planning process, the need to consider cultural issues was identified early. An historic property need not *always* be preserved unchanged. It may not have significant values, and the SHPO may require only recording the plans and archival quality photographs. Or, a structure may be relocated if the SHPO agrees. Even an archaeological site may be safely preserved under a paved area, so long as it is recorded and agreed to in the MOA. Since the significance of an archaeological site often lies in its research potential, adverse effects may be mitigated by partial (or complete) excavation and evaluation of the site before proceeding with the action. It is when such issues are not recognized early in the process that the most severe problems arise. Allow your agency to make the best use of its time and funds by raising these questions at the earliest stages of the project planning. This is the time when the lowest cost is associated with modifications in project planning.

13.7   Ecorisk

What is ecorisk?

Ecological risk assessment, often called *ecorisk* for short, is a systematic process for analyzing the risk, or likelihood of adverse effects, to the ecology of an area in response to human activities. The activities may be either contemplated (proposed) or ongoing. An ecorisk analysis provides a decision maker with a reasoned set of environmental information to use for effective resource management, to either avoid adverse impacts or minimize adverse effects. The essence of ecorisk assessment is its scientific approach to developing multiple lines of ecological evidence through multivariate analyses or other comprehensive analytical processes.

In 1998 the EPA issued a set of guidelines to cover the ecorisk process and applications. The EPA guidelines (EPA, 1998a) define ecorisk as follows:

Ecological risk assessment is a process that evaluates the likelihood that adverse ecological effects may occur or are occurring as a result of exposure to one or more stressors. The process is used to systematically evaluate and organize data, information, assumptions, and uncertainties in order to help understand and predict the relationships between stressors and ecological effects in a way that is useful for environmental decision-making. An assessment may involve chemical, physical, or biological stressors, and one stressor or many stressors may be considered.

The EPA *Terms for Environment* (EPA 1998b) includes the following definitions:

Ecological risk assessment: The application of a formal framework, analytical process, or model to estimate the effects of human action(s) on a natural resource and to interpret the significance of those effects in light of the uncertainties identified in each component of the assessment process. Such analysis includes initial hazard identification, exposure and dose-response assessments, and risk characterization.

Risk management: The process of evaluating and selecting alternative regulatory and non-regulatory responses to risk. The selection process necessarily requires the consideration of legal, economic, and behavioral factors.

As is the case with the environmental analyses done under a NEPA review, ecorisk is but one tool available to the resource manager to make informed decisions. Other factors outside the environmental arena that may be of interest to the manager include things such as social parameters, technical considerations, cost, or legal considerations (CENR, 1999).

Ecological risk assessment is a comparatively recent term, having come into common use only since the late 1980s (Calabrese and Baldwin, 1993). As discussed elsewhere in this book (see Chapter 1), the United States slowly became aware of the insidious social, economic, and environmental consequences of pollution and other types of adverse impacts in the last third of the past century. Before then, the adverse effects of industrial pollution or urban development were considered to be part of the necessary cost of doing business, or "not my problem." However, starting in the 1960s the federal government and state agencies began to become aware that adverse impacts upon human health and welfare from industrial development were not inevitable, but could be forecast, managed, and in some cases avoided or mitigated before they occurred. Close on the heels of the awareness of adverse impacts on human health was the growing awareness of the consequences of human activities on the environment and the sometimes-hidden interrelationship between human activities and environmental impact.

This new awareness of the relationships between human activities and environmental effects was reflected in the cornerstone of environmental legislation, NEPA. NEPA, drafted in the 1960s and signed into law on January 1, 1970, established (NEPA, Sec. 2)

...a national policy which will encourage productive and enjoyable harmony between man and his environment; to promote efforts which will prevent or eliminate damage to the environment and biosphere and stimulate the health and welfare of man; to enrich the understanding of the ecological systems and natural resources important to the Nation.

In the 1980s, in response to NEPA and other environmental laws, researchers began to use risk assessment as an analytical technique to predict and describe the risks of human activities on human health. In parallel with the application of risk assessment to human health, scientists began to use similar techniques to analyze the risk to ecological parameters of value. As the understanding of risk to human health from chemical or physical stressors evolved, researchers began to apply similar risk assessment techniques to understanding the risk of human activities on specific factors of the environment (Calabrese and Baldwin, 1993). Terms for similar environmental assessment processes are

- Ecological risk assessment
- Hazard assessment
- Cumulative ecological assessment
- Environmental impact analyses
- Bioassessments
- Ecotoxicity analyses (ecotoxicology)
- Habitat evaluation

In contrast with assessing the effects of human activities on the environment at large, as is generally done in a NEPA review, the ecorisk technique generally is applied to the estimation of risk to specific species or to ecosystem attributes imposed by specific chemical pollutants (Suter, 1993). Ecorisk analyses tend to focus on those environmental receptors that serve as an indicator species or a measure of ecosystem health, such as certain amphibians; those that are valued as a national or regional resource, such as threatened and endangered species; or those that are economically important, such as agricultural products (CENR, 1999).

Over the past decade analytical techniques have improved, such as development of faster computers and better modeling codes, and our understanding of toxicological effects and causal relationships has improved. This has allowed for a more precise estimation of the effects of stressors on specific ecological receptors. However, in many cases toxicological effects on species of concern are still not well understood, in terms of either cause and effect, exposure pathways, physiological response to specific stressors, or reaction variation within the species. This hinders the ability of researchers to give resource managers precise predictive information to support risk-based decisions. Ecorisk techniques allow the risk assessor to estimate the degree of uncertainty inherent in risk predictions and acknowledge situations where data may be incomplete or limited. The challenge for ecorisk assessors and resource managers in the decades to come will be the inclusion of better science in the environmental decision-making process in order to achieve sustainable systems of resource management and use (NCSE, 2000).

Uses of ecorisk analyses

Ecorisk is often associated with environmental remediation work such as pollution prevention, waste management, environmental restoration, and environmental cleanup. However, ecorisk also can be applied to more traditional impact assessment or ecological management areas such as protecting threatened and endangered species, managing wildlife habitat, improving agricultural or silvicultural practices, and managing fisheries (EPA, 1998a). Ecorisk can also provide a credible scientific basis for the environmental impact analyses performed under a NEPA review, or a floodplains and wetlands assessment. However, because of limitations in the knowledge base discussed above (e.g., limited understanding of ecotoxicology), the results of ecorisk assessments are often most useful for helping scientists focus on problem areas of greatest concern, and defining the need for and path of future investigation.

Ecorisk can be integrated into an agency's decision-making process, such as baseline surveys, environmental monitoring, and implementing decisions. A resource manager may want to use an ecorisk screening process to help guide fieldwork plans. If faced with limited resources (and most agencies are always under time, budget, or personnel constraints), a resource manager may want to concentrate baseline field studies or data collections in areas of greatest interest instead of examining a large, amorphous area. In such cases the manager may want to use the ecorisk process to identify specific, smaller areas of greatest concern to a species in question or a specific resource management decision. Ecorisk can provide a means to identify and acknowledge data gaps, which in turn can guide and influence ecological research.

Ecorisk analyses also can be applied to environmental monitoring. Environmental monitoring is the process of sampling and analyzing specific environmental media (such as soil, water, or plants) for evidence of contaminant levels over time. Environmental monitoring thus provides a measure of actual environmental impacts (as opposed to

NEPA analyses, for example, which forecast or predict the environmental impacts that would be expected). Ecorisk techniques can be applied to design a study methodology for environmental monitoring, determine uncertainties within the study area parameters, and serve as an analytical basis to determine the severity or consequence of measured impacts to the environmental media of concern.

Ecorisk analyses can be used to prioritize resource management needs, which may in turn help a manager prioritize decisions to be implemented first, and help establish what risks can be effectively and economically mitigated. Ecorisk analyses also can serve as a framework to monitor the effectiveness of risk-based resource management decisions over time.

Many federal and some state agencies have developed their own approaches to considering ecorisk (CENR, 1999). For example, the Department of Energy's Oak Ridge National Laboratory, in Tennessee, has developed extensive detailed methodologies to use to guide remediation decisions and management of the large Oak Ridge Reservation (Suter, et.al. 1995). These methods have been widely adopted at DOE sites, particularly those sites that are subject to CERCLA Superfund requirements.

Ecorisk analysis process

Ecological risk assessment provides a structured means to define an environmental problem, design a process to address that problem, use scientific means to collect and analyze data, and develop information that is of use to the resource manager. It provides a process for organizing and analyzing data, assumptions, and uncertainties to evaluate the likelihood of adverse ecological effects (EPA, 1998a). This is especially useful for complex ecosystem problems that include many variables.

Ecological risk assessments are similar to human health risk assessments and use the same general steps:

- Identify hazards
- Identify receptors
- Assess exposure
- Determine response
- Characterize risk
- Identify mitigation measures

The EPA describes the risk assessment process as having three phases: (1) problem formulation, (2) analysis, and (3) risk characterization. Following the risk assessment, the risk manager will take three more steps: (1) decide upon a course of action, (2) implement the decision, and (3) evaluate the actual effect of the decision over time (EPA, 1998a). The risk manager is charged with providing good management decisions, while the risk assessor is charged with providing good science.

Because the value of the ecorisk process is its grounding in science, the first step, problem formulation, is very important. It is through this step that the risk assessor and the resource manager agree on the actual environmental value that is to be protected, and link the analysis to the management decision facing the resource manager. The manager may want to gain additional input from other agencies or the general public to help define the entity that is important and is potentially at risk, similar to the information collecting aspect of the NEPA scoping process. Interaction between the risk assessor and the risk manager is critical to ensure that the information generated during the assessment process is relevant to the decisions the risk manager needs to make. The risk assessor must define an endpoint for the analysis, determine what conceptual models, if any, will be used, develop an analysis plan to be followed, define the value or species that is potentially at risk, and define the past, current, or potential human activities that may lead to the risk (and the related chemical release, or pollution). Once the assessor knows what chemical or toxic constituents may be present and what species are at risk, the assessor must determine possible exposure pathways for that species, the ecological effects of the chemical, and how it interacts with ecological receptors. In some cases, physical stressors such as heat, noise, or light may be of interest instead of chemical stressors. During this planning stage, the EPA suggests that a risk assessor define data quality objectives, which are the levels of confidence and certainty needed for the management decisions at hand and public confidence. Data quality objectives help determine the boundaries of a study as well as evaluate the quantity and quality of data necessary for the study; by looking at alternative methods, the assessor will optimize the analytical design (EPA, 1998a).

Through the ecorisk analysis phase the risk assessor will characterize the exposure of the receptor to the stressor, and characterize the ecological effects. The steps of this phase must be accomplished interactively. The assessor will define the source of the stressor and its distribution within the area of study, identify ecosystem receptors of concern, and identify areas where the stressor may come into contact with the ecological receptor. In the second step of this phase, the assessor will characterize the relationship between stressors and ecological effects, and analyze whether the projected or observed response is actually a response to a given stressor. After considering cause and effect, the assessor will perform an uncertainty analysis of the risk information, and then prepare summary profiles that describe exposure and the stressor-response relationship.

In the risk characterization phase, the risk assessor will estimate the risk of the stressor's having an adverse effect on the receptor. The assessor may rely on complex computer models or empirical field observations. After integrating information on exposure and the stressor-response profile from the second phase, the assessor will describe the risk by presenting the evidence and discussing how the determination of ecological adversity was derived. Any measure of ecological adversity must be technically defensible (science-based) and meaningfully reflect management goals. As a final step of the risk characterization, the assessor will write a report to communicate the findings of risk to the risk manager, other interested parties, and the public at large.

The ecorisk process is as follows (EPA, 1998a):

Phase 1: Problem formulation

- Evaluate goals and select assessment endpoints.
  - What is the study area?
  - What is at risk (endpoints)?
  - What stressors are present?
- Prepare the conceptual model.
  - Computer codes.
  - Empirical data.
- Develop an analysis plan.
  - Data needs and data gaps.
  - Data quality objectives.
  - Data collection design and methods.
  - Time frames.
  - Data analysis process.

Phase 2: Analyses

- Characterize exposure to stressors.
  - Source of stressors.
  - Distribution of stressors in the environment.
  - Occurrence of ecological receptors.
  - Distribution (in time and space) of ecological receptors.
  - Interface (contact or co-occurrence) between stressors and receptors.
- Characterize relationship between stressors and ecological effects.
  - Stressor-response relationship (cause and effect).
  - Uncertainty analysis.
  - Summary profiles.

Phase 3: Risk characterization

- Estimate risk through integration of exposure and stressor-response profiles.
  - Indicate the degree of confidence in the risk estimates.
- Describe risk and adverse ecological effect.
  - Cite evidence to support the risk estimate.
  - Describe and interpret the adversity of ecological effects.
- Prepare ecorisk report.
  - State results.
  - Provide major assumptions and uncertainties.
  - Identify reasonable alternative interpretations.
  - Separate scientific conclusions from policy judgments.

Although early application of ecorisk was as a tool for determining remediation decisions, over the past 10 years the application of ecorisk as an analytical and management tool has moved beyond consideration of environmental remediation or cleanup. For example, ecorisk is very suitable as an analysis tool for habitat management considerations, especially in relationship to an urban or industrial setting. This has implications for assessing and mitigating the effects of contaminants on the habitat or food supply of threatened or endangered species. The multivariate analyses of ecorisk can be used to simulate diet and other exposure pathways, predict habitat ranges, postulate toxicity response, generate information about hazards, and cast cumulative effects from multiple contaminants found within the range of a given animal (Gonzales et al., 1998).

At heart, an ecorisk analysis is simply an exercise in environmental problem solving, and provides a way to correlate toxicological and ecological information to estimate the probability of risk of damage to the environment (Bartell et al., 1992). An ecorisk approach provides a way to organize and analyze assumptions and data, and identify analytical uncertainties. Because the relationship of stressors, ecological receptors, and susceptibility to adverse effect is complex, and because our knowledge of ecosystems is very incomplete, most researchers make use of various types of multivariate analyses and probabilistic risk analyses. Different researchers have developed various computer codes, sometimes integrated with geographic information system analyses (see Chapter 7), as tools to manipulate and array large quantities of spatial and receptor-specific data. For example, integration of spatial information contained in a geographic information system with species information and contaminant information can be analyzed by a computer to produce a three-dimensional array of risk to a specific species over its home range. (Gallegos and Gonzales, 1999).

The EPA envisions ecorisk as an iterative process (EPA, 1998a). Under this repeated, or tiered, approach, less complex assessments are employed first, using a greater level of conservatism and accepting a higher level of uncertainty. After this initial screening process, the risk researcher probably will be able to determine that some areas do not require additional analysis for managers to make reasonable risk-based decisions. However, where results of the initial screening process indicate an area of greater concern, or where the researcher is aware of a contaminant of greater concern, the researcher can employ progressively more complex and realistic assessments with a lower uncertainty. This will give more precise estimates of risk in these smaller focus areas. By using a tiered approach, the researcher and risk manager can relatively quickly screen out areas of marginal interest and concentrate time and resources on areas of greater risk or greater impact consequence.

Uncertainty and risk analysis

Through an ecorisk analysis, an assessor estimates cause and effect—the interaction of a stressor with an ecological receptor. However, these analyses are merely estimates, and in some cases, for a variety of reasons, the assessor may be much more certain that a given receptor will respond as predicted in the risk analysis. In other cases, the assessor will face a high degree of statistical uncertainty regarding the outcome of the risk analysis.

One of the identifying hallmarks of ecorisk assessments, as opposed to other types of ecological assessments, is the central premise of identifying the likelihood that a specific adverse impact will occur, and the degree of uncertainty (or certainty) in that estimate (Suter, 1993). For example, if a decision maker is certain that a specific event will occur (that is, the probability of occurrence is known to be 1), or is certain that it will not occur (that is, the probability of occurrence is known to be 0), there are no issues of risk uncertainty. However, this simplistic example would not be expected to be the case, particularly with multivariate situations. An uncertainty analysis allows the risk assessor to address the uncertainty (or certainty) associated with estimating both the likelihood of occurrence and the scale (magnitude, intensity, or duration) of the impact. In practice, for the purposes of an ecorisk analysis the risk assessor often assumes the probability of occurrence to be 1.

Uncertainty is inherent in all risk estimates, and may arise from several sources. Uncertainty can be introduced by a researcher or may arise from unknowns, inaccuracies, or natural variability (stochasticity) (Suter, 1993; Cohrssen, and Cavello, 1985). It can also arise from the uncertainties in other, related, sampling or statistical analyses (experimental variability). These uncertainties affect all stages of the risk assessment, such as hazard identification, exposure assessment, and dose-response assessment. Uncertainty is generally expressed as a probability distribution or a confidence interval, or determined through a sensitivity analysis. Sometimes a researcher merely acknowledges that some degree of uncertainty exists, and does not try to quantify the degree of uncertainty. Pinpointing the sources of uncertainty can identify areas where more scientific information is needed. Currently in the field of ecorisk assessment, one of the most problematic sources of uncertainty is the uncertainty that a given organism or group of organisms in nature will respond to a specific contaminant in the same way as a test species (which may be very different from the receptor of interest) responded under laboratory conditions.

For example, if a resource manager has a need to control flooding in a drainage channel, a proposal might be to build a flood retention structure. Aside from engineering considerations, there might be a question of where to place the structure to minimize risk to wildlife. The ecorisk assessor may help by providing a multivariate risk analysis. The analysis might correlate the 100-year flood level, animal species that would be affected in the event of a flood, and sediment contaminants that might be translocated during a flood event. Although by definition a 100-year flood is that which statistically would be expected once every hundred years, in actuality, although statistically unlikely, a 100-year flood can occur in any year or for several years in a row. This is an example of statistical uncertainty. Although the researcher may find out from a wildlife biologist what species are known to inhabit the floodplain, neither expert can predict with certainty what species might be present on the day of the hypothetical flood. This is an example of natural variability. The researcher may ask an ecologist to identify sediment areas with contaminants, but the ecologist may not know with certainty the exact location, extent, concentration, or toxicity of the contaminants. This is an example of experimental error, or the difference between the actual attributes and the measured attributes. If the ecologist thought a contaminant was present but in fact it was not, this would be an example of an inaccuracy. If the ecologist did not know or did not postulate that a contaminant was present, this would be an example of an unknown. These individual uncertainties can be combined mathematically to produce an overall estimate of uncertainty. This information would help the resource manager understand the variability of the risk analysis.

This text does not attempt to provide an in-depth explanation of statistical parameters. A more complete discussion of definitions and measurement of uncertainty, error, and confidence can be found in essentially any basic textbook on statistics or experimental design.

Implementing risk-based decisions

Risk management and risk assessment are two distinct activities, though interrelated. The reason that ecorisk analyses are done is to help make better management decisions by providing scientifically based risk-receptor correlations. In order for this to occur, the decision maker must be aware that a risk analysis has been done and of the probable environmental consequences of the decisions to be made. In this way the decision maker can consider the risk of impacts from various courses of action, how certain this risk might be to occur in a given case, and how some courses of action may lessen the risk to a greater degree than other courses of action. This is referred to as risk management, or making a risk-based decision. The risk assessor and risk manager must be aware that they play different roles in the process. While a risk assessor is generally interested in the science of data acquisition and risk analysis, the risk manager is often more interested in balancing the ecorisk analysis with other policy, regulatory, or social considerations and may not be in a position to focus on scientific nuance or technical details (CENR, 1999).

The steps in risk-based decision making are as follows (EPA, 1998a):

Step 1: Make a risk-based decision
- Review risk characterization report
- Review options and mitigation measures
- Balance risk analysis information against other decision factors
- Select a course of action
- Communicate the decision to interested parties

Step 2: Implement the decision
- Translate decision and mitigation measures into project engineering or design
- Put mitigation measures into place
- Carry out actions

Step 3: Monitor decision over time
- Was decision properly implemented?
- Did the mitigations reduce risk?
- Is ecological recovery occurring?
- Is another tier of assessment warranted?

At some point, the risk manager will make some decision regarding some aspect of resource management. A decision, however well intentioned or well thought out, is meaningless unless implemented. For example, a forest manager may decide to protect the habitat of migratory birds, but this decision cannot be effectively implemented without knowing what species of migratory birds would be expected to be present in a given area, when they might be there, what type of habitat they prefer, the reaction of specific bird species to environmental stressors such as noise or toxic chemicals, the consequence, or threat, to the habitat areas that might be caused by planned activities, and the response of the species to degradation or decrement of habitat. If the decision maker is aware of these factors, and understands the relationships between the bird species and their ecosystem, the manager can take proactive measures to ensure that the bird habitat is protected, or at least not unduly degraded. Absent this information on the birds and their environment, the manager may take no actions that serve to protect the bird habitat, or take actions that are inappropriate or ineffective in protecting the habitat. In either case, although the manager decided in principle to protect the habitat, the decision was not implemented and the habitat was not protected.

An ecorisk analysis is done to help a resource manager make a decision on a specific proposal. The risk analysis is one piece of information, but not the only piece of information, that is available to the decision maker for consideration during the decision-making process. To ensure that resource decisions are in fact risk-based decisions, the ecorisk assessor must communicate the findings of the risk assessment and mitigation recommendations to the manager in sufficient detail and at the most appropriate time for inclusion in the decision-making process. This is generally done through the ecorisk report prepared at the end of the risk characterization, which is the final phase of the ecorisk assessment.

The ecorisk report may also suggest options that may be employed to either protect the species analyzed or lessen adverse impacts to the species. For example, an ecorisk report on a proposed remediation activity might find that a specific amphibian species is best served if no cleanup is undertaken, because cleanup might resuspend contaminants in surface water. However, the decision maker may wish to proceed with cleanup because of public pressure. The ecorisk report then may suggest ways to lessen the risk to the species through mitigation measures, such as sediment traps that would reduce runoff and lessen the chance that contaminants would reach the surface water.

Once the resource manager has selected a course of action, the specifics of the decision must be incorporated into project design or engineering practices. The ecorisk report may be a tool to communicate the specifics of the risk-based decision to project engineers or construction personnel. The best-thought-out decisions and mitigation measures will be ineffective if the person wielding a chainsaw or driving a bulldozer does not understand that specific groves of trees must be left untouched or that specific areas of soil must be left undisturbed.

The responsibility of the resource manager does not end when the construction project is finished or remediation completed. Assuming that the risk-based decisions and mitigations have been successfully carried out, the wise manager will monitor the situation over time to determine if the actions taken were truly effective in reducing risk to the environmental receptors. Because in many cases the relationship between stressor and receptor is not well understood, and because the ecorisk process gives the manager a way to make risk-based decisions in the face of missing or incomplete information or other forms of uncertainty, the resource specialist and the manager may not know if a planned mitigation measure or course of action will actually be effective in reducing risk over time. The ecorisk assessor can suggest a statistically valid environmental sampling plan that will yield contaminant monitoring information over time to corroborate or disprove whether the action taken did in fact reduce adverse effect.

Relationship to other environmental assessment processes

Ecorisk is an analytical tool that has application to many legal or regulatory drivers. This country has enacted a suite of federal laws that address protecting the environment from toxic chemicals. Ecorisk analyses may be helpful in addressing the assessment requirements of these laws. In particular, the Comprehensive Environmental Response, Compensation, and Liability Act of 1980 (CERCLA, or the "Superfund Act") requires both a human health risk assessment and an ecological risk assessment (Bartell et al., 1992). For other laws, while an ecorisk analysis is not required, it provides a scientific basis for disclosing adverse environmental impacts, and the uncertainties associated with the analysis where appropriate, such as impact assessments performed under NEPA or various state environmental disclosure laws.

Some of the environmental laws (see Chapter 2) that require or are facilitated by an ecorisk analysis are

- Comprehensive Environmental Response, Compensation, and Liability Act (CERCLA)
- Natural Resource Damage Assessment (NRDA) provisions of CERCLA and the Oil Pollution Act of 1990
- Resource Conservation and Recovery Act (RCRA)
- Endangered Species Act (ESA)
- Toxic Substances Control Act (TSCA)
- Federal Insecticide, Fungicide, and Rodentcide Act (FIFRA)

CERCLA, passed in 1980, requires retrospective evaluations of the effect of past contamination in a given area, and a determination as to whether or not the environment would be best served if cleanup or remediation is undertaken. Under CERCLA, risk managers are required to protect human health and the environment, and to comply with applicable, relevant, and appropriate requirements. The law sets up a process to identify Superfund sites, or significantly polluted areas, that require remediation, and by regulation establishes a National Priority List to prioritize polluted sites and the National Contingency Plan, or the Superfund Regulations, at 40 CFR 300, to facilitate cleanup. Under this law a lead agency is identified to be in charge of assessment, remediation, or cleanup of a given site, while the EPA retains overall authority for the process. The lead agency conducts a baseline risk assessment during the remedial investigation/feasibility study in order to characterize current and potential threats to human health and the environment. For human health risk analyses, the EPA looks for quantifiable levels of acceptable risk; however, the agency has not established comparable quantifiable risk goals for ecological risk. For any given site, the risk manager must determine whether contaminants on the site present an unacceptable risk to important resources; if they do, whether the site should be cleaned up or whether the remedy would do more harm than good; and if the site is to be cleaned up, how to select a cost-effective response that will provide adequate protection. Ecorisk can be used for each of these questions. It can help characterize baseline risk to determine whether a cleanup should be considered. The analysis can be used to derive levels of concentration of contaminants that would no longer pose an unacceptable risk to the environment. Ecorisk can help evaluate alternative approaches for remediation to determine which would be most effective, and whether any would increase the risk instead of decreasing it. Following cleanup, ecorisk analyses can be helpful in determining a monitoring plan to establish whether the remedy was truly effective in reducing risk. The EPA recommends that a tiered approach be used to allow the assessor to quickly identify and eliminate sites that are not at risk, or to identify sites that are known with great certainty to be at risk. Following this screening process, the assessor can concentrate on the remaining sites and perform a greater depth of analysis focused on the areas at question. This graded approach reduces analytical costs, and provides for more information to be generated to reduce uncertainties where needed (CENR, 1999).

The NRDA provisions of CERCLA and the Oil Pollution Act identify the need for assessing injury and damage to natural resources caused by spills or other means. The Department of the Interior has promulgated regulations for conducting damage assessments as provided in CERCLA (43 CFR 11), and the National Oceanic and Atmospheric Administration has promulgated regulations for conducting damage assessments as provided in the Oil Pollution Act (15 CFR 990). Damage assessments are conducted to calculate the monetary cost of restoring five types of natural resources—air, surface water, groundwater, biotic, and geologic—from injury that results from releases of hazardous substances or discharges of oil. Under the National Contingency Plan regulations (40 CFR 300), natural resource trustees with specific trust responsibilities over natural resources can claim injury in the event of resource damage; trustees are defined to include applicable states or Native American tribes, and five federal agencies—the Departments of Agriculture, Commerce, Defense, Energy, and Interior. Damage is evaluated by identifying the functions or services provided to the public by the resources and quantifying the monetary loss to the trustee from the reduction in service as a result of the discharge and the cost of restoration. For example, a spill of oil into a body of surface water might cause damage (injury) to a fishery if there is an economic loss (DOE, 1995). Ecorisk is applicable to determining the link between exposure to contaminants and adverse effect (injury); therefore, it is of use in establishing (or disproving) a causal relationship between a release or spill and natural resource damage. Ecorisk is not effective in assessing other aspects of natural resource damage, such as the determination of injury or monetary loss. The ecorisk technique is designed to evaluate baseline ecological risk, and often depends on an evaluation of specific sensitive species as an indication of risk to the ecosystem as a whole; however, in a damage assessment, the monetary damage may be related to loss of a different species, such as a game species, that is not the more sensitive indicator identified in the ecorisk analysis. If damage or injury has occurred, sometimes the trustee may want to evaluate restoration options to determine which might be most effective to bring a damaged resource back into service or to the condition it would be expected to have if the damage had not occurred. Ecorisk is well suited as an analytical technique to help the resource manager weigh restoration options (43 CFR 11; CENR, 1999).

The RCRA, as amended by the Hazardous and Solid Waste Amendments, requires federal resource managers to develop corrective actions where releases of hazardous waste or constituents have occurred, and to investigate and clean up contamination in areas designated as solid waste management units where waste products have been disposed. Like CERCLA, the actions required under RCRA are largely retrospective in that they focus on existing contamination (in contrast to laws such as NEPA, which forecast or predict future impacts of proposed actions). The remedial actions are carried out by the relevant lead agency, while the overall authority for the process rests with the EPA. The application of ecorisk to RCRA cleanup actions is essentially the same as for CERCLA cleanup actions, and in general follows the iterative process set forth in the EPA ecological risk assessment guidelines (EPA, 1998a).

The ESA and its implementing regulations (50 CFR 402) establish a consultation process to conserve the ecosystems that threatened or endangered species require, and to take measures to assist in recovery of these species. The Department of the Interior lists threatened and endangered species periodically, and the listed species change over time as new species are added and recovered species are delisted. The ability to effectively model risk to threatened and endangered species is central to the success of the act. Risk assessment comes into play in two distinct arenas: the risk to a species of extinction, and the management of related ecosystem resources to reduce the risk of contamination as a stressor to the species. These concepts, and the use of risk analyses in predicting the likelihood of species extinction, are outlined in a report of the National Research Council (NRC, 1995; CENR, 1999). Under the ESA, federal agencies must disclose when their proposed actions may affect listed threatened or endangered species, and if so must enter into consultation with the U.S. Fish and Wildlife Service, an agency of the Department of the Interior. Resource managers may have to consider the toxic effect of past contamination as well as predict the impact on listed species from planned or proposed actions. Ecorisk analyses can be of use for either of these types of impacts.

The TSCA, passed in 1976, regulates certain industrial chemicals such as solvents, polymers, adhesives, coatings, and plastics, but not chemicals such as pesticides or pharmaceuticals that are covered by other regulations. Manufacturers or importers of new chemicals are required to submit a premanufacture notification to the EPA. The agency must sift through a large number of applications (about 2000 per month) in a relatively short time (each application must be reviewed within 90 days). To facilitate its decisions, the agency applies a tiered ecorisk approach: If initial screening suggests a level of risk, the agency does a more detailed analysis. Ecorisk is a useful way for the agency to apply scientific information, determine the likelihood of ecotoxic effect to ecosystems even when there is little toxicity data, and compare potential ecological effect with potential exposure concentrations. The agency's experience under this act demonstrates that ecorisk can be conducted with minimal toxicity and exposure data, and that regulatory decisions can be made quickly using the best data available at that time (CENR, 1999).

The FIFRA regulates pest control substances such as herbicides, fungicides, rodenticides, and biological agents, and imposes certain labeling requirements on packages of these materials. The EPA is responsible for reviewing and registering existing and proposed new pesticides. The law is a cost-benefit statute, and the agency must make a regulatory determination that use of the pesticide would cause no unreasonable adverse effects to human health or the environment, while weighing other factors as well. Potential ecological effects of pesticides often can be mitigated by controlling application frequency, dose, area, or type of use, or imposing other restrictions. As is the case with TSCA, the agency uses a tiered approach: If initial screening, using a conservative approach, indicates a level of risk, the agency uses more refined means to identify and characterize that risk. Ecorisk provides the risk manager with a science-based method to integrate hazard and exposure assessments into a characterization of risk (CENR, 1999).

Summary

In summary, ecorisk (SETAC, 1997) is a resource management tool that can be used to

- Identify and prioritize the greatest ecological risks
- Allow decision makers to consider the consequences of various potential management actions
- Facilitate identification of environmental values of concern
- Identify critical knowledge gaps

Ecorisk provides an effective means to identify and characterize stressors caused by human actions and the ecological receptors that would be adversely affected by those stressors, and perform multivariate analyses on the interaction between stressors and receptors to characterize the risk to the receptors. It serves as a tool for resource managers to make risk-based decisions; this information is balanced against the other considerations before the risk manager, such as economic, regulatory, or social concerns. Through an ecorisk analysis, the manager can identify critical knowledge gaps, and identify and pursue research needs. Ecorisk is a scientific tool, and is based on statistical and other mathematical modeling techniques. To date, beyond assessing the potential for impact, ecorisk assessments have been especially useful to help focus further ecological research by identifying the most problematic contaminants, species, and geographical areas of concern.

Ecorisk, however, has limitations. Currently, ecorisk analyses are limited by a shortage of toxicological information that accurately conveys how species will respond in nature. Ecorisk is not effective for weighing trade-offs among different types of impacts, such as ecological impacts against cost considerations. It does not provide an analysis of every type of environmental or ecological impact; instead, it is focused on specific species of concern, which may have been selected because they serve as a measure of ecological sensitivity. Ecorisk does not provide a venue to make or document a final decision.

The value of ecorisk rests with its ability to provide a comprehensive scientific framework for ecological analysis. This allows a resource manager to make reasoned risk-based decisions to allow for improved resource management.

13.8 How Are These Contemporary Issues Different?

The seven issues presented fall into two broad categories. First are two topics, endangered species and cultural resources (especially historic preservation), where there is clear legislation, the publication of binding regulations, and an administering agency of the government responsible for compliance. They are "must comply" topics when, and if, the resources discussed exist in the project area. Each has several internal processing steps which apply, and outside consultation which is mandated in the regulations. The major pitfalls within these areas are in not remembering to consider them soon enough in the project planning (and environmental assessment) schedule. Compliance may require many months which were not allowed for.

The other five topics, global warming, acid rain, deforestation, biodiversity, and ecorisk, are all "must discuss" issues where they are relevant. At this time there are no laws or regulations generally covering any specific *compliance* for these issues. CFCs are a minor exception. They are all, however, "big ticket" scientific controversies at the present time. The technical reviewers of your document are likely to find fault with it if you do not raise the issue, where relevant, as a *contributory* element to environmental damage. Some actions may have a minor positive effect on one or more of these areas; others may be minor negatives. Unless the major premise of the agency's action is the alleviation of one of these problems, an in-depth discussion is not fruitful. The acknowledgement, however, of the *relevance* of the controversy to your proposed action and the acknowledgement of those effects which *do* relate to these topics is usually adequate coverage.

13.9 Discussion and Study Questions

1 What is your personal belief about the existence of the greenhouse effect? Do you believe its presence has been proven? Do you believe that it is contributing significantly to global warming? Discuss the evidence for and against these propositions.

2 Explain how postulated temperature increases of tiny fractions of a degree per year could eventually cause major changes in our way of life. What would be some of the most apparent consequences?

3 Discuss the dichotomy between the industries, states, and countries believed to be contributing most strongly to acid rain generation and those most affected by the phenomenon.

4 Where are the forests about which environmentalists are most concerned? Where are those about which industry is most concerned? Do you see any conflict in these views of the world? What changes in local lifestyle would have to come about if tropical forests were to be managed for continuous production? What changes would there have to be in the lifestyle of the developed countries?

5 Why was recognition of the concept of a threatened (as opposed to endangered) species an important advance in conservation legislation?

6 How does a biological assessment differ from an environmental assessment? How are they similar? When and where do they fit together within the NEPA context?

7 How does concern about ecological biodiversity differ from concern about endangered species? In what ways are they driven by similar concerns?

8 Do you believe that biodiversity should become a mandated part of every EA and EIS through amendments to NEPA? Discuss why or why not.

9 Discuss some of the way(s) in which biodiversity, species endangerment, the effects of acid rain, and tropical deforestation are interrelated.

10 Discuss the way(s) in which the NHPA and the ESA are similar to each other. How are they similar to the NEPA? How do they differ from the basic concept of NEPA?

11 Discuss Native American concerns about tribal artifacts and items once buried with the dead. Do you believe the problems which their return has caused to museums and collectors serve a legitimate purpose? Can you propose alternative means to accommodate these concerns? What would they entail?

12 How does the ecological risk assessment process compare to the analytical process used in an EIS? What types of environmental attributes used in a NEPA review lend themselves to ecorisk analysis, and what types do not? What aspects of ecorisk assessment are not useful for a NEPA review?

13.10   Further Readings

Abrahamson, Dean E., ed. *The Challenge of Global Warming*. Washington, D.C.: Island Press, 1989.
Falkowski, P., et al. "The Global Carbon Cycle: A Test of Our Knowledge of Earth as a System." *Science*, 290:291–296, 2000.

Glass, James A. *The Beginnings of a New National Historic Preservation Program: 1957 to 1969.* Nashville, Tenn.: American Association for State and Local History, 1990.

Gould, Roy. *Going Sour: Science and Politics of Acid Rain.* Boston: Birkhauser, 1985.

Gradwohl, Judith, and Russell Greenberg. *Saving the Tropical Forests.* Washington, D.C.: Island Press, 1988.

Intergovernmental Panel on Climate Change. Houghton, J. T., G. J. Jenkins, and J. J. Ephraums, eds. *Climate Change: The IPCC Scientific Assessment.* New York: Cambridge University Press, 1990.

Marcoux, Alain. *Population and the Environment: A Review and Concepts for Population Programmes. Part III: Population and Deforestation.* Food and Agriculture Organization, June 2000.

Matthews, John R., and Charles J. Moseley, eds. *The Official World Wildlife Fund Guide to Endangered Species of North America.* Washington, D.C.: Beacham Pub., 1990.

Miller, Kenton, and Laura Tangley. *Trees of Life: Saving Tropical Forests and Their Biological Wealth.* Washington, D.C.: World Resources Institute, 1989.

Prance, Ghillean T., ed. *Tropical Rain Forests and the World Atmosphere.* Boulder, Colo.: Westview Press, 1986.

Regens, James L., and Robert W. Rycroft. *The Acid Rain Controversy.* Pittsburgh, Penn.: University of Pittsburgh Press, 1988.

Smith, George S., and John E. Ehrenhard, eds. *Protecting the Past.* Boca Raton, Fla.: CRC Press, 1991.

U.S. Fish and Wildlife Service. "ESA Basics: Over 25 Years of Protecting Endangered Species." June 1998. http://www.fws.gov/r9endspp.html.

Epilogue

NEPA was written, almost 35 years ago, first as a fundamental policy statement. It provided the nation with an overarching environmental policy: "to create and maintain conditions under which man and nature can exist in productive harmony, and fulfill the social, economic, and other requirements of present and future generations of Americans" (43 USC 4331). In order to carry out the new policy, the second tier of the law provided that federal planners and decision-makers consider environmental ramifications alongside other pertinent factors such as cost and technical issues. As a third tier, the framers of the new law provided a new action-forcing device—preparation of documented assessments—to lay the process open to public scrutiny and assist federal agencies in determining environmental impacts of proposed actions.

Was NEPA successful in achieving its purpose? What can we now say about the real essence of NEPA? Did we, as a nation, work to achieve the law's policy goal to "prevent or eliminate damage to the environment and biosphere and stimulate the health and welfare of man?" Did we "enrich the understanding of the ecological systems and natural resources important to the Nation?" Or do we now view NEPA merely as "the National Environmental Impact Statement Act"—an analytical nuisance and a roadblock to real progress?

We can look back and agree that the quality of our national environment is better than it was thirty or forty years ago, thanks to laws such as NEPA, the Clean Air Act, and the Clean Water Act. Most federal agencies, and many states, routinely consider the environment when taking new actions. The public has learned the value of NEPA as a means to gain information, and as a conduit into the federal decision making process. The ecology sciences have made progress into providing a better understanding of how natural processes work, and databases of environmental attributes are more extensive than in the past. But these reflect more the secondary and tertiary processes of NEPA rather than the underlying policy. Given the fact that the emphasis has been, by and large, on the assessment process rather than the national policy, is NEPA a success? We think so.

After NEPA had been in effect for 25 years, CEQ conducted a study of the experience of federal agencies in performing NEPA reviews. The CEQ study found five factors critical to successful NEPA implementation: strategic planning, public information and input, interagency coordination, an interdisciplinary and place-based approach to decision making, and a science-based and flexible management approach for project implementation (CEQ, 1997). These can be seen as manifestations of NEPA's success.

One may ask, from a pragmatic point of view, what is it that agencies undertaking environmental review need to do? This book has discussed basic concepts associated with environmental assessment and has provided guidance in preparing various types of environmental assessment documents. The text includes international perspectives, methods for assuring effective public participation, contemporary environmental issues, energy use, and other environmental concerns. The issues are many. However, we recognize that with the flood of information ranging from procedural requirements, scientific information, and consideration of contemporary concerns, the essential purpose of NEPA may be lost or obscured. What can we say about the real essence of the NEPA process?

CEQ answered this question most profoundly when it stated, "NEPA is much more than environmental impact statements and environmental assessments. It is an eloquent and inspiring declaration which, well before the term 'sustainable development' became widely used, called for the integration of our varied aspirations as a society. NEPA is a tool with tremendous potential to help build community and to strengthen our democracy" (CEQ, 1997).

Building community and strengthening democracy are tall orders. How nations use their environmental resources to meet citizens' aspirations, economic needs, and social mores may ultimately determine not only the sustainability of their society but also what types of social, economic, and political institutions evolve as a result. A society that continues to ignore environmental concerns to obtain short-term economic, political, or military objectives is likely to create institutions that will not remain viable in the long run. Arguably, nations should recognize that there is indeed a global dimension to resource use, and should strive to meet basic social and economic needs while preserving, protecting, and enhancing the environment. Other countries recognize the United States' lead in this regard, and the assessment process prescribed by NEPA can be a tool to help them achieve these aims.

Beyond the familiar analytical review process, we see the future of NEPA as helping achieve the fundamental national policy of sustaining our environment. Although the focus of NEPA has often been the sometimes-voluminous documents generated by agencies to assess environmental impact, this is not the essence of NEPA. As is stated in the CEQ regulations, "NEPA's purpose is not to generate paperwork—even excellent paperwork—but to foster excellent action." The assessments and disclosures mandated by NEPA are merely the path to reach "excellent action." The authors of this book hope that in some small way this text will contribute to the reader's ability to navigate this path.

National Environmental Policy Act

This appendix presents the full text of the National Environmental Policy Act of 1969, enacted as Public Law 91-190. It was signed on January 1, 1970. The act was amended by PL 94-52, PL 94-83, and PL 97-258.

The National Environmental Policy Act of 1969, as amended

(Pub. L. 91-190, 42 U.S.C. 4321-4347, January 1, 1970, as amended by Pub. L. 94-52, July 3, 1975, Pub. L. 94-83, August 9, 1975, and Pub. L. 97-258, § 4(b), Sept. 13, 1982)

An Act to establish a national policy for the environment, to provide for the establishment of a Council on Environmental Quality, and for other purposes.

Be it enacted by the Senate and House of Representatives of the United States of America in Congress assembled, That this Act may be cited as the "National Environmental Policy Act of 1969."

Purpose

Sec. 2[42 USC § 4321]. The purposes of this Act are: To declare a national policy which will encourage productive and enjoyable harmony between man and his environment; to promote efforts which will prevent or eliminate damage to the environment and biosphere and stimulate the health and welfare of man; to enrich the understanding of the ecological systems and natural resources important to the Nation; and to establish a Council on Environmental Quality.

TITLE I
CONGRESSIONAL DECLARATION OF
NATIONAL ENVIRONMENTAL POLICY

Sec. 101 [42 USC § 4331]. (a) The Congress, recognizing the profound impact of man's activity on the interrelations of all components of the natural environment, particularly the profound influences of population growth, high-density urbanization, industrial expansion, resource exploitation, and new and expanding technological advances and recognizing further the critical importance of restoring and maintaining environmental quality to the overall welfare and development of man, declares that it is the continuing policy of the Federal Government, in cooperation with State and local governments, and other concerned public and private organizations, to use all practicable means and measures, including financial and technical assistance, in a manner calculated to foster and promote the general welfare, to create and maintain conditions under which man and nature can exist in productive harmony, and fulfill the social, economic, and other requirements of present and future generations of Americans.

(b) In order to carry out the policy set forth in this Act, it is the continuing responsibility of the Federal Government to use all practicable means, consistent with other essential considerations of national policy, to improve and coordinate Federal plans, functions, programs, and resources to the end that the Nation may—

1. fulfill the responsibilities of each generation as trustee of the environment for succeeding generations;

2. assure for all Americans safe, healthful, productive, and aesthetically and culturally pleasing surroundings;

3. attain the widest range of beneficial uses of the environment without degradation, risk to health or safety, or other undesirable and unintended consequences;

4. preserve important historic, cultural, and natural aspects of our national heritage, and maintain, wherever possible, an environment which supports diversity, and variety of individual choice;

5. achieve a balance between population and resource use which will permit high standards of living and a wide sharing of life's amenities; and

6. enhance the quality of renewable resources and approach the maximum attainable recycling of depletable resources.

(c) The Congress recognizes that each person should enjoy a healthful environment and that each person has a responsibility to contribute to the preservation and enhancement of the environment.

Sec. 102 [42 USC § 4332]. The Congress authorizes and directs that, to the fullest extent possible: (1) the policies, regulations, and public laws of the United States shall be interpreted and administered in accordance with the policies set forth in this Act, and (2) all agencies of the Federal Government shall—

(A) utilize a systematic, interdisciplinary approach which will insure the integrated use of the natural and social sciences and the environmental design arts in planning and in decisionmaking which may have an impact on man's environment;

(B) identify and develop methods and procedures, in consultation with the Council on Environmental Quality established by title II of this Act, which will insure that presently unquantified environmental amenities and values may be given appropriate consideration in decisionmaking along with economic and technical considerations;

(C) include in every recommendation or report on proposals for legislation and other major Federal actions significantly affecting the quality of the human environment, a detailed statement by the responsible official on—

   (i) the environmental impact of the proposed action,

  (ii) any adverse environmental effects which cannot be avoided should the proposal be implemented,

 (iii) alternatives to the proposed action,

 (iv) the relationship between local short-term uses of man's environment and the maintenance and enhancement of long-term productivity, and

  (v) any irreversible and irretrievable commitments of resources which would be involved in the proposed action should it be implemented.

Prior to making any detailed statement, the responsible Federal official shall consult with and obtain the comments of any Federal agency which has jurisdiction by law or special expertise with respect to any environmental impact involved. Copies of such statement and the comments and views of the appropriate Federal, State, and local agencies, which are authorized to develop and enforce environmental standards, shall be made available to the President, the Council on Environmental Quality and to the public as provided by section 552 of title 5, United States Code, and shall accompany the proposal through the existing agency review processes;

(D) Any detailed statement required under subparagraph (C) after January 1, 1970, for any major Federal action funded under a program of grants to States shall not be deemed to be legally insufficient solely by reason of having been prepared by a State agency or official, if:

(i) the State agency or official has statewide jurisdiction and has the responsibility for such action,

(ii) the responsible Federal official furnishes guidance and participates in such preparation,

(iii) the responsible Federal official independently evaluates such statement prior to its approval and adoption, and

(iv) after January 1, 1976, the responsible Federal official provides early notification to, and solicits the views of, any other State or any Federal land management entity of any action or any alternative thereto which may have significant impacts upon such State or affected Federal land management entity and, if there is any disagreement on such impacts, prepares a written assessment of such impacts and views for incorporation into such detailed statement.

The procedures in this subparagraph shall not relieve the Federal official of his responsibilities for the scope, objectivity, and content of the entire statement or of any other responsibility under this Act; and further, this subparagraph does not affect the legal sufficiency of statements prepared by State agencies with less than statewide jurisdiction.

(E) study, develop, and describe appropriate alternatives to recommended courses of action in any proposal which involves unresolved conflicts concerning alternative uses of available resources;

(F) recognize the worldwide and long-range character of environmental problems and, where consistent with the foreign policy of the United States, lend appropriate support to initiatives, resolutions, and programs designed to maximize international cooperation in anticipating and preventing a decline in the quality of mankind's world environment;

(G) make available to States, counties, municipalities, institutions, and individuals, advice and information useful in restoring, maintaining, and enhancing the quality of the environment;

(H) initiate and utilize ecological information in the planning and development of resource-oriented projects; and

(I) assist the Council on Environmental Quality established by title II of this Act.

Sec. 103 [42 USC § 4333].   All agencies of the Federal Government shall review their present statutory authority, administrative regulations, and current policies and procedures for the purpose of determining whether there are any deficiencies or inconsistencies therein which prohibit full compliance with the purposes and provisions of this Act and shall propose to the President not later than July 1, 1971, such measures as may be necessary to bring their authority and policies into conformity with the intent, purposes, and procedures set forth in this Act.

Sec. 104 [42 USC § 4334].   Nothing in section 102 [42 USC § 4332] or 103 [42 USC § 4333] shall in any way affect the specific statutory obligations of any Federal agency (1) to comply with criteria or standards of environmental quality, (2) to coordinate or consult with any other Federal or State agency, or (3) to act, or refrain from acting contingent upon the recommendations or certification of any other Federal or State agency.

Sec. 105 [42 USC § 4335].   The policies and goals set forth in this Act are supplementary to those set forth in existing authorizations of Federal agencies.

TITLE II

COUNCIL ON ENVIRONMENTAL QUALITY

Sec. 201 [42 USC § 4341].   The President shall transmit to the Congress annually beginning July 1, 1970, an Environmental Quality Report (hereinafter referred to as the "report") which shall set forth (1) the status and condition of the major natural, manmade, or altered environmental classes of the Nation, including, but not limited to, the air, the aquatic, including marine, estuarine, and fresh water, and the terrestrial environment, including, but not limited to, the forest, dryland, wetland, range, urban, suburban and rural environment; (2) current and foreseeable trends in the quality, management and utilization of such environments and the effects of those trends on the social, economic, and other requirements of the Nation; (3) the adequacy of available natural resources for fulfilling human and economic requirements of the Nation in the light of expected population pressures; (4) a review of the programs and activities (including regulatory activities) of the Federal Government, the State and local governments, and nongovernmental entities or individuals with particular reference to their effect on the environment and on the conservation, development and utilization of natural resources; and (5) a program for remedying the deficiencies of existing programs and activities, together with recommendations for legislation.

Sec. 202 [42 USC § 4342].   There is created in the Executive Office of the President a Council on Environmental Quality (hereinafter referred to as the "Council"). The Council shall be composed of three members who shall be appointed by the President to serve at his pleasure, by and with the advice and consent of the Senate. The President shall designate one of the members of the Council to serve as Chairman. Each member shall be a person who, as a result of his training, experience, and attainments, is exceptionally well qualified to analyze and interpret environmental trends and information of all kinds; to appraise programs and activities of the Federal Government in the light of the policy set forth in title I of this Act; to be conscious of and responsive to the scientific, economic, social, aesthetic, and cultural needs and interests of the Nation; and to formulate and recommend national policies to promote the improvement of the quality of the environment.

Sec. 203 [42 USC § 4343].

(a) The Council may employ such officers and employees as may be necessary to carry out its functions under this Act. In addition, the Council may employ and fix the compensation of such experts and consultants as may be necessary for the carrying out of its functions under this Act, in accordance with section 3109 of title 5, United States Code (but without regard to the last sentence thereof).

(b) Notwithstanding section 1342 of Title 31, the Council may accept and employ voluntary and uncompensated services in furtherance of the purposes of the Council.

Sec. 204 [42 USC § 4344].   It shall be the duty and function of the Council—

1. to assist and advise the President in the preparation of the Environmental Quality Report required by section 201 [42 USC § 4341] of this title;

2. to gather timely and authoritative information concerning the conditions and trends in the quality of the environment both current and prospective, to analyze and interpret such information for the purpose of determining whether such conditions and trends are interfering, or are likely to interfere, with the achievement of the policy set forth in title I of this Act, and to compile and submit to the President studies relating to such conditions and trends;

3. to review and appraise the various programs and activities of the Federal Government in the light of the policy set forth in title I of this Act for the purpose of determining the extent to which such programs and activities are contributing to the achievement of such policy, and to make recommendations to the President with respect thereto;

4. to develop and recommend to the President national policies to fos ter and promote the improvement of environmental quality to meet

the conservation, social, economic, health, and other requirements and goals of the Nation;

5. to conduct investigations, studies, surveys, research, and analyses relating to ecological systems and environmental quality;

6. to document and define changes in the natural environment, including the plant and animal systems, and to accumulate necessary data and other information for a continuing analysis of these changes or trends and an interpretation of their underlying causes;

7. to report at least once each year to the President on the state and condition of the environment; and

8. to make and furnish such studies, reports thereon, and recommendations with respect to matters of policy and legislation as the President may request.

Sec. 205 [42 USC § 4345]. In exercising its powers, functions, and duties under this Act, the Council shall—

1. consult with the Citizens' Advisory Committee on Environmental Quality established by Executive Order No. 11472, dated May 29, 1969, and with such representatives of science, industry, agriculture, labor, conservation organizations, State and local governments and other groups, as it deems advisable; and

2. utilize, to the fullest extent possible, the services, facilities and information (including statistical information) of public and private agencies and organizations, and individuals, in order that duplication of effort and expense may be avoided, thus assuring that the Council's activities will not unnecessarily overlap or conflict with similar activities authorized by law and performed by established agencies.

Sec. 206 [42 USC § 4346]. Members of the Council shall serve full time and the Chairman of the Council shall be compensated at the rate provided for Level II of the Executive Schedule Pay Rates [5 USC § 5313]. The other members of the Council shall be compensated at the rate provided for Level IV of the Executive Schedule Pay Rates [5 USC § 5315].

Sec. 207 [42 USC § 4346a]. The Council may accept reimbursements from any private nonprofit organization or from any department, agency, or instrumentality of the Federal Government, any State, or local government, for the reasonable travel expenses incurred by an officer or employee of the Council in connection with his attendance at any conference, seminar, or similar meeting conducted for the benefit of the Council.

Sec. 208 [42 USC § 4346b].   The Council may make expenditures in support of its international activities, including expenditures for: (1) international travel; (2) activities in implementation of international agreements; and (3) the support of international exchange programs in the United States and in foreign countries.

Sec. 209 [42 USC § 4347].   There are authorized to be appropriated to carry out the provisions of this chapter not to exceed $300,000 for fiscal year 1970, $700,000 for fiscal year 1971, and $1,000,000 for each fiscal year thereafter.

The Environmental Quality Improvement Act, as amended

(Pub. L. No. 91-224, Title II, April 3, 1970; Pub. L. No. 97-258, September 13, 1982; and Pub. L. No. 98-581, October 30, 1984.)
42 USC § 4372

(a) There is established in the Executive Office of the President an office to be known as the Office of Environmental Quality (hereafter in this chapter referred to as the "Office"). The Chairman of the Council on Environmental Quality established by Public Law 91-190 shall be the Director of the Office. There shall be in the Office a Deputy Director who shall be appointed by the President, by and with the advice and consent of the Senate.

(b) The compensation of the Deputy Director shall be fixed by the President at a rate not in excess of the annual rate of compensation payable to the Deputy Director of the Office of Management and Budget.

(c) The Director is authorized to employ such officers and employees (including experts and consultants) as may be necessary to enable the Office to carry out its functions; under this chapter and Public Law 91-190, except that he may employ no more than ten specialists and other experts without regard to the provisions of Title 5, governing appointments in the competitive service, and pay such specialists and experts without regard to the provisions of chapter 51 and subchapter III of chapter 53 of such title relating to classification and General Schedule pay rates, but no such specialist or expert shall be paid at a rate in excess of the maximum rate for GS-18 of the General Schedule under section 5332 of Title 5.

(d) In carrying out his functions the Director shall assist and advise the President on policies and programs of the Federal Government affecting environmental quality by—

1. providing the professional and administrative staff and support for the Council on Environmental Quality established by Public Law 91-190;

2. assisting the Federal agencies and departments in appraising the effectiveness of existing and proposed facilities, programs, policies, and activities of the Federal Government, and those specific major projects designated by the President which do not require individual project authorization by Congress, which affect environmental quality;

3. reviewing the adequacy of existing systems for monitoring and predicting environmental changes in order to achieve effective coverage and efficient use of research facilities and other resources;

4. promoting the advancement of scientific knowledge of the effects of actions and technology on the environment and encouraging the development of the means to prevent or reduce adverse effects that endanger the health and well-being of man;

5. assisting in coordinating among the Federal departments and agencies those programs and activities which affect, protect, and improve environmental quality;

6. assisting the Federal departments and agencies in the development and interrelationship of environmental quality criteria and standards established throughout the Federal Government;

7. collecting, collating, analyzing, and interpreting data and information on environmental quality, ecological research, and evaluation.

(e) The Director is authorized to contract with public or private agencies, institutions, and organizations and with individuals without regard to section 3324(a) and (b) of Title 31 and section 5 of Title 41 in carrying out his functions.

42 USC § 4373. Each Environmental Quality Report required by Public Law 91-190 shall, upon transmittal to Congress, be referred to each standing committee having jurisdiction over any part of the subject matter of the Report.

42 USC § 4374. There are hereby authorized to be appropriated for the operations of the Office of Environmental Quality and the Council on Environmental Quality not to exceed the following sums for the following fiscal years which sums are in addition to those contained in Public Law 91-190:

(a) $2,126,000 for the fiscal year ending September 30, 1979.

(b) $3,000,000 for the fiscal years ending September 30, 1980, and September 30, 1981.

(c) $44,000 for the fiscal years ending September 30, 1982, 1983, and 1984.

(d) $480,000 for each of the fiscal years ending September 30, 1985 and 1986.

42 USC § 4375

(a) There is established an Office of Environmental Quality Management Fund (hereinafter referred to as the "Fund") to receive advance payments from other agencies or accounts that may be used solely to finance—

1. study contracts that are jointly sponsored by the Office and one or more other Federal agencies; and

2. Federal interagency environmental projects (including task forces) in which the Office participates.

(b) Any study contract or project that is to be financed under subsection (a) of this section may be initiated only with the approval of the Director.

(c) The Director shall promulgate regulations setting forth policies and procedures for operation of the Fund.

B

Attribute Descriptor Package

The environmental characteristics, or attributes, defined and described in this appendix represent a selected set of elements designed to be used in the environmental assessment process. They may be too generalized for many analyses, but too specific for others. Use them as a guide in classifying the environment into factors which may be affected by the actions of your agency. The manner in which they are incorporated into an analysis is shown in Appendix C.

It should be noted that these descriptions are intended to give the reader an overview of each attribute in the context of its role in impact assessment. None of the descriptions should be considered complete, as indeed, many of the individual subject areas themselves form the basis for complete texts. It is anticipated that familiarity with these 49 attributes can serve to expedite interdisciplinary studies, which frequently encounter difficulties due to lack of communication between disciplines. This communication problem can be overcome when the participants attain some understanding of each other's terminology, problems, and difficulties in achieving solutions to those problems.

Air

Air attributes are factors that indicate the quality of the air. Basically, two kinds of environmental factors relate to air quality. They relate to

Structural elements of the environment

Inputs to or emissions from human activities

Factors relating to the structural elements of the air environment are stability, temperature, mixing depth, wind speed, wind direction, humidity, precipitation, pressure, and topography. On the other hand, factors relating to inputs from human activity are dust, fumes, gases, vapors, mists, smoke, soot, and compounds of arsenic, aluminum, etc.

Nine attributes may be utilized in describing the impact of human activities on the air environment:

Diffusion factor

Particulates

Sulfur oxides

Hydrocarbons

Nitrogen oxides

Carbon monoxide

Photochemical oxidants

Hazardous toxicants

Odors

The first attribute, the diffusion factor, is related to the structural elements of the environment; the remaining attributes are related to the emissions from human activities.

Diffusion factor

Definition of the attribute.   Diffusion factor is an attribute that is related to various atmospheric and topographic aspects of the environment. For example, vertical temperature structure affects movement of air in the atmosphere. Wind structure in a region determines the scavenging action in the environment as well as the impact of inversions. Topography may change temperature and wind profiles because of the combined effects of surface friction, radiation, and drainage. Valleys are more susceptible to stagnation and to air pollution than are flatlands or hill slopes. The mixing depth, in fact, also determines the intensity of air pollution in a given region. The status of stability or instability of the atmosphere determines to what extent air pollution can build up in a given region. Humidity and pressure also affect the diffusion rate of a given pollutant emitted to the atmosphere. In addition, precipitation is an important scavenger element that can clean up pollutants in the air.

Together, all of the above environmental factors determine the diffusion factor in a given region.

Activities that affect the attribute.   Generally, most human activities will not affect the diffusion factor. However, since research has shown the possibility of certain activities affecting the weather and other related meteorological factors, it is necessary to consider such activities that are now known (about which limited information is available) which may affect the diffusion factor. For instance, artificial methods for generating storms, seeding clouds, and research and testing of these new and powerful methods can, and will, cause changes in the diffusion factor.

Source of effects. As indicated above, impacts of certain specialized activities can have a major effect on the diffusion factor. Weather modification, in terms of cloud seeding, hail suppression, or alternate forms may affect precipitation patterns and other atmospheric attributes. The effects must be examined on a case-by-case basis, and where details of these activities are classified, for security reasons, it is not possible to provide detailed information on their potential impacts.

Variables to be measured. Variables to be measured to determine the diffusion factor are many. The major ones are stability, mixing depth, wind speed, precipitation, and topography. Various measures of each of these variables will indicate the extent and nature of the diffusion factor in a given region.

How variables are measured. Generally, data on stability, mixing depth, wind speed, direction, and precipitation are collected by meteorological survey stations of the National Weather Service. Data on these attributes are readily available from the Weather Service offices across the country. Topography data can be obtained from the United States Geological Survey (USGS) maps of largest available scale.

Data sources. Primary sources of data for the variables that define diffusion factors are the National Weather Service and the USGS; both have offices in most major cities throughout the country and web sites.

Skills required. Collection and analysis of such data require a sophisticated meteorological background. Persons with specialized training in meteorology and trained technicians are required to collect and develop information relating to these variables.

Instruments. A full-scale meteorological laboratory is needed to monitor the selected attributes that define the nature and extent of the air diffusion factor.

Evaluation and interpretation of data. The diffusion factor can be classified into three or more major ratings. For example, the diffusion factor can be high, medium, or low. The high rating represents an environmental quality (EQ) value of 1.0; the medium rating represents an EQ value of 0.5; and the low (or poor) rating represents an EQ value of 0.

The environmental impact of selected activities on the diffusion factor is measured by the change in diffusion factor ratings. When a diffusion factor changes only a small amount and its rating remains unaltered, the impact is considered insignificant. When the change in the diffusion factor rating is altered by one step (e.g., between high and medium or medium and poor), the impact is considered to be moderate. When a change in the diffusion factor rating occurs through two steps (e.g., between high and poor), the impact is treated as significant.

Geographical and temporal limitations.   There can be substantial variation in the diffusion factor, spatially and temporally, depending upon variations in the determinant variables. It is known, for instance, that wind speed, precipitation, stability, and mixing depth change with time and location in a given region. These variations, therefore, alter the diffusion factor accordingly.

Mitigation of impact.   Generally, the impact of most activities on diffusion has not been adequately defined. The mitigation techniques are also not well established.

Secondary effects.   The diffusion factor can be related to land-use patterns in the vicinity of air pollution sources. The prevailing wind direction can render certain land areas aesthetically undesirable or otherwise environmentally unacceptable during specific times or seasons, if the air pollution is not eliminated or satisfactorily dispersed.

Other comments.   Research is needed to identify potential activities, their impacts, and the mitigation strategies relating to potential impacts on the diffusion factor. Also, a mathematical model is needed to relate all of the determinant variables to the diffusion factor. This will help establish a suitable relationship between variables and the diffusion factor.

References.   de Nevers, N. *Air Pollution Control Engineering,* 2d ed. New York: McGraw-Hill, 2000.

Particulates

Definition of the attributes.   Particulates exist in the form of minute separate particles of solid and liquid suspended in the air. They may be of organic or inorganic composition.

Particulates are finely divided solid and liquid particles suspended in the ambient air. They range from over 100 micrometers (μm) to less than 0.01 μm in diameter. Particulates of smaller size (less than 10 μm) suspended in air can scatter light and behave like a gas. These smaller particulates are called aerosols.

Activities that affect the attribute. Many human activities generate particulates that are emitted to the air. These include construction, operation, maintenance, and repair ctivities; transportation; and industrial activities. Examples of subactivities are site preparation; demolition, removal, and disposal; excavation; concrete construction; operation and maintenance of aircraft; operation and maintenance of automotive equipment; use of construction equipment; use of explosives; mineral extraction; foundry operation; manufacturing; noninitiating high explosives; and use of transportation vehicles.

Source of effects. In general, the atmosphere naturally contains some level of particulate matter. Emissions resulting from various activities, are released to the atmosphere causing a higher concentration of particulates. Particulates can cause increased mortality and morbidity in the exposed population by aggravating diseases such as bronchitis, emphysema, and cardiovascular diseases. Particulates can soil clothes and buildings and can cause serious visibility problems. Steel and other metal structures can be corroded as a result of exposure to particulates and humidity. Property values and psychic welfare of people can be undermined.

Variables to be measured. Particulate concentration is generally measured as the concentration of all solid and liquid particles averaged over a period of 24 h. For purposes of impact assessment, particulate concentration is measured as the average annual arithmetic mean of all 24-h particulate concentrations at a given location.

How variables are measured.* Particulate concentrations are usually measured by the high-volume method. The air is drawn into a covered housing, through a filter, by a high-low blower at a rate of 35 to 64 ft³/min. The particles, ranging from 100 to 0.1 μm in diameter, are ordinarily collected on fiberglass filters. The concentration of suspended particulates is then computed by measuring the mass of collected particulates in the volume sample in micrograms per cubic meter.

Data source. Sources of data are generally state pollution control departments, county air pollution control offices, multicounty air

---

*Variable measurement methods are continually modified by federal and state regulatory agencies. Readers should consult the latest requirements. For air pollution, these requirements are found at 40 CFR 50.

pollution control offices, or city air pollution control offices. High-volume samplers may be installed to monitor particulates from specific operations.

Skills required. Basic paraprofessional training in mechanical or chemical engineering with special training in operating high-volume samplers is adequate to collect particulate concentration data. Specialized supervision is needed to ensure that data are properly collected and analyzed.

Instruments. The apparatus used for sampling particulate concentration is called a high-volume air sampler. The sampler is installed in a shelter to protect it against extremes of temperature, humidity, and other weather conditions. It has a filter medium with a collection efficiency of about 99 percent for particles of 0.3 μm in diameter.

Evaluation and interpretation of data. The primary effects of particulates on environmental quality range from visibility problems to health impairments. Visibility problems occur at concentrations as low as 25 μg/m^3. As the concentration of particulates increases to about 200 μg/m^3, human health begins to be affected. The concentration levels mentioned above refer to 24-h average annual concentration. Particulate concentration of less than 25 μg/m^3 is also considered less desirable for the environment, since it provides condensation nuclei upon which fog and cloud droplets settle. From these considerations, a particulate value function was developed, based on a 24-h average annual concentration, as shown in Fig. B.1.

The determination of environmental impact of proposed activities on particulate level is measured by the change in particulates concentration. When the particulates concentration changes to the extent that its rating remains unaltered (e.g., high-quality air remains high quality), the impact may be considered insignificant. When a change in particulates rating occurs through two steps (e.g., between high quality and low quality, and vice versa), the impact is treated as significant; a one-step change is considered moderate.

The particulate value function (Fig. B.1) is used for rating air quality in terms of high, moderate, and low quality, based on 24-h annual geometric mean. For a given value of 24-h annual geometric mean particulates concentration on the horizontal axis, a point on the curve identifies the environmental quality rating from the vertical axis of Fig. B.1 (e.g., 130 μg/m^3 indicates a moderate quality of 0.4).

Geographical and temporal limitations. The concentration of particulates does not remain constant over the entire spatial extent of a given region. Also, it will not remain constant over time. As such, substantial spatial

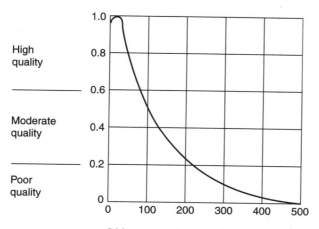

Figure B.1 These value functions are provided for conceptual evaluation of air quality impact. It should be noted that any time air quality standards established by the governing regulatory agency are exceeded, the impact is significant.

and temporal variations in the concentration of particulates can be expected. It is generally claimed that the impact of particulates on the environment and on humans depends on the total amount of exposure over the entire year. Spatial variations can be accounted for by analyzing minuscule units of urbanized regions. This requires extensive calculations based on a diffusion model or a large-scale monitoring program. Since the use of a large-scale monitoring network is infeasible in most situations, the problem can be adequately addressed using diffusion models to predict air quality values over the entire spatial area.

Mitigation of impact. Particulate pollution impacts can be mitigated by means of four major alternatives:

Reduction in particulate emission from sources

Reduction or removal of receptors from the polluted areas

Particulate removal devices such as cyclones, settling chambers, impactors, scrubbers, electrostatic precipitators, and bag houses

Use of protected controlled environment (e.g., oxygen masks, enclosed shopping malls)

A combination of the first three alternatives should be considered to provide an optimal strategy for the mitigation of particulate pollution impacts.

Secondary effects. Particulate emissions are associated with problems of human health—increased mortality and morbidity in the exposed population. In addition to these direct effects, particulates also cause numerous secondary impacts. Particulates soil clothes and structures, resulting in economic loss. In addition to visibility problems and increased accident risk, aesthetic considerations reduce property values and undermine the general psychic welfare of people. Steel and other metal structures can be corroded as a result of exposure to particulates and humidity. Water quality from storm runoff and vegetation can also be deteriorated by particulate matter present.

Other comments. Particulates are present even in the cleanest air at the most remote locations uncontaminated by humans. Sources of particulate pollution relate to activities such as construction; industrial operation; operation, maintenance, and repair work; and transportation. Automobile emissions are only a minor source of particulate pollution.

Sulfur oxides

Definition of the attribute. Sulfur oxides are common air pollutants generated primarily by combustion of fuel. Solid and liquid fossil fuels contain a high degree of sulfur in the form of inorganic sulfides and organic sulfur compounds.

Sulfur oxides are usually a combination of sulfur dioxide, sulfur trioxide, sulfuric acid, and sulfurous acid. Combustion of fossil fuels normally produces about 30 parts sulfur dioxide for 1 part sulfur trioxide. Sulfur dioxide is the most dominant portion of the sulfur oxides concentration; as such, the sulfur oxides attribute is defined in terms of the sulfur dioxide parameter.

Sulfur dioxide is a nonflammable, nonexplosive, transparent gas with a pungent, irritating odor. The concentration of this gas in parts per million (ppm) measures the magnitude of sulfur oxides pollution in a given region.

Activities that affect the attribute. Many human activities use fossil fuels. Coal- and oil-fired furnaces, fossil-fueled electric generating plants, and industrial uses of fossil fuels appear to be major generators of sulfur dioxide pollution. In addition, operation of various facilities can cause significant sulfur dioxide pollution. Construction work and transportation also create a minor sulfur dioxide problem from the operation of diesel engines.

Sources of effects. The effects of sulfur dioxide pollution can be high morbidity; increased mortality; increased incidence of bronchitis, respiratory diseases, and emphysema; and general deterioration of health. It can also cause increased corrosion of metals, chronic plant injury, excessive leaf dropping, and reduced productivity of plants and trees. The effect of sulfur dioxide pollution in the presence of particulates can result in synergistic impacts on the environment. Synergistic impacts of sulfur dioxide in the presence of nitrogen dioxide have also been noted. For example, even low levels of sulfur dioxide, in combination with other contaminants such as particulates, aggravate symptoms of asthma, bronchitis, and emphysema. SO_2 levels as low as 0.25 ppm may cause attacks in asthmatics participating in exercise (Dickey, 1999).

Variables to be measured. The primary variable that measures the extent of the sulfur oxides problem is expressed by the 24-h annual arithmetic mean concentration of sulfur dioxide present in the ambient air. This variable is used to predict the potential impact of sulfur oxides on the environment.

Here, the use of one variable is not entirely adequate. Concentration of particulates, ozone, and nitrogen oxides affects the impacts of sulfur oxides. However, to take advantage of the simplification, only one variable has been used.

How variables are measured. The sulfur dioxide concentration is commonly measured by the pararosaniline method. In principle, sulfur dioxide is absorbed from air in a solution of potassium tetrachloromercurate (TCM). The resulting complex is added to pararosaniline and formaldehyde to form an intensely colored acid solution which is analyzed spectrophotometrically. The spectrophotometric analysis is a colorimetric method in which the concentration of sulfur dioxide absorption is measured by the intensity of the color produced in the resulting acid solution. The method is recommended by the Environmental Protection Agency in the National Primary and Secondary Ambient Air Quality Standards published in 40 CFR 50.4 and 50.5. Test methods are in 40 CFR 50 Appendix A.

Data source. Air quality measurements for sulfur oxide are made by air quality monitoring programs established by state pollution control agencies, the federal Environmental Protection Agency, and county, regional, multicounty, or city air pollution control agencies. Generally, the data are compiled annually and are published with summaries by the state agency responsible for air quality monitoring.

Skills required. The skills required for measuring sulfur dioxide concentration in air can be developed by special technician-level training imparted at a technical school or as part of an on-the-job training program. Technician-level training in mechanical and chemical engineering is adequate to develop the necessary skills to operate a monitoring and recording system for sulfur dioxide.

Instruments. The instruments required for monitoring sulfur dioxide concentration are

All-glass midget impinger

Air pump

Air flowmeter

Spectrophotometer

Evaluation and interpretation of data. A review of the literature indicates that the minimum sulfur dioxide concentration for vegetation damage is 0.03 ppm. A sulfur dioxide concentration less than 0.03 ppm should be considered a characteristic of a safe environment. As concentration increases, more damage will be done to the vegetation and materials. Visibility of the atmosphere is also impaired. At a concentration of 0.2 ppm of sulfur dioxide, increased mortality rates are observed. This situation should reflect a value function of zero. Based on these considerations, a value function was developed for sulfur oxide, as shown in Fig. B.2.

The determination of environmental impact of proposed activities on sulfur dioxide level is measured by the change in sulfur dioxide concentration. When the sulfur dioxide concentration changes to the extent that its rating remains unaltered (i.e., high-quality air remains high-quality, and so on), the impact is considered insignificant. If the change in sulfur dioxide concentration is such that its rating changes by one step (i.e., from high quality to moderate quality, etc.), the impact is treated as moderate. Furthermore, when a change in sulfur dioxide rating occurs through two steps (i.e., from high quality to low quality, and vice versa), the impact is treated as significant.

The sulfur dioxide value function (Fig. B.2) is used for rating air quality in terms of high, moderate, and low quality based on the 24-h annual arithmetic mean. For a given value of 24-h annual geometric mean sulfur dioxide concentration on the horizontal axis, the environmental quality rating can be read for the horizontal axis in Fig. B.2.

Geographical and temporal limitations. Concentration of sulfur dioxide does not remain constant over the entire spatial extent in a given region. Also, it will not remain constant over time. As such, substantial spatial and temporal variations in the concentration of sulfur dioxide on the environment and on humans depends on the total amount

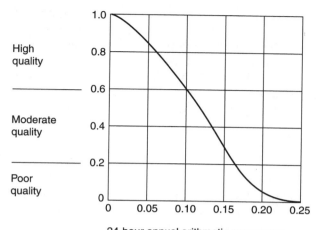

Figure B.2   Sulfur dioxide value function.

of exposure over the entire year. Spatial variations can be accounted for by taking minuscule units of urbanized regions for purposes of analysis. This requires extensive calculations based on a diffusion model or a large-scale monitoring program. Since the use of a large-scale monitoring network is infeasible in most situations, the problem can be adequately addressed using diffusion models to predict air quality values over the entire spatial region.

Mitigation of impact.   The impacts can be mitigated by means of four major alternatives or a combination thereof:

Reduction in sulfur dioxide emissions from sources

Reduction or removal of receptors from the polluted areas

Gas removal devices using absorption (liquid as a medium), adsorption (molecular sieve), and catalytic converters

Use of protected, controlled environment, such as oxygen masks, enclosed athletic facilities, etc.

Secondary effects.   Secondary effects of sulfur oxides include economic and resource loss through damage to material surfaces and vegetation, water quality deterioration through the natural "cleansing" of the atmosphere through precipitation, and aesthetic and general welfare quality reduction that accompanies the degradation of a vital resource. Land-use patterns and community needs may be affected in localized areas with point-source emissions.

Other comments. Sulfur dioxide is generally harmful to the health and welfare of a community. Its impact can be substantially increased by the presence of suspended particulates due to the synergistic relationship of the two pollutants. Despite this, the value function is based only on the concentration of sulfur dioxide. This is done to simplify the value function. However, the impacts have been adjusted for the concentrations of particulates that generally accompany given levels of sulfur dioxide in the ambient air.

Hydrocarbons

Definition of the attribute. Hydrocarbon is a general term used for several organic compounds emitted when organic materials such as petroleum fuels are burned. Automobile exhaust accounts for over half of the complex mixture of hydrocarbons emitted to the atmosphere; the remaining hydrocarbons arise from natural sources like decomposable organic matter on land, swamps, and marshes; hydrocarbons haze from plants and forest vegetation; geothermal areas; coal fields, natural gas, and petroleum fields; and forest fires. Usually, hydrocarbons consist of methane, ethane, propane, and derivatives of aliphatic and aromatic organic compounds.

The hydrocarbons attribute is defined as the total hydrocarbon concentration (THC) present in the ambient air. Hydrocarbons are organic compounds consisting of carbon and hydrogen; their concentration is measured in parts per million by volume or in micrograms per cubic meter of air. For most U.S. cities, except Los Angeles, the peak hydrocarbon concentration occurs between 6 and 9 a.m.

Activities that affect the attribute. Many activities emit high levels of hydrocarbons into the environment. For example, industrial operations, home heating, and vehicle operations involve substantial combustion of fuel, causing hydrocarbon emissions due to inefficient combustion processes. Gasoline and diesel engines are used for purposes of construction, operation, maintenance, repair, and transportation. In addition, many industrial activities have petroleum and petrochemical operations that emit high levels of hydrocarbons. Areas with natural vegetation and forests also generate high levels of hydrocarbon concentration.

Source of effects. Hydrocarbons are of concern primarily for their role in the formation of photochemical oxidants and smog. Direct health effects of gaseous hydrocarbons in the ambient air have not been demonstrated. Health effects occur only at high concentrations (about 1000 ppm or more) that interfere with oxygen intake. Hydrocarbons in the atmosphere have been found to cause lacrimation, coughing, sneezing, headaches, laryngitis, pharyngitis, and bronchitis, even at low concentrations. In addition, hydrocarbons may cause breathing problems and eye irritation. In combination with nitrogen oxides, hydrocarbon impacts can be significantly increased.

Variables to be measured.   The variable expressing the impact of hydrocarbons is measured by the 3-h average annual concentration of ambient hydrocarbons, expressed in parts per million. The time concentration is measured from 6 to 9 a.m., at which time peak hydrocarbon concentration is expected to occur in most U.S. cities except Los Angeles.

Nitrogen oxide variables interact synergistically with the concentration of hydrocarbons. Nitrogen oxides combined with hydrocarbons generate oxidants causing smog. The impact of smog is significantly greater than that of hydrocarbons alone. However, for purposes of simplicity, nitrogen oxides are treated as a separate variable.

How variables are measured.   There are two different methods of analysis for the total hydrocarbons:

Flame ionization method

Spectrophotometric method

The EPA, in its national primary and secondary ambient air quality standards document, has recommended use of the hydrogen flame ionization method to measure total hydrocarbon concentration. The flame ionization technique uses a measured volume of ambient air delivered semicontinuously (about 4 to 12 times per hour) to a hydrogen flame ionization detector (FID). A sensitive electrometer detects the increase in ion concentration which results from the interaction of the hydrogen flame with a sample of air contaminated with organic compounds such as hydrocarbons, aldehydes, and alcohols. The ion concentration response is approximately proportional to the number of organic carbon atoms in the sample. The FID serves as a carbon atom counter.

The measurement can be made by two modes of operation:

A complete chromatographic analysis showing continuous output from the detector

Programming the system to display selected output from the detector

The latter is adequate for recording hydrocarbons system concentration values from 6 to 9 a.m. only. See 40 CFR 50 Appendix E for test methods.

Data sources. Hydrocarbon data are generally collected by state air quality monitoring programs. Other potential sources include the federal EPA and city or county monitoring agencies.

Skills required. Basic paraprofessional training in mechanical or chemical engineering with special training in operating air pollution samplers is adequate to collect data relating to hydrocarbons. Specialized supervision is needed to ensure that the instruments are correctly operated and recorded. This requires either experienced personnel or experienced consultants specializing in air quality monitoring.

Instruments. Instruments used for measuring hydrocarbons are the following:

Commercial THC analyzer

Sampler introduction system (including pump, flow control, valves, automatic switching valves, and flow meter)

In-line filter (a binder-free glass-fiber with a porosity of 3 to 5 μm)

Stripper or per column (the column should be repacked or replaced every 2 months of continuous use)

Oven (containing analytical column and analytical converter)

The instruments are installed and connected in accordance with the manufacturer's specifications.

Evaluation and interpretation of data. The extent of hydrocarbon impact is measured by the degree to which it affects smog intensity. Hydrocarbon criteria are, therefore, keyed to the 6 to 9 a.m. average annual concentration. At low concentrations, hydrocarbons are relatively harmless and unimportant. The quality of the environment deteriorates rapidly as conditions for smog development approach (i.e., 0.15 to 0.25 ppm). A sharp decrease in environmental quality is noted within this range. Above a 0.25-ppm hydrocarbon concentration, the value function gradually levels off to zero, since the marginal impact of increases in hydrocarbons concentration is small. The value function is thus a flat S curve. On the basis of these considerations, the hydrocarbons value function shown in Fig. B.3 was developed.

The determination of environmental impact of proposed activities on hydrocarbon levels is measured by the change in hydrocarbon concentration. When the hydrocarbon concentration changes to the extent that its rating remains unaltered (e.g., high-quality air remains high quality), the impact is considered insignificant. When the change in hydrocarbons concentration is such that its rating changes by one step (e.g., between high quality and moderate quality), the impact is treated as

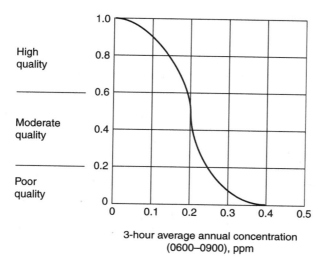

3-hour average annual concentration
(0600–0900), ppm

Figure B.3 Hydrocarbons value function.

moderate. When a change in hydrocarbons rating occurs through two steps (e.g., from high quality to low quality, and vice versa), the impact is considered significant.

The hydrocarbons value function (Fig. B.3) is used for rating air quality in terms of high, moderate, and low quality based on the 3-h average annual concentration. For a given value of 3-h average annual concentration on the horizontal axis, a point on the curve identifies the environmental quality rating from the vertical axis of Fig. B.3 (e.g., 0.3 ppm indicates a poor quality of 0.15).

Geographical and temporal limitations. Concentration of hydrocarbons does not remain spatially constant in a given region. Also, it will not remain constant over time. As such, a substantial spatial and temporal variation in the concentration of hydrocarbons can be expected. It is generally claimed that the impact of hydrocarbons on the environment and humans depends upon the total exposure during peak periods. Spatial variations can be accounted for by taking small units of urbanized regions for analysis. This requires extensive calculations based on a diffusion model or a large-scale monitoring program. Since the use of a large-scale monitoring network is infeasible in most situations, the problem can be adequately addressed using diffusion models to predict air quality values over the entire spatial area.

Mitigation of impact. There are four major strategies for the mitigation of impacts of hydrocarbons on the environment. These are

Control of motor vehicle emissions

Control of stationary source emission (including evaporation, incineration, absorption, condensation, and material substitution)

Reduction or removal of receptors from polluted areas

Creation of a controlled environment to avoid pollution (including use of oxygen masks)

These strategies can be used in an optimal combination in order to get the best results from an abatement program.

Secondary effects.   Production of hydrocarbons beyond acceptable levels may result in secondary impacts through reduction of property values, shifts in land-use patterns, and adverse effects on vegetation. Increased accident occurrence can accompany the reduction in vision and other human health aspects.

Other comments.   Hydrocarbon concentration is one of the parameters that defines the extent of smog development in an environment. In selecting attributes, the ozone parameter was avoided, since the formation of ozone is determined by the interaction of hydrocarbons and nitrogen oxides in the presence of sunlight. The environment receives many different kinds of hydrocarbon emissions; as such, these emissions are an important indicator of environmental impact.

Nitrogen oxides

Definition of the attribute.   Many nitrogen oxides are found in the urban environment. The most important are nitric oxide (NO) and nitrogen dioxide (NO_2). In addition, nitrous oxide (N_2O) is another oxide of nitrogen present in the atmosphere in appreciable concentration. The term NO_2 is often used to represent the composite atmospheric concentration of nitrogen oxides in the environment.

Nitrogen oxides are emitted by exhausts from high-temperature combustion sources. They result from the reaction of nitrogen with oxygen; with hydrocarbons they produce photochemical smog. Nitrogen oxide concentrations are measured in parts per million by volume.

Activities that affect the attribute.   Many human activities generate nitrogen oxides which are emitted to the air. Industrial operations; research, development, and testing operations; operation and maintenance of motor vehicles; and stationary combustion sources (like power plants, natural gas burners, diesel-operated construction machineries) are some of the sources of nitrogen oxides. However, a large portion of nitrogen oxides is produced by natural sources, such as bacterial action in forests, swamps, and parks.

Source of effects. There is very little documented information on the health effects of nitrogen oxides at concentrations normally found in ambient air. The human threshold for sensing the odor of nitrogen dioxide is about 0.12 ppm. Data from human and animal studies indicate that nitrogen oxides have adverse effects on human health. Nitrogen dioxide is about four times more toxic than nitric oxide.

In addition, nitrogen oxides can affect vegetation, causing acute (chronic) injury to leaves as well as to productivity of certain plants. Nickel alloys are subject to corrosion in the presence of nitrogen oxides; synthetic fibers fade and white clothes yellow in the presence of nitrogen oxides.

Variables to be measured. The variable measuring the extent of pollution from nitrogen oxides is the average annual concentration of nitrogen oxides in the ambient air. The nitrogen oxides level is measured in parts per million (ppm).

Other variable factors that might interact with nitrogen oxides are hydrocarbons and particulates. These variables are considered separately in defining air quality impacts, even though they interact synergistically.

How variables are measured. Nitrogen dioxide is the only atmospheric nitrogen oxide which can be measured directly with current techniques.* Measurement of nitrogen oxides, therefore, must rely on some type of converter that oxidizes nitric oxide to nitrogen dioxide.

The reference method for the determination of nitrogen dioxide is the Griess-Saltzman technique, modified by the EPA. It is a 24-h continuous sampling method. In principle, nitrogen dioxide–contaminated air is bubbled through a sodium nitrite. The nitrite concentration in the sample solution is measured colorimetrically by the reaction of an exposed absorbing agent with phosphoric acid, sulfanilamide, and NEDA solution.

Data sources. Sources of data are generally state pollution control departments and county, multicounty, or city air pollution control offices. They can also install monitoring samplers at critical distances from emission sources to determine the level of nitrogen oxides generated by the particular activities.

---

*Nitrogen oxides pollution is measured by the concentration of nitrogen dioxide expressed in terms of annual arithmetic mean concentration.

Skills required. Basic paraprofessional training in mechanical or chemical engineering, with special training in operating air quality sampling devices, is adequate to collect data relating to nitrogen oxides. Specialized supervision is needed to ensure that the data are properly collected and analyzed. Specialized supervision should include personnel or experienced consultants trained in the field of air quality monitoring.

Instruments. Nitrogen dioxide is measured with an apparatus consisting of the following components:

Absorber tubes

Probe with membrane filter, glass funnel, and trap

Flow-control device with a calibrated 27-gauge hypodermic needle and a membrane filter protection

Air pump capable of maintaining a flow of 0.2 l/min and a vacuum of 0.7 atmosphere

Calibration equipment

Evaluation and interpretation of data. Generally, nitrogen oxides concentrations below 0.05 ppm (on average annual basis) do not pose health problems. Exposure above this level can be correlated with a higher incidence of acute respiratory problems. At levels higher than those normally present in ambient air (i.e., about 0.05 ppm), nitrogen dioxide acts as a toxic agent. Based on these considerations, a nitrogen dioxide value function has been developed, as shown in Fig. B.4.

The determination of environmental impact of proposed activities on nitrogen oxides level is measured by the change in nitrogen oxides (NO_x) concentration. When the NO_x concentration changes to the extent that its rating remains unaltered (e.g., high-quality air remains high-quality), the impact is considered insignificant. When the change in NO_x concentration is such that its rating changes by one step (e.g., between high quality and moderate quality), the impact is treated as moderate. When a change in NO_x ratings occurs through two steps (e.g., from high quality to low quality, and vice versa), the impact is considered significant.

The nitrogen oxides value function (Fig. B.4) is used for rating air quality in terms of high, moderate, and low quality, based on average annual concentration. For a given value of average annual concentration on the horizontal axis, a point on the curve identifies the environmental quality rating from the vertical axis of Fig. B.4 (e.g., 0.1 ppm indicates a poor quality of 0.1).

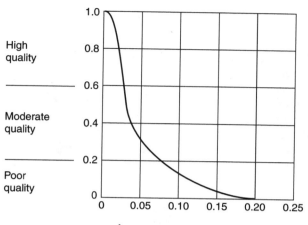

Figure B.4 Nitrogen oxides value function.

Geographical and temporal limitations. Concentration of nitrogen dioxide does not remain constant over the entire spatial extent in a given region. Also, it will not remain constant over time. As such, substantial spatial and temporal variations in the concentration of nitrogen dioxide can be expected. It is generally claimed that the impact of nitrogen dioxide on the environment and humans depends on the total amount of exposure over the entire year. Spatial variations can be accounted for by analyzing small units of urbanized regions. This requires extensive calculations based on a diffusion model or a large-scale monitoring program. Since the use of a large-scale monitoring network is infeasible in most situations, the problem can be adequately addressed using diffusion models to predict air quality values over the entire spatial area.

Mitigation of impact. There are five major strategies for the mitigation of impacts of nitrogen dioxide on the environment. These are

Control of motor vehicle emissions

Control of stationary-source emissions (including incineration and evaporation)

Reduction or removal of receptors from polluted areas

Gas removal devices using absorption (liquid as a medium), adsorption (molecular sieves), and catalytic converters

Creation of a controlled environment to avoid pollution (such as the use of oxygen masks)

These strategies can be used in an optimal combination to get the best results from an abatement program.

Secondary effects. Secondary effects due to nitrogen oxide production include economic losses ranging from damage to vegetation to deterioration of building materials. Shifting land-use patterns, reduced property values, and increased accident occurrence can accompany the formation of smog and other direct effects.

Carbon monoxide

Definition of the attribute. Carbon monoxide (CO) is the most widely distributed and most commonly occurring air pollutant. The majority of atmospheric CO is produced by the incomplete combustion of carbonaceous materials used for fuels for vehicles, space heating, industrial processing, and the burning of refuse.

Activities that affect the attribute. All activities that involve the combustion of organic materials are sources of CO. In addition, industrial operations contribute to the CO burden in the air. CO is also formed by explosions and the firing of weapons, and it can occur naturally.

Source of effects. Adverse health effects on humans have been observed for exposures of 8 h or more at CO concentrations of 12 to 17 mg/m^3 (10 to 15 ppm). Adverse health effects consist of impaired time interval discrimination, physiologic stress on heart patients, etc.

Variables to be measured. The concentration of CO is measured in micrograms per cubic meter. The variable measuring the extent of carbon monoxide pollution is the maximum 8- and 1-h concentration.

How variables are measured. The reference method for the continuous measurement of carbon monoxide is nondispersive infrared spectrometry. The measurement technique is based on the absorption of infrared radiation by carbon monoxide. By comparing absorption of infrared radiation passing through a reference cell and a test cell electronically, the concentration of CO in the test cell can be measured.

Instruments are available that measure in the range of 0 to 58 mg/m^3. The sensitivity is 1 percent of full-scale response per 0.6 mg CO/m^3 (0.5 ppm). See 40 CFR 50 Appendix F for test methods and 40 CFR 50-11 for standards.

Data sources. The sources of data are generally the State Pollution Control Department, the County Air Pollution Control Office, or the City Air Pollution Control Office. Monitoring equipment can be installed at critical locations near specific operations to determine the level of carbon monoxide generated.

Skills required. A basic paraprofessional training in mechanical or chemical engineering with special training in operating the air quality instruments is adequate to collect data relating to carbon monoxide. Specialized supervision will be needed to ensure that the data are properly collected and analyzed. Specialized supervision should include trained and experienced personnel or experienced consultants in the field of air quality monitoring.

Instruments. Instruments recommended for measuring carbon monoxide are

Commercial nondispersive infrared spectrometer

Sample introduction system (including pump, flow control valve, and flowmeter)

In-line filter (use a filter with a porosity of 2 to 10 μm to trap large particles)

Moisture controller (refrigeration units or drying tubes)

The instruments are installed and connected in accordance with the manufacturer's specifications.

Evaluation and interpretation of data. Generally, carbon monoxide does not pose a health problem to the general public. Continuous exposure to CO concentrations of 10 to 15 ppm, however, can cause impaired time interval discrimination. CO levels of 30 ppm have caused physiologic stress in patients with heart disease, while concentrations of 8 to 14 ppm have been correlated with increased fatality rates in hospitalized myocardial infarction patients.

Geographical and temporal limitations. The concentration of carbon monoxide does not remain constant over the entire spatial extent in a given region. Also, it will not remain constant over time. As such, substantial spatial and temporal variations in the concentration of carbon monoxide can be expected. It is generally claimed that the impact of carbon monoxide on the environment and humans depends on the total amount of exposure over the entire year. The spatial variations can be accounted for by taking small units of urbanized regions for purposes of analysis. This would require extensive calculations based on a diffusion model or a large-scale monitoring program. Since the use of a large-scale monitoring network is infeasible in most situations, the problem can be adequately addressed using diffusion models to predict air quality values over the entire spatial area.

Mitigation of impact.   There are three major strategies for the mitigation of impact of carbon monoxide on the environment. These are

Control of motor vehicle emissions

Control of stationary source emission

Reduction or removal of receptors from polluted areas

Secondary effects.   Presently identifiable specific secondary impacts due to increased carbon monoxide emissions are those related to human health effects—economic loss, increased accident rate, etc. Long-term secondary effects on the ecosystem due to increased carbon monoxide levels are not yet understood.

Photochemical oxidants

Definition of the attribute.   Products of atmospheric reactions between hydrocarbons and nitrogen oxides which are initiated by sunlight are called photochemical oxidants. The product of these reactions which is most commonly found and measured in the atmosphere is *ozone*. Other oxidants of interest include peroxyacetyl nitrate (PAN) and acrolein. Atmospheric measurement techniques measure the net oxidizing properties of atmospheric pollutants and report these photochemical oxidant concentrations as equivalent ozone concentration. Photochemical oxidants can be found anywhere hydrocarbons and nitrogen oxides interact in the presence of sunlight.

Activities that affect the attribute.   All activities that generate oxides of nitrogen and hydrocarbons simultaneously contribute to the generation of photochemical oxidants. Industrial activities and the operation and maintenance of motor vehicles and stationary combustion sources are major sources of nitrogen oxides and hydrocarbons. In addition, many other activities have petroleum and petrochemical operations that emit high levels of hydrocarbons.

Source of effects.   The data from animal and human studies are sparse and inadequate for determining the toxicological potential of photochemical oxidants. Injury to vegetation is one of the earliest manifestations of photochemical air pollution. The oxidants can cause both acute and chronic injury to leaves. Leaf injury has occurred in certain sensitive species after a 4-h exposure to 100 µg/m³ (0.05 ppm) total oxidants. Photochemical oxidants are known to attack certain materials. Polymers and rubber are important materials that are sensitive to photochemical oxidants.

Variables to be measured.   The concentration of ozone is measured in micrograms per cubic meter, as a maximum hourly average concentration. The standard is found in 40 CFR 50.9, and the test method in 40 CFR 50 Appendix D.

How variables are to be measured.   Since ozone is the major constituent contributing to photochemical oxidants, it is used as the reference substance in reporting levels of photochemical oxidants.

Ambient air and ethylene are delivered simultaneously to a mixing zone, where the ozone in the air reacts with the ethylene to emit light, which is detected by a photomultiplier cell. The resulting photocurrent is amplified and displayed on a recorder. The range of most instruments is from 0.005 ppm to greater than 1 ppm of ozone. The sensitivity is 0.005 ppm of ozone.

Data sources.   The sources of data are generally the State Pollution Control Department, the County Air Pollution Control Office, or the City Air Pollution Control Office. They can also install monitoring equipment at critical locations near their operations to determine the level of photochemical oxidants generated by activities.

Skills required.   A basic paraprofessional training in mechanical or chemical engineering with special training in operating the air quality instruments is adequate to collect data relating to photochemical oxidants. Specialized supervision will be needed to ensure that the data are properly collected and analyzed. Specialized supervision should include trained and experienced personnel or experienced consultants in the field of air quality monitoring.

Instruments.   Instruments for carrying out photochemical oxidant measurements include

Detector cell

Air flowmeter capable of controlling air flows between 0 and 1.5 l/min

Ethylene flowmeter capable of controlling ethylene flows between 0 and 50 ml/min

Air inlet filter capable of removing all particles greater than 5 μm diameter

Photomultiplier tube

High-voltage power supply (2000 V)

Direct current amplifier and a recorder

Evaluation and interpretation of data.   Photochemical oxidants are keyed to the 6 to 9 a.m. concentration values. At low concentrations, photochemical oxidants do not pose a problem. The quality of the environment, however, rapidly deteriorates as conditions for smog development approach (i.e., hydrocarbon concentration of 0.15 to 0.25 ppm). The values of the oxidant levels during the early morning determine the intensity of the oxidants to be expected later in the day. After sunset, the oxidant concentrations are reduced to low levels.

Geographical and temporal limitations.   The concentration of photochemical oxidants does not remain constant over the entire spatial extent in a given region. Also, it will not remain constant over time. As such, a substantial spatial and temporal variation in the concentration of photochemical oxidants can be expected. It is generally claimed that the impact of photochemical oxidants on the environment and humans depends on the total amount of exposure during the peak periods. The spatial variation can be accounted for by taking small units of urbanized regions for purposes of analysis. This would require extensive calculations based on a diffusion model or a large-scale monitoring program. Since the use of a large-scale monitoring network is infeasible in most situations, the problem can be adequately addressed using diffusion models to predict the air quality values over the entire spatial area.

Mitigation of impact.   All strategies for mitigating hydrocarbons and oxides of nitrogen are applicable to photochemical oxidants.

Secondary effects.   Sensitivity of plants to photochemical oxidants results in economic loss, as well as other secondary impacts on ecological balance. Other economic loss occurs with material deterioration and reduced property values.

Hazardous toxicants

Definition of the attribute.   Many kinds of hazardous air pollutants may be released to the environment. Some of these toxic elements or compounds are arsenic, asbestos, barium, beryllium, boron, cadmium, chromium, copper, lead, molybdenum, nickel, palladium, titanium, tungsten, vanadium, zinc, zirconium, radioactive wastes, mercury, and phenols. These toxic substances at certain concentrations may cause serious damage to the health and welfare of an exposed community.

Hazardous toxicants are substances like asbestos, beryllium, mercury, and other harmful elements and their compounds. Exposure to these toxicants can cause serious health hazards and diseases. These health impairments can result in increased mortality, morbidity, susceptibility to diseases, and loss of productivity.

Activities that affect the attribute. Hazardous toxicants may be generated by human activities such as construction; operation, maintenance, and repair of existing systems; industrial operations; research, development, and testing operations; and demolition of structures. For example, the surfacing of roadways with asbestos tailings can cause serious asbestos hazards.

The manufacture of clocks, cord, wicks, tubing, tape, twine, rope, thread, cement products, fireproofing and insulating materials, friction products, paper, mill board, felt, floor tile, paints, coatings, caulks, adhesives, sealants, and plastics may produce visible emissions of asbestos. Also, construction emissions produce substantial amounts of asbestos dust.

Source of effects. Hazardous toxicants can create serious health hazards and diseases of a chronic nature. For instance, exposure to asbestos dust at high concentrations and for longer durations can cause asbestos and bronchial cancer. In addition, asbestos is a cause of mesotheliomas; tumors; and membrane, intestine, and abdomen cancers. Most asbestos diseases have a latency period of 30 years.

Today, research has failed to establish an emission limit or concentration range above which asbestos dust can be harmful to human health. The EPA, however, recommends that no visible emissions be permitted from asbestos-generating activities.

Beryllium is another hazardous air pollutant which can seriously affect human health. Its effects are acute and chronic lethal inhalation, skin and conjunctival effects, cancer induction, and other beryllium diseases. The lowest beryllium concentration producing a beryllium disease was found to be greater than $0.01 \, \mu g/m^3$. At a concentration of $0.10 \, \mu g/m^3$ or above, the majority of exposed persons will develop beryllium diseases.

Variables to be measured. The variable measuring the extent of impact of a specific hazardous toxicant varies with the toxicant. For example, beryllium concentrations are required not to exceed 10 g over a 24-h period, while radionuclides emissions must not exceed the amount that would cause a member of the public to receive a specified dose equivalent per year.

How variables are measured.   There are many different methods of measuring various hazardous toxicants. See 40 CFR 61 for the method required for each toxicant.

Data sources.   Only a few city, county, regional, and state agencies monitor hazardous toxicants and emissions. Monitoring of selected hazardous toxicants is occasionally done by the EPA in cooperation with state or local agencies for selected periods at critical locations. Such monitoring is done only when a special hazardous toxicant is identified in a given region. Data on toxicant monitoring are available from state and local air pollution control agencies when collected.

Skills required.   Skills required for various hazardous toxicants measuring techniques are not well defined in the literature and require specialized supervision for use. Specialized consulting services are needed to implement these measurement techniques.

Instruments.   Complex sampling trains have to be designed on a case-by-case basis for each hazardous toxicant in the environment. The full range of instrumentation necessary for measurement of each hazardous toxicant is described in some of the standard documents mentioned above.

Evaluation and interpretation of data.   There are no well-defined value functions available for the hazardous toxicants identified in the environment. Generally, for each hazardous toxicant, it is possible to establish the upper and lower concentration limits of acceptability for the environment. The upper limit of acceptability is called the permissible level, the excess of which is considered highly undesirable and damaging to human health. On the other hand, the lower concentration limit of acceptability is called the desirable level, below which concentrations the quality of air can be considered acceptable; that is, the value function equals 1.

Emission limits have not been established for all the known hazardous toxicants. The EPA has established standards for major hazardous toxicants. These are found in 40 CFR 61.

The environmental impact of proposed activities on hazardous toxicant level is measured by the change in the hazardous toxicant concentration (HTC). When the HTC changes to the extent that its rating remains unaltered (e.g., high-quality air remains high-quality), the impact is considered insignificant. When a change in HTC is such that the rating changes by one step (e.g., between high quality and moderate quality), the impact is treated as moderate. When a change in HTC rating occurs through two steps (e.g., from high quality to poor quality or vice versa), the impact is considered significant.

Geographical and temporal limitations. Concentrations of hazardous toxicants do not remain constant over the entire spatial extent in a given region. Also, they will not remain constant over time. As such, substantial spatial and temporal variations in the concentrations of hazardous toxicants can be expected. It is generally claimed that the impact of hazardous toxicants on the environment and humans depends on the total amount of exposure over the entire day. Spatial variations can be accounted for by analyzing small units of urbanized regions. This requires extensive calculations based on a diffusion model or a large-scale monitoring program. Since the use of a large-scale monitoring program is infeasible in most situations, the problem can be adequately addressed using diffusion models to predict air quality values over the entire spatial area.

Mitigation of impact. There are five major strategies for the mitigation of impacts resulting from hazardous toxicants:

Use of materials that do not generate hazardous toxicants

Use of processes that do not generate hazardous toxicants

Avoiding or reducing activities that generate hazardous toxicants

Removal of hazardous emissions

Moving people from contaminated areas

Secondary effects. Secondary effects of hazardous toxicants include the economic losses which accompany lowered health standards and decreased productivity. Deterioration of water quality may result as these toxicants are cleansed from the air by natural processes. Effects on plant and animal life (aquatic and terrestrial) would vary with the toxicants and the levels present.

Other comments. Hazardous toxicants are powerful damaging agents for a community. Any industry can ill afford to be negligent about such emissions. Any attempt on the part of industry to compel communities to endure dangerous levels of toxicants resulting from its activities should be strongly discouraged. The use of this parameter will help to identify potential hazardous toxicant problems resulting from various operations.

Odors

Definition of the attribute. Industrial malodors are generally considered harmless, even though they frequently cause loss of personal and community pride, loss of social and economic status, discomfort, nausea, loss of appetite, and insomnia. It is true that odor effects on human health and welfare have been recognized only recently, and it seems that very little attention has been given in the literature to this air contaminant.

Malodors are generally caused by organic and sulfur compounds. The resulting odor characteristics are described by commonly accepted odor descriptors. Some common odor descriptors and their odor contaminants are indicated in Table B.1. For each odor contaminant, a concentration can be defined for which there can be no perception of the odor by a panel of individuals. This concentration is generally known as olfactory threshold or odor threshold. The odor thresholds of a few selected gaseous sulfur compounds in the air are shown in Table B.2.

The odor intensity is a measure of the stimulus resulting from the olfactory sensation of a given concentration of odorant. According to the Weber-Fechner law, odor intensity increases only logarithmically with the increase in concentration of the odor.

Activities that affect the attribute. In general, industrial operations; research, development, and testing operations; and operation and maintenance activities are potentially capable of emitting odor contaminants to the air.

Specific examples include metallurgical, chemical, petroleum, and food processing operations, feedlots, and burning activities.

Sources of effects. Malodors can affect both health and welfare of a community. These effects result from the loss of personal and community pride, reducing property values, tarnishing silver and paints, corroding steel, reducing appetite, producing nausea and vomiting, causing headache, and disturbing sleep, breathing, and olfactory sensations. These result in significant impacts, causing major public concern.

Variables to be measured. There are two major variables that measure the extent of odor problems. First, the average annual concentration of selected odor contaminants in parts per million (ppm) by volume is a useful measure of the extent of odor pollution at a given receptor point in a community. Second, the odor intensity, determined by an "odor jury" consisting of a panel of eight persons, is another measure of odor problems. The odor intensity scale has the following levels:

Levels	Descriptors
0	No odor
1	Odor threshold (or very slight odor)
2	Slight odor
3	Moderate odor
4	Strong odor

The concentration and intensity variables are used interchangeably for odor measurements.

TABLE B.1   Selected Malodors and Contaminants

Chemical compound or type material	Commonly accepted description of odor types
Acetylaldehyde	Fruity
Acetic acid	Vinegar
Acetone	Nail polish remover
Acetylene	Ethereal, garlic
African fiber	Musty, sour
Banana oil	Nail polish remover
Burnt protein	Burnt toast, scorched grain
Cannery waste	Rotten egg
Carbon disulfide	Rotten egg
Carbon tetrachloride	Cleaning fluid
Cresol	Creosote
Decayed fish	Rendering
Dimethyl sulfide	Rotten vegetables
Enamel coatings	Fatty linseed oil
Fatty acids	Grease, lard
Fermentation	Yeast or stale beer
Foam rubber curing	Sour sulfides
Gas house	Gas odors
Hydrogen sulfide	Rotten egg
Indole	Rest room
Iodoform	Iodine
Medicinal	Iodoform
Methyl ethyl ketone	Nail polish remover
Mercaptans (methyl)	Rotten cabbage
Oils: castor, coconut, soya, linseed	Rancid grease
Phenolic	Carbolic acid
Phenolic resins	Carbolic acid
Pig pen	Waste lagoons
Pyridine	Acrid, goaty
Septic sewage	Rotten egg
Skatole	Rest room
Sludge drying	Burnt grain
Sulfur dioxide	Irritating, strong, suffocating

SOURCES: Weisburd and Post.

TABLE B.2   Typical Odor Recognition Thresholds

Compound	ppm by volume
Acetaldehyde	0.2
Acetic acid	1
Acetone	100
Acrolein	0.2
Acrylonitrile	20
Allyl chloride	0.5
Ammonia	50
Aniline	1
Benzene	5
Benzyl chloride	0.05
Benzyl sulfide	0.002
Bromine	0.05
Butyric acid	0.001
Carbon disulfide	0.2
Carbon tetrachloride	20
Chloral	0.05
Chlorine	0.3
o-Cresol	0.001
Dimethylacetamide	50
Dimethylformamide	100
Dimethyl sulfide	0.001
Diphenyl sulfide	0.005
Ethanol	10
Ethyl acrylate	0.0005
Ethyl mercaptan	0.001
Formaldehyde	1
Hydrochloric acid	10
Hydrogen sulfide	0.0005
Methanol	100
Methylene chloride	200
Methyl ethyl ketone	10
Methyl isobutyl ketone	0.5
Methyl mercaptan	0.002
Methyl methacrylate	0.2
Monochlorobenzene	0.2
Nitrobenzene	0.005
Perchlorethylene	5
Phenol	0.05
Phosgene	1
Phosphine	0.02
Pyridine	0.02
Styrene	0.05
Sulfur dichloride	0.001
Sulfur dioxide	0.5
Toluene	5
Trichloroethylene	20
p-Xylene	0.5

SOURCE: Corbitt (1990).

How variables are measured. Two distinct methods for measuring malodors are

Scentometer method

Odor judgment panel

A scentometer can be used to measure ambient odor intensities when traveling through dusty areas. Strong, constant odors are measured by a scentometer over a square mile of area. It is a useful routine surveillance device that can identify threshold levels, possible odor problem areas, patterns of peak odor intensity, etc., over a given region.

On the other hand, an odor judgment panel can be used to verify the source of an unidentified odor, odor intensity, and damage potential of a given odor.

Data sources. The federal government has not yet established standards for potential odorants. No systematic monitoring and data collection are done with regard to odorants or odor contaminants by state and local agencies. Only in isolated cases will it be possible to find data on odor contaminants for selected periods and monitoring stations operated by state or local agencies.

Skills required. The use of a scentometer requires at least a technician level training and about a year's experience in using the equipment. The odor panel approach does not require any specific qualifications or formal training. It requires careful selection of juries based on olfactory sensation, and continuous training of the jurors to develop proper perception of different types of odors.

Instruments. The scentometer is the only equipment that is required in the first method of measuring odor problems. The second method (i.e., odor panel approach) does not require any equipment whatsoever.

Evaluation and interpretation of data. An environment with no odor at all is considered to be an ideal environment, with an environmental quality value of 1.0. Odor threshold concentration represents a tolerable level of odor contamination in the air; as such, it has an environmental quality value of 0.6. The value function falls rapidly with the occurrence of slight odor, and to 0 with a strong odor. Based on the above considerations, the value function for various odorants is presented in Fig. B.5. For practical purposes, the odor threshold of any odorant is the odorant concentration that can be detected only by 5 to 10 percent of the panelists. The slight odor is detected by about 20 to 25 percent of the panelists. The moderate odor is detected by about 40 percent of the panelists, and the strong odor is detected by about 100 percent of the panelists.

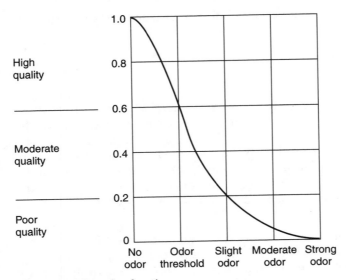

Figure B.5 Odor value function.

Geographical and temporal limitations. Concentration of malodors does not remain constant over the entire spatial extent in a given region. Also, it will not remain constant over time. As such, substantial spatial and temporal variations in the concentration can be expected. Spatial variations can be accounted for by analyzing small units of urbanized regions. This requires extensive calculations based on a diffusion model or a large-scale monitoring program. Since the use of a large-scale monitoring network is infeasible in most situations, the problem can be adequately addressed using diffusion models to predict air quality values over the entire spatial area.

The environmental impact of proposed activities on odor level is measured by the change in odor intensity. When the odor intensity changes to the extent that its rating remains unaltered (e.g., high-quality air remains high-quality), the impact is considered insignificant. When the change in odor intensity is such that its rating changes by one step (e.g., between high quality and moderate quality), the impact is treated as moderate. When a change in odor rating occurs through two steps (e.g., from high to low quality, or vice versa), the impact is considered significant.

The odor value function (Fig. B.5) is used for rating air quality in terms of high, moderate, and low quality, based on measured odor intensity. For a given value of odor intensity on the horizontal axis, a point on the curve can be found which identifies the environmental quality rating from the vertical axis of Fig. B.5 (e.g., odor intensity of slight odor indicates a moderate quality of greater than 0.2).

Mitigation of impact. The many different methods of abating potential impacts of odorous contaminants include

Dilution of odorant (dilution can change the nature as well as the strength of an odor)

Odor counteraction or neutralization (certain pairs of odors in appropriate concentrations may neutralize each other)

Odor masking or blanketing (certain weaker malodors may be suppressed by a considerably stronger good odor)

Reduction in odor emissions

Removal of receptors from polluted areas and/or from downwind odor path

Fatigued olfactory odor perception (certain levels of odor can be tolerated as a result of perception fatigue due to long-term exposure to the odor)

Planning can help establish optimal combinations of these mitigation alternatives to ensure that the best solution is made available to a community.

Secondary effects. Additional effects of malodors include the lowering of socioeconomic status, damaging community reputation, discouraging capital investment in a community, and discouraging tourism. Effects on the ecosystem and animal populations are not well understood.

References

Abron, Lilia A., and Robert A. Corbitt. "Air and Water Quality Standards" in *Standard Handbook of Environmental Engineering*. New York: McGraw-Hill, 1990.

Control Techniques for Particulate Emissions from Stationary Sources, Vol. 1, U.S. Environmental Protection Agency, EPA 450/3-81/005a, 1982.

Corbitt, Robert A. "Air Quality Control" in *Standard Handbook of Environmental Engineering*. New York: McGraw-Hill, 1990.

Dickey, J. H. "No Room to Breathe: Air Pollution and Primary Care Medicine." Physicians for Social Responsibility, Washington, D.C., www.psr.org/breathe.htm, 2001.

Englund, H. M., and Seymour Calvert. *Handbook of Air Pollution Technology.* New York: Wiley, 1984.

EPA regulations, standards, and test methods relevant to air pollution can be found in these chapters of the *Code of Federal Regulations:*
 40 CFR 50—*National Primary and Secondary Ambient Air Quality Standards*
 40 CFR 53—*Ambient Air Monitoring Methods*
 40 CFR 60—*Standards of Performance for New Stationary Sources*
 40 CFR 61—*National Emission Standards for Hazardous Air Pollutants*

Evaluation of Control Technologies for Hazardous Air Pollutants—Appendices. U.S. Environmental Protection Agency, EPA 600/7-86/0096, 1985.

Godish, Thad. *Air Quality.* Chelsea, Mich.: Lewis Publishers, 1990.

Majumdar, S. K., et al., eds. *Air Pollution: Environmental Issues and Health Effects.* Easton, Pa.: Pennsylvania Academy of Science, 1991.

Handbook—Control Technologies for Hazardous Air Pollutants. U.S. Environmental Protection Agency, EPA 625/6-86/014, 1986.

Samet, Jonathan M., and John D. Spengler, eds. *Indoor Air Pollution: A Health Perspective.* Baltimore: Johns Hopkins University Press, 1991.

Water

Water assumes different meaning and significance to different people. A particular definition depends, in large measure, on the personal uses to which water is put by the definer. Water may be considered an absolute necessity to sustain life and a necessary resource for all economic activity by some, and yet a refuge for biological pests and nuisances by others.

Pollution of water may be defined as a reduction in water quality by activities causing an actual hazard to public health or impairment of beneficial use of water.

The water environment is an intricate system of living and nonliving elements. Physical, chemical, and biological factors influencing water quality are so interrelated that a change in any water quality parameter triggers other changes in a complex network of interrelated variables. Often it is difficult to categorize the nature of these interrelationships that may result from human activity and influence on the entire water system.

To simplify analysis in the area of water, attributes of similar nature have been grouped together. This grouping was done with the following objectives. The list of selected attributes should be

1. As compact as possible

2. Equally applicable to surface and groundwater quality

3. Representative of comprehensive water quality indicators

4. Measurable in the field

5. Relevant to the spectrum of major activities

6. Capable of being measured on a project scale

Self-purification of natural waters.   All natural waters have the capability to assimilate certain amounts and types of waste without apparent effect upon the environment. The process by which self-purification is achieved is different for surface water and groundwater systems. Processes associated with both types of water systems are briefly described below.

Surface water system.   Some minor degradation of surface water quality may be overcome by the natural capacity of water bodies for withstanding certain insults. Such natural waste assimilation capacity is a result of dilution, sedimentation, flocculation, volatilization, biodegradation, aeration, aging, and uptake by organisms. The effects of relatively small amounts of waste are mitigated, and the water system recovers itself. If the waste load exceeds waste assimilation capacity, even for a short period, the effects may be devastating. The process of self-purification in surface waters is a complex phenomenon.

Groundwater system. Pollution of groundwater systems is a serious problem. Fortunately, contaminants typically must travel through the soil column before reaching the water table. Many soils have the capacity to mitigate manifold types of wastes. The processes by which waste is removed or purified in the soil column are aerobic and anaerobic decomposition, filtration, ion exchange, adsorption, absorption, etc. The process of dilution also reduces the concentration of contaminants.

Certain contaminants undergo significant removal during movement of water through the soil (unless the groundwater is directly contaminated by fissure cracks, leaks, abandoned or improperly constructed wells, pipes, or holes). Examples of such contaminants are microorganisms, organic matter, and turbidity. Dissolved solids, gases, and colloids are also important in assessing potential groundwater pollution. These contaminants, as discussed later, cause taste, odor, and physiological effects.

When groundwater becomes contaminated, remediation may be both difficult and expensive. Due to the relatively low flow rates characteristic of groundwater systems, pollutants are not readily diluted and thus tend to remain localized problems for extended periods of time. There may also be a considerable lag in time before pollution becomes noticeable in a groundwater system. As a result, today's activity may result in a significant impact only after several years, and today's problems are often traced to discharges of 10 to 30 years ago.

Description of selected water attributes. Fourteen attributes are used to define potential effects on water from the basic activities associated with various programs. These attributes, in three major categories, physical, chemical, and biological, are outlined below.

I. Physical
    A. Aquifer safe yield
    B. Flow variation
    C. Oil
    D. Radioactivity
    E. Suspended solids
    F. Thermal pollution
II. Chemical
    A. Acid and alkali

    B. Biochemical oxygen demand (BOD)
    C. Dissolved oxygen (DO)
    D. Dissolved solids
    E. Nutrients
    F. Toxic compounds

III. Biological

    A. Aquatic life
    B. Fecal coliforms

Table B.3 is a summary table indicating the 14 water quality attributes, conditions contributing to each, and a useful scale of impacts.

Aquifer safe yield

Definition of the attribute. The safe yield of an aquifer is exceeded when the rate of withdrawal surpasses the rate of recharge. Aquifer safe yield describes the general availability of the total groundwater system to supply water for human uses without the ultimate depletion of the aquifer. Aquifer safe yield incorporates physical attributes of the aquifer, which are porosity, permeability, transmissibility (which is permeability times thickness of the aquifer), and the storage coefficient.

Activities that affect the attribute. Many human activities affect the aquifer yield. The aquifer safe yield (available water resource) may decrease due to overpumping or by restricting this movement of water into or through the aquifer. During overpumping, as a result of turbulence in the well bore, fine-grained material moving near the well may cause a decrease in water movement toward the well. Land-use patterns may significantly reduce the water percolation into the ground. Also, improper waste injection may cause clogging of the formation due to suspended solids or bacterial action.

    Leaching of landfills may also clog the pores. All these factors decrease transmissibility of an aquifer and result in decreased aquifer safe yield. In regions dependent upon groundwater for water supplies, a decrease in safe yield could be highly undesirable. Lowering of the water table may cause public controversy, even in regions almost wholly dependent upon surface waters as a water supply. In coastal regions, uncontrolled water pumping from the ground may reverse the normal seaward gradient of the water table and permit salt water to move inland and contaminate the aquifer.

    Many activities may increase water availability due to increased water entering the system, which may result in raising of the water table accompanied by increased aquifer safe yield. Examples of such activities are water impoundment and reservoir construction and changes in topography to increase percolation. High water table is often accompanied by waterlogging problems in soils and water problems during excavation.

Sources of effects. As discussed above, many activities may upset the aquifer yield by directly or indirectly altering physical factors such as permeability, porosity, and ground surface conditions. The effects may be damaging and reduce potential groundwater resources.

Variables to be measured; how variables are measured. Maximum safe yield is measured in thousands of acre-feet of water withdrawn in a unit of time (usually a year); the method of measurement is based upon several techniques which all utilize extensive pumping tests.

Evaluation and interpretation of data. Knowledge concerning the relationship between degree of change in aquifer safe yield and environmental impact is extremely limited. It would not be possible at this time to make any quantitative judgment. However, since the reasonable environmental goal is to minimize the impact, a qualitative judgment can be made relating to deviation from the natural condition. Table B.3 summarizes five degrees of environmental impact based upon the qualitative judgments, and Table B.4 provides National Primary Drinking Water Standards.

Geographical and temporal limitations. Impacts related to aquifer safe yield are most likely to occur in areas with (1) high dependency on groundwater for supply, (2) a high water table, or (3) significant seasonal precipitation and subsequent infiltration. Local U.S. Geological Survey offices and state water agencies are excellent data sources for groundwater information.

Mitigation of impact. All activities likely to change the physical nature of the aquifer, to affect land surface runoff and percolation, and, in general, to increase or decrease water availability to the aquifer should be carefully controlled. Included are changing land-use patterns, landfilling, lagooning, reservoir construction, deep well injection, and pumping rate modifications. Complete tests should be made to investigate the existing groundwater hydrology, and correctional techniques should be selected to minimize adverse effects. These may relate to land slope and topography; surface area; reservoir, lagoon, and landfill lining; and deep well injection. Pumping rates may be adjusted to minimize the impact. Artificial recharge may also be employed.

TABLE B.3 Selected Attribute and Environmental Impact Categories

Selected attributes	Observed condition	Environmental impact category*				
		1 (most desirable)	2	3	4	5 (least desirable)
Physical Aquifer safe yield†	Changes occurring in physical attributes of aquifer (porosity, permeability, transmissibility, storage coefficient, etc.)	No change	No change	Slight change	Significant change	Extensive change
Flow variation‡	Flow variation attributed to activities; (Q_{max}/Q_{min})	None	None	Slight	Significant	Extensive
Oil§	Visible silvery sheen on surfaces, oily taste and odor to water and/or to fish and edible invertebrates, coating of banks and bottom or tainting of attached associated biota	None	None	Slight	Significant	Extensive
Radioactivity§	Measured radiation limit, 10^{-7} microcurie/ml	Equal to or less	Equal to or less	Exceed limit	Exceed limit	Exceed limit
Suspended solids‡	1. Sample observed in a glass bottle	Clear	Clear	Fairly clear	Slightly turbid	Turbid
	2. Turbidity in net transfer units	3 or less	10	20	50	100
	3. Suspended solids mg/l	4 or less	10	15	20	35
Thermal discharge‡	Magnitude of departure from natural condition, C°	0	2	4	6	10
Chemical Acid and alkali§ pH units	Departure from natural condition,	0	0.5	1	1.5	2

		1	2	3	5	10
BOD§	mg/l	1	2	3	5	10
DO‡	% saturation	100	85	75	60	Low
Dissolved solids§	mg/l	500 or less	1000	2000	5000	High
Nutrients‡	Total phosphorus mg/l	0.02 or less	0.05	0.10	0.20	Large
Toxic compounds§	Concentration mg/l	Not detected	Traces	Small	Large	Large
Biological						
Fecal coliform§	Number per 100 ml	50 or below	5000	20,000	250,000	Large
Physical						
Aquatic life‡	Green algae	Scarce	Moderate quantities in shallows	Plentiful in shallows	Abundant	Abundant
	Gray algae	Scarce	Scarce	Scarce	Present	Plentiful
	Delicate fish; trout, grayling	May be plentiful	Plentiful	Probably absent	Scarce	Absent
	Coarse fish; chub, dace, carp, roach	May be present	Plentiful	Plentiful	Scarce	Absent
	Mayfly naiad, stonefly nymph	May be plentiful	Plentiful	Scarce	Absent	Absent
	Bloodworm, sludge worm, midge larvae, rat-tailed maggot, sewage fly larvae and pupa	May be absent	Scarce	May be present	Plentiful	Abundant

*Environmental impact category: category 1 indicates most desirable condition; category 5 indicates extensive adverse condition. Because all attributes are related to environmental quality between 0 and 1, it is possible to compare different attributes and five categories on a common base. Each category is equivalent to approximately 20 percent of overall environmental quality. In the physical sense, water quality for five categories will be very clean, clean, fairly clean, doubtful, and bad. Environmental impact may be adverse or favorable. Adverse impact will deteriorate the environmental quality while favorable impact will improve the quality. Proper signs and weights must be used to achieve overall effects.

†Applies to groundwater systems only.

‡Applies to surface water systems only.

§Applies to both the groundwater and surface water.

TABLE B.4 2000 National Primary Drinking Water Regulations, U.S. EPA

Inorganic chemicals	Maximum contaminant level goal (MCLG) (mg/l)	Maximum contaminant level (MCL) (mg/l)	Contaminants	
			Potential health effects from ingestion of water	Sources of contaminant in drinking water
Antimony	0.006	0.006	Increase in blood cholesterol; decrease in blood glucose.	Discharge from petroleum refineries; fire retardants; ceramics; electronics; solder
Arsenic	None	0.05	Skin damage; circulatory system problems; increased risk of cancer.	Discharge from semiconductor manufacturing; petroleum refining; wood preservatives; animal feed additives; herbicides; erosion of natural deposits
Asbestos (fiber > 10 micrometers)	7 million fibers per liter (MFL)	7 MFL	Increased risk of developing benign intestinal polyps.	Decay of asbestos cement in water mains; erosion of natural deposits
Barium	2	2	Increase in blood pressure.	Discharge of drilling wastes; discharge from metal refineries; erosion of natural deposit
Beryllium	0.004	0.004	Intestinal lesions.	Discharge from metal refineries and coal-burning factories; discharge from electrical, aerospace, and defense industries
Cadmium	0.005	0.005	Kidney damage.	Corrosion of galvanized pipes; erosion of natural deposits; discharge from metal refineries; runoff from waste batteries and paints
Chromium (total)	0.1	0.1	Some people who use water containing chromium well in years excess of the MCL over many could experience allergic dermatitis.	Discharge from steel and pulp mills; erosion of natural deposits

Contaminant	MCLG	MCL or TT	Potential health effects from exposure	Sources of contaminant in drinking water
Copper	1.3	Action level = 1.3; TT*	Short-term exposure: Gastrointestinal distress. Long-term exposure: Liver or kidney damage. Those with Wilson's Disease should consult their personal doctor if their water systems exceed the copper action level.	Corrosion of household plumbing systems; erosion of natural deposits; leaching from wood preservatives
Cyanide (as free cyanide)	0.2	0.2	Nerve damage or thyroid problems.	Discharge from steel/metal factories; discharge from plastic and fertilizer factories
Fluoride	4	4	Bone disease (pain and tenderness of the bones); children may get mottled teeth.	Water additive which promotes strong teeth; erosion of natural deposits; discharge from fertilizer and aluminum factories
Lead	0	Action level = 0.015; TT*	Infants and children: Delays in physical or mental development. Adults: Kidney problems; high blood pressure.	Corrosion of household plumbing systems; erosion of natural deposits
Inorganic mercury	0.002	0.002	Kidney damage.	Erosion of natural deposits; discharge from refineries and factories; runoff from landfills and cropland
Nitrate (measured as nitrogen)	10	10	"Blue baby syndrome" in infants under six months: life-threatening without immediate medical attention. Symptoms: Infant looks blue and has shortness of breath.	Runoff from fertilizer use; leaching from septic tanks, sewage; erosion of natural deposits
Nitrate (measured as nitrogen)	1	1	"Blue baby syndrome" in infants under six months: life-threatening without immediate medical attention. Symptoms: Infant looks blue and has shortness of breath.	Runoff from fertilizer use; leaching from septic tanks, sewage; erosion of natural deposits
Selenium	0.05	0.05	Hair or fingernail loss; numbness in fingers or toes; circulatory problems.	Discharge from petroleum refineries; erosion of natural deposits; discharge from mines
Thallium	0.0005	0.002	Hair loss; changes in blood; kidney, intestine, or liver problems.	Leaching from ore-processing sites; discharge from electronics, glass, and pharmaceutical companies

TABLE B.4 2000 National Primary Drinking Water Regulations, U.S. EPA (*Continued*)

			Contaminants	
Organic chemicals	MCLG (mg/l)	MCL (mg/l)	Potential health effects from ingestion of water	Sources of contaminant in drinking water
Acrylamide	0	TT*	Nervous system or blood problems; increased risk of cancer.	Added to water during sewage/wastewater treatment
Alachlor	0	0.002	Eye, liver, kidney, or spleen problems; anemia; increased risk of cancer.	Runoff from herbicide used on row crops
Atrazine	0.003	0.003	Cardiovascular system problems; reproductive difficulties.	Runoff from herbicide used on row crops
Benzene	0	0.005	Anemia; decrease in blood platelets; increased risk of cancer.	Discharge from factories; leaching from gas, storage tanks and landfills
Benzo(a)pyrene	0	0.0002	Reproductive difficulties; increased risk of cancer.	Leaching from linings of water storage tanks and distribution lines
Carbofuran	0.04	0.04	Problems with blood or nervous system; reproductive difficulties.	Leaching of soil fumigant used on rice and alfalfa
Carbon tetrachloride	0	0.005	Liver problems; increased risk of cancer.	Discharge from chemical plants and other industrial activities
Chlordane	0	0.002	Liver or nervous system problems; increased risk of cancer.	Residue of banned termiticide
Chlorobenzene	0.1	0.1	Liver or kidney problems.	Discharge from chemical and agricultural chemical factories

Contaminant			Potential Health Effects	Sources
Dalapon	0.2	0.2	Minor kidney changes.	Runoff from herbicide used on row crops
1,2-Dibromo-3-chloropropane (DBCP)	0	0.0002	Reproductive difficulties; increased risk of cancer.	Runoff/leaching from soil fumigant used on soybeans, cotton, pineapples, and orchards
o-Dichlorobenzene	0.6	0.6	Liver, kidney, or circulatory system problems.	Discharge from industrial chemical factories
p-Dichlorobenzene	0.075	0.075	Anemia; liver, kidney, or spleen damage; changes in blood.	Discharge from industrial chemical factories
1,2-Dichloroethane	0	0.005	Increased risk of cancer.	Discharge from industrial chemical factories
1,1-Dichloroethylene	0.007	0.007	Liver problems.	Discharge from industrial chemical factories
cis-1,2-Dichloroethylene	0.07	0.07	Liver problems.	Discharge from industrial chemical factories
trans-1,2-Dichloroethylene	0.1	0.1	Liver problems.	Discharge from industrial chemical factories
Dichloromethane	0	0.005	Liver problems; increased risk of cancer.	Discharge from pharmaceutical and chemical factories
1,2-dichloropropane	0	0.005	Increased risk of cancer.	Discharge from industrial chemical factories
Di(2-ethylhexyl)adipate	0.4	0.4	General toxic effects or reproductive difficulties.	Leaching from PVC plumbing systems; discharge from chemical factories
Di(2-thylhexyl)phthalate	0	0.0006	Reproductive difficulties; liver problems; increased risk of cancer.	Discharge from rubber and chemical factories.

TABLE B.4 2000 National Primary Drinking Water Regulations, U.S. EPA (*Continued*)

Organic chemicals	MCLG (mg/l)	MCL (mg/l)	Contaminants	
			Potential health effects from ingestion of water	Sources of contaminant in drinking water
Dinoseb	0.007	0.007	Reproductive difficulties.	Runoff from herbicide used on soybeans and vegetables
Dioxin (2,3,7,8-TCDD)	0	0.00000003	Reproductive difficulties; increased risk of cancer.	Emissions from waste incineration and other combustion; discharge from chemical factories
Diquat	0.02	0.02	Cataracts.	Runoff from herbicide use
Endothall	0.1	0.1	Stomach and intestinal problems.	Runoff from herbicide use
Endrin	0.002	0.002	Nervous system effects.	Residue of banned insecticide
Epichlorohydrin	0	TT*	Stomach problems; reproductive difficulties; increased risk of cancer.	Discharge from industrial chemical factories; added to water during treatment process
Ethylbenzene	0.7	0.7	Liver or kidney problems.	Discharge from petroleum refineries
Ethylene dibromide	0	0.00005	Stomach problems; reproductive difficulties; increased risk of cancer.	Discharge from petroleum refineries
Glyphosate	0.7	0.7	Kidney problems; reproductive difficulties.	Runoff from herbicide use
Heptachlor	0	0.0004	Liver damage; increased risk of cancer.	Residue of banned termiticide
Heptachlor epoxide	0	0.0002	Liver damage; increased risk of cancer.	Breakdown of heptachlor

	MCLG	MCL	Potential health effects	Sources of contaminant
Hexachlorobenzene	0	0.001	Liver or kidney problems; reproductive difficulties; increased risk of cancer.	Discharge from metal refineries and agricultural chemical factories
Hexachlorocyclo-pentadiene	0.05	0.05	Kidney or stomach problems.	Discharge from chemical factories
Lindane	0.0002	0.0002	Liver or kidney problems.	Runoff/leaching from insecticide used on cattle, lumber, gardens
Methoxychlor	0.04	0.04	Reproductive difficulties.	Runoff/leaching from insecticide used on fruits, vegetables, alfalfa, livestock
Oxamyl (Vydate)	0.2	0.2	Slight nervous system effects.	Runoff/leaching from insecticide used on apples, potatoes, and tomatoes
Polychlorinated biphenyls (PCBs)	0	0.0005	Skin changes; thymus gland problems; immune deficiencies; reproductive or nervous system difficulties; increased risk of cancer.	Runoff from landfills; discharge of waste chemicals
Pentachlorophenol	0	0.001	Liver or kidney problems; increased risk of cancer.	Discharge from wood preserving factories
Picloram	0.5	0.5	Liver problems.	Herbicide runoff
Simazine	0.004	0.004	Problems with blood.	Herbicide runoff
Styrene	0.1	0.1	Liver, kidney, or circulatory system problems.	Discharge from rubber and plastic factories; leaching from land-fills
Tetrachloroethylene	0	0.005	Liver problems; increased risk of cancer.	Discharge from factories and dry cleaners

TABLE B.4 2000 National Primary Drinking Water Regulations, U.S. EPA (*Continued*)

Organic chemicals	MCLG (mg/l)	MCL (mg/l)	Contaminants — Potential health effects from ingestion of water	Sources of contaminant in drinking water
Toluene	1	1	Nervous system, kidney, or liver problems.	Discharge from petroleum factories
Total trihalomethanes (THMs)	None	0.1	Liver, kidney, or central nervous system problems; increased risk of cancer.	Byproduct of drinking water disinfection
Toxaphene	0	0.003	Kidney, liver, or thyroid problems; increased risk of cancer.	Runoff/leaching from insecticide used on cotton and cattle
2,4,5-TP (Silvex)	0.05	0.05	Liver problems.	Residue of banned herbicide
1,2,4-Trichlorobenzene	0.07	0.07	Changes in adrenal glands.	Discharge from textile finishing factories
1,1,1-Trichloroethane	0.2	0.2	Liver, nervous system, or circulatory problems.	Discharge from metal degreasing sites and other factories
1,1,2-Trichloroethane	0.003	0.005	Liver, kidney, or immune system problems.	Discharge from industrial chemical factories
Trichloroethylene	0	0.005	Liver problems; increased risk of cancer.	Discharge from petroleum refineries
Vinyl chloride	0	0.002	Increased risk of cancer.	Leaching from PVC pipes; discharge from plastic factories
Xylenes (total)	10	10	Nervous system damage.	Discharge from petroleum factories; discharge from chemical factories

TABLE B.4 2000 National Primary Drinking Water Regulations, U.S. EPA (*Continued*)

Radionuclides	MCLG (mg/L)	MCL (mg/L)	Contaminants — Potential health effects from ingestion of water	Sources of contaminant in drinking water
Beta particles and photon emitters	None	4 millirems per year	Increased risk of cancer.	Decay of natural and manmade deposits
Gross alpha particle activity	None	15 picocuries per liter (pCi/l)	Increased risk of cancer.	Erosion of natural deposits
Radium 226 and Radium 228 (combined)	None	5 pCi/l	Increased risk of cancer.	Erosion of natural deposits

Microorganisms	MCLG (mg/L)	MCL (mg/L)	Potential health effects from ingestion of water	Sources of contaminant in drinking water
Giardia lamblia	0	TT*	Giardiasis, a gastroenteric disease.	Human and animal fecal waste
Heterotrophic plate count (HPC)	N/A	TT*	HPC has no health effects, but can indicate how effective treatment is at controlling microorganisms.	N/A
Legionella	0	TT*	Legionnaire's disease, commonly known as pneumonia.	Found naturally in water; multiplies in heating or air conditioning systems

TABLE B.4 2000 National Primary Drinking Water Regulations, U.S. EPA (Continued)

			Contaminants	
Microorganisms	MCLG (mg/l)	MCL (mg/l)	Potential health effects from ingestion of water	Sources of contaminant in drinking water
Total coliforms (including fecal coliform and *E. Coli*)	0	No more than 5.0% samples total coliform-positive in a month	Used as an indicator that other potentially harmful bacteria may be present.	Human and animal fecal waste
Turbidity	N/A	TT*	Turbidity has no health effects but can interfere with disinfection and provide a medium for microbial growth. It may indicate the presence of microbes.	Soil runoff
Viruses (enteric)	0	TT*	Gastroenteric disease.	Human and animal fecal waste

*Treatment technique (TT). An enforceable procedure or level of technical performance which public water systems must follow to ensure control of a contaminant.

SOURCE: EPA, Office of Water, July 2000.

NOTE: MCLG is the level of a contaminant in drinking water below which there is no known or expected risk to human health. "None" indicates no goal has been established. MCLGs are non-enforceable goals. MCL is the highest level of a contaminant that is allowed in drinking water and is set as close to the MCLG as possible. MCLs are enforceable standards.

Secondary effects. Alterations in aquifer safe yield can be related to other attributes in terms of secondary impacts. Aside from being a community need, a safe, dependable water supply is necessary for community and regional economic stability. It can affect land-use patterns as well, since it is a factor in domestic, industrial, and agricultural requirements.

Flow variations

Definition of the attribute. The velocity of flow and rate of discharge are extremely important to aquatic organisms in a number of ways, including the transport of nutrients and organic food past those organisms attached to stationary surfaces; the transport of plankton and benthos as drift, which in turn serve as food for higher organisms; and the addition of oxygen to the water through surface aeration. Silts are moved downstream, and sediments may be transported as bed load. These, in turn, are often associated with major nutrients, such as nitrogen and phosphorus, which may be released at some point downstream.

Natural flow variations are, therefore, critical factors governing the type of ecological system that will develop and survive in a given watercourse. If the pattern of stream flow variation is changed markedly from what is natural, subsequent disruption of the natural ecology may result.

Activities that affect the attribute. Major activities that may influence stream flow include water resource projects and changing the ground surface and topography for different types of land-use projects. This may include site clearing, earthwork and borrowing, paving of land areas, and building construction. Other activities include modification of vegetation, which can lead to altered runoff patterns, and water use changes in withdrawal and return flow rates.

Sources of effects. Water resource projects may be for flood control (that reduces high flows), power generation (that minimizes low-flow conditions), or any desired use that alters the flow pattern of the stream. The land-use project alters the runoff, percolation, and evaporation in the drainage basin. These changes may increase or decrease the runoff. Other attributes affected by such activities are suspended solids and nutrients in the watercourses; they may, in turn, affect the population of photosynthetic organisms and, thus, the food chain. Direct flow variations are caused by fluctuating municipal, industrial, and/or agricultural demands and the return flows from these users.

Variables to be measured; how variables are measured. Flow measurement is relatively simple. Many types of automatic flow measurement devices that can be installed in a selected reach of a watercourse are commercially available. The typical unit of flow measurement is cubic feet per second (ft^3/s). Velocity measurement may be accomplished by current meters, which register in feet per second (ft/s).

Evaluation and interpretation of data. If flow variations are rapid and extensive, more disruption to natural ecology results. However, due to lack of information, classification of water cannot be made on the basis of qualitative measurement. Five degrees of environmental impact are summarized in Table B.3. This classification is based upon qualitative or observed conditions.

Special conditions. Flow variations become most significant at the extreme conditions—low flows and high flows. Under low-flow conditions, the natural assimilative capacity of a given stream is greatly reduced, and the adverse effects of natural and human-induced waste loads are most critical. At high flows, physical damage due to flooding and inundation of vegetation becomes a major concern.

Geographical and temporal limitations. Low-flow considerations are of importance on all streams; however, they warrant particular consideration in areas which typically experience prolonged periods of drought. These periods frequently coincide with summer, when biological activity is high and dissolved oxygen content in streams is at a minimum, thus compounding the significance of the problem.

Impacts associated with high-flow conditions are most likely to occur in areas and climates with conditions conducive to flooding.

Mitigation of impact. All activities such as land-use projects and water impoundment and operation should be given consideration to minimize flow variations from the mean natural flow.

Secondary effects. Human-induced flow variations may have secondary impacts in ecology, land-use patterns, and the socioeconomic realm. Many species of plant and animal life are sensitive to flow variations and require specific ranges of flow conditions. Floodplain development can be a function of the degree of control over flow variations. Economic losses are felt through flooding of agricultural as well as built-up areas, and adverse psychological effects are apparent when there are threats of flood.

Oil

Definition of the attribute.   Oil (i.e., petroleum) slicks are barely visible at a concentration of about 25 gal/mi^2. At about 50 gal/mi^2, an oil film is about 3.0×10^{-6} in thick and is visible as a silvery sheen on the surface. Oil is destructive to aquatic life in the following ways:

Free oil and emulsions may coat and destroy algae and plankton.

Heavy coating may interfere with the natural processes of reaeration and photosynthesis.

Water-soluble fractions may exert a direct toxic action.

Settleable oil substances may coat the bottom, destroy benthic organisms, and interfere with spawning areas.

Activities that affect the attribute.   Major activities responsible for oil pollution include bilge and ballast waters discharged from ships; oil refinery wastes; industrial plant wastes, such as oil, grease, and fats, and lubrication of machinery; gasoline filling stations; bulk stations; and accidental spills.

Sources of effects.   Oil may reach natural waters by direct discharge or by surface runoff. Direct discharge may occur from bilge and ballast waters or by accidental spill from barges or tankers. Indirect oil release may occur from surface runoff or storm sewers or combined sewer overflows. In all cases, damage could be severe and long lasting. Water quality parameters affected by oil discharge are dissolved oxygen, general appearance, and taste and odor.

Variables to be measured; how variables are measured.   Dissolved or emulsified oil or grease is extracted from water by intimate contact with various organic solvents. The results are expressed in milligrams per liter oil or grease. Other measurements are qualitative and include (1) visible oil slick, (2) oily taste and odor in fish and edible invertebrates, and (3) coating of banks and bottom or tainting of associated biota. Quantitative measurement of oil and grease is by extraction in a separating funnel with either trichlorotrifluoroethane or petroleum ether. The technique is used as routine analysis in water and wastewater analysis.

Evaluation and interpretation of data.   Due to lack of information, classification of water cannot be made on the basis of quantitative measurement or concentrations. Five degrees of water impacts are summarized in Table B.3 on the basis of qualitative or observed conditions.

Mitigation of impact. Oil pollution can be minimized by controlling all direct discharge into natural waters. Surface runoff from oil-handling areas should be treated for oil separation before discharge into the environment. If oil wastes are combined with sanitary sewage, oil separation will be necessary at the wastewater treatment facility. Lagooning of oil wastes and land disposal of oily sludges should be restricted in order to avoid possible contamination of the groundwater system.

Secondary effects. Secondary effects of oil discharges are manifested through impacts on aquatic ecology and waterfowl, economic loss through decreased recreational desirability, and lowered property values if the discharges become frequent. Increased activity in exploration, production, and transportation of petroleum can increase controversy, divide communities, alter land-use patterns, and indirectly affect public and private land markets—whether or not any actual spills take place.

Radioactivity

Definition of the attribute. Ionizing radiation, when absorbed in living tissue in quantities substantially above that of natural background, is injurious. It is, therefore, necessary to prevent excessive levels of radiation from reaching any organism, such as humans, fish, or invertebrates.

Activities that affect the attribute. Human activities responsible for radiation hazards are application of nuclear methods in power development, industrial operation, medical laboratories, research and development, nuclear weapons testing, and radiation warfare. In all applications, radioactive substances may be released accidentally, by inadequately planned and controlled activity, or by disposal of radioactive wastes.

Sources of effects. Radioactivity, once released to the aquatic environment, may (1) remain in solution or in suspension, (2) precipitate and settle to the bottom, or (3) be taken up by plants and animals. Immediately upon introduction of radioactive materials into the water, the wastes may become diluted by dispersion or may become concentrated by the process of biological magnification.*

Variables to be measured; how variables are measured. The measure of radioactivity is the curie (Ci), the quantity of any radioactive material in which the disintegrations per second are 3.7×10^{10}; this is a large

---

*Biological magnification, or bioaccumulation, is a process in which some substances become concentrated as they pass through the food chain.

amount of radioactivity. Two smaller units, the microcurie (μCi, 10^{-6} Ci) and the picocurie (pCi, 10^{-12} Ci or 2.22 disintegrations per minute), are often used. Radioactive waste can be diluted in water to below the allowable limit. The allowable limit of radiation in natural water is taken as 10^{-7} μCi/ml when the activity is caused by an unknown mixture of beta- and gamma-emitting isotopes.

Measurement techniques are not difficult because radiation-counting equipment of high sensitivity and stability is commercially available.

Evaluation and interpretation of data.   It is difficult to determine the long-term effects of radiological wastes upon aquatic life. For this reason, and as a practical matter, radioactivity exceeding the allowable limit of 10^{-7} μCi/ml may be considered detrimental to human health and aquatic life. Five classes of water impacts are given in Table B.3.

Special conditions.   Special precautions should be taken to prevent radioactive materials from entering ground or surface waters to be used for drinking water supply, fish production, or recreation. Table B.4 provides national primary drinking water standards that should be consulted for water that is to be used as a drinking water supply.

Mitigation of impact.   Release of radioactive wastes from radiation facilities must be monitored and controlled. Radioactivity in sewage after treatment is likely to be concentrated in sludge; thus sludge disposal becomes a difficult problem. Therefore, waste containing radioactivity should be treated separately by means of dewatering procedures, and solids or brine should be disposed of by special care (deep well injection or containment). Fallout of radioactive dust will induce radioactivity in surface runoff, the treatment of which is a difficult task. All efforts should, therefore, be made to minimize release of radioactivity into the environment.

Secondary effects.   While it is generally understood that aquatic organisms are relatively tolerant of radioactive materials, little is known of the mechanism of concentrating radioactive elements by these organisms and the effect this might have on human or other consumer organisms. Although there have been few cases of actual radioactive contamination of water resources, the fear of such contamination actually has had much greater impact. These fears have resulted in controversy and altered land-use patterns and have had other socioeconomic effects.

Suspended solids

Definition of the attribute.   Suspended solids are solids contained in water which are not in solution. They are distinguished from dissolved solids by laboratory filtration tests. Suspended solids comprise settleable, floating (specific gravity lower than water), and nonsettleable (colloidal suspension) components. These may contain organic (volatile suspended solids) or inert (nonvolatile) substances. Turbidity may be caused by a wide variety of suspended materials, ranging in size from colloidal particles to a coarse dispersion, depending upon the turbulence and light scattering properties of suspended materials.

Suspended solids are perhaps of greatest significance from the standpoint of aesthetics. Natural waters may contain wide variations of suspended solids. These may be due to clay, silt, silica, organic matter, microorganisms, or sewage. Suspended solids may be undesirable in many ways. In public water supplies, turbid water is difficult and costly to filter. Disinfection may require higher chemical dosages if the water is turbid. Also, excessive suspended solids can be harmful to fish and other aquatic life by coating gills, blanketing bottom organisms, reducing solar radiation intensity, and, thus, affecting the natural foodchain. In stream pollution-control work, all suspended solids are considered to be settleable solids, because eventually (by bacterial decomposition and chemical flocculation) those solids are deposited.

Activities that affect the attribute.   Activities directly responsible for suspended-solids release are dredging, wastewater discharge, construction of hydraulic structures, and gravel washing. Activities that indirectly affect suspended solids result from land use: site clearing, surface paving, building construction, landscaping, and mine tailings. All change the surface runoff pattern, which, in most cases, increases the storm flow. Suspended-solid load in the surface runoff may change considerably due to erosion. Also, flow variations in streams may change the bed load and solids transport.

Source of effects.   As discussed above, many activities will increase or decrease the suspended-solid condition in natural waters. It may be mentioned that many times this effect may be temporary. For example, dredging may increase suspended solids during operation. After completion of dredging, the channel may become deeper and wider; thus, dredging may actually reduce velocity and encourage settling. Likewise, many other activities, such as construction, site clearing, and excavation, may have effects that should be evaluated as long- or short-term.

Many water quality attributes may be affected by change in suspended-solid condition. These include dissolved oxygen (DO) (due to increase in photosynthesis), nutrient enrichment, and direct deleterious effect to fish and other aquatic life by coating gills or blanketing bottom organisms, for example.

Variables to be measured; how variables are measured. Readily set-tleable suspended solids are measured in milliliters per liter of settled water. Suspended solids are measured by filtering a sample through a membrane filter or a glass fiber mat in a Gooch crucible. Turbidity is measured in net transfer units equivalent to the interference to light transmission caused by 1 mg/l of a standard suspension.

Many types of commercially available instruments can continuously measure and record the turbidity and the suspended solids in water. They rely upon transmission, diffraction, and absorption of light through a standard light path.

Evaluation and interpretation of data. Water quality is considered lower with increasing turbidity and suspended solids. Table B.3 summarizes the five classes of water impact, based upon turbidity, suspended solids, and visual consideration.

Mitigation of impact. The impact due to suspended solids may be min-imized by controlling discharge of wastes that contain suspended solids; this includes sanitary sewage and industrial wastes. Also, all activity that increases erosion or contributes nutrients to water (thus stimulating algae growth) should be minimized.

The gravel-washing activity, mine tailings, and anything causing dust may be controlled by utilizing available technology.

Secondary effects. Increase in suspended solids content may have sec-ondary impact on socioeconomic attributes as a result of loss of pro-ductivity (e.g., decline in fish harvest) and reduction in various recreation-oriented activities. Additionally, increased costs to remove suspended solids for domestic or industrial water use may occur as a result. Long-term effects include siltation of reservoirs, reducing use-ful capacity, and filling of marsh areas in estuaries, reducing produc-tive habitat.

Thermal discharge

Definition of the attribute. Temperature is a prime regulator of natural processes within the water environment. In addition to affecting the rate of chemical reactions, it governs physiological function in organ-isms and, acting directly or indirectly in combination with other water quality constituents, affects aquatic life with each change. Water tem-perature controls spawning and hatching, regulates activity, and stim-ulates or suppresses growth; it can kill when the water becomes heated or chilled too suddenly. Colder water generally suppresses development; warmer water generally accelerates activity.

Activities that affect the attribute. Human activities affecting the attribute are discharges with temperatures above or below that of the receiving waters. Heated discharge may result from sources such as thermal power generation, heavy machine operations, and industrial operations.

Cold water discharges may result from flows from large, deep reservoirs.

Sources of effects. Heated wastes, when discharged into the water environment, raise the temperature of the water. The extent to which the temperature is raised depends upon the quantity of waste heat discharged and the amount of diluting water available. As water temperature increases, the solubility of oxygen decreases. Furthermore, the accelerated biological activity imposes higher oxygen demand. The net result is a decrease in DO level, which can reach critical levels.

Water released from lower depths of stratified reservoirs may be significantly lower in temperature and DO content than would prevail in normal ambient stream conditions. Thus, release depths can have a pronounced effect upon the aquatic life below reservoirs.

Variables to be measured; how variables are measured. Temperature measurement is simple and accurate. Many types of automatic temperature recording devices are commercially available. Measurement scale is either degrees Celsius or degrees Fahrenheit. Prediction of the effects of projects on ambient water temperatures is a complex problem which may be addressed through the use of mathematical models for mixing and heat exchange in the aquatic environment.

Evaluation and interpretation of data. In environmental quality assessment, the temperature effects are best handled in terms of the magnitude of departure from the natural conditions. Table B.3 summarizes five classes of water impacts, based upon temperature rise above natural conditions. Allowable departures from ambient temperatures may vary with location, so state water quality regulations should be consulted.

Geographical and temporal limitations. Fogging problems may be associated with warm water discharges in cold regions or under special climatic conditions.

Mitigation of impact. Cooling towers can be used to convert once-through systems into closed systems. A very efficient way is to utilize treated wastewaters (such as sewage, industrial wastes, or stored surface runoffs) as cooling water makeup. Many industrial plants are consider this type of closed system. Chromium may be recovered from cooling tower blowdown before treatment and disposal of tower blowdown. Selective withdrawal may be employed to control the temperature of water released from stratified reservoirs.

Secondary effects.   Effects on the aquatic environment resulting from temperature alteration may, in turn, have other biophysical and socioeconomic consequences. Increased heat to water bodies accelerates evaporation, and thus the suspended solids content of the water. This and other impacts on the biological activity may alter the aesthetic and recreational desirability of a given area. Depending upon the circumstances, these effects may be of a positive or a negative nature. On the one hand, heat addition may speed the eutrophication process and reduce recreational use. In other instances, this effect has increased recreational benefit through increased productivity. Aquaculture, or "fish farming," has been investigated as a possible beneficial secondary result of heated discharges.

Acid and alkali

Definition of the attribute.   Acid and alkaline wastes discharged into waters may change the natural buffer system. The pH of the water may significantly change, depending upon the extent of acid or alkali discharged. Change in the pH of natural water is hazardous for fish and other aquatic life. Below a pH of 5.0 and above 9.0, fish mortalities may be expected.

Activities that affect the attribute.   Activities which may contribute acid and alkali waste to the environment are industrial wastes such as pickling liquors, tanning, metal finishing, food processing, accidental spills of chemicals, and mining operations.

Sources of effects.   Acid and alkali wastes can be extremely damaging to aquatic life. Toxicity due to the solubility or precipitation of heavy metals is increased by synergism. Also, the capacity of natural waters to assimilate organic wastes is significantly reduced by these wastes.

Variables to be measured; how variables are measured.   The pH is considered to be an important measure of environmental quality. High pH reflects an alkaline situation, and low pH reflects an acid condition (a neutral solution has a pH equal to 7.0).

The pH measurement is simple. Many types of continuous measuring and recording instruments are commercially available for this purpose.

Evaluation and interpretation of data. Since the natural pH of aquatic ecosystems varies from one locale to another, the best measure of pH is in terms of departure from natural levels. Table B.3 summarizes five classes of water impacts, based upon pH departure from the normal. It has been assumed that both positive and negative departures are equally damaging to the environment. This may not be strictly true in normal cases, but due to lack of evidence, such assumptions may be considered valid.

Special conditions. In some cases, alkaline or acid wastes actually may help to balance a pH problem. Acid mine drainage is an example of a problem which would be neutralized by an alkaline discharge; however, these are more productively controlled through treatment processes.

Secondary effects. Secondary effects of impacts on the acidity or alkalinity of waters follow as a result of any condition that deteriorates the quality of water. Social and economic losses in terms of reduced productivity, decline in recreational benefits, and additional costs of treatment to correct problems related to pH are a few examples.

Biochemical oxygen demand (BOD)

Definition of the attribute. BOD of water is a direct bioassay to measure the amount of oxygen required to biologically degrade the organic material present. It is, thus, an indication of the amount of DO that will be depleted from water during the biological assimilation of organic pollutants. The BOD test is widely used to determine the organic strength of sewage and industrial wastes in terms of oxygen that would be required to oxidize organics if these wastes were discharged into natural waters in which aerobic conditions exist. The test is one of the most important in stream pollution-control activities. By its use, it is possible to determine the degree of organic pollution in natural waters at any time. This test is also of prime importance in regulatory work and in studies designed to evaluate purification capacity of receiving bodies of water.

Activities that affect the attribute. Activities associated with normal municipal and industrial operations may contribute to BOD wastes. These human activities (e.g., sanitary sewage; wastewaters from hospitals, food-handling establishments, and laundry facilities; and floor washing from shops) constitute BOD wastes. If all wastes are collected by a network of sewers to a central location, adequate treatment must be provided to minimize impact upon the surface water system. If cesspools, septic tanks, and soakpits are utilized, groundwater in the vicinity may become adversely affected.

Sources of effects.   The discharge of wastes containing organic material imposes oxygen demand in the natural body of water and reduces the DO level. If wastewaters are treated, the combined sewer overflows and surface runoffs may also exert effects under wet weather conditions. All parameters directly or indirectly related to DO also affect the organic waste assimilation. These parameters include depth of water, velocity of flow, temperature, and wind velocity (see section on DO for general discussion).

Variables to be measured; how variables are measured.   BOD values are generally expressed as the amount of oxygen consumed (mg/l) by organisms during a five-day period at 20°C. Several other parameters, such as chemical oxygen demand (COD) and total organic carbon (TOC), are also used to represent the organic matter in water and wastewater. The COD value indicates the total amount of chemically oxidizable material present and is normally greater than BOD. TOC is a measure of bound carbon. Both these tests are closely related to BOD and are used in water and wastewater monitoring programs.

Routine BOD measurements are made in laboratories by dilution techniques; results are obtained in five days. Some modifications of BOD tests may require less time. COD measurements take a few hours, and TOC takes only seconds. Several types of instruments which measure TOC on a continuous basis are commercially available.

Evaluation and interpretation of data.   Table B.3 indicates five classes of water: very clean, clean, fairly clean, doubtful, and bad, depending upon the BOD of water. It may be mentioned, however, that this classification must be used on relative terms. As an example, a sluggish stream, reservoir, or lake may show undesirable conditions at BOD of 5 mg/l, whereas a swift mountain stream may easily handle 5 mg/l of BOD without significant deleterious effects.

Mitigation of impacts.   All wastes containing organic material should be processed by treatment methods. The treatment methods may include biological and/or chemical processes. Also, several types of packaged treatment units are commercially available that can be installed for desired applications.

Secondary effects.   By virtue of the biologic and aesthetic effects of BOD on aquatic environments, secondary impacts are manifested in terms of additional impacts on aesthetics, reduced recreational benefits, and costs to alleviate the direct consequences of BOD on waters scheduled for reuse. The success of land-use planning efforts in areas where water is an integral part of the planning effort (e.g., recreational areas and industrial siting) is dependent upon the quality of those waters. BOD is a parameter of importance.

Dissolved oxygen (DO)

Definition of the attribute. Almost all living organisms depend upon oxygen in one form or another for their metabolic process. Aerobic organisms require DO and produce innocuous end products. Anaerobic organisms utilize chemically bound oxygen, such as that from sulfates, nitrates, and phosphates, and the end products are odorous. For a diversified warm-water biota, including game fish, DO concentration should remain above 5 mg/l. Absence of DO will lead to the development of anaerobic conditions with odor and aesthetic problems. In surface waters, DO is measured frequently to maintain conditions favorable for the growth and reproduction of fish and other favorable aquatic life.

Activities that affect the attribute. The activities discussed in BOD also apply to DO. Other activities that may influence DO include site preparation demolition, dredging, and excavation, all of which may cause turbidity and nutrient release. Routine operation, such as operation and maintenance of aircraft, watercraft, and automotive equipment, may cause oil release. Oil film interferes with the natural process of reaeration.

Sources of effects. Discharge of all organic wastes will lower the DO in receiving waters. A shallow, swift mountain stream can assimilate large quantities of organic wastes without deleterious effects. This is because swift-moving streams have greater capacity for natural reaeration and for preventing deposition of organic materials at the stream bed. In a sluggish stream or reservoir, small amounts of BOD released may cause relatively large adverse effects. The solubility of oxygen in water decreases with increases in temperature and dissolved salts (in fresh water, solubility of oxygen at $0°C$ is 14.6 mg/l, and at $35°C$, it is 7 mg/L). Biological activity is also increased at higher temperatures, and thus the rate of DO utilization from natural waters is significantly increased. Therefore, BOD wastes discharged into natural waters have more pronounced effects during summer months, when the water is warm. Thus, water quality parameters, such as temperature, dissolved salts, depth and velocity of the stream, wind velocity, and natural reaeration, are all interdependent. Also, in nutrient-rich bodies of water, due to algae bloom, the DO level may reach supersaturation during sunny days. At night, however, the DO level drops considerably, due to lack of photosynthesis. High turbidity in water may also interfere with photosynthesis by reducing the depth of light penetration. Oil slicks may reduce the natural reaeration process, too. Therefore, nutrients, algae, sunny days, turbidity, and oil slicks are all interdependent parameters.

Variables to be measured; how variables are measured.   The unit of DO measurement is milligrams per liter. It can be measured by titration techniques using the azide modification method. Many commercially available DO meters can be used for DO measurement.

Evaluation and interpretation of data.   The oxygen requirements for fish vary with species and age. Cold water fish require higher oxygen concentration than do the coarse fish (carp, pike, eel). It may be stated that the 3- to 6-mg/l range is the critical level of DO for nearly all fish. Below 3 mg/l, further decrease in DO is important only insofar as the development of local anaerobic conditions is concerned; the major damage to fish and aquatic life will already have occurred. Above 6 mg/l, the major advantage of additional DO is as a reservoir or buffer to handle shock loads of high-oxygen-demanding waste loads. Table B.3 indicates five classes of water according to DO levels.

Geographical and temporal limitations. Typically, the most critical DO problems occur in summer, when biological activity is high and saturated DO content is low.

Mitigation of impact.   The methods are the same as those given for BOD.

Secondary effects.   Secondary impacts are the same as those listed for BOD.

Dissolved solids

Definition of the attribute.   High amounts of total dissolved solids (TDS) are objectionable because of physiological effects, mineral tastes, or economic effects. TDS is the aggregate of organic and inorganic compounds, such as carbonates, bicarbonates, chlorides, sulfates, phosphates, nitrates, and other salts of calcium, magnesium, sodium, potassium, and other substances. All salts in solution change the physical and chemical nature of the water and exert osmotic pressure; the magnitude of the change is, to a large extent, dependent upon the total salt concentration (salinity), measured as the sodium index.

Activities that affect the attribute.   Major areas which may contribute to TDS include mining and quarrying, municipal and industrial waste disposal, brine disposal, lagooning, landfilling of solid wastes, and accidental spill of chemicals.

Source of effects. Major activities listed above may cause release of salts either directly or indirectly into the natural water system. Direct release includes discharging the waste laden with salts into the water system. Indirect release may be due to runoff from the affected land or seepage from filled areas. Landfill seepage or leaching may affect groundwater quality, and, if groundwater feeds the water courses, surface water may be affected as well.

As a result of salt discharge, many water quality parameters will be affected. DO will decrease as a result of high salinity. High quantities of salts give a mineral taste. Sulfates and chlorides are associated with corrosion damage. Sulfate in water has a laxative effect. Nitrate plus nitrite causes methemoglobinemia (blue baby disease).

Variables to be measured; how variables are measured. Total dissolved solids is determined after evaporation of a sample of water and its subsequent drying at 103°C in an oven. This includes "nonfilterable residue." The results are expressed in milligrams per liter TDS.

Evaluation and interpretation of data. For reasons of palatability and unfavorable physiological reaction, a limit of 500 mg/l TDS in potable water has been recommended. Highly mineralized waters are also unsuitable for many industrial applications. Irrigation crops are highly sensitive to salt concentrations; waters containing over 2000 mg/l are of marginal value for irrigation use, and waters containing 3000 mg/l are unsuitable. The upper limits for some freshwater fish are as high as 5000 mg/l. In such cases, the sodium index is used to estimate the total salt concentration and its effects on osmotic pressure. Based upon TDS, the five impact classes are summarized in Table B.3.

Special conditions. The amount of dissolved ionic matter in a sample may often be estimated by multiplying the specific conductance by an empirical factor. After the empirical factor is established, for a comparatively constant water quality, specific-conductance measurement will yield TDS. Specific-conductance measurement is relatively simple and is a measure of a water's capacity to convey an electric current at 25°C. Specific conductance is expressed as microohms per centimeter.

Mitigation of impact. Wastes containing high TDS are difficult to treat. Recommended treatment methods include removal of liquid and disposal of residue by controlled landfilling to avoid any possible leaching of the fills. Deep well injection has been used for disposal of brine. All surface runoffs around mines or quarries should be collected and concentrated. The brine may be disposed of by deep well injection or other means acceptable to water quality control authorities.

Secondary effects. Effects on irrigated crop land (reduced productivity and economic loss) constitute perhaps the most significant secondary impacts due to TDS. Other effects include those on health, where drinking waters are concerned, and those on economics and land use, where industrial and municipal consumption are to be considered.

Nutrients

Definition of the attribute. *Eutrophication* is a term meaning enrichment of waters by nutrients through either human-induced or natural means. Present knowledge indicates that the fertilizing elements most responsible for eutrophication are phosphorus and nitrogen. Inorganic carbon (CO_2), iron, and certain trace elements are also important. Eutrophication results in an increase in algae and weed nuisances and an increase in larvae and adult insects. Dense algal growths (blooms) may form surface water scums and algae-littered beaches. Water may become foul smelling when algae cells die; oxygen is used in decomposition, and fish kills result. Filter-clogging problems at municipal water treatment plants and taste and odor in water supplies may all be due to the dense algal population and by-products resulting from its subsequent decay.

Activities that affect the attribute. Sewage and sewage effluent contain a generous amount of the nutrients necessary for eutrophication. Treated or untreated sewage discharge will contribute to nutrients in receiving waters. Mining, tunneling, blasting, and quarrying into phosphate rocks may cause increased phosphorus from surface runoff. Dredging of waterways will release the storehouse of nutrients contained within the mud bottom; as a result, the water will become enriched during and soon after the dredging operation. Many other activities may enrich the natural waters. These include drainage from cultivated agricultural lands, surface irrigation return flows, logging and sawmilling, deposition of dead trees and leaves, and growth of natural organisms.

Source of effects. Nutrients released from many activities (described above) will cause aquatic plant problems, turbidity, taste, and odor; cause reservoir and other standing waters to collect nutrients and to store a portion of these within consolidated sediments (once nutrients are combined within the ecosystem of receiving waters, their removal by natural process is very slow); and induce excessive weed growth, which will eventually block waterways or turn lakes into swamps.

As a result of nutrients released into natural waters, many water quality parameters will be affected directly or indirectly. Some of these effects are turbidity, due to excessive algae growth—then, when algae cells and other plants die, oxygen is used in decomposition and the DO level declines, causing fish kill; rapid decomposition of dense algal growths, giving rise to odors and hydrogen sulfide gas that create strong citizen disapproval; and serious water treatment problems, caused by color, taste, and odor.

Variables to be measured; how variables are measured. Phosphorus, nitrogen, carbon, iron, and trace metals all act as nutrients. Growth of aquatic plants is governed by the law of minimum (i.e., any nutrient, out of a broad array of materials required for growth and development, governs the growth if it is present in a limiting concentration). In natural waters, phosphorus is present in limiting amounts and commonly governs the rate of normal plant growth.

Phosphorus occurs in natural waters and in wastewaters almost solely in the form of phosphates. These forms are commonly classified into orthophosphates, condensed phosphates (pyro-, meta-, and polyphosphates), and organically bound phosphates. These phosphates may occur in the soluble form, in particles of detritus, or in the bodies of aquatic organisms. Because the ratio of total phosphorus to that form of phosphorus readily available for plant growth is constantly changing and ranges from 2 to 17 times or greater, it is desirable to establish limits on the total phosphorus, rather than the portion that may be available for immediate plant use.

Phosphate analysis embodies two general procedural steps: (1) conversion of the phosphorus form of interest to soluble orthophosphate, and (2) colorimetric determination of soluble orthophosphates. The result may be expressed as milligrams per liter phosphorus (mg/l P).

Evaluation and interpretation of data. Although the concentration of inorganic phosphorus that will produce problems varies with the nature of the aquatic environment and the levels of other nutrients, most relatively uncontaminated lake districts are known to have surface waters that contain 0.001 to 0.003 mg/l total phosphorus as P (they are nutrient deficient). Above 0.02 mg/l P, one gets into a region of potential algae bloom. Above 0.1 mg/l P, water is excessively enriched. Table B.3 categorizes five classes of waters, based upon total P contact.

Geographical and temporal limitations. Since algae growth is temperature-dependent, adverse effects due to eutrophication in northern climates are more pronounced in summers. In southern climates, the effects are felt over the entire year, with fall overturn in stratified reservoirs being the most critical time.

Mitigation of impact. Once nutrients are combined within the ecosystem of the receiving waters, their removal is tedious and expensive. In a lake, reservoir, or pond, phosphorus is removed naturally only by overflow, by insects that hatch and fly out of drainage basins, by harvesting a crop (such as fish), and by combination with consolidated bottom sediments.

The most desirable method to mitigate impact is to treat wastewater to a desired phosphorus level before discharge into the environment. Also, all activities mentioned above should be performed under controlled conditions.

Secondary effects. Various adverse secondary impacts occur with advanced stages of eutrophication, including a decline in recreational benefits, effects on land use, and the economic losses that normally accompany any deterioration in water quality.

Toxic compounds

Definition of the attribute. Wastes containing concentrations of heavy metals (mercury, copper, silver, lead, nickel, cobalt, arsenic, cadmium, chromium, etc.), either individually or in combination, may be toxic to aquatic organisms, and thus have a severe impact on the water community. Other toxic substances include pesticides, ammonia-ammonium compounds, cyanides, sulfides, fluorides, petrochemical wastes, and other organic and inorganic species. A severely toxic substance will eliminate aquatic biota until dilution, dissipation, or volatilization reduces concentration below the toxic threshold. Less generally, toxic materials will reduce the aquatic biota, except those species that are able to tolerate the observed concentration of the toxicant. Because toxic materials offer no increased nutrient supply, such as discussed for organic wastes, there is no increase seen in the population of those organisms that may tolerate a specific concentration.

Activities that affect the attribute. Many human activities may contribute to release of toxic compounds into the environment. These include waste discharged from maintenance and repair shops and from industrial operations. Wastes that are particularly likely to contain toxic compounds result from electroplating, galvanizing, metal finishing, and cooling tower blowdown. Other activities which may contribute to toxic chemicals are mining, accidental spills of chemicals, chemical warfare, and leaching of landfills containing toxic compounds.

Source of effects. Chemicals released into the environment may affect surface water or groundwater systems by direct discharge of wastes containing toxic compounds or from surface runoff which may come in contact with toxic material left as residue over the ground surface.

Variable to be measured; how variables are measured. The spectrum of toxic materials is extremely large and highly diverse in terms of effects. Measurement may be expressed as μg/l of the specific compound under consideration. For a group of toxic compounds, it should be pointed out that possible synergistic or antagonistic interactions among mixed compounds may cause different effects from those associated with the respective toxic compounds considered separately.

Bioassay is an important tool in the investigation of these wastes, because results from such a study indicate the degree of hazard to aquatic life of particular discharges; interpretations and recommendations can be made from these studies concerning the level of discharge that can be tolerated by the receiving aquatic community.

A typical basic bioassay might consist of a 96-h exposure of an appropriate organism, in numbers adequate to assure statistical validity, to an array of concentrations of the substance, or mixture of substances, that will reveal (1) the level of pollution that will cause irreversible damage to 50 percent of the test organisms, and (2) the maximum concentration causing no apparent effect on the test organisms in 96 h.

Evaluation and interpretation of data. The bioassay may indicate the concentration at which toxic compounds will not cause an apparent effect upon the test organism. However, long-term effects of toxic compounds having more subtle changes, such as reduced growth, lowered fertility, altered physiology, and induced abnormal patterns, may have more disastrous effects on the continued existence of a species. Also, the biological magnification and storage of toxic residue of polluting substances and microorganisms may have another serious aftereffect. For all these reasons, and as a practical matter, toxic compounds, if they could be detected in natural waters by modern water quality analysis methods, may render water undesirable for propagation of healthy aquatic life. The five classes of water, based upon toxic compounds, are given in Table B.3.

Special conditions. Synergistic action may magnify toxic effects under special conditions (e.g., under an increased temperature or a low dissolved oxygen situation).

Mitigation of impact. All wastes containing toxic chemicals should be monitored and controlled. Those released into sanitary sewers should be carefully regulated so that such release does not affect the treatment process. Also, after dilution, effluent concentration should not exceed the desired level. Runoffs from chemical-handling areas should also be considered, to the extent that pollution is expected. If necessary, suitable treatment may be given to all contaminated runoffs.

Secondary effects. While toxic compounds have a primary effect on lower organisms in the aquatic environment, secondary effects may be felt all through the food chain, with human health as a final major consideration. Procedures to remove these compounds, once they have been released to the aquatic environment, may be nonexistent or, at best, extremely expensive. Failure to remove them or to prevent their initial entry may degrade the water quality, with ensuing effects on aesthetics, economics, and biophysical relationships.

Aquatic life

Definition of the attribute. Organisms in any community exist in a dynamic state of balance, in which the population of each species is constantly striving to increase. However, population is maintained at a fluctuating level determined by food supply, predators, chemical characteristics of the water, and physical variables. Since these factors vary greatly, several types of communities exist in balance. Any human-created pollution tends to upset the natural state of balance. This may cause abundance of a few types of organisms, while others may decline or completely disappear. Because of some variation in response among species to conditions of existence within the environment, and because of inherent difficulties in aquatic invertebrate taxonomy, ecological evaluation of the total organism community is the acceptable approach in water pollution–control investigation.

Activities that affect the attribute. All activities discussed above (with various water attributes) affect aquatic life to some degree. Change in an aquatic community depends upon the type and extent of pollution.

Source of effects. Discharge of organic wastes (sewage) tends to lower the natural DO and to eliminate DO-sensitive organisms. Thermal discharge affects the normal life cycle of many organisms. Toxic wastes will reduce the aquatic biota, except those species that are able to tolerate the observed concentration of the toxicant. In general, changes in any attributes, whether they are physical or chemical, will influence the aquatic life.

Variables to be measured; how variables are measured. For aquatic life interpretation, field observations are indispensable. However, many of the biological parameters cannot be evaluated directly in the field. The specific nature of a problem and the reasons for collecting samples will dictate those aquatic communities of organisms to be examined, and those, in turn, will establish sampling and analytical techniques. The following communities and types of organisms are considered: plankton, periphyton, macroinvertebrates, macrophytes, and fish. Sampling and identification techniques are based upon routine biological sampling and analysis methods. Readers are referred to the latest edition of *Standard Methods for Examination of Water and Wastewater.*

Evaluation and interpretation of data.   Based upon the most common aquatic life in natural waters, five classes of water are given in Table B.3.

Mitigation of impact.   See all water quality attributes for mitigation of impact upon aquatic life.

Secondary effects.   Economic and recreational benefits may be affected as a result of adverse impacts on aquatic life. Loss of productivity reduces fishing harvest, and decline in recreational activity produces additional economic loss.

Fecal coliforms

Definition of the attribute.   Water acts as a vehicle for the spread of disease. All sewage-contaminated waters must be presumed potentially dangerous. The presence of coliform organisms in water is regarded as evidence of fecal contamination, as their origin is in the intestinal tract of humans and other warm-blooded animals. They are also found in soil and water which has been subjected to pollution by dust, insects, birds, and small and large animals. Fecal coliforms, per se, are not as proximate a hazard to water supplies as they once were, but are still utilized as a surrogate for pathogens in general. However, the test continues to retain importance because of water-contact recreational usage of water, and of implications that viral diseases can be transmitted through fecal contamination of water supplies. Indirect routes, such as the contamination of foods with fecally contaminated irrigation water and accumulation of contaminants by oysters, clams, and mussels from fecally contaminated marine waters, continue to be areas of concern.

In the 1990s, the most common of all coliforms, *Escherichia coli* (*E. coli*), moved from being a benign component of the intestinal flora to being classified as a dangerous killer. This is the result of several outbreaks of an especially toxic strain named *E. coli* O157:H7. Originally found in, and still most commonly associated with, undercooked hamburgers, it caused several deaths in the Pacific northwest. Since that time, *E. coli* O157:H7 has been found on fruits and vegetables, cheeses, hot dogs, poultry, and a wide variety of other foods. The common element in plant materials appears to be surface contamination with untreated water containing the organism, which now appears to be widespread in surface waters.

Activities that affect the attribute.   The activities discussed in BOD and DO also apply to this attribute.

Source of effects.   See BOD and DO attributes.

Variables to be measured; how variables are measured.   Two methods are used for determining the presence of coliform organisms: the multiple-tube fermentation technique and the membrane filter technique. The results of multiple-tube fermentation techniques are expressed as most probable number (MPN), based upon certain probability formulas. The results of membrane filter tests are obtained by actual count of coliform colonies developed over membrane filter. In both cases, the estimated coliform density is reported in terms of coliform per 100 ml. The equipment used is the type commonly needed in routine microbiological study.

Evaluation and interpretation of data.   Present water quality criteria restrict the use of water, depending upon fecal coliform density. The desirable criterion for surface water supply is fecal coliform less than 20 per 100 ml, and for recreational use (including primary contact recreation), the recommended value is 200 per 100 ml. Based upon the coliform density, five classes of water are summarized in Table B.3.

Mitigation of impact.   See attributes BOD and DO.

Secondary effects.   Quantification of the presence of fecal coliforms in recreational waters results in a classification by permissible use. This classification restricts not only the use of the waters, but also the economic benefits which might be obtained from those waters. Effects on shellfish harvests are other economic impacts which may result from fecal contamination.

References

Canter, Larry W. *Ground Water Pollution Control.* Chelsea, Mich.: Lewis Publishers, 1986.

Corbitt, R. A. *Standard Handbook of Environmental Engineering,* 2d ed. New York: McGraw-Hill, 1999.

Eckenfelder, W. Wesley. *Industrial Water Pollution Control.* New York: McGraw-Hill, 2000.

Nemerow, N. L. *Stream, Estuary, Lake and Ocean Pollution.* New York: Van Nostrand Reinhold, 1991.

Odum, E. P. *Fundamentals of Ecology.* Philadelphia: W. B. Saunders.

Sawyer, C. N. *Chemistry for Environmental Engineering,* 4th ed. New York: McGraw-Hill, 1994.

Suess, Michael J., ed. *Examination of Water for Pollution Control: A Reference Handbook.* New York: Pergamon Press, 1982.

Thomann, R. V., and J. A. Mueller. *Principles of Surface Water Quality Modeling and Control.* New York: Harper and Row, 1987.

Viessman, W., and M. Hammer. *Water Supply and Pollution Control.* Menlo Park: Addison-Wesley, 1998.

EPA regulations governing water quality standards and test methods can be found in the *Code of Federal Regulations,* as listed below:

40 CFR 125—*Criteria and Standards for the National Pollutant Discharge Elimination*

40 CFR 129—*Toxic Pollutant Effluent Standards*

40 CFR 131—*Water Quality Standards*

40 CFR 133—*Secondary Treatment Regulation*

40 CFR 136—*Guidelines Establishing Test Procedures for the Analysis of Pollutants*

40 CFR 141—*National Primary Drinking Water Regulations*

40 CFR 142—*National Secondary Drinking Water Regulations*

40 CFR 401–471—*Effluent Guidelines and Standards for Various Industries*

Land

As with other resources, land is not available in unlimited quantities. Because of this, it is becoming increasingly recognized in this country, and in other countries with less of an endowment of land resources, that land use must be properly planned and controlled. CEQ regulations recognize this need for the rational management of land resources, and, because the price system does not allow rational allocation of land, the CEQ has provided for a specific consideration of the relationship of a changed pattern in land use to the existing pattern. Therefore, land is being treated in much the same manner as our other scarce natural resources, air and water.

To consider these factors requires comprehensive consideration of existing and projected land capabilities and land-use patterns. The most significant elements of the land use question have been collapsed into three attributes:

Erosion

Natural hazards

Land-use patterns

Erosion

Definition of the attribute. Erosion is defined as the process through which soil particles are dislodged and transported to other locations by the actions of water and/or wind. The two most common forms attributable to water are sheet erosion, in which the upper surface of the soil is more or less evenly displaced, and gully or rill erosion, in which the downward-cutting action of the overland flow of water results in linear excavations deep into the soil horizon. While the latter type of erosion is often more spectacular to the eye, loss of uniform layers of topsoil through sheet erosion is the more serious of the two. Wind erosion is similar to sheet erosion in that very small soil particles containing plant nutrients and organic matter are the ones that are carried away, leaving coarse and less productive material.

Soils of almost all types are held in place by vegetative cover and its associated root system. Removal of this cover exposes the soil to the erosive forces of water and wind. Erosion is intensely destructive. First, the site itself may be denuded of its most productive topsoils and/or may be gullied to the extent that it becomes almost totally unproductive, often to the point of posing a physical barrier to other activities. Second, the streams and lakes which receive the attendant sediment loads may be affected. The landscape, after erosive forces have been at work, is barren and aesthetically unappealing.

Activities that affect the attribute. Activities that affect the extent and rate of erosion are those associated in any way with removal or reestablishment of vegetative cover. Some of these are land clearing for construction, road building, or other cut-and-fill operations; timber harvesting or vegetative suppression by herbicide application; controlled burning; reforestation or afforestation; strip mining; agricultural activities; off-road vehicular traffic; and large animal grazing.

Source of effects. Land clearing and mechanized off-road activities strip land of its vegetative cover, organic surface material, and root structures which formerly protected the soil, thereby opening it up to direct attack by wind and water. Timber harvesting, application of herbicides, and controlled burning can result in the removal of a sufficient quantity of organic surface material and vegetative cover to cause an increase in the intensity of rainfall and wind movement at the soil surface. Conversely, reforestation and afforestation can reintroduce a vegetative canopy and root structure which—over time—can reduce the intensity of these erosive forces and result in a buildup of organic surface material. Road building and other cut-and-fill activities lay bare previously vegetated soil, alter natural drainage patterns, change the gradient of slopes, and create somewhat unconsolidated fill areas upon which vegetative cover is often not immediately reestablished. The stripping away of vegetative shrub and ground cover in semiarid areas by overgrazing is one of the most widespread causes of wind erosion. If grazing rights are withdrawn, or native grazing animals are fenced out of and/or removed from overgrazed areas, seeding of native grasses can accelerate the return of vegetative cover and reduce erosion potential.

Variable to be measured.   Major variables affecting erosion are soil composition or texture, degree of slope, uninterrupted length of slope, nature and extent of vegetative cover, and intensity and frequency of exposure to the eroding forces. The interaction of these variables is complex and difficult to measure directly. Magnitude of the impact is also directly dependent on the extent of the affected area.

Soil texture is determined by the percentage of its sand, silt, and clay components. Generally accepted textural classes in order of decreasing particle size (coarse to fine) are

Sand	Sandy clay loam
Loamy sand	Clay loam
Sandy loam	Silty clay loam
Loam	Sandy clay
Silt loam	Silty clay
Silt	Clay

While such a statement is subject to contradiction on a specific site, finer-textured soils are usually more susceptible to water erosion. Sandy soils and granulated clays are those most easily eroded by wind.

Water erosion increases with the length and steepness of slope. A general rule is that if the length of slope is doubled, soil loss from erosion will increase by a factor of 1.5. The relationship between degree of slope (gradient) and erosion potential can be specified in general terms as follows:

10 percent $\geq$ highly erodible

10 percent = moderately erodible

2 percent $\leq$ slightly erodible

The erosion hazard depends upon the intensity and frequency of rain and wind storms. While the amount of yearly rainfall is important, of greater significance is the force with which it strikes the ground, volume in a given time, and return frequency of intense storms. The impact of wind varies with velocity, direction, and soil moisture content.

The difference in types of vegetative cover and the extent of each also affect erosion potential. A mature forest with a heavy overstory (leaf or needle) cover, an understory of trees with less dense leaves, scattered ground vegetation, and a heavy layer of decaying organic matter will protect the soil from wind and water to a greater extent than will brush and sparse ground cover found in arid and semiarid areas. These are extremes—pasture and cultivated cropland fall somewhere between.

Before proceeding further, some informed judgment should be made as to whether these variables are operative to a degree and in sufficient combination to warrant the rather extensive calculations to be described next. If necessary, an agronomist or agricultural engineer from the local office of the Natural Resources Conservation Service (NRCS) could assist in making this initial assessment.

How variables are measured.   Most soil loss or soil erosion equations are based upon models that represent interrelationships among the variables just discussed. One such model developed for agricultural cropland (*Rainfall Erosion*, USDA Handbook 282), but subject to modification for other vegetative types is

$$A = RKLSCP$$

where A = Computed soil loss per unit area (acre)
    R = Rainfall factor
    K = Soil erodibility factor
    L = Slope-length factor
    S = Slope gradient factor
    C = Crop management factor (or relative vegetation cover)
    P = Erosion control practice factor.

While the techniques of arriving at numbers to represent the various factors are adequately described in a handbook which should be available from the local NRCS office, it would be helpful to have the expert advice of a team of NRCS agronomists, hydrologists, and agricultural engineers in applying it to a specific site. Soil loss should be computed both with and without the project. This will provide a comparison for analysis.

The area affected should be outlined on a map overlay or geographic information system (GIS) at an appropriate scale. By using a planimeter (map overlay) or computer calculations and with the assistance of an engineer, the number of acres affected can be determined. Total soil loss with and without the project can then be calculated by multiplying the soil loss per acre, as previously obtained from the model, by the number of acres involved.

Evaluation and interpretation of data.   Overall magnitude of the impact can be represented by the percentage of change in total soil loss as calculated above. If a more sophisticated analysis appears to be warranted, this quantitative figure can be tempered by a further evaluation that takes into account change in soil fertility (productive capacity) and the impact of changes in sediment load in streams that drain the affected area. This kind of analysis could best be done by an interdisciplinary team of economists, agronomists, engineers, and ecologists.

Special conditions.   If the land were productive for agricultural crops or forest products, the economic and ecological impacts might be greater than if it were relatively infertile.

Geographical and temporal limitations.   While there are few areas in the United States where the potential for at least moderate erosion does not exist, the most severe erosion has occurred in the Appalachian area of the Southeast, in the Great Plains, and in some desert and semiarid areas of the Southwest. The major temporal limitation on erosion involves the time of year when the soil is exposed and the length of time it remains exposed relative to the time of year that intense rain and windstorms are likely to occur.

Mitigation of impact.   It is much easier to prevent erosion before it begins than it is to arrest it or restore the land afterward. The environmental impact of soil erosion can best be mitigated by removing vegetative cover only from the specific site on which construction is to take place and by disturbing the vegetation in adjacent areas as little as possible. Construction, land management, or mining activities that result in the soil being laid bare could be scheduled in such a way that some type of vegetative cover appropriate to the site could be established prior to the onset of intense rain or windstorms. If grass is to be seeded, a mulch of straw will help to protect the soil from less extreme erosive forces until vegetative and root development begin. Natural drainage patterns can often be maintained by preparing sodded waterways or installing culverts. Steep slopes can be terraced, thereby effectively reducing the length of slope. Catch basins built near construction sites can reduce the quantity of eroded soil particles reaching free-flowing streams or lakes.

Secondary effects.   Secondary effects of erosion include increased sediment loads in streams which may clog reservoirs and fill large areas of bays and estuaries. These sediment loads also affect aquatic life through such mechanisms as the covering of fish eggs and spawning areas, coating of gills, and retarding light penetration, which, in turn, reduces the photosynthetic process necessary for aquatic plant production. As the aquatic environment is degraded, the results are losses in areas of aesthetic, recreational, and economic benefits. Other economic losses include adverse effects on land-use suitability, crop production reduction, and frequent filter replacement due to increased particulate materials in the air.

Other comments.   If the erosive effects resulting from an activity are not confined but spill over into adjacent private lands (sediment deposition), or if severely eroded land is visible from public highways, after-the-fact controversy over the project may develop. This is especially true if these considerations are not directly addressed in the environmental assessment/impact statement and if the mitigation possibilities are not discussed and evaluated.

Natural hazards

Definition of the attribute. Natural hazards are those occurrences brought about by the forces of nature that cause discomfort, injury, or death to humans; damage or destroy physical structures and other real or personal property; change the physical character of land, water, and air; and damage or destroy the plant and animal life of the affected area. The severity and frequency of occurrence of floods, earthslides, and wildfires may be influenced by various activities. Other natural hazards, such as earthquakes and hurricanes, may cause greater personal and physical damage than would be the case if human activities were located in areas other than those where these natural events occur with some frequency and severity.

Activities that affect the attribute. Some activities that often have an impact on the frequency and magnitude of natural hazards are construction, land management, land use, agriculture, and industrial development. These activities do not affect the natural processes that are the root causes of hazards—intense rain or wind storms, the geologic structure and soil and bedrock properties of an area, or lightning strikes from the thunderstorms. Rather, it is the destructive nature of the results of these occurrences that human activities influence.

Source of effects. The effects of construction activities on the destructive potential of natural hazards are quite diverse. Land clearing, which precedes most kinds of construction, lays bare the soil surface, a condition conducive to increased volumes of water runoff and increased sediment loads in streams—both of which tend to cause increases in flood heights and return frequencies, the two greatest determinants of flood damage. Paving large areas with asphalt and concrete—often done for parking lots and outdoor storage areas—reduces infiltration of water into the soil, thereby increasing runoff and the peak volume of water that streams are required to carry. The building of structures such as dams and levees, as well as stream channelization to reduce flood levels, may greatly modify the flow regimes of natural water courses, which, in turn, may result in the diversion of floodwaters to previously unflooded areas.

The probable incidence of earthslides may be increased by road construction activities if natural shear stresses in the earth are increased, excessive pore pressure developed, or rock and soil strata exposed by road cuts. Failure may be induced by blasting, changes in slope, greater overburden, etc. Earthslides can block streams and cause a backup of water, which, in turn, can result in upstream damage due to a gradual rise in water level and extensive downstream damage due to the rapid release of water when the slide is overtopped or eaten away. Earthslides also destroy vegetation, increase sediment loads in streams, and disrupt transportation routes. On the positive side, road construction in remote areas can reduce potential wildfire damage by permitting more rapid access by firefighting crews and equipment.

Land management includes activities such as timber harvest, reforestation and afforestation, herbicide application, and controlled burning. Timber harvest can create at least temporary increases in runoff volume and sediment loads as a result of the removal of some of the vegetation cover and the disturbance of the soil surface by tractors and other mechanized equipment. Rehabilitation of eroded areas by reforestation, afforestation, or seeding decreases runoff and sediment loads. Timber harvest on steep slopes can result in landslides which disturb the soil horizon to the extent that natural tree regeneration will not take place. When vegetation killed by herbicides and logging debris left after timber-harvest operations dries to the point where the plant material will ignite easily and burn with considerable intensity, lightning strikes are more likely to cause fires that are difficult to control and may do great damage. Conversely, controlled burning can reduce the incidence and destructive potential of wildfire by creating a low-temperature blaze that consumes the dry underbrush and organic matter on the forest floor without damaging mature timber. (This favorable impact of controlled burning should not overshadow the fact that it may adversely affect other environmental attributes, such as vegetative diversity, wildlife populations, and erosion.)

Land-use considerations that dictate where certain projects will be located often have a decided impact on natural hazards. Any physical structure (building, bridge pier, or temporary bridge) that occupies a portion of the floodway (the stream channel carrying the normal water flow) or is situated on the floodplain (that area covered by flood waters when a stream overflows its banks) will restrict the flow of water and decrease the volume which the floodplain can accommodate at a particular level, thereby increasing flood heights both upstream and downstream. Permitting homes or other structures to be located in floodplains poses the possibility of increased physical damage to the structures and loss of life to their occupants. During a flood, portable or temporary structures can damage other structures by direct impact or lodge in the stream channel in such a way as to form a temporary dam, raising flood levels behind them. Siting housing areas on brush or forest land subject to wildfire can increase the damage potential to life and property. The same is essentially true of any structures placed near known fault lines in active earthquake zones or in coastal or inland river areas subject to frequent wind and water damage from hurricanes.

Variables to be measured. Each type of hazard has its own set of variables that influence frequency of occurrence and severity. For floods, changes in volume of the overland flow of water, changes in sediment deposits in stream channels, or alteration of the floodplain cause variation in flood height and resultant damage levels. Baseline data can sometimes be obtained from gauging stations that record the magnitude of the increased stream flow resulting from runoff associated with the storm. Changes in the infiltration rate of water cause changes in the volume of surface runoff from overland flow and in the amount of sediment carried into the stream. The resulting change in return frequencies of certain levels of flooding is the critical determinant of impact.

Earthslide-prone areas are those characterized by unstable slopes and land surfaces which—because of a history of actual occurrences, geology, bedrock structure, soil, and climate—present a significant hazard potential. The variables here are the extent to which soil and rock strata are exposed to wetting, drying, heating, and cooling processes; the slope gradient of the cut, which exposes the relevant stratum; and changes in internal earth stresses caused by surface or subsurface loadings (e.g., blasting, heavy machinery operation, and installation of footings and foundations).

The variables associated with wildfire are changes in flammability of the organic matter on the ground (duff) and the areal extent of the activity. Changes in wind velocity near the ground, depth of the duff, and moisture content of the duff influence its capacity to support combustion and the intensity with which it will burn once ignited. The size of the activity, in terms of changes in the volume or area of standing timber, in the number and value of physical facilities, and in the number of people housed or working in areas susceptible to wildfire, influences both the probable incidence of wildfire and the magnitude of the resultant damage.

How variables are measured. Few, if any, of the variables associated with baseline data on natural hazards are subject to measurement by the layperson. It is even more difficult to project changes in the variables over time as a result of specific activities.

For floods, the assistance of an expert hydrologist is required to relate rainfall intensity (rate over time), infiltration capacity of the soil (the maximum rate at which soil in a given condition can absorb water), overland flow (rainfall excess that reaches stream channels as surface runoff) and its effect on channel depth, and the resulting increase in flow rate over time (hydrograph) which would yield a certain flood height and attendant damage level. The major variable—change in soil infiltration capacity—is influenced by such diverse and interrelated factors as interception of rain by trees and buildings, depth of surface detention of water and thickness of saturated soil layer, soil moisture content, compaction due to machines and animals, microstructure of the soil, vegetative cover at or near the surface, and temperature. The nearest district office of the Corps of Engineers or the U.S. Geological Survey should be able to assist in obtaining baseline data and in projecting the effects of various activities on flood heights and return frequencies.

The relative tendency of an area to have earthslides is not subject to simple measurement; the forces which cause an earthslide and the extent of their interactions are extremely complex. To the expert geologist, the type of geologic structure common to the area, the type of bedrock, soil structure, height of water table, type of surface material, degree of natural slope, and past history indicate whether an area is prone to earthslides. Such general information can often be obtained from the U.S. Geological Survey, from state geologists, or from local universities. If the area is prone to earthslides, an engineering analysis should be made to determine whether physical changes that result from the activity are likely to increase or decrease the probable incidence with which slides may occur. The services of both a soil and a civil engineer would be required for a thorough analysis.

Baseline data on the conditions and occurrence of wildfire should be available directly from the nearest office of the U.S. Forest Service or from the state forester's office. These records usually include or can be correlated with other data relating to the thickness of the duff, the relative humidity, number of days since the last rain, wind velocity, and other local factors, which in combination give the fire danger ratings and could assist in projecting the change the activity would have on the previously identified specific variables (i.e., wind velocity near the ground, depth of duff, and moisture content of the duff). Any change in the area (acres) susceptible to wildfire should also be measured. This can be done with before and after overlays of the area prepared from maps, aerial photographs, or site plans. Through using GIS or a planimeter, the size of the area for each can be determined. The assistance of an engineer may be required to make this calculation.

Evaluation and interpretation of data. For flood hazards, the magnitude of the impact of a change in infiltration capacity of the soil and the attendant change in rate of surface runoff on flood stage height and of return frequency needs to be evaluated. A more sophisticated analysis could relate the change in flood height and return frequency to potential dollar losses or losses of human life, taking into account existing structures that might be affected, as well as any new ones to be located in the floodplain. This analysis could probably best be made by insurance underwriters associated with the National Flood Insurers Association or by the Federal Insurance Administration of the U.S. Department of Housing and Urban Development, who have maps of flood hazard areas.

Evaluation of changes in potential incidence of earthslides is less straightforward. The areas where earthslides are most likely to occur should be evident from the previously recommended engineering analysis. The impact of a slide in a particular area could be calculated in terms of the dollar value of physical damage to structures, loss of life, and the ecological damage to watercourses and vegetation. A team of engineers, geologists, and insurance underwriters could develop risk factors associated with changes in the potential incidence of earthslides.

Just as with other natural hazards, wildfire has two aspects to be separately evaluated—the change in potential incidence and the amount of damage that might result from an occurrence. Again, the considerations are complex and not amenable to one-dimensional evaluation and interpretation. The change in incidence is related to change in flammability and areal extent of the duff, to greater or lesser numbers of people in the area, to the nature of the proposed activity, and to measures taken to prevent or reduce wildfire damage. A team of foresters and fire insurance underwriters should be able to develop risk factors associated with the change in potential incidence and intensity of wildfire and then estimate property damage or the loss of life that might result, both with and without the project.

Special conditions. If increases in flood heights and frequencies are likely to adversely affect floodplains where extensive industrial, commercial, or residential development already exists; if increased incidence of earthslides is likely to damage population areas and/or cause severe ecological damage; or if residential or prime-timber-producing areas are subjected to higher risks of damage from wildfire—particularly if any of the effects are felt outside the confines of the activity—then controversy over the projected magnitude of the impacts is almost certain to develop. In such instances, an interdisciplinary team of qualified professionals is needed to develop and substantiate these projections. Actions having such consequences may be regulated by state law.

Geographical and temporal limitations. Geographical limitations on natural hazards have to do with observed frequencies of occurrence (e.g., hurricanes are most likely to affect Gulf and Atlantic coastal areas, earthslides are unlikely to occur in areas of relatively flat terrain, and earthquakes occur more frequently and with greater severity along known geological fault lines). While floods and wildfire can occur almost anywhere, the frequency and severity of lightning storms in mountain regions of the western states increase the incidence of wildfire in that geographical area. There are some general temporal limitations for natural hazards: Wildfire is most likely to occur in the summer and fall, when the moisture content of living vegetation and the duff is lowest; floods of greatest severity occur with certain predictability in the spring, but flash floods can take place at almost any time of the year; the hurricane season is considered to be summer and early fall, and earthslides of various types most often occur in the winter and spring. Temporal limitations do not seem to apply to earthquakes.

Mitigation of impact. Primary mitigation techniques for hurricanes and earthquakes center around the avoidance of areas where these hazards occur with sufficient frequency and intensity to cause severe damage and the use of proofing techniques in the construction of physical facilities. Proofing techniques include the use of "floating" foundations and height restriction in earthquake zones and increased foundation height, wall strength, and roof support in areas periodically subject to hurricanes.

The frequency and/or severity of flooding can be held to a minimum by prohibiting any construction activity or land use that restricts the flow of water in natural channels or that reduces the floodplain area that retains overflow waters during times of flooding. Generally speaking, all forms of temporary structures should be banned from the floodplain, and all permanent structures should be raised to a height above the level which flood waters can be expected to reach, on average, once every 100 years (100-year flood). No temporary dwelling units—mobile homes and the like—should be permitted in the floodplain. Increases in surface runoff can be mitigated by disturbing the existing vegetation and natural contour of the land as little as possible. Installation of underground drainage structures helps to reduce sediment loads (overland flow is reduced), but not total runoff volume.

Earthslides can be mitigated by avoiding areas with a high probability of incidence or those where proposed activity will significantly increase their probability. Engineering plans can be drawn to reduce the area of exposed strata subject to earthslides, reduce the inclination of slope of earth cuts on fills below what might otherwise be acceptable, provide physical support for exposed soil or rock faces, concentrate or distribute—as appropriate—the weight loadings of foundations to areas or strata better able to support that weight, use small charges for blasting, and restrict the movement of heavy machinery during the construction phase.

The effects of wildfire can be mitigated by clearing fire lanes in strategic locations and building restricted-access roads into areas having a high probability of wildfire incidence. Removal of live vegetative cover, which permits the drying forces of wind and sun to interact more directly with the duff, should, if possible, be avoided. In timber-harvest operations, the removal from the woods of as much of the total tree as is commercially possible to use will reduce the amount of vegetative logging debris left to contribute to depth and flammability of the duff. Restrictions on the use of areas during periods of high fire danger is another type of mitigation technique. Also, buildings should be sited (on the prevailing downwind slope) and roads constructed (more than one access and egress point) so as to minimize physical damage and loss of life if a wildfire should occur.

Secondary effects. Activities that increase the risk of occurrence of natural hazards also have secondary impacts on various social and economic factors. General feelings of security and well-being may be reduced by the increased threat of potential disaster. These psychological effects would be experienced most severely by individuals who believe their lives and property would be affected, should the disaster occur. Economic effects also could result in the forms of increased insurance premiums or changes in property values as hazard risks increase.

Other comments. The impact of human activities in areas subject to hurricanes and earthquakes has not been treated in detail. The most appropriate measure of impact in such cases is the change in the number of people and in the dollar value of physical facilities exposed to these hazards as a result of the activity.

Land-use patterns

Definition of the attribute. Land-use patterns are natural or imposed configurations resulting from spatial arrangement of the different uses of land at a particular time. Land-use patterns evolve as a result of (1) changing economic considerations inherent in the concept of highest and best use of land, (2) imposing legal restrictions (zoning) on the uses of land, and (3) changing (zoning variances) existing legal restrictions.

The critical consideration is the extent to which any changes in land-use patterns resulting from an action are compatible with existing adjacent uses and are in conformity with approved or proposed land-use plans. Where a conflict or inconsistency exists (between a proposed action and the objectives and specific terms of an approved or proposed federal, state, or local land-use plan, policy, or control), the should describe the extent to which the agency analysis reconciled its proposed action with the plan, policy, or control, and the reasons why the agency has decided to proceed notwithstanding the absence of full reconciliation.

Activities that affect the attribute.   Changes involving transportation systems (roads, highways, airports, etc.), water resources projects, industrial expansion, and changes in the working or resident populations are examples of activities likely to induce changes in the pattern of land use and create compatibility problems with adjacent uses. The building of new, or the expansion of existing, facilities through a program of land acquisition would be an activity likely to result in a conflict with approved or proposed federal, state, regional, or local land-use plans. If such a conflict exists, it is quite possible that a compatibility problem with adjacent uses will also emerge. Recreational opportunities and second home or resort area development are other areas where land-use conflicts are evolving.

Source of effects.   Activities involving land acquisition will conform or conflict with approved or proposed federal, state, regional, and local land-use plans in relation to whether such plans exist at all, their detail, and the specific use of the acquired land. For example, if an agency purchased land for the construction of an office building in an area specifically designated for residential use by an approved zoning ordinance, there would be a direct conflict with a land-use plan. Conversely, if the land were purchased as a site for the construction of family housing units, there would be no apparent conflict.

In terms of changes in land-use compatibility patterns, increased or decreased noise levels could have a decided impact. If an industrial-type activity is established at a location that was previously administrative in nature, the attendant increase in rail and truck traffic, particularly if routed near or through residential areas adjoining the site, could result in increased noise levels that might be incompatible with the existing use. Even greater noise problems affecting land-use compatibility patterns arise when activities involve airfield expansion or construction or propose modification of flight patterns.

Military installation closings resulting in the working and resident populations being reduced almost to zero would usually have a decided impact on the land-use patterns of nearby private property. These changes might not be easily perceived at first. Residential and commercial areas would remain, but their intensity of use would probably be sharply curtailed. Portions of such areas might eventually revert to a lower use, the structures possibly razed, and the land permitted to return to open space or some nonintensive form of agriculture. The issue of compatibility with adjacent uses might arise if the use revision took place in a random and essentially uncontrolled fashion.

Large increases in a project-related labor force at a given location would almost certainly have repercussions on land-use patterns in the area. An example would be the introduction into nearby areas of residential structures that are basically unsuited for such development. Mobile home parks or high-density apartment complexes might be sited adjacent to the approach pattern of aircraft runways on what was previously agricultural land. This could come about if a variance to zoning ordinances was granted by some local governments in an attempt to encourage population growth in their political jurisdictions.

Activities which influence changes in land-use patterns certainly do not always do so adversely. There may be compatibility conflicts in the existing land-use pattern which would be ameliorated by other activities. An influx of people (with an appreciation of planning) into an area having no comprehensive zoning ordinances or land-use plans could result in the formulation and adoption of such ordinances or plans. Over time, this could result in more compatible land uses in the area surrounding the activity.

Variables to be measured. Compatibility of use between one parcel of land and adjacent properties involves variables such as type and intensity of use (residential, commercial, industrial, transportation, agricultural, mineral extraction, and recreational, and sub-breakdowns within each that reflect use intensity), population density, noise, transportation patterns, prevailing wind direction, buffer zones, and aesthetics. For example, a high level of residential/transportation land-use compatibility would be evident where a single-family home is set back 30 ft from a two-lane street having a traffic volume of 20 cars per hour which travel at an average speed of 25 miles per hour. Conversely, considerable incompatibility would exist if the same house is set back the same distance—with no intervening barriers—from a four-lane highway with a traffic volume of 2000 vehicles per hour, the majority of which travel at 55 or more miles per hour.

Conformity of a proposed new use of land with approved or existing land use plans is determined by whether a plan exists for the area in question, and if so, whether the proposed use conforms with the ones permitted in the plan. This is a very straightforward relationship unless attempts are made to correlate use/plan conflicts with variances under which precedents for change may have been set.

How variables are measured. Because the constraints that influence compatibility vary widely with the types of land use involved and the spatial arrangement of one with another, variables (such as traffic flow; population density; noise levels; depth, width, and area of buffer zones; and constituents and quantity thereof in air, water, solid effluents) are subject to physical measurements by engineering and planning professionals. Even aesthetic qualities are subject to a somewhat objective measurement by landscape architects. With respect to compatibility of use, however, measurement alone does not indicate the magnitude of the impact. It is the relationship of these variables to one another in the context of their specific spatial arrangement that determines compatibility.

Measurement of variables reflecting conformity with a land-use plan is essentially a yes-no proposition. A plan with which the proposed use can be compared either exists or does not. If a plan exists, the proposed use either conforms or conflicts with its provisions. In practice, the assistance of a spatial planner/zoning expert would probably be required if the proposed use is complex or if the plan is couched in legal terminology. Land-use plans may be prepared at all levels of local government: by incorporated towns and municipalities, townships, and counties; by regional planning agencies (for agencies in specific areas, refer to Regional Councils Directory, published periodically by the National Association of Regional Councils); by state departments of planning, development, and natural resources (for specific state-by-state information, refer to *A Summary of State Land Use Controls,* published by Land Use Planning Reports); and by federal land management agencies, such as the Bureau of Land Management, the National Park Service, the Bureau of Reclamation, the Corps of Engineers, the Tennessee Valley Authority, and the Department of Energy.

Evaluation and interpretation of data. Discussion of the variables involved in land-use compatibility attempts to convey the idea that there is no simple way to relate these variables and arrive at a compatibility index. While planning standards exist, the way they are applied in practice varies considerably from one political entity to another, from one geographical area to another, and with the types of existing and proposed uses. The assistance of a city and regional planner with a background in the spatial arrangement of land uses would be essential in measuring and analyzing interactions among variables and, subsequently, in interpreting the results in terms of the relative compatibility of the uses.

For reasons of continuity, evaluation and interpretation of whether a proposed use of certain parcels of land conforms or conflicts with existing or proposed land-use plans was included in the previous discussion on the measurement of variables.

Geographical and temporal limitations. There appear to be no geographical limitations directly influencing the compatibility of adjacent uses of land. On the other hand, geographical boundaries of political entities govern the areal extent of the particular land-use plans which the activity may affect.

Temporal considerations relate to the problem of projecting how land-use patterns are likely to evolve as a result of a proposed activity. The period of analysis usually used is the expected beneficial lifespan of the project.

Mitigation of impact. Compatibility between adjacent land uses can best be assured by providing an open-space buffer zone between the proposed activity and nearby properties where any significant degree of incompatibility is likely to result. The width/depth/area of this buffer zone should not be excessive, since to make it so could be construed as an inefficient use of land. As for mitigating the impact of changes in existing uses among adjacent off-post parcels of land likely to evolve as a result of the proposed activity, officials of affected local political entities and regional, state, and federal agencies could be apprised at an appropriate time of the projected impacts. They would then have the opportunity to change existing, or enact new, land-use plans.

Mitigation of conflicts between a proposed use of land and proposed existing land-use plans can best be accomplished during the planning stage. Obviously, it would be most desirable from an environmental standpoint to locate the activity where no conflict in use would exist. If this is not feasible, discussions could be held with representatives responsible for the plans, with a view toward resolving the conflict through the granting of a zoning variance or plan modification. Even if no satisfactory agreement can be reached, the fact that such discussions were initiated and conducted in good faith might have a positive impact on any future controversy or litigation.

Secondary effects. Just as direct impacts on many geophysical and socioeconomic attributes induce effects on land use, direct effects on land use result in secondary effects on other biophysical and socioeconomic attributes. Transportation projects, for example, may concentrate air or ground traffic, with resultant increases in levels of air pollution and noise production. Population shifts result in changes in demand for utilities (water supply, sewage treatment, electricity, etc.), and affect wholesale and retail markets and community services (police, fire, schools, etc.). In essence, land-use designation can be related to all areas—air, water, land, ecology, sound, human, economic, and resources.

Other comments. On the surface, it would appear that proposed land-use plans, policies, or controls, as well as those which generally address land use without supportive legal instruments (ordinances, laws, administrative rules), would not be as binding—or taken into account to the same degree—as would those specifically and carefully drawn, officially enacted or promulgated, and having the support of legal precedent. However, the language of the previously quoted CEQ regulations is rather unequivocal. For an impact assessment/statement, no differentiation is made between approved and proposed plans, policies, and controls. The charge is still to examine the conformance of the proposed action to the plan or policy.

References

Control of Erosion and Sediment Deposition from Construction of Highways and Land Development. U.S. Environmental Protection Agency, 1971.
Dodson, R. *Storm Water Pollution Control,* 2d ed. New York: McGraw-Hill, 1998.
Environmental Protection Agency. "Sediment and Erosion Control: An Inventory of Current Practices," W-278. April 20, 1990.
Rainfall Erosion Losses for Cropland. Agricultural Research Service, USDA Agricultural Handbook No. 282.

Ecology

The characteristics of the human environment are intimately related to the nonhuman ecology that surrounds us. Problems that affect lower-level elements in the ecological system may ultimately affect humans. For example, the accumulation of pesticides and heavy metals in lower levels of the ecological system may be harbingers of dangerous levels of these materials in humans.

In addition, despite progress that we have made in providing for our needs, the total ecological balance of the environment is crucial to the viability of our species. For this reason, species diversity and balance must be maintained. Convincing evidence exists that species diversity in an ecosystem is closely related to the stability of that system, with increasing species diversity indicating an increased ability of the ecosystem to resist disturbance and stress. Evaluation of impacts on a given ecological system should include an assessment of the effect of proposed alterations of the environment on species diversity, based on existing information or on special field studies.

The attributes that have been identified to describe the "ecology" resource are

Large animals

Predatory birds

Small game

Fish, shellfish, and waterfowl

Field crops

Listed species

Natural land vegetation

Aquatic plants

Large animals (wild and domestic)

Definition of the attribute. Large animals are those, both wild and domestic, that weigh more than about 50 lb when fully grown. Common wild animals falling into this category are deer, bear, elk, and moose. Domesticated animals of this size include horses, sheep, cattle, swine, and goats.

Activities that affect the attribute. Since most large animals (except for some which are quite rare, i.e., cougars, wolves, etc.) are browsers or grazers, activities having the greatest effect upon them are those which diminish the animals' vegetative food supply or otherwise make inhospitable to them all or portions of the area over which they range. Examples of such activities are construction of new facilities (roads, fences, buildings, etc.), military training exercises, and encroachment into wildlife habitat by vehicular traffic or recreation activities. In the western states, encroachment on the unfenced open rangelands may have an equivalent effect on domestic livestock.

Source of effects. Vegetative and other forms of cover—for traveling, eating and watering, sleeping, breeding, and rearing of young—are required by all wild animals if they are to thrive in an area. Construction activities which result in the clearing of underbrush by burning or other physical means can reduce the available range over which large animals forage. Likewise, application of herbicides can reduce both cover and food, unless utilized in programs specifically designed to increase cover and food. Acquisition of new land for various activities, if such land was previously used for the grazing of domestic livestock, can reduce the total area available for that purpose in a particular locality, resulting in human social and economic effects.

Noise can cause large wild animals to leave or avoid a particular area. Fencing can restrict the movement of animals, either denying them access to food and water areas or keeping them penned within an area smaller than that required for their well-being.

Variable to be measured. The most direct variable is animal population. The type (species) and number of large animals should be determined. To arrive at the magnitude of the impact on the population, the change in the amount (acres) of land suitable for large animal habitat must be determined. A relative measure of the increased noise generated by extensive human intrusion into wild, remote areas formerly ventured into only by hunters or herders should be made. Intense and prolonged noise-generating activity can sufficiently change the habits of large animals to cause them to vacate an area, at least temporarily, until human activities are reduced or the animals become accustomed to them. Adjacent areas can be stressed by having to temporarily support greater populations.

How variables are measured. A census of large animal populations can be made by direct observation. If small, the entire area can be censused. If large, counts can be taken on random plots and projected over the total area of suitable habitat. Good observational and outdoor skills are required for many direct counts. In some areas of fairly open terrain, skilled photointerpreters can take the census of large animals from aerial photographs. If direct observation is not practical or possible (lack of skilled people, large area, nature of the habitat or animal species), a local wildlife biologist affiliated with a federal or state wildlife agency should be consulted for his or her estimate of the population (numbers of domestic animals should be available from ranchers using the land). Wildlife specialists are professionally qualified to judge how noise and other nondestructive activities of humans and vehicles affect the use of an area by large animals.

The change in acreage of a particular habitat type can be obtained from before and after overlays prepared from aerial photographic prints, project plans, or maps. Through using GIS or a planimeter, the size of these areas can be determined with the assistance of an engineer or surveyor. While a simple proportional relationship can be made between the large animal population and acres of available habitat, it would be helpful to have a wildlife biologist review the calculations and determine the relative effect of the seasonal variation, etc.

Evaluation and interpretation of data. The increase or decrease of the large domestic animal population of an area can be interpreted on the basis of the resulting change in annual income. A more subjective evaluation must be made for wild animals. The number lost or gained relative to the number originally in the area is the most critical element. If any of these wild animals prey on smaller animals, the effect of the increase or decrease in that population should be considered. Not to be overlooked are the aesthetic value of large wild animals and the economic dividends which accrue to an entire region if the animals are subject to hunting. Neither of these two values can be readily quantified, and any judgment of their significance must remain highly subjective. Activities which may adversely affect hunting access or success can be extremely controversial.

Special conditions. If there is a long tradition of grazing rights for domestic livestock and these rights are to be withdrawn, the impact of the activity could become controversial—particularly if these rights had previously been exercised by Native American tribes. If any of the wild animals are considered to be endangered or threatened—regionally, nationally, or internationally—a reduction in their numbers as a result of some activity, particularly habitat alteration, would likely result in controversy and legal compliance problems. (The attribute discussion for listed species goes into greater detail on this subject.)

Geographical and temporal limitations. Concern about domestic animals and associated grazing rights is of significance primarily in the western United States, where extensive rights to use federal lands for this purpose still exist. As already noted, the impact of a particular activity on wild animals may be short-term, occurring only during the construction or direct activity period, when people and equipment intrude most heavily on the animal's home range. Also, especially in alpine and high-plains areas, large animals have both a summer and a winter range: a factor in determining their presence in, or absence from, an area. The impact of the reduction in summer range would likely not be as severe, for example, as would be a reduction in winter range.

Mitigation of impact. The impact of activities on large animals can best be mitigated by intruding as little as possible on their habitat. If such animals use the area where the activity will take place, the activity should be concentrated to the maximum extent possible in those parts of the area which they least often frequent. During the planning phase of an activity, an attempt should be made to avoid extending into the home range of large wild animals. If this is not feasible, the activity should be completed as quickly as possible at a time when the animals are not present, and regular and sustained use of the area over time should be minimized. If land acquisition is necessary and a choice is possible, a productive range used by large domestic and/or wild animals should be avoided.

Secondary effects. Economic interests resulting from hunting-related business and aesthetic qualities supported by the presence of wild animal species may be affected as a result of impacts on large animals. Other secondary impacts may occur if natural predator-prey balances are upset by the activity. Undesirable effects may also result if some aspect of a project, such as extensive irrigated landscaping, *attracts* large populations of the animals.

Other comments. If the activity impinges upon the range of large wild animals that have previously been hunted in the particular area and if the activity will result in either closing that area to hunting or a reduction in the number of such animals available for annual harvest, fish and game clubs are likely to oppose the activity.

Predatory birds

Definition of the attribute. Birds of prey are flesh eaters and obtain their food primarily by hunting, killing, and eating small animals, other birds, and fish. Common birds in this group (orders Falconiformes and Strigiformes) are hawks, owls, and vultures. Less common are eagles, ospreys, and some of the falcons. The California condor is quite rare.

Activities that affect the attribute. Since birds of prey nest primarily in trees—sometimes in areas remote from human habitation—cutting of mature timber stands or the selected removal of individual overmature or noncommercial trees could result in a disproportionate reduction in their numbers. Burning of brush or grasslands, applications of herbicides and pesticides, and the use of poisoned bait in animal-control programs are other activities that could directly affect the survival of predatory birds. Activities resulting in intrusions by persons into or near nesting areas could affect these birds, particularly eagles, ospreys, condors, and some types of falcons that are less tolerant of humans. Historically, widespread use of chlorinated hydrocarbon pesticides caused populations of many of these species to become endangered. This has become less critical following the ban on most uses for these materials.

Source of effects. The removal of nesting trees as a part of any general land-clearing program preceding construction activities or the selected removal of such trees in a forest management "sanitation" cutting could destroy unhatched eggs or cause the death of birds too young to survive outside the nest. If suitable nesting habitat is not available elsewhere in the vicinity, adult birds may disappear from an entire area. Burning of brush and grasslands destroys the habitat and large numbers of the prey species (small animals) on which predatory birds depend for food. Similarly, application of defoliants could reduce the food and cover available for small animals and birds, with a consequent reduction in their numbers. The reproductive capacity of these and many other birds may be reduced if sufficient quantities of DDT and other chlorinated hydrocarbon insecticides are concentrated in the food. As an example, eggshells can be weakened to the point where they break before the young are ready to emerge. Direct killing of predatory birds can result from their eating of poisoned bait (portions of animal carcasses) intended for coyotes, cougars, and other flesh-eating animal predators. Extensive outdoor activities resulting in the visibility of humans and the noise of vehicles and heavy equipment use, if conducted intensively over an extended period of time or at frequent intervals, could cause birds to desert their nests. If the activity is sustained over a long enough period of time, adult birds may leave the area permanently.

Variables to be measured.   The number and types of birds of prey that nest and/or capture their food within the affected area should be determined. The change in the amount of available habitat (nesting and/or feeding) must be ascertained to estimate the numbers of birds which the existing habitat will support once the activity is completed or the project becomes operational.

How variables are measured.   While a direct census of common birds of prey is possible in areas of limited size, the observational and general outdoor skills and the time required make it most impractical to conduct one. A usable population figure could be best obtained from a local wildlife biologist affiliated with a federal or state wildlife agency. A biologist of the Natural Heritage Program, Audubon Society, or similar private wildlife conservation organization may also be able to provide accurate counts of the less common species and the locations of their nesting and feeding areas. The change in acreage of nesting and feeding habitats can be obtained from before and after overlays prepared from aerial photographic prints, mosaics, or topographic maps. Through using GIS or a planimeter, the size of these areas can be determined with the assistance of an engineer or surveyor. For the more common species of hawks and owls, the nesting and feeding habitats can be combined, and a direct proportion established between the bird population and the number of acres of available habitat.

The relationship between available habitat and the generally larger, less numerous predatory birds is less direct and more subjective. If such species as the bald eagle, golden eagle, osprey, peregrine falcon, or California condor are present in an area, it will be necessary to solicit the opinion of expert wildlife biologists in determining what portion of the existing population would remain after the activity was completed. If the species are threatened or endangered, consultation with the U.S. Fish and Wildlife Service is mandatory.

Evaluation and interpretation of data. The change in numbers of common birds of prey, as related to a particular activity and location, is an overall indicator of the change in habitat quality for other birds and animals within the area. Any substantial reduction in the numbers and types of hawks and owls would be generally indicative of a widespread adverse ecological impact. As with the large wild animal attribute, any reduction of the less common species of avian predators could be expected to bring forth objections from private conservation organizations, as well as from federal and state agencies charged with their management and protection.

Special conditions. If any of the predatory birds of the area are considered to be threatened or endangered—regionally, nationally, or internationally—any effects on their habitat resulting from the activity would be controversial and subject to USFWS review. (The attribute discussion covering listed species goes into greater detail on this subject.)

Geographical and temporal limitations. Most of the large, less common birds of prey have very restricted geographical ranges. Maps showing these ranges are contained in most field guides to bird identification. A review of such range maps would reveal whether or not these species are likely to be found in the activity area. Special attention should be given to any short-term activities that might disturb the birds during their nesting season. The regional office of the USFWS is the definitive authority on these matters.

Mitigation of impact. The potential detrimental impact of human activities on the avian predator population can best be mitigated by locating the activity at places not considered a part of the habitat essential for the survival of these birds. This is best accomplished during the site-selection planning stage of a project, rather than after a specific site has been chosen. Unless operational considerations are absolutely overriding, the habitat of the large, uncommon species should not be disturbed at all. Regular or sustained intrusions of workers or equipment into nesting areas should be avoided to the maximum possible extent, especially while eggs are being incubated by the adults and until the young have left the nest. No known nests should be destroyed by the sanitation cuttings of individual noncommercial trees.

Secondary effects. Secondary impacts from an increase or decrease in predatory birds may be observed in the populations of animals upon which these birds prey. These animals may, in turn, have economic benefit through hunting-related business, or they may play significant roles in other ecologic relationships.

Other comments. If the existing habitat of the bald eagle, golden eagle, osprey, peregrine falcon, California condor, or other threatened or endangered species is affected by an activity, the resultant controversy is likely to be intense, prolonged, and acrimonious.

Small game

Definition of the attribute. Small game includes both upland birds and animals which, as adults, weigh less than about 30 pounds, and many are commonly hunted for sport. Some small game species falling into this category are rabbit, squirrel, raccoon, quail, grouse, and pheasant.

Activities that affect the attribute. Since most small game animals and upland birds are relatively tolerant of humans, the activities most damaging to them are those which physically destroy their habitat (area in which all welfare factors such as food, cover, water, and space required for their survival and propagation are present in sufficient quantity and diversity). Land-clearing activities for buildings, road construction, etc., are most often the ones that significantly and adversely affect small game. Conversely, such game can be expected to return to formerly built-up areas, which, when abandoned, revert to native vegetation. Distribution of poisoned baits used in rodent and predator control and use of herbicidal defoliants can also reduce small game populations, as can the use of certain pesticides.

Source of effects. The removal of native vegetation from, or the rearrangement of topography and surface features by the grading of, an area denies small game the kinds of habitats they require. Without the food and cover provided by vegetation and irregular surface features, populations of small game diminish rapidly. Conversely, they will often quickly return to abandoned areas given over to native vegetation. If poisoned bait is used, only a few small game animals and birds are likely to be affected, except in winter, when food is scarce and populations are at their annual minimum. Herbicides temporarily destroy small game habitat, and repeated applications can cause the permanent abandonment of an area. Persistent chemicals are accumulated in body tissues through ingestion of residues with food and water.

Variables to be measured. The small game population of the area to be affected by the activity must be censused. Once this is accomplished, the number of acres of existing habitat must be determined, as well as the amount by which it will increase or decrease over time as a result of the activity. The relationship between these variables and the attribute is fairly straightforward; the carrying capacity (wildlife population an area can support indefinitely without habitat degradation) is increased or decreased in direct proportion to the amount of available habitat remaining. While the quality of small game habitat existing before, and available after, the completion of an activity is an important variable, it is very difficult to quantify and will not be specifically discussed. It will, however, enter into subjective evaluations and judgments.

How variables are measured. While an accurate census of small game is difficult to make, usable estimates of the number of different species per acre of habitat can often be obtained from local wildlife biologists affiliated with federal or state wildlife agencies.

The change in acreage of small game habitat can be obtained from before and after habitat overlays prepared from aerial photographic prints, mosaics, or topographic maps. Through using GIS or a planimeter, the size of these areas can be determined with the assistance of an engineer or surveyor. A direct proportion can then be established between the small game population and the number of acres of suitable habitat.

Evaluation and interpretation of data. The relative importance of a change in the small game population of an area is a very subjective judgment. If habitat is to be destroyed, significance should be attached to the relative amount and quality available in adjacent areas, as well as to the relative amount and quality of total habitat under control that will remain after the activity is completed.

Special conditions. If the activity will cause a significant reduction in the available small game habitat in an area subject to heavy hunting, the impact will likely be controversial to both sports enthusiasts and, to a lesser extent, economic interests in the area. This could happen if a prime small game hunting area is fenced and placed off limits to the general public, or if its quality is to be heavily degraded.

Geographical and temporal limitations. While it is unlikely that any small game species would fall into the endangered or threatened category nationally, certain ones, such as grouse, woodcock, and turkey, might be rare in some states or local areas. Many outdoor activities during nesting season often destroy eggs, the result of which may be a significant reduction of the game-bird population for one or more years.

Mitigation of impact. Activities affecting small game can best be mitigated by disturbing the vegetative cover and altering the physical contour of the land as little as possible. Selecting areas of poorer habitat quality and preserving prime areas will reduce the severity of the activities' impact on the small game population. Opening of large areas to the general public during certain periods of the small game hunting season (opening day and holidays, when hunting pressure is particularly heavy) will also tend to ameliorate the withdrawal of areas formerly freely available to public hunting.

Secondary effects. Economic interests resulting from hunting-related business and aesthetic qualities supported by the presence of wildlife may be affected as a result of impacts on small game. Other secondary impacts may occur if natural ecological predator-prey balances are upset by the activity. Most changes in small animal populations are the result of, not the cause of, secondary effects.

Fish, shellfish, and waterfowl

Definition of the attribute. Fish are cold-blooded, water dwelling animals that obtain oxygen through a gill system. They inhabit saltwater and freshwater bodies and streams and vary widely in size. Common species are minnows, sunfish, trout, bass, pike, salmon, and tuna.

Shellfish are aquatic animals that have an exoskeletal shell rather than an internal vertebrate structure of backbone and ribs. Common freshwater and saltwater species are mussels, crayfish, clams, oysters, shrimp, crabs, and lobsters.

Waterfowl are birds which frequent and often swim in water, nest and raise their young near water, and derive at least part of their food from aquatic plants, animals, and insects. Ducks and geese are the most familiar waterfowl. Because of similar habitat requirements, the generally protected swans, herons, cranes, pelicans, and gulls are also included here. The whooping crane is a frequently cited example of an endangered species that falls into the waterfowl category.

Activities that affect the attribute. Since fish, shellfish, and waterfowl depend directly upon good-quality water for all or some facets of their existence, activities which affect water quality and water level have the greatest impact upon their well-being. Examples of particularly damaging activities are dredging, stream channelization, construction that exposes mineral soil and subsoil subject to erosion, disposal of untreated or insufficiently treated sewage in water courses, permitting toxic materials to drain into water courses without collection and treatment, disposal of industrial cooling water in the ocean or in streams and lakes, application of pesticides—the residue of which may drain into water courses, draining of swamps or potholes, building of water-level control structures such as dams or dikes, and disposal of hazardous wastes at sea.

Source of effects. Dredging can temporarily displace the bottom organisms on which these categories of wildlife feed and can destroy spawning grounds. Stream channelization results in the removal of native vegetation which supports the insects eaten by fish. In addition, alteration of flow and substrate characteristics resulting from stream channelization can be as harmful as loss of vegetation. Certain species of fish are affected by even small amounts of solid material suspended in the water, a condition resulting from dredging or soil erosion. Some species of fish and shellfish are affected by siltation, which can both cut off their oxygen supply and reduce the availability of food. A special case should be pointed out with respect to release of trapped pollutants when bottom sediments are disturbed, as through dredging. Heavy metals, such as mercury, and persistent pesticides, such as DDT and its derivatives, even though their use and disposal has been prohibited for decades, may be reintroduced into the food chain. Analysis of the bottom sediments prior to removal is now routine.

Discharge of insufficiently treated sewage may introduce disease-causing bacteria and viruses and reduce the oxygen content of the water—the life-support system upon which fish and shellfish are totally dependent. Insufficiently treated sewage also introduces nutrients which accelerate plant growth and decay in the water, often affecting the quantity of available fish habitat by further reducing the oxygen supply. Toxic materials, such as oils spilled or draining into water courses, cause the feathers of waterfowl to no longer shed water, bringing about death from exposure. Toxic materials, such as mercury, can eventually be so concentrated in the food chain that fish are no longer safe for people to eat. Other toxic materials can cause the outright death of fish by damaging their gills and preventing them from extracting oxygen from water.

The acidity level of water, if too high (pH 5 or less) or too low (pH 11 or greater), can cause similar gill damage. Increases in water temperature often cause sport fish to abandon the area to less-desirable species of so-called rough fish such as carp. Rapid fluctuations in water temperature can kill fish outright. Pesticide residues draining into water courses and concentrating through the food chain may eventually become present in sufficient quantities in fish to cause their reproductive capacity and the survivability of young to be impaired. Pesticides can become even more concentrated in the tissues of fish-eating birds and animals.

Draining of swamps or potholes is very detrimental to waterfowl, as it is in these bodies of water that reproduction, nesting, and the rearing of young take place. The artificial raising and lowering of water levels is often beneficial to wildlife habitat, if done at times consistent with needs for food and nesting cover. However, since changing of water levels is most often a flood-control requirement, fish and waterfowl habitat can be drastically affected by changes not in consonance with their needs. Depending on the lethality of the material, hazardous waste disposed of at sea could cause the destruction of all aquatic life in both the immediate area and other areas where the waste is transported by ocean currents.

Variables to be measured. The detailed variables to be measured for fish are the same as those identified in the attribute descriptions involved with surface water quality. Some of these important variables are dissolved oxygen content, coliform bacteria levels, acidity levels (pH), heavy metal concentrations, and pesticide concentrations detrimental to fish life.

While many substances (petroleum products, hydrogen sulfide, copper, and other metals) can taint shellfish and make them unpalatable for reasons of odor, taste, or color, it is the pathogenic bacteria and viruses which they take up from the surrounding water which may render them unfit for human consumption. Measurements of coliform bacteria present in the water provide a standard for determining when oysters, clams, and mussels can be safely eaten.

The main variable to be measured for waterfowl is change in available habitat. Quantity of suitable nesting habitat—which equates roughly to the length of shoreline—is a heavy determinant of waterfowl population on a year-by-year basis, but it is difficult to relate the two exactly. Winter habitat is also important but more difficult to quantify and to relate to increases or decreases in waterfowl.

How variables are measured. As indicated, measurements of water quality variables are discussed under surface water quality. Some acceptable general standards for maintaining a healthy aquatic fish habitat are that dissolved oxygen content should not fall below 5 mg/l and that pH level should be maintained in the 6 to 9 range. The latest EPA or state water quality standards which specify limits for pollutants within various water use categories, including aquatic life, should be followed.

While the standards for fish also apply generally to shellfish, coliform bacteria count is the important variable to be measured. Criteria for water from which shellfish are harvested are contained in the U.S. Public Health Service Manual, *Sanitation of Shellfish Growing Areas.*

General standards for coliform bacteria are that the median most probable number (MPN) must not exceed 70 per 100 ml.

If the length of existing shoreline or its character is altered so as to render it unsuitable for waterfowl nesting, the amount of change should be determined. Before and after overlays of suitable wildlife nesting habitat along shorelines should be prepared from aerial photographs, project plans, or maps. Through using GIS or a map measurer, shoreline length can be determined. The amount of change between present and future habitat can then be calculated. Additionally, the change in number of individual bodies of water between the two overlays should be noted. Once these data are obtained, they are still not directly convertible to change in the number of pairs of nesting waterfowl the habitat can support. This is a subjective judgment which only an expert wildlife biologist (e.g., from the U.S. Fish and Wildlife Service) can make.

Evaluation and interpretation of data.   Although it is difficult to directly relate small alterations in water quality to changes in fish and shellfish populations, changes in fecundity, population counts, and growth rates are often sensitive indicators of such alterations. Therefore, an attempt should be made to assess population changes that might result from proposed alteration of the environment. If water quality is degraded or improved to the point where commercial fishing activities are affected, the change in annual revenues derived from this source can be determined. If the change in water quality affects species associated with sportfishing, the number of miles of streams affected would provide some measure of the significance of the impact. If a prime sportfishing area is involved, economic gains or losses to businesses deriving a part of their income from those who fish might be an important consideration. Estimates of the effects of such changes might be obtained from federal or state wildlife agencies.

Changes in quantity of nesting habitat would definitely affect the number of waterfowl available for annual harvest. However, the effect is felt more in areas where waterfowl are hunted than where they nest. An expert from the U.S. Fish and Wildlife Service could provide an insight into the extent of the resulting environmental (ecological and economic) impacts.

Special conditions.   If activities will cause a significant reduction in the length of streams or areas of coastal waters suitable for sportfishing or in the amount of waterfowl nesting habitat, the impact will likely be controversial to anglers. This could happen if even small stretches of renowned trout streams were to be affected of if prime fishing waters were placed off limits to the general public. Commercial interests would most likely oppose any intrusion into prime fish and shellfish areas or any reduction in the annual catch/harvest. If any waterfowl that are considered to be threatened or endangered—regionally, nationally, or internationally—use the activity area for nesting, migration stopover, or feeding, significant controversy would probably result. (The attribute discussion covering listed species goes into greater detail on this subject.)

Geographical and temporal limitations.   The only geographical limitations on fish and shellfish relate to particular types found in the activity area and whether coastal estuaries or open sea areas are involved. Temporal considerations are those involved with conducting the activity during spawning, migration, or harvest seasons.

The critical region of migratory waterfowl nesting habitat is generally considered to be in the northern tier of states, Canada, and Alaska. This would, however, not be true of nonmigratory waterfowl associated with estuaries and seacoasts. Activity which would disturb waterfowl during the nesting season and while the young are being reared would be most damaging.

Mitigation of impact.   Impacts upon fish and shellfish populations can be mitigated by restricting the input of polluting substances into fresh water bodies, estuaries, and the open sea. This can best be accomplished by ensuring that wastewater treatment facilities of suitable capacity and design are constructed so as to be in operation by the time it is anticipated that waste products from the proposed project will be generated. If soil erosion is a problem, construction activities should be scheduled at times of the year when intense rainfall is least likely to occur. Protective measures such as catchment and retention basins and silt fences may be effective.

Impacts on waterfowl from an activity can best be mitigated by disturbing the land-water interface in the area as little as possible. Vegetation along water courses should not be cleared indiscriminately. Neither should potholes or swamps be drained unless absolutely necessary for successful completion of the activity. Additionally, when a part of the activity involves water-level control, changes in such levels should be programmed—to the extent it is possible to do so—in a way that will only minimally disturb nesting and feeding habitat. These considerations for the natural environment will help to ensure that waterfowl habitat available for nesting and feeding is not appreciably diminished in either quantity or quality. If habitat is permanently lost, the USFWS may require compensatory development of new waterfowl habitat.

Secondary effects.   Economic interests resulting from business related to hunting and commercial and sportfishing activities would be affected by impacts to fish, shellfish, and waterfowl. Other secondary impacts may occur if natural ecological relationships are upset by the action.

Other comments. Water quality and fish and shellfish habitat go hand in hand. Any substantial degradation of the former will have a decided impact on the fish and shellfish populations relative to both quality and number. All aquatic oxygen-using (aerobic) organisms will be affected to some degree by decreases in water quality. The effect of an activity on fish and shellfish is a general indicator of the impact on the entire water environment.

Field crops

Definition of the attribute. Field crops are those commercially cultivated for the primary purpose of providing food and fiber for people and food for domestic livestock. Common field crops include corn, wheat, cotton, soybeans, and truck produce (tomatoes, melons, and table vegetables).

Activities that affect the attribute. Since almost all land highly suitable for field crops is in private ownership, acquisition of that land—for whatever project purpose—would take it out of agricultural production. Acquisition of prime agricultural lands for nonagricultural purposes is likely to have the greatest impact on field crops. Reservoir construction and operation, along with various runoff control projects, may affect the flooding regime of large areas of field crops. Application of herbicides on land adjacent to an agricultural area planted in field crops would have a more localized impact.

Source of effects. Diversity, both human and natural, is an important and valuable characteristic of ecosystems. If the area previously given over to field crops is to be built upon (the most likely reason for acquiring the relatively flat land that field crops usually occupy) or is to be used extensively for nonagricultural activities, the vegetative diversity will be reduced. Wildlife could also be affected, as many game and nongame animals and birds obtain some food and cover from field crops. If the acquired land is later allowed, through successive vegetative stages, to revert to the natural climax type of area, the impact might be ecologically beneficial.

Reservoirs and impoundments may raise groundwater levels, flooding root systems and severely damaging crops. Other flow diversions may decrease probability of flood damage.

Herbicides applied by aerial spraying might carry onto adjacent agricultural land, killing crops with which they come in contact. While the area might be relatively small, the resulting damage could be highly controversial.

Variables to be measured. The main variables to be measured are the number of acres of land now given over to field crops which would be taken out of production, as well as the percentage of that land which would be permitted to revert to natural vegetation. Since field crops and natural vegetation are both ecologically important, an assumption can be made that if one type is not unduly created at the expense of the other, each is of equal significance. In practice, row crops are not as valuable, per acre, as natural vegetation for purposes of wildlife habitat. The measure of ecological impact would then be determined by the loss of productive vegetative cover.

How variables are measured. Specific acreages of field crop land to be taken out of production by a land acquisition program could be measured directly, but it would be easier to obtain figures from local offices of the Farm Services Agency (FSA) of the U.S. Department of Agriculture. Land previously used for crops but permitted to revert to natural vegetation should be depicted on an overlay prepared to scale from a project plan, a map, or an aerial photograph. Through using GIS or a planimeter, the size (acreage) of that area can be determined with the assistance of an engineer or surveyor. The percentage of crop land reverting to natural vegetation could then be derived.

Only a general estimate is possible when determining field crop acreages that might be damaged by herbicidal spraying. Some of the variables involved would be the kind of application system used, wind direction and velocity, and state of crop development. However, if these variables are reduced to an assumption that 500 ft is the maximum distance into the field that the herbicide could produce crop damage, the other variable involved would be the linear measure of crop land directly adjoining the area where the herbicide is to be applied. Area affected and potential economic cost could then be calculated. The FSA should be able to provide acreage yields and selling prices of various field crops which might be affected. Herbicide programs are always controversial, and are probably best avoided.

Evaluation and interpretation of data. The magnitude of the impact of the change in land use that results from crop land acquisition is related to the percentage of that land which will continue to support natural vegetation. The greater the percentage of field crop land that is built upon or otherwise taken out of vegetation production, the greater the impact. For crops damaged by application of herbicides, a measure of impact could be made by comparing the dollar loss of the destroyed crops to the annual value of that crop in the country or area concerned. Again, the greater the percentage of dollar loss is of the total crop value, the greater the impact.

Special conditions.   If the crop land is especially productive relative to other crop land in the general area, or if the crop grown upon that land is of very high value, the impact may be greater than what would otherwise be anticipated.

Geographical and temporal limitations.   Because of climate and soil or other requirements, some field crops—particularly specialty crops (avocados are a good example)—can be grown only in a very limited geographical area. If the crop land to be acquired is in such an area, a significant reduction in the local or even national output of that crop might result. On the other hand, there are vast areas in the western United States where conditions are not suitable for the cultivation of field crops. Land acquisition activity in those areas would not affect this attribute.

Herbicidal damage to field crops is greatest when the plant is growing fastest (spring), before the vegetative product (corn ears, grain kernels, bean pods) has matured. However, this is generally the time when herbicides will be used, because they have the greatest suppressive effect on vegetation at which they are directed. This is most severe when crop lands are intermixed with forest or rangeland where herbicide programs are most commonly proposed. This is a form of land-use conflict.

Mitigation of impact.   The detrimental impact of acquiring productive field crop land can best be mitigated by locating the activity in an area where very little land is given over to field crop production or where the farming enterprise is of marginal economic value. Some additional mitigation in the form of trade-offs is possible if a large portion of the crop land is allowed to revert to natural vegetation. This would be possible in buffer areas acquired to shield private lands from the activities of a major development project.

Mitigation of the impact of herbicidal applications could take the form of cutting vegetation and applying the herbicide directly to the stumps in those areas where field crops are directly adjacent to the activity. Further mitigation of impact is possible if the stump application of herbicides is done at a time when the adjacent field crops are vegetatively dormant. If spraying is a preferred method of vegetative suppression, it should be done at times when wind velocity is low and wind direction is such that the possibility of the herbicide carrying into the field crop area is minimal. This may be regulated by state law or local regulation.

Secondary effects.   Loss of field crop production could have significant effects on local or regional economic stability, particularly if the loss is of long-lasting or permanent duration. Other effects on land prices and farm product availability could result. Such effects are also highly controversial.

Other comments. If significant economic loss will result from the acquisition of crop land and its removal from agricultural production, farmers' organizations will be likely to actively oppose the project.

Listed species

Definition of the attribute. Federally listed species (including those categorized as threatened and endangered) include all forms of plant and animal life whose rates of reproduction have declined to the point where their populations are so small they are in danger of disappearing or may soon decrease to this level. Listed species are classified as such on both a regional and a national basis. A species classified as threatened within a state may occur only in limited numbers at very few locations within that state but be relatively common in other states. National endangered species are those found only in very small numbers or near extinction throughout the United States. Lists of threatened and endangered animal and plant species are published periodically by the Fish and Wildlife Service, U.S. Department of the Interior* (see 50 CFR 17 Subpart B). Examples of more commonly known listed species are the timber wolf, grizzly bear, southern bald eagle, California condor, and whooping crane. Less commonly known listed species include the black-footed ferret, key deer (Florida), Devil's Hole pupfish, Florida kite, Nene goose, and Delmarva fox squirrel. While animal species are the ones most often in the public eye, there are many species of plants that qualify for listing, but few are well known to the general public, and all are of very restricted distribution. See also Sec. 13.4.

Activities that affect the attribute. These activities are basically the same, depending on the animal or plant species involved, as those mentioned under the large animals, predatory birds, and natural land vegetation attributes. Refer to those sections if a listed species' habitat is located within the geographic area that a specific action will affect. Threatened and endangered species are also discussed in the NEPA context in Chapter 13 of this book.

Source of effects. The source of the effects of various activities on listed species of animals and plants is essentially the same as those listed for

---

*The "list" maintained by the Secretary of the Interior is literally a list of those species determined to be threatened or endangered, thus the origin of the name, listed species.

large animals, predatory birds, and natural land vegetation. Refer to those attributes if the habitat of a listed species is located within an area where the effects of a particular action will be felt. Consultation with the USFWS will be required.

Variables to be measured; how variables are measured. The variables to be measured and the method of doing so are highly dependent upon the particular species of plant or animal affected. While the information contained in similar parts of the attribute discussion for large animals, predatory birds, and national land vegetation could serve as a general guide in the case of listed species, the assistance of an ecological team of wildlife biologists, zoologists, botanists, and plant physiologists in accumulating relevant data would be almost a necessity.

Evaluation and interpretation of data. This function can be adequately carried out only by a group of professional ecologists familiar with the myriad details associated with the listed species itself, its place in the ecosystem, and the nature of the particular habitat which is to be affected. Logically, this team of ecologists should be the same group responsible for collection of the basic data on which the evaluation is to be based. However, an additional critical review of their conclusions by an eminent ecologist might help to ensure public acceptability of those findings. The various state natural heritage programs will also be interested parties.

Special conditions. Any potential effect at all to either a listed species or any special interest species is extremely sensitive, and likely to be controversial. This controversy may be the basis of public opposition or legal challenges or both. Federally listed species are specifically protected by the Endangered Species Act, and strong objections are also certain to be raised by the Fish and Wildlife Service. The project may trigger controversy even if the proposal is not directed toward the listed species. Secondary consequences, if predictable, are adequate to trigger the provisions of the ESA. This is clearly an area where extreme care must be taken every time a listed species is affected in any manner by a proposed action. Many states have protective legislation as well, which will cover species of local importance. Compliance with these regulations must also be considered mandatory, or the proponent will suffer the cost of acrimonious public relations in addition to legal penalties.

Geographical and temporal limitations. Clearly, species never found within the sphere of effects of the proposed action cannot be involved. Many nonresident species may, at some time of the year, however, utilize other lands and waters for migration, breeding, or feeding. A brief action during the period when a listed species is not present may be permissible, although coordination with the USFWS is mandatory even in this case. Planners should consult with knowledgeable wildlife professionals for every proposed project, period. The state natural heritage programs maintain records of presence of listed species, as do state wildlife and natural resources agencies and the USFWS. All will provide basic information readily at no cost.

Mitigation of impact. The primary way to mitigate the impact of activities on listed species is to avoid any disruption—physical or biological—of their habitats which might result in a decrease in their populations. While it would appear to be less damaging to disturb the habitat of a species classified as threatened by a state than of one that is federally listed or as threatened rather than endangered, these trade-offs are usually not feasible. It is best to avoid disturbing the known habitat of any listed species. The rulings (opinions) of the USFWS will state what accommodations are required to meet legal requirements.

Secondary effects. As indicated above (in special conditions), many secondary effects occur along with the impacts on threatened species. In addition to their human and aesthetic interest value, some of these species may be of significant importance to the dynamic aspect of the ecosystems in which they are found. Their use as indicators of overall environmental quality should not go unnoticed.

Other comments. If any activity has the potential of adversely affecting the populations of any listed species, naturalist and wildlife groups are almost certain to vigorously oppose it in public hearings and/or in court. The opinion of the USFWS must, by law, be sought prior to taking any action.

If any question exists as to the presence of a listed species—either intermittently or year-round—in the area of a project, USFWS wildlife biologists or botanists should be called upon to verify that presence and to give a preliminary assessment of the impact of the activity on the population of this species.

Natural land vegetation

Definition of the attribute. Natural land vegetation is that which uses soil (as opposed to water) as its growth medium and which is not the subject of extensive cultural practices by humans. Included in this category are a number of diverse groups of plants, including trees, shrubs, grasses, herbs, ferns, and lichens.

Activities that affect the attribute.   Any activities that affect the land surface will affect the vegetation that grows upon it. Timber-harvest operations, land-clearing activities prior to construction, burning, application of herbicides, off-road vehicular traffic, and the application of artificial paving materials are some activities that can cause adverse impacts on natural vegetation. Abandonment of a project site can result in natural vegetation becoming reestablished through a series of successional stages.

Source of effects.   Timber-removal operations employing inappropriate forest management methods can reduce the possibility of reestablishing fully stocked stands of the same species. Without the protection of the forest canopy, shrubs and other plants left after timber removal may weaken and become prime targets for disease and insects. Land-clearing activities can cause the outright destruction of natural vegetation, and resulting soil erosion can inhibit its reestablishment. Improper use of herbicides can result in the destruction of nontarget species of natural vegetation and can disrupt the overall stability of the ecosystem. Mechanized military field training destroys lower vegetative forms outright, and the resultant soil compaction and erosion—each in its own way—can inhibit their reestablishment. Paving can deny native vegetation to large areas for extended time periods. As previously indicated, a reduction in the magnitude of activity at a particular installation or its closing can encourage the reestablishment of native vegetation.

Variables to be measured.   The variables to be measured are the number of acres of native vegetation existing before and after the activity, as well as any significant vegetative changes that may develop. A reduction in an area given over to native vegetation can result in increased soil erosion, a decrease in soil fertility, and a decrease in quality and quantity of wildlife habitat. It can also accelerate the invasion of weeds and other undesirable pest species. Reintroduction of native vegetation can—over time—have the opposite effect. Successional change in vegetative type is slow, however, and the least-desirable plant types are the first to become reestablished on a site after a major clearing activity.

How variables are measured.   The change in acreage of natural vegetation can be obtained from before and after overlays of vegetative types. The before overlay can best be prepared from recent aerial photographs. A photointerpreter can assist in differentiating and plotting the major vegetative types. In this way, the total area of vegetation cover can be ascertained along with the subareas in each of the major types. The after-activity overlay should be prepared at the same scale, using the project plan to outline areas of existing natural vegetation which will be affected. The remaining total acreage in native vegetation by major type should then be determined. These calculations of acreage can best be done through using GIS or a planimeter, information from a remote-sensing system. The percentage of original native vegetation remaining—both total and by major type—is then derived.

Evaluation and interpretation of data. The magnitude of the impact of the activity on natural vegetation can be determined from the percentages previously given. However, the specific changes and types of vegetation which would result from the activity can be projected only by a botanist or forester intimately familiar with the local area. Even more difficult to interpret objectively are the aesthetic considerations involved.

Special conditions. Destruction of natural vegetation in the particularly fragile ecosystems that exist under extremely adverse environmental conditions—such as tundra and desert—can have greater impact than in an area with a more moderate climate. They may also require many more years to recover—40 to 50 years or longer.

Geographical and temporal limitations. The only geographical limitations of impacts on this attribute occur in those rare areas of desert and bare rock where no native vegetation exists. Much greater damage may occur when the soils are wet, or when annual vegetation has not become established. Fire, including prescribed burning, may be too destructive when conditions are dry.

Mitigation of impact. The best way to mitigate the impact of activities on natural vegetation is to design the project so as to restrict the area affected. Examples of other mitigation possibilities are to restrict land-clearing activities to the absolute minimum, apply ecologically sound management practices in timber harvest and timber stand improvement, confine vehicular activities to designated areas and restrict expanding them into new areas, apply vegetation suppression techniques of controlled burning and herbicide application only when other methods are not feasible, and use crushed stone rather than asphalt or concrete for surfacing parking areas.

Secondary effects. In addition to economic gains from timber harvesting, natural land vegetation provides habitat for wildlife species; recreational areas for hunting, camping, and other pursuits; and countless other resources of both aesthetic and material nature, including reduction of erosion and runoff from storms.

Other comments.   If activities result in the destruction of unique areas of natural vegetation, opposition can be anticipated from local and national naturalist organizations. These natural areas are usually well known locally and are often cataloged at the state level by departments of natural resources or the natural heritage programs. Any activity that would alter these unique and rare areas of natural vegetation should be avoided to the same extent as one involving the habitat of a threatened species of wildlife.

Aquatic plants

Definition of the attribute.   Aquatic plants are those whose growth medium is primarily water, though they may be rooted in bottom sediments. They include free-floating plants such as phytoplankton, all surface and submerged rooted plants, and swamp and marsh vegetation whose roots are periodically or permanently submerged in water. Aquatic plants are essential elements in the food web.

Activities that affect the attribute.   Activities which cause changes in water level or water quality parameters have the greatest impact on aquatic plants. Examples of particularly damaging activities are dredging, stream channelization, construction that exposes mineral soil and subsoil subject to erosion, disposal of untreated or insufficiently treated sewage in water courses, disposal of cooling waters in oceans and in streams or lakes, draining swamps and marshes, and building of water-level-control structures such as dams or dikes.

Sources of effects.   Dredging can temporarily—and sometimes for long periods—displace rooted and bottom-dwelling aquatic plants. Stream channelization intentionally removes all stream-side vegetation. Erosion can cause increased sediment loads sufficient to restrict the sunlight on which aquatic plants depend for photosynthesis. The discharge of insufficiently treated sewage into the water courses induces excessive aquatic plant growth. Increases in temperature also tend to accelerate aquatic plant growth, particularly algae. The draining of swamps and marshes reduces the area in which aquatic plants can survive. Changes in water level can cause the destruction of aquatic plants, either by exposing their roots to the drying influence of sunlight and air or by flooding to levels which deny air to bank- or marsh-dwelling species for long periods of time.

Variables to be measured. The essential variable is the change in amount of water area suitable for the growth of aquatic plants. There are two elements to this variable: changes in water surface areas and changes in those elements of water quality which accelerate or restrict plant growth. Any changes in the kind of vegetation and its productivity can influence all other organisms that depend upon it for food.

How variables are measured. The only direct measurement that can be readily made is the quantity of total aquatic plant habitat available before and after the activity. This can be done by an expert photointerpreter, who should prepare before and after overlays from large-scale aerial photography. Through using GIS or a planimeter, the acerage can be calculated and the change in total available aquatic habitat derived. Infrared photography or remote sensing are especially useful.

The quality of the water habitat existing before and after the project can be ascertained only by intensively examining the aquatic plant life, measuring the various water quality parameters affecting plant growth, projecting changes in water quality that will result from the activity, and projecting the changes in aquatic plant habitat that will follow. This is a complex procedure which can best be accomplished with the assistance of an interdisciplinary team of biologists, botanists, zoologists, ecologists, and engineers, and should be a part of the water quality studies.

Evaluation and interpretation of data. Since it is not usually possible to measure directly the change in the quality and quantity of aquatic plant life, only a very imprecise measure of impact can be obtained from the change in water area. Generally, if the percentage change in aquatic plant habitat exhibits a value greater than 20, an attempt should be made to measure the qualitative change as well. Further, if changes in the water quality parameters measured or projected under surface water attributes indicate increased nitrogen and phosphorus concentrations, increased water temperature, decreased water flow, or high sediment loads, the advice of ecological experts should be sought relative to the extent of the impact on aquatic life.

Special conditions. If the change in quantity, quality, or type of aquatic vegetation will result in waters being rendered unfit for swimming or will cause a reduction in the gamefish or commercial fish populations, greater controversy over the project is likely to result, since the impact will be more directly felt by the general public. Dredging and filling of wetlands and water areas is regulated by federal law and by many states as well. Permits will be required before actions take place.

Geographical and temporal limitations. The only geographical limitations on aquatic vegetation are the particular types native to certain areas. Temporal considerations do not appear to be significant, except in that explosive growth of algae and floating plants (blooms) is most common during warm weather.

Mitigation of impact. Impacts on aquatic plant life can best be mitigated by minimizing the input of nutrients, erosion products, and heat into water bodies. This can be accomplished by assuring that wastewater treatment facilities of appropriate size are constructed so as to be in operation by the time the increased amount of nutrients is scheduled to be generated. In addition to catchment, catch basins can be constructed to permit the settling out of suspended solids prior to the runoff water reaching natural water bodies. (The attribute discussion covering erosion goes into greater detail on this subject.) Additionally, construction activities can be scheduled at times of the year when intensive rainfall is least likely to occur. Cooling water can be processed or stored in artificial ponds until the difference in temperature between it and the receiving water is more nearly equal.

Swamps and marshes should not be proposed to be drained unless such action is absolutely necessary for the successful completion of the activity. Artificial changes in water level should be minimized and programmed during the fall and winter, when the plants are dormant. If herbicides are used to suppress excessive aquatic plant growth, they should be applied selectively and in amounts that will reduce the undesirable species but not kill all aquatic plants. Their use is controlled by federal and state law.

Secondary effects. Since aquatic plants are essential elements in the food web, adverse impacts to these elements will also be reflected in impacts to higher-order consumers (fish, animals, and humans). Excessive growth of aquatic plants, on the other hand, can choke waterways and recreational areas, with resultant induced reduction of economic, social, and aesthetic benefits. Algal blooms may affect the taste and odor of public water supplies.

Other comments. Water quality and quantity are directly related to the suitability of water bodies for desirable aquatic plant growth. Introduction of pollutants will reduce plant productivity and plant species diversity and result in an aquatic plant community composed predominantly of pollution-tolerant forms. This will, in turn, have a decided impact on the fish populations that inhabit the waters. Changes in the food web can have impacts throughout the ecosystem, but these are often not completely understood. Some contaminants may accumulate in aquatic plants and be further taken up by birds, other animals, and humans.

Sound

The level of sound (noise) is an important indicator of the quality of the environment. Ramifications of various sound levels and types may be reflected in health (mental and physical) and/or in aesthetic appreciation of an area. Because of the important consequences of a too-noisy environment, this resource, sound, is examined separately rather than under various other resource categories.

The sound (noise) in an environment is indicated by many attributes, but some important ones are

Physiological effects

Psychological effects

Communication effects

Performance effects

Social behavior effects

Physiological effects

Definition of the attribute.   Noise can affect the physiology of the human body in three important ways:

Internal bodily systems

Hearing threshold

Sleep pattern

Internal bodily systems are defined as those physiological systems essential for life support, that is, cardiovascular (heart, lungs, vessels), gastrointestinal (stomach, intestines), neural (nerves), musculoskeletal (muscles, bones), and endocrine (glands). Noise stimulation of nerve fibers in the ear may indirectly harmfully affect these systems. High-intensity noise (e.g., artillery fire, jet aircraft takeoff) constricts the blood vessels, increases pulse and respiration rates, increases tension and fatigue, and can cause dizziness and loss of balance. However, these effects are generally temporary and, to some extent, adaptation does occur. The process of adaptation is in itself indicative of an alteration in body functions and is therefore undesirable. High noise levels can also reduce precision of coordinated movements, lengthen reaction time, and increase response time, all of which can result in human error (Miller, 1971).

Hearing threshold is defined as the lowest sound level or loudness of a noise that can be heard. The lower the sound level that can be heard, the lower the hearing threshold. If the sound level necessary for a noise to be heard (or the hearing threshold) is higher than normal for a person, then hearing loss or partial deafness is indicated. Noise can cause temporary or permanent hearing loss (i.e., an increase in the hearing threshold) and can cause ringing in the ears (tinnitus). Hearing loss can be temporary, in that the ear recovers relatively soon after the termination of the noise. Over time, the recovery may be incomplete, and a permanent loss results. Hearing loss of any degree is serious because accidents can occur if warning signals, commands, etc., cannot be heard or understood. In addition, hearing loss is undesirable from social, economic, psychological, and physiological points of view.

Sleep pattern is defined as a natural, regularly recurring condition of rest, and is essential for normal body and mental maintenance and recuperation from illness. Noise can affect the depth, continuity, duration, and recuperative value of sleep. The disruption or lack of sleep results in irritability, often irrational behavior, and the desire for sleep. Even a shift in the depth of sleep can result in fatigue. Also, while suffering or recovering from illness, rest and sleep are essential to health and recovery. Thus, it is important for noise to be kept at a minimum, or at least constant, during night hours.

Activities that affect the attribute.   Most human activities cause some level of noise, but the most serious impacts result from the following.

Construction.   Construction projects create noise through the use of vehicles, construction equipment, and power tools. The noise affects the operators, personnel, and communities near the site, and the people near transportation routes to the site.

Work activities.   The operation of most types of aircraft and surface vehicles, machinery, and power-generating equipment will generate noise. Maintenance and repair produce noise through the use of all types of tools and when a number of noise sources are operating at the same time in the same general area (e.g., a vehicle repair shop).

Military training.   Training courses and exercises which use any type of vehicle, weapon, power tools, appliances, and machinery create noise for operators and military and civilian personnel, and, in large-scale exercises, can affect nearby civilian communities. High-performance aircraft and heavy weapons use are the most intrusive.

Industrial plants.   The machinery and tools contained in these plants are a significant source of noise to the personnel and, if noise levels are sufficiently high, can affect the nearby community.

Source of effects. The sources of noise which affect this attribute include the following.

Military equipment. Missiles and artillery of all types, including small arms, have extremely high noise levels and can severely affect the hearing threshold of their operators in addition to disrupting civilian activities.

Vehicles. Vehicles in the air, on the ground, or on water are important noise sources which affect the operator, other personnel, and the community. Examples include the following:

Aircraft on and around commercial airports and military air bases significantly affect the community, particularly sonic booms or night operation, which can affect sleep patterns.

Large vehicles such as trucks, buses, and armored vehicles can affect the hearing threshold of the operators and passengers.

Most vehicles, when operated at night, can affect sleep patterns.

Construction equipment. These types of equipment, which include vehicles and power tools, have high noise levels which can affect hearing thresholds of operators and site personnel. Pile drivers are notorious offenders.

Machinery. Machinery in industrial plants, where noise levels are high and continuous, can significantly affect operator hearing thresholds.

Variables to be measured. The important variables of noise which affect this attribute are its loudness, duration, and frequency content. As the loudness and/or duration increase, the effects of noise on the body increase. The internal bodily systems are increasingly under stress, the hearing threshold increases to the point where permanent damage (called noise-induced hearing loss) can occur, and sleep becomes increasingly impossible. Noises which contain high frequencies or contain, or are, pure tones are more damaging and disturbing than those which do not.

Finally, the impulsivity of a noise is important. An impulsive noise is highly intense and short in duration (generally less than 1 sec), for example, artillery or small arms fire. Recommended noise measures and their explanation are provided in Table B.5.

How variables are measured. The loudness of noise is measured in terms of decibels (abbreviated as dB). Decibels are measured by using a sound-level meter. Normally, loudness is measured with a sound-level meter incorporating an "A"-weighted electronic network. The resulting measure is called dBA or dB(A). Most of the evaluation criteria are

given in dB(A) units. The intensity of an impulsive noise may be difficult to read visually from a sound-level meter alone, due to the very
short duration of the noise. To determine the intensity of an impulsive
noise, special equipment, such as a true integrating noise monitor, is
needed and C weighting is used. Some of the pioneering work on com-
pulsive noise has been done at the U.S. Army Construction
Engineering Research Laboratories, including developing a true integrating noise monitor for measuring impulsive noise levels and community response to high-amplitude impulse noise (Schomer, 1977,
1985, 1986, 1991).

The frequency content of noise is more difficult to measure, and complex equipment is required. Subjectively, however, high-frequency content and pure tones are recognizable (assuming the observer's hearing
is normal). For example, noises with high frequencies have a whine (jet
aircraft), screech (certain machinery), clank or clink, squeal, squeak,
whistle, whine, or ping, or simply a tone. Noises with these characteristics are more annoying and disturbing to people and, at high loudness
levels, more damaging. In general, subjective evaluations are not
acceptable except to support or verify objective measurements.

Measurement for the physiological attribute, for existing situations,
should be taken at the expected position of the human body with
respect to the noise source(s). The sound-level meter should be placed
where the body or people are or will be located. When the noise source
is active, several readings should be taken and averaged (see discussion in geographical and temporal limitations section).

In those situations where the noise source is in the future and thus
cannot be measured directly, which is often the case for environmen-

TABLE B.5   Noise Measurement for Environmental Assessment

Type of environment	Type of criteria	Recommended measures
General audible noise	Hearing loss potential	A-weighted L_{dn}*
	Health and welfare effects on people	A-weighted L_{dn}
	Environmental degradation/effects on structures and animals	A-weighted L_{dn}
High-amplitude impulsive noise blasts, artillery, helicopters, pile drivers	Structural damage	Peak pressure/peak acceleration
	Annoyance	C-weighted L_{dn}

*L_{dn} = Day/night average sound levels; a measure of the noise environment over a 24-h day
with a 10-dB penalty applied to nighttime (10 p.m. to 7 a.m.) activities (Goff and Novak, 1977).

tal assessment, analytical models need to be used for estimating noise levels (*CHABA Guidelines,* 1977; *Highway Noise,* 1971; Goff and Novak, 1977).

Finally, to measure hearing loss or hearing thresholds of an individual, an audiometer should be used by a trained person certified as an audiometric technician under the supervision of a physician or an audiologist. It is important to measure an individual's hearing before he or she is subjected to noise sources so that a baseline audiogram can be prepared. This audiogram can then be used for future reference and comparisons with later tests.

Evaluation and interpretation of data. The following criteria can be used to determine if the noise source will affect the body in any manner. The intensity and duration of the noise at the body should not exceed the values given in Table B.6.

Special conditions. The most serious noise impacts on this attribute are

Partial hearing loss caused by artillery or small arms fire

Partial hearing loss to equipment operators caused by construction equipment or vehicles

Sleep loss

Noise-induced hearing loss due to artillery, small arms fire, or combat vehicles is not uncommon in the military. Military personnel exposed to these noise sources should have their hearing checked periodically and wear protective equipment at all times.

Noise sources of any type should not be located near schools, hospitals, or homes for the aged. Night operations should be isolated from these places and from any areas where people are sleeping.

Geographical and temporal limitations. Any activities and noise sources should be geographically located so as to minimize their impact on communities and populations. Isolation of the activity can be accomplished by geographical distance and/or placement with natural barriers (vegetation, hills, or mountains).

Noise sources affect people differently during the day. During the day people expect noise levels to be normal, but during the evenings, when outdoor events, family activities, rest, television watching, etc., take place, noise levels are expected to be much less. At night, of course, noise sources are not expected to be active. Similarly, during weekends, noise sources should not be active. The use of the 10-dB night penalty reflects this expectation.

TABLE B.6   Noise Exposure Intensity and Duration Levels

	Intensity	Duration
Internal bodily system	85 dB(A)*	Any
Hearing threshold (continuous sound, if	80 dB(A)†	16 h
sound of intermittent summation is	85 dB(A)	8 h
required; use meters especially designed	90 dB(A)	4 h
for this purpose or contact audio	95 dB(A)	2 h
engineers or sound specialists)	100 dB(A)	1 h
	105 dB(A)	30 min
	110 dB(A)	15 min
	115 dB(A)	7.5 min
	>115 dB(A)	Never
Hearing threshold‡		
(Impulsive sound)	140 dB (at ear)†	100 μsec
Sleep pattern		
(Causes awakening)	55–60 dB(A)	Any
(Causes shift in sleep)	35–45 dB(A)	Any

*These dB(A) levels may be changed to reflect different distances between the noise source and measurements by applying the rule of subtracting or adding 6 dB(A) per doubling or halving of distance. For example, if the estimate is given as 90 dB(A) measured at 50 feet and the actual distance between the noise source and the personnel is 100 feet, the dB(A) can be estimated to be 84 dB(A). Noise sources which are "line" sources, such as trains and heavy streams of traffic, reduce in noise level 3 dB(A) per doubling of distance.
†American Conference of Government Industrial Hygienists (ACGIH), 1973.
‡Operators of artillery and small arms can be expected to receive higher intensity levels.

In terms of measurement, variables should be measured or projected at various geographical distances and directions from the source until criterion values are reached to determine the extent of the noise. In addition, it is important to measure noise from transportation routes and flight patterns through communities and the airfield. Also, the variables should be measured at various times during the day, evening, and night to determine the worst and best noise conditions.

Mitigation of impact.   The optimal method of reducing sound level is, of course, to reduce the noise being produced by the source. Since this method can be difficult or expensive to use on existing noise sources, the techniques of isolation and insulation are often used. If these techniques fail to reduce noise levels sufficiently, then the use of ear protective devices is recommended.

To reduce noise levels at the source requires engineering solutions. These solutions may include damping, absorption, dissipation, and deflection methods. Common techniques involve constructing sound enclosures, applying mufflers, mounting noise sources on isolators, and/or using materials with damping properties. Redesigning the mechanical operation of noise sources may be necessary. Performance specifications for noise represent a way to ensure that the procured item is controlled.

When an individual is exposed to steady noise levels above 85 dB(A), in spite of the efforts made to reduce noise level at the source, hearing conservation measures should be initiated.

The federal government has promulgated three regulations that relate to controlling noise at the source. These noise regulations have been issued by the General Services Administration, the Environmental Protection Agency, and the Department of Labor.

General Services Administration. The General Services Administration issued construction-noise specifications effective July 1, 1972, for earth-moving, materials-handling, stationary, and impact equipment (see Table B.7). They require that all on-site equipment used by a contractor while under contract with the General Services Administration have A-weighted sound level requirements [dB(A)] measured 50 ft from the equipment. For example, a tractor, regardless of type, must not exceed 80 dB(A) while in operation on the site at a distance of 50 ft. Noise violations result in a cancellation of the contract. Construction equipment that exceeds these levels would require some type of engineering noise control, and "quiet" equipment has been made available for many purposes.

Environmental Protection Agency. Under provisions of the Noise Control Act of 1972, the EPA is required to promulgate noise-emission standards for four new product categories:

Construction equipment

Transportation equipment

Motor or engine

Electrical or electronic equipment

In addition, all railroad and motor carriers engaged in interstate commerce will be subject to noise-emission requirements. Furthermore, any product adversely affecting the public health or welfare must be labeled with the specific sound level (see Noise Control Act, Chapter 2).

Department of Labor. Noise exposure criteria have been established by the Department of Labor under provision of the Occupational Safety and Health Act. To meet the provisions of this act, a hearing conservation program must be initiated for protecting noise-exposed personnel; emphasis should be placed on engineering noise control. Hearing protective devices must be issued to the workers, but only as an inter-

TABLE B.7   General Services Administration Construction-Noise Specifications

Equipment	Effective dates	
	July 1, 1972	January 1, 1975
Earthmoving		
Front loader	79	75
Backhoes	85	75
Dozers	80	75
Tractors	80	75
Scrapers	88	80
Graders	85	75
Trucks	91	75
Pavers	89	80
Materials handling		
Concrete mixers	85	75
Concrete pumps	82	75
Cranes	83	75
Derricks	88	75
Stationary		
Pumps	76	75
Generators	78	75
Compressors	81	75
Impact		
Pile drivers	101	95
Jackhammers	88	75
Rock drills	98	80
Pneumatic tools	86	80
Other		
Saws	78	75
Vibrators	76	75

NOTE: Equipment to be employed on the site shall not produce a noise level exceeding the following limits of dB(A) at a distance of 50 ft from the equipment under test in conformity with the Standards and Recommended Practices established by the Society of Automotive Engineers, Inc., including SAE Standard J 952 and SAE Recommended Practice J 184.

im measure while engineering solutions are being planned. In practice, the "interim" has been over 20 years in many industries.

Other mitigation methods include isolation and insulation. The noise source and personnel or structures can be isolated from one another by distance. (The intensity of noise decreases at an approximate rate of 6 dB per doubling of distance.) Another method is to build barriers between the noise source and personnel. Increasing insulation in structures will also reduce inside noise levels.

Ear protective devices can be used to shield and protect individuals from noise. Occupational health programs have emphasized the proper fitting and issuing of hearing protective devices (i.e., earplugs or earmuffs) to noise-exposed personnel as an essential element of a hearing conservation program.

Secondary effects. Exposure to high noise levels appears to have potentially detrimental effects on worker performance and accident rates and absenteeism in industry. In addition, this exposure can cause general stress. Continued noise production can lead to land-use changes, with associated socioeconomic and biophysical ramifications.

Psychological effects

Definition of the attribute. Noise can affect an individual's mental stability and psychological response (annoyance, anxiety, fear, etc.).

Mental stability refers to the individual's ability to function mentally or act in a normal manner. The mental well-being of an individual is essential for personal maintenance and efficiency. It is generally agreed that noise does not cause mental illness but may aggravate existing mental or behavioral problems. Noise predominately causes psychological responses such as anger, irritability, increased nervousness, and, most of all, annoyance. It is the annoyance reaction which can cause individual and community outcry and lawsuits against noise sources such as airports, aircraft, and highway transportation.

Activities that affect the attribute. Many activities can cause annoying and unacceptable noise. The most serious are discussed below.

Construction. Construction projects create noise through the use of vehicles, construction equipment, and power tools. The noise affects operators, personnel, and communities near the site, and those people near transportation routes to the site.

Work activities. The operation of most types of air/surface vehicles, machinery, and power-generating equipment will generate noise. Maintenance and repair produce noise through the use of all types of tools and when a number of noise sources are operating at the same time in the same general areas (e.g., a vehicle repair shop).

Military training. Training courses and exercises which use any type of vehicle, weapon, power tools, appliances, or machinery create noise for operators and military and civilian personnel, and, in large-scale exercises, can affect nearby civilian communities.

Industrial plants. The machinery and tools contained in these plants are a significant source of noise to the personnel, and, if noise levels are sufficiently high, can affect nearby communities.

Source of effects. The sources of noise which affect this attribute include the following.

Military equipment.   Missiles and artillery, including small arms, have extremely high noise levels which can disturb and annoy personnel and nearby communities.

Vehicles.   Vehicles in the air, on the ground, or on water are significant sources of noise which annoy and disturb operators, personnel, and nearby communities. In particular, aircraft around airports and military bases can disturb and annoy both base personnel and communities. Some individuals living directly beneath flight paths experience anxiety and fear from the aircraft noise. Additionally, they may find they must stop their work and mental processes due to passing aircraft, which, in turn, produces annoyance reactions.

Construction equipment.   These types of equipment, which include vehicles and power tools, have high noise levels which annoy operators, workers, and nearby community citizens.

Variables to be measured.   The important variables of noise which affect this attribute are loudness, duration, and frequency content. As loudness and duration increase, psychological stress, annoyance, anger, and irritability also increase. In terms of frequency content, people are generally more annoyed by high frequencies and pure tones. The frequency content of a noise source also gives the sound an identity. Certain noises are annoying, disturbing, or fear-producing (to some people) because of their identity (e.g., sirens, jackhammers, horns, motorcycles, aircraft, buzzers, trucks, backfires, gunshots, and air compressors).

Noises which have very high noise levels but very short durations (called impulsive noises), such as gunshots, vehicle backfires, and sonic booms, startle people. These individuals not only are annoyed, but express feelings of fear and anxiety, and their activities (particularly sleep) are severely interrupted.

How variables are measured.   The measurement of loudness, duration, frequency content, and impulsivity is discussed under physiological effects.

Evaluation and interpretation of data.   It is difficult to establish a single set of criteria, due to the variety of acoustical and social factors. In addition to the intensity or loudness and duration of noise, other acoustical considerations involve pattern, occurrence, and the noise source itself. Social variables, such as demographic characteristics, personality type, and predisposition to nervousness, must be considered.

While the spectral content and temporal patterns of noise pressure levels are important, as general criteria, ambient noise levels exceeding 55 dB(A) during the day or 45 dB(A) during the night will disturb and annoy some people. Figure B.6 describes the community reaction to intrusive noise as a function of normalized day/night sound equivalent level. Table B.8 provides a summary of human response to selected (55, 65, and 75 dB) day/night sound equivalent levels.

Special conditions. While environmental noise alone probably does not produce mental illness, the continual bombardment of noise on an already depressed or ill person cannot be helpful. Certainly it interferes with sleep, producing irritability and other tensions. Comparative studies of persons living adjacent to London's Heathrow Airport with others living in a quieter environment revealed that among those living in the noise environment there was a significantly higher rate of admission to mental hospitals (EPA, 1972).

Another medical discovery was the effect of noise on unborn babies. Previously, unborn babies were thought to be insulated from the noise stress of the outside world, but now physicians believe that external noises can trigger changes in fetuses (EPA, 1972).

Study of steelworkers indicated that those working in a noisy environment are more aggressive, distrustful, and irritable than workers in

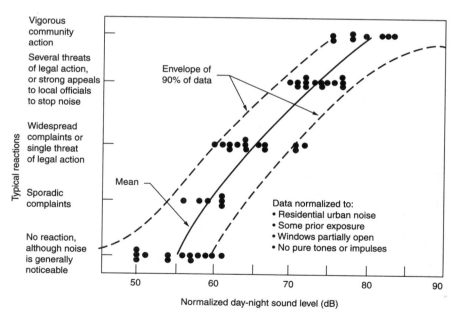

Figure B.6 Community reaction to many types of intrusive noise as a function of normalized daylight sound equivalent level. (From Community Noise, NTID 300.3 U.S. EPA, 1971).

TABLE B.8 Summary of Human Response to Selected Sound Levels

Type of effect	Magnitude of effect		
	L_{dn} = 55 dB	L_{dn} = 65 dB	L_{dn} = 75 dB
Speech—indoors	100 percent sentence intelligibility (average) with a 5-dB margin of safety	99 percent sentence intelligibility (average) with a 4-dB margin of safety	Sentence intelligibility (average) less than 99 percent
—outdoors	100 percent sentence intelligibility (average) at 0.35 m 99 percent sentence intelligibility (average) at 1.0 m 95 percent sentence intelligibility (average) at 3.5 m	100 percent sentence intelligibility (average) at 0.1 m 99 percent sentence intelligibility (average) at 0.35 m 95 percent sentence intelligibility (average) at 1.2 m	100 percent sentence intelligibility not possible at any distance 99 percent sentence intelligibility (average) at 0.1 m 95 percent sentence intelligibility (average) at 0.35 m
Average community reaction	None; 7 dB below level of significant "complaints and threats of legal action" and at least 16 dB below "vigorous action" (attitudes and other nonacoustical factors may modify this effect)	Significant; 3 dB above level of significant "complaints and threats of legal action" but at least 7 dB below "vigorous action" (attitudes and other nonacoustical factors may modify this effect)	Very severe; 13 dB above level of significant "complaints and threats of legal action" and at least 3 dB above "vigorous action" (attitudes and other nonacoustical factors may modify this effect)
High annoyance	5 percent, depending on attitude and other nonacoustical factors	15 percent, depending on attitude and other nonacoustical factors	37 percent, depending on attitude and other nonacoustical factors
Attitudes toward community	Noise essentially the least important of various factors	Noise is one of the most important adverse aspects of the community	Noise is likely to be the most important of all adverse aspects of the community

SOURCE: From *Guidelines for Preparing Environmental Impact Statements on Noise*. Report of CHABA Working Group Number 69 (Committees on Hearing and Bioacoustics, February 1977).

a quieter environment (EPA, 1972). These studies show that it is very important to keep noise levels as low as possible in communities near hospitals, mental institutions, homes for the aged, and any place where people may be particularly annoyed or placed under stress by noise.

Geographical and temporal limitations. See discussion under physiological effects.

Mitigation of impact. Mitigation procedures relevant to the attribute are discussed under physiological effects.

Secondary effects. Continued noise production can lead to land-use changes, with the associated socioeconomic and biophysical ramifications.

Communication effects

Definition of the attribute. Noise can affect face-to-face and telephonic communication, and, during extremely high levels of intensity, visual impairment has been reported.

Aural face-to-face communication, or the ability to give and receive information, signals, messages, or commands, without instrumentation, is an essential activity. The temporary interference with or interruption of communication during phases of human activity can be annoying, and occasionally hazardous, to personal well-being. Interference occurs when background or ambient noise levels of the environment are of sufficient intensity to mask speech, making it inaudible or unintelligible. Noise that interferes with communication can be dangerous, particularly when a message intended to alert a person to danger is masked or when a command is not heard or understood. More commonly, however, noise is annoying because it disrupts the communication process.

Telephonic communication, or the ability to give and receive information through telephones, headsets, receivers, etc., is also an important activity. Noise affects this type of communication in the same way as face-to-face (i.e., it causes annoyance and disruption). However, due to the insulation effect of the telephone or headsets and control over the volume of the incoming or outgoing signals, higher levels of loudness or intensity can be tolerated.

Activities that affect the attribute. Many activities generate noise sufficient to interfere with aural communication.

Construction. Construction projects create noise through the use of vehicles, construction equipment, and power tools. Noise levels are high enough to affect all types of communication, particularly for the operator and personnel in the general construction area.

Work activities.   The operation of most types of vehicles, machinery, and power-generating equipment will create noise at levels which will interfere with communication of operators, personnel in the area, and communities. Communication in and near operating maintenance and repair shops will also be affected by the noise generated by tools and vehicles.

Military training.   Training exercises that use air, land, and water vehicles; weapons; and machinery create noise levels sufficient to interfere with communication between military personnel.

Industrial plant activities.   Machinery and power tools contained in industrial plants are a significant source of noise affecting communication within the plant.

Source of effects.   The sources of noise which affect this attribute include the following.

Vehicles.   Vehicles in the air, on the ground, or on water are important noise sources that affect the communication of operators and the community. Examples include the following:

Aircraft on and around airports and military air bases significantly affect communication, particularly in airport operations and in community areas directly beneath flight paths.

Large vehicles generate very high noise levels and can affect communication between the operators and other personnel in operating areas.

Other vehicles, particularly when operating in groups, affect communication near highways and other routes.

Construction equipment.   This equipment also has high noise levels and affects the intelligibility of communication at the construction site. Transportation or routes to the site may also generate noise levels that interfere with communication near the routes.

Military equipment.   Weapons of all types (including small arms) have extremely high noise levels and can interrupt face-to-face communication. During large-scale activities, even telephonic communication can become difficult. Weapons achieve noise levels which may be sufficient to momentarily distort vision.

Machinery.   Machinery located in industrial plants where many machines are operating continuously can severely affect the attitudes and irritability of workers within the plant.

Variables to be measured. The important variables of noise which affect face-to-face communication are loudness of the ambient noise level and the distance between the speaker and the listener. As the loudness increases, masking of the speech increases and speech intelligibility and discriminability decrease. Also, as the distance between speaker and listener increases, the speech becomes more difficult to hear and to understand, and annoyance and frustration rise.

In telephonic communication, the noise variable of concern is the loudness of the background noise level.

As these variables increase, the speaker raises his or her voice to overcome the masking. Of course, the voice reaches a point where it strains and cannot overcome masking, and communication becomes impossible. In addition, the strain of shouting—and of trying to hear—is both fatiguing and frustrating in any situation, and may lead to inefficiency.

How variables are measured. The variable loudness can be measured or projected in dB(A) units, as specified in the physiological effects section. The distance between the speaker and the receiver should be measured in feet.

Evaluation and interpretation of data. The impact of noise on face-to-face communication can be evaluated by using the chart in Fig. B.7. Enter the side of the chart at the expected dB(A) noise level and the bottom at the expected average distance between speaker and listener. If the intersection of the two values falls above the Area of Nearly Normal Speech Communication, then speech communication is being adversely affected.

In one-to-one personal conversation, the distance from speaker to listener is usually about 5 ft; nearly normal speech communication can proceed in noise levels as high as 66 dB(A). Many conversations involve groups; for this situation, distances of 5 to 12 ft are common and the intensity level of the background noise should be less than 60 dB(A). At public meetings and outdoor training sessions, distances between speaker and listener are often about 12 to 30 feet, and the sound level of the background noise should be kept below 55 dB(A) if nearly normal speech communication is to be possible (Miller, 1971).

In telephonic communication, background intensity levels above 65 dB become increasingly intrusive (see Fig. B.8).

Special conditions. There are special areas where communication which should not be disturbed takes place. These areas include training and testing areas, schools, churches, libraries, theaters, offices, hospitals, and research laboratories. Noise sources including air and land

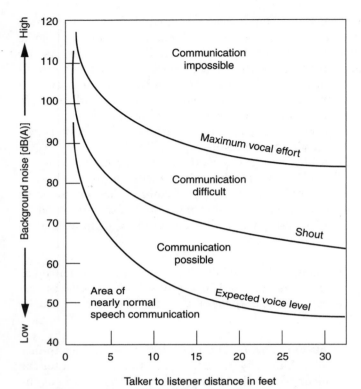

Figure B.7 A simplified chart that shows the quality of speech communication in relation to sound levels of noise [dB(A)] and the distance between the speaker and the listener (Miller, 1971).

transportation should be isolated from these communication-sensitive areas or the areas should be well insulated against external noise.

Geographical and temporal limitations.   See discussion under physiological effects.

Mitigation of impacts.   To ensure intelligible communication, the noise sources and the personnel need to be isolated or insulated from one another. The special areas (see special conditions), where communication is especially sensitive, should be well isolated and insulated against external noise. When it is unavoidable to have personnel who must communicate near high noise levels, special communication devices should be used (e.g., headsets).

Secondary effects.   As communication becomes more and more difficult to accomplish, impacts on psychological and performance effects

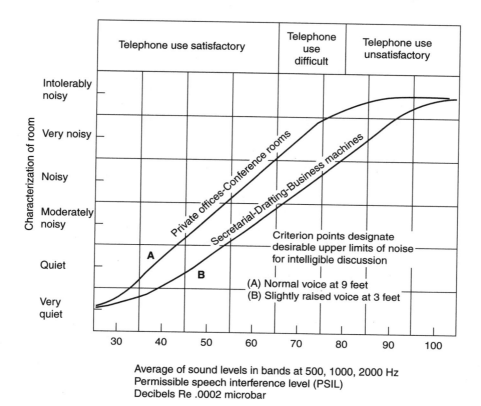

Average of sound levels in bands at 500, 1000, 2000 Hz
Permissible speech interference level (PSIL)
Decibels Re .0002 microbar

Figure B.8 Rating chart for office noise. The data were determined by an octave-band analysis and correlated with subjective tests (Peterson, 1980).

may occur. Continued difficulties may lead to land-use changes, precipitating a series of socioeconomic and biophysical ramifications.

Performance effects

Definition of the attribute. Noise can affect the ability of humans to perform mechanical and mental tasks. Noise can adversely affect performance through

The increase in muscular tension that can interfere with movement

The lapse in attention or a diversion of attention from the task at hand

The masking of needed auditory signals

The startle response to high-intensity noises

Mechanical tasks can range from simple mechanical assembly to more complex tasks. Lower-order tasks, such as mechanical assembly or manual routine-type activity, are least influenced by noise. However, tasks of this nature are altered in three essential ways by high-intensity noise. Although work output remains fairly constant, worker errors can increase (Miller, 1971), judgment of time intervals can become distorted, and a greater effort is necessary to remain alert (Kryter, 1985). Noise is most likely to affect the performance of tasks which are quite demanding and/or require constant alertness (U.S. Department of Commerce, 1970).

Mental tasks, such as problem solving and creative thinking, are more affected by noise. Higher-order tasks requiring greater mental facilities (although dependent on the individual) are generally disrupted by lower noise intensities than mechanical tasks. It is important, therefore, to keep noise at a minimum in and near office areas (Fig. B.8).

When a task (mental or mechanical) requires the use of auditory signals, speech or nonspeech, noise at any intensity level sufficient to mask or interfere with the perception of these signals will interfere with the performance of the task (Miller, 1971).

Activities that affect the attribute.   The most important activities that can reach noise levels sufficient to affect performance are discussed below.

Construction.   Construction projects create noise through the use of vehicles, construction equipment, and power tools. The noise levels are certainly high enough to affect mental tasks. The mechanical tasks at the site are generally considered highly physical and probably would be unaffected by the noise levels. Construction adjacent to occupied buildings is a common source of problems.

Work activities.   The operation of all types of vehicles and machinery will create noise levels sufficient to affect human performance, primarily through distraction.

Military training.   Training exercises which use vehicles, weapons, and machinery create noise levels high enough to distract people.

Industrial plant activity.   Machinery and power tools contained in industrial plants are a significant source of noise which affects mental and mechanical performance.

Source of effects.   The sources of noise that can affect performance include the following.

Weapons.   Weapons of all types generate noise levels which can interfere with and interrupt mental and complex, precise mechanical tasks.

Vehicles. Vehicles of all types are significant noise sources which can interfere with and interrupt task performance. In particular, aircraft can disrupt the mental tasks of large segments of communities.

Machinery. Machinery and power tools in industrial areas create high noise levels that could affect some complex and precise mechanical tasks.

Construction equipment. Noise from this equipment can affect the mental tasks of personnel in the area.

Variables to be measured. The important variable of noise which affects task performance is loudness. As the loudness of noise increases, the effects of the noise on performance increase. First, mental tasks are affected; then, as the loudness further increases, complex and precise mechanical tasks become affected.

How variables are measured. Loudness of noise is measured in terms of decibels. A detailed discussion of how to measure or project decibel levels can be found under physiological effects.

Evaluation and interpretation of data. In addition to the information about the impact of noise discussed earlier, Tables B.9 and B.10 provide criteria for the relative compatibility of mental tasks and various land uses.

Special conditions. Special areas where mental tasks take place should not be disturbed. These areas include offices, conference areas, schools, indoor training areas, libraries, and research laboratories. In terms of mechanical tasks, it is difficult to be specific. Wherever complex, precise, and demanding mechanical tasks are performed, the environment should be protected from high-intensity noise sources.

Geographical and temporal limitations. See discussion under physiological effects.

Mitigation of impact. The optimal method of reducing sound levels is through noise source reduction, isolation, or insulation. Methods of achieving noise reduction are discussed under physiological effects.

Secondary effects. Continued exposure to noise may lead to land-use changes, precipitating a series of socioeconomic and biophysical impacts.

Social behavior effects

Definition of the attribute. Social behavior refers to the individual's ability to mentally function in a normal manner on an interpersonal

TABLE B.9   Recommended Outdoor Criteria for Various Land Uses

	CHABA		JSPM	
	L_{dn}	L_{eq}	L_{dn}	L_{eq}
Residential	55		64	
Hospitals	55		64	
Hotel, motel	60		64	
School/outdoor teaching areas		55		64
Church		60		64
Office buildings		70		69
Theater		70		69
Playground, active sports		70		74
Parks		60		69
Special purpose outdoor areas		*		*

*Outdoor amphitheaters or other critical land uses requiring special consideration should not allow new intruding noise to exceed a level 5 dB below the present L_{eq}.

†Guidelines for Preparing Environmental Impact Statements on Noise. Draft Report of CHABA Working Group Number 69 (Committees on Hearing and Bioacoustics, February 1977).

‡Planning in the Noise Environment, Joint Services Planning Manual (1977).

L_{dn}—Day/night average sound levels; a measure of the noise environment over a 24-hour day with a 10-dB penalty applied to nighttime (10 p.m. to 7 a.m.) activities.

L_{eq}—Equivalent sound levels; level of a constant sound with the same sound energy as a time-varying sound level.

TABLE B.10   Noise Criteria for Mental Tasks

Type of room	Maximum permissible level (measured when room is not in use)
Small private office	45 dB(A)*
Conference room	35–40 dB(A)
Secretarial offices (typing)	60 dB(A)
School rooms	30–35 dB(A)
Reading	40–45 dB(A)
Meditation	40 dB(A)
Studying	40–45 dB(A)
Individual creative activity	40–45 dB(A)

*These figures are general guidelines. For specification purposes, Noise Criterion (NC) curves which cover the different frequency bands should be used (Beranek, 1960).

basis. Under certain conditions within communities, interpersonal relationships are altered when noise is of sufficient intensity. Areas of socialization may become restricted due to noise exposure. Outdoor areas are first to be affected, thus limiting socialization to residential interiors. Patterns of entertainment become confined and restricted.

When one or more methods of basic auditory communication (face-to-face or telephonic) are masked, the channels for social interaction become limited. These results, in turn, affect personal attitudes and create annoyance.

Activities that affect the attribute. Many activities generate noise levels which could interfere with social behavior.

Construction. Construction projects generate sufficient noise to interfere with the social behavior of personnel and communities located near the site. In particular, new transportation routes to the site will introduce new noise levels to people living nearby, which, in turn, can adversely affect social behavior.

Work activities. The operation of aircraft and surface vehicles and other equipment increases the noise level of the outside environment, where much socialization takes place. Social behavior inside structures can also be affected by these activities, if their noise levels are extreme.

Military training. Training exercises which use vehicles, weapons, and machinery create noise levels sufficient to interfere with socialization between personnel. Noise from large-scale training exercises can also affect the community.

Industrial plant. Machinery and activities in the plant are a significant source of noise which affects social behavior between personnel and in nearby communities.

Source of effects. Most noise sources are capable of influencing social behavior in community environments, particularly outdoors. In and around military air bases and airports, weapons, construction equipment, aircraft, helicopters, and ground transportation vehicles generate noise levels sufficient to interfere with interpersonal communication, thus affecting and limiting social behavior on the base and in surrounding communities. Social behavior inside structures is probably affected mostly by machines, heating and air-conditioning units, operating appliances, research equipment, or external noise sources with very high levels, such as aircraft, trucks, and construction equipment.

Variables to be measured. The important variables affecting this attribute are the same as those discussed under communication effect. As communication becomes difficult or impossible, social behavior and interpersonal relationships become limited, especially in the outdoor environment.

How variables are measured. See discussion under communication effects and physiological effects.

Evaluation and interpretation of data. Evaluation techniques, as outlined under communication effects and psychological effects, should be applied to this attribute for both outdoor and indoor environments.

Special conditions. Social behavior is important to people. Being able to socialize with friends, neighbors, and members of the family is an essential human activity. Constant interruption of these activities can only result in frustration and annoyance. Consequently, it is important to consider the impact of continuous or highly repetitive noise sources such as aircraft, weapons, and vehicles on social behavior in surrounding communities.

Geographical and temporal limitation. Socialization is normally expected to take place in the community and in and around living and entertainment areas. These areas then should be measured (inside and out) to determine if any noise sources are affecting social behaviors. One of the most prevalent noise sources is from transportation, both from routes through communities and from flight paths. These, too, must be measured.

Socialization occurs most often in the evening, in the early night hours, and on weekends. Measurement of noise during these periods should be emphasized for this attribute.

Mitigation of impact. Social behavior is primarily affected by noise sources which create high noise levels in the outdoor environment. The mitigation techniques of source control, isolation of the sources from the community, or creation of barriers would be useful.

Secondary effects. Continued exposure to noise may lead to land-use changes, precipitating a series of socioeconomic and biophysical impacts.

References

American National Standard. *Methods for Measurement of Sound Pressure Levels*. American National Standards Institute, New York, ANSI S1.13-1971.

Beland, R. D., and T. P. Mann. *Aircraft Noise Impact Planning Guidelines for Local Agencies*. U.S. Department of Housing and Urban Development, November 1972, NTIS-PB-213020.

Beranek, L. L. *Noise Reduction*. New York: McGraw-Hill, 1960.

Fay, Thomas H., ed. *Noise & Health*. New York: New York Academy of Medicine, 1991.

Foreman, John E. K. *Sound Analysis and Noise Control*. New York: Van Nostrand Reinhold, 1990.

Goff, R. J., and E. W. Novak. *Environmental Noise Impact Analysis for Army Military Activities: User Manual*. USA-CERL Technical Report N-30, 1977.

Harris, David A., ed. *Noise Control Manual: Guidelines for Problem-Solving in the Industrial/Commercial Acoustical Environment*. New York: Van Nostrand Reinhold, 1991.

Jain, R. K., et al. *Environmental Impact Study for Army Military Programs.* Technical Report No. IRD 13, U.S. Army Construction Engineering Research Laboratory, Champaign, Ill., December 1973.

Kryter, K. D. *The Effects of Noise on Man.* New York: Academic Press, 1985.

Miller, J. D. *Effects of Noise on People.* U.S. Environmental Protection Agency, Washington, D.C., December 31, 1971, PB-206723.

Noise and Conservation of Hearing. U.S. Army, Technical Bulletin Med. 251, March 1972.

Noise Effects Handbook: A Desk Reference to Health and Welfare Effects of Noise. U.S. Environmental Protection Agency, EPA 550/9-82/106, 1981.

Peterson, A. P. G., and E. E. Gross. *Handbook of Noise Measurement.* West Concord, Mass.: General Radio Company, 1980.

Roberts, John, and Diane Fairhall, eds. *Noise Control in the Building Environment.* Brookfield, Vt.: Gower Technical, 1988.

Schomer, P. D. "Assessment of Community Response to Impulsive Noise," *Journal of the Acoustical Society of America,* Vol. 77, No. 2, 1985, pp. 520–535.

Schomer, P. D. "Evaluation of C-weighted Ldn for Assessment of Impulse Noise," *Journal of the Acoustical Society of America,* Vol. 62, No. 2, 1977, pp. 386–399.

Schomer, P. D. *Handbook of Acoustical Measurements and Noise Control,* Chap. 50, "Community Noise Measurements," 2d ed. New York: John Wiley and Sons, 1991.

Schomer, P. D. "High-Energy Impulsive Noise Assessment," *Journal of the Acoustical Society of America,* Vol. 79, No. 1, 1986, pp. 182–186.

Suter, Alice H., and John R. Franks. *A Practical Guide to Effective Hearing Conservation Programs in the Workplace.* U.S. Dept. of Health and Human Services, Public Health Service, Centers for Disease Control, National Institute for Occupational Safety and Health, 1990.

U.S. Army Military Standard. *Noise Limits for Army Material.* MIL-STD-1474 (MI), 1 March 1973.

U.S. Department of Commerce, Panel on Noise Abatement. *The Noise Around Us.* September 1970, COM 71-00147.

U.S. Environmental Protection Agency. *Noise Pollution.* August 1972, U.S. Government Printing Office, Washington, D.C., 5500-0072.

U.S. Environmental Protection Agency. *Public Health and Welfare Criteria for Noise.* July 27, 1973.

U.S. Environmental Protection Agency. *Report to the President and Congress on Noise.* February 1972, U.S. Government Printing Office, Washington, D.C., 5500-0040.

Human Aspects

A critical aspect of the human environment is characterized by the way in which we interact with other people and the natural environment. Owing to the complexity of our activities and interrelationships, it is difficult to identify general parameters that describe the condition of human resources. The attributes that have been identified for this purpose are obviously not completely descriptive of all human activities and may appear to miss many important issues. Nevertheless, these few attributes have been chosen because, for most projects envisioned, they will be able, if applied, to capture the major human (or community) elements of environmental impact.

Because of their generality, the attributes are difficult to measure and define, and as a general rule, an adequate assessment of impact on human resources will probably have to be undertaken by persons with special expertise in this area (sociologists or psychologists). One possible division of the attributes that should be examined is

Lifestyles

Psychological needs

Physiological needs

Community needs

Lifestyles

Definitions of the attribute. This attribute refers to the many social activities of humans. Such activities often take on structural characteristics which eventually cause them to be organizations. The make-up of organizations may vary, depending upon the characteristics, interests, and objectives of the organization population. Some common bases for these organizations are racial, ethnic, political, religious, and occupational. Another perspective of this attribute occurs in the form of informal interaction between friends, relatives, and coworkers.

Activities that affect the attribute. The major classifications of activities affecting this attribute include those which affect employment and job security, standards of living, community development, and recreational opportunity. Examples of these include population migration, transportation projects, large construction projects, water resources projects, and industrially related activities. A number of minor activities falling within these major categories could affect this attribute.

Sources of effects. Changes or impacts that occur in this attribute will be dependent upon changes that occur in the population. For example, in a community where various activities have been established, the out-migration of a large portion of the population could disrupt a number of both formal and informal activities. Examples of some of these activities are community athletic teams, schools, and church groups. Individual, informal interactions are also to be considered in this attribute.

The following example is given to illustrate how a significant change in population can cause changes in this attribute. If the population were predominantly elderly people, the type of activities they would be involved in might include hobby clubs, craft clubs, and card-playing clubs. If a large portion of this population were to migrate out of the area, the stability of some of these activities might be affected. Likewise, if many more people of the same age group, with the same interests, moved into the community, the stability of these groups might be strengthened. Also, if the population mix were changed significantly, perhaps by an influx of many more younger people, the stability of such groups might be threatened, or, more importantly, present members of such groups might *feel* threatened.

Variables to be measured. The variables to be measured for this attribute cannot be precisely identified. The purpose in considering this attribute is to identify those instances in which a noticeable change will occur that will affect many people. The objective is to identify general changes in social activities and practices which will be caused by the proposed action.

How variables are measured. The variables in this attribute cannot be precisely measured. One approach that can be taken to "measure" changes in this attribute would involve making a survey of the area to determine the number and kinds of social organizations and activities that exist before the proposed action takes place. Then, having determined the changes expected to occur in the population and the characteristics of that population, impacts on organizations and activities can be predicted in terms of how they may grow, desist, or experience noticeable alterations.

Some persons who might be good sources for predicting and interpreting impacts in this attribute are leaders and participants in the organizations, coaches of local athletic teams, community social and recreation leaders, and local political leaders.

Evaluation and interpretation of data. Interpretation of the impacts or changes in this attribute must be performed by the impact assessment, which should then be analyzed in conjunction with the opinions and assessment of the people mentioned above.

Geographical and temporal limitations. Usually, a geographical area larger than that of the immediate community must be included in the analysis, because impacts often occur outside of the area following changes within the community. The residence and work locations of those who take part in the activities that are considered a part of this attribute must be included in the geographic area.

The analysis should include a summarization of this attribute before the proposed action takes place and the anticipated changes that will result from the proposed activity. In addition to these considerations, to get a more realistic view of the total impacts, consideration must be given to what the condition of this attribute would be if the present situation, or the previous situation, were to continue, as well as to what changes would occur normally, without the proposed action.

Mitigation of impact. Although impacts to this attribute cannot be completely mitigated (with the exception of postponing the proposed action indefinitely), the effect of anticipated impacts could be lessened simply by forewarning participants that such changes are expected to occur.

This will enable the organizations and participants in informal activities to prepare themselves to adjust to expected impacts.

Secondary effects.   Socioeconomic changes frequently may result in secondary or indirect impacts on the biophysical environment. These impacts need to be identified and assessed.

In the case of social changes, there may be environmental effects on air, water, land, etc., from increasing or decreasing the population in an area. Additional people cause increased demand for water, sewage treatment, and power; they require new housing, which takes land, shopping centers, and schools; and they require transportation, which increases traffic congestion and degrades air quality.

Psychological needs

Definition of the attribute.   This attribute refers to the needs of human beings that can be distinguished from the physiological needs and relate primarily to emotional stability and security. Although this attribute could relate to such factors as instincts, learning processes, motivation, and behavior, these factors are not included in this attribute because of the difficulty in relating changes in outside factors to changes in these factors. Emotional stability and security are, therefore, the only two psychological needs that are considered in this attribute.

Activities that affect the attribute.   The major classifications of activities that are likely to affect this attribute are essentially the same as those affecting lifestyles.

Source of effects.   Changes in the degree of emotional stability and feelings of security within the individuals affected could occur from a number of activities. For example, in major construction or industrial project activities, it may be necessary for some people to be moved from their homes or businesses. Even though it is difficult to anticipate the effect of such relocation, experience has shown that such activities almost always have negative effects on the people involved. These effects will vary in their degree of permanency.

Also, when the proposed activity would involve increasing or decreasing the number of jobs or other opportunities (e.g., recreational) in an area, it can be assumed that such activities will either increase or decrease feelings of security, particularly for those who are directly affected by the change in job availability.

Feelings of concern for physical security may be affected by fear for personal safety from crime elements or from natural or human-induced disasters (e.g., from nuclear power plants or industrial facilities).

Variables to be measured. Although no specific variables are identified for this attribute, a general feeling of the degree to which the psychological needs of individuals and communities are being met can be obtained.

Evaluation and interpretation of data; how variables are measured. Data concerning impacts of this attribute must be obtained from several sources. One source would be detailed plans of the proposed activities and identification of groups who might be affected in such ways as in the example above. This information could then be given to psychologists, who could best anticipate and interpret the changes that will occur as a result of the proposed activity. Impacts in this attribute cannot be measured, but can be identified as to whether the impacts are potentially beneficial or disruptive.

Other information may be obtained from personal surveys, or by consulting local counselors, clergy, and law enforcement officials. A good public involvement plan will assist in data acquisition (see Chapter 11).

Geographical and temporal limitations. The geographical area for this attribute must be that which contains people who believe they would be affected by it. This, therefore, could include areas both within and outside of the immediate community.

The time limits for this attribute would be the same as for the other attributes in this section. The "before" time period should be that time shortly before instigation of the proposed activity. The "after" time period should include that time immediately after the proposed activity has been completed.

Mitigation of impact. Some adverse impacts might be averted by including, in the proposed activity funds, an action plan that would permit assistance for those people who would be affected. For example, when a number of jobs are to be eliminated, a service could be set up in which those people who would be without jobs could obtain assistance in locating jobs in other areas. In problems caused by relocation, some program of assistance could be instituted in which people could be aided in finding housing and business locations similar to those they now have.

Fears for personal safety may be alleviated through planned safety programs coordinated with public interest groups.

Secondary effects. See description under lifestyles.

Other comments. Even though impacts which may occur in this attribute are difficult to identify, measure, and evaluate, the attribute is included in the impact assessment process because it is very important. Therefore, it is necessary to attempt to identify situations where such impacts might occur, even if only the possibility of potential impacts can be identified, with very little interpretation or evaluation. This attribute is useful, at least in trying to anticipate where impacts may occur and in identifying situations for which mitigation procedures may have to be planned and included in the proposed activity.

Physiological systems

Definition of the attribute. This attribute refers to anything that is a part of a person's body or that plays a part in a bodily function and is, therefore, related to physical health and well-being. It includes both individual parts (organs) and systems, such as the transport, respiratory, circulatory, digestive, skeletal, and excretory systems. All parts of the human body that contribute to its effective, efficient functioning are included in this attribute.

Activities that affect the attribute. Major classifications of activities that can affect this attribute include construction; operational activities; military training and mission change; industrial; and research, development, testing, and evaluation. Any activity that can harm or threaten the efficient functioning of any part of the human body must be considered in light of its effect on this attribute.

Source of effects. The possible sources of impacts in this attribute are many. They range from activities performed in a laboratory to construction activities that might impair the safety of individuals working in the area. This attribute considers any hazards that may impair the health or safety of any individual.

Variables to be measured. There is not a list of variables that can be measured for this attribute. The purpose of this attribute is to identify potential sources of harm to people. Therefore, detailed elements and implications of the proposed activity must be examined to determine if any of those activities may be potentially harmful. A public involvement program (Chapter 11) will be valuable in identifying these factors.

How variables are measured; evaluation and interpretation of data. It would be helpful to rely upon the knowledge and skill of people who are familiar with the kinds of harm considered in this attribute that can occur. It is suggested that physicians be contacted and given a description of the proposed activity. The seriousness of the potential impacts can then be determined through professional opinion.

Special conditions. It must be determined how many persons will be affected by the expected impacts. Although the impacts are not considered slight even if they affect only a few, it may safely be said that seriousness will increase as the number of affected people increases.

Mitigation of impact. Anticipated impacts in this attribute can be mitigated by taking whatever precautionary measures are necessary to avoid the impact. This may take the form of including in the proposed activity specific safety practices and protective devices.

Secondary effects. Effects on physiological systems can also affect psychological needs, and may have additional economic ramifications if a significant number of workers or production is affected.

Community needs

Definition of the attribute. This attribute refers to some of the many services that a community requires. It included such things as housing; water supply; sewage disposal facilities; utilities such as gas, electricity, and telephone; recreational facilities; and police and fire protection. The nature of change or impact that occurs in this attribute as a result of the proposed activity will be very much dependent upon the type of change that is expected to occur in the population as a result of this proposed activity.

Activities that affect the attribute. Major classifications of activities that are likely to affect this attribute are essentially the same as those affecting lifestyles.

Source of effects. As changes in population and characteristics of the population occur, the needs or services required for that population will change, too. For example, in the general activity category of construction, a temporary force of construction workers may be required to perform the activity. If the construction workers and their families must settle in an area until the construction is completed, these workers and their families will require particular services, such as those mentioned in this attribute. Likewise, when they leave the community, the demand for these services will have been lessened, or perhaps even dissolved, thus leaving the community with a supply of services that is no longer needed, but for which public debt has been incurred.

Also, in industrial development activities, a number of people may be brought into an area on a permanent basis, and the community may find itself unprepared to provide the services and needs to this permanent addition to the population. Also, impacts can occur as a result of a change in military mission or a change in the number of training activities taking place on a particular military base. These impacts may result from fewer numbers of people requiring the services that have already been designed to serve a greater number of people. For example, a community may find itself with an oversupply of houses or have to decrease the number of personnel required for such activities as police and fire protection.

In these and other activities, there are particular subactivities that relate directly to the provision of some of these services. Therefore, any proposed activity that has to do with the provision of such services should be investigated as to the impact that will occur.

Variables to be measured.   For the impact assessment procedure, variables that should be measured are those which will indicate services in the community that are available as well as what services are needed. The community should be surveyed in order to determine (1) the change in population and the characteristics of that population, (2) the number of houses and apartments available to meet the needs of the population if there will be an increase, (3) the number of homes supplied with water and sewage disposal and other facilities, (4) the number of personnel on the police force and the fire department, and (5) the number of acres of land devoted to recreational activities and the number of recreational activities available in the area.

How variables are measured.   Communities should be surveyed to determine what services are now available. For example, a survey should be made to determine the number of dwelling units (houses, apartments, and trailers, for example) that are available and the number of those served with adequate water, sewage, and utility service. The availability of recreational facilities can be determined by noting the number of acres devoted to recreational usage and the number of recreational activities available. The number of police and fire protection personnel should be determined to indicate the level of service now available to the population.

Various sources can be utilized for obtaining this information. Planning agencies often have information on all of these services. Police and fire department personnel are sources which can give an indication of the adequacy of these kinds of services.

After this information is obtained, it will be necessary to relate the present conditions to the change in population that is anticipated from the proposed activity. If the population will increase, it must be determined if there are enough facilities and services available to serve the incoming population. On the other hand, if there will be a decrease in population or an outmigration, the services provided by the community must be considered in light of the oncoming decrease in demand.

Perhaps other uses can be made of those services no longer in demand in their usual functions.

Evaluation and interpretation of data. There are no standardized means of interpreting the above-mentioned variables. For the purpose of an impact assessment, when anticipated changes in population of an area will cause serious problems in the services needed by the population, the situation must be further studied for the impact statement. Expert judgment may be useful in determining when a serious problem will exist, given an immigration or outmigration of a significant number of people in the community.

Geographical and temporal limitations. The geographical area to be considered in this attribute will vary, depending upon the proposed activity. The area to be considered will depend upon where the affected population resides and works. Therefore, any area where people who will be affected by the services discussed herein reside or work must be considered in the determination of impact.

In determining the impact that occurs within this attribute, the analysis must be done for the area before and after the proposed activity is instituted. It is suggested that the "before" time period incorporate those conditions that exist or can be anticipated to exist shortly before the proposed activity is instituted. It is also suggested that the "after" time period be that time period shortly after the proposed activity or project has been completed and is in full operation.

Mitigation of impact. Impacts in this attribute can be mitigated by including in the planning process for the proposed activity a plan for providing the services that have been identified as being needed or proposing alternative uses that can be made of services that will no longer be needed as such by the population.

Secondary effects. See description under lifestyles.

Economics

The potential impact on the economic structure of changes resulting from project activities stems primarily from the direct effect of purchases of goods and services for project activities and the indirect effects arising from goods and services purchased from payrolls. These effects may be summarized by reference to three major attributes that reflect impact on industrial and commercial activities, the local government, and the individual. These attributes are as follows:

Regional economic stability

Public sector revenue and expenditures

Per capita consumption

Regional economic stability

Definition of the attribute. This attribute indicates the ability of a region's economy to withstand severe fluctuations, or the speed and ease an economy demonstrates in returning to an equilibrium situation after receiving a shock. This is an *ex post* definition, whereas a surrogate *ex ante* definition is the diversity of a regional economy or the degree of homogeneity of the region's economic activities in contributing to the gross regional product. The more diverse an economy is and the more closely related it is to growth areas of the national economy, the more stable it is likely to be.

Activities that affect the attribute. Any activity that results in some input or output relationship with a local business or individual has an impact on the growth and stability of the regional economy. Direct purchases would have an effect, as would indirect purchases through payrolls. Proposed increases *or* decreases in either area are important causes of effects.

Source of effects. The severity of a change in stability is directly proportional to the degree of dependence of the regional economy on one affected business sector for income and employment. Thus, if one or a few industries or firms dominate a region's economy (measured by the share of gross regional product or proportion of total employment), that region is highly sensitive to factors affecting those industries. Hence, activities that decrease the industrial diversity in an area are reducing the stability of the region, especially when the key industries are locally important but declining nationally. "One-factory" towns and smaller cities dependent on a military installation are common examples.

Variables to be measured. Effects on the regional economy are indicated by the percentage of total regional economic activity affected by the activity. For example, if 25 percent of all retail sales in a county stem from agency personnel purchases, significant impacts can be anticipated from a change in personnel. Likewise, the agency's direct purchase of labor or other materials from the local economy should be examined as a percentage of local economic activity.

How variables are measured. Considerable ingenuity must be exhibited by the individual who is measuring impact on regional economic stability. Variables to be examined would include employment in economic activity related to specific activities. Production and income variables might also be examined.

Evaluation and interpretation of data.   There are no rules that would enable one to determine whether or not a given change is small or large. Instead, judgment must be exercised, with explicit reference to the basis for judgment. This approach would enable any reviewer to evaluate the facts and, perhaps, to disagree with the judgment. At least, full consideration of the issues and the rationale for a conclusion will have been given.

Special conditions.   Stability and, perhaps, growth are two goals of a regional economy. They are usually incompatible because, in the long run, some specialization is required if a growth rate higher than that for the rest of the country is to be realized. Therefore, the unique or special characteristics of the regional economy must be considered. An economy with an agricultural base, for example, might be much more severely affected by the withdrawal of agricultural land for use in a project than if agricultural land were to be withdrawn from use in an industrial-based economy.

Geographical and temporal limitations.   The same geographical and temporal limitations that exist for the per capita consumption attribute are applicable here.

Mitigation of impact.   Mitigation of negative effects can be achieved in one of two ways: Either increasing the demand for the output of high-growth industries in the region, or changing the distribution of demand for the output of different firms so that the resulting employment redistribution approximates more closely the situation at the national level (taking into account the potential for regional specialization).

Secondary effects.   Economic changes frequently result in secondary or indirect impacts on the biophysical environment. These impacts need to be identified and assessed.

In the case of economic effects, programs or actions that add or reduce revenue in an area will result in additional or decreased population and new economic activity in local communities. This may take the form of new or fewer retail outlets (stores, garages, etc.), increased or decreased service-oriented businesses, and land-use changes as new home developments, shopping centers, etc., are created or requirements for them are reduced. Most of these activities will have a secondary impact on air, water, and land attributes.

Public sector revenue and expenditures

Definition of the attribute. This attribute is an expression of the annual per capita revenues and expenditures of local and state governments and associated agencies in the region under study. Changes in this variable can be interpreted as a measure of the change in economic well-being of the public sector.

Activities that affect the attribute. Changes in the economic, social, or physical conditions of the area due to project activities may result in changes in public sector revenues and expenditures. The effects would be felt primarily through changes in employment, industrial or manufacturing activities, and the acquisition or release of real estate by agency action.

Source of effects. Tax receipts are directly affected by changes in personal income. For major federal projects, payments from the federal government to local governments to compensate for increased local expenses also may occur. Changes in land usage and, therefore, assessed value also affect revenues collected.

Numerous changes in the costs of services (and therefore in requirements for public expenditures) occur in such areas as education, transportation, public welfare, health, utilities, and natural resources as the direct result of an activity or indirectly through employment changes caused by the activity.

Variables to be measured. One measure of an impact is the average annual revenues and expenditures of the relevant government and its agencies in a defined geographic region over the lifetime of the project, assuming the project or activity has been undertaken, minus the same measure over the same time span, but assuming the activity has not been undertaken (and everything else remains the same). In lieu of the annual average, one recent year may be chosen arbitrarily and the change in annual net revenue computed for that year.

Another set of variables would be a comparison, on a function-by-function basis, of the expenditures necessary to provide adequate public services with and without the project.

How variables are measured. The geographic extent of the affected public sector must be defined *a priori*, usually as a local (town, city, or county) or state government. Changes in revenues and expenditures must then be estimated on an item-by-item basis. Tax revenue changes may be estimated as described in the section on measurement of variables in the per capita consumption attribute. Local sales tax rates should be used in lieu of state rates where pertinent, and the state or local income tax rates should be used in place of the composite national rate as described in the per capita consumption attribute. Effective state income tax rates can be found in publications pertaining to specific states or local areas, such as the *State and Metropolitan Area Data Book* released periodically by the U.S. Department of Commerce, or similar local or regional publications. Corporate tax receipts for local areas are generally not important.

The change in gasoline tax receipts is determined by calculating the percentage change in the number of vehicles in the area. This implies that the tax rate, the per mile gasoline consumption for each vehicle, and the total mileage per vehicle are constant. The percentage change in vehicles may also be assumed to be proportional to the change in personal income. Independent estimates may be made through interviewing automobile dealers or by multiplying population changes times a factor representing cars per capita. The preliminary value for motor fuel tax receipts can be based on the annual *Statistical Abstract of the United States*, taking the figures for state tax collections and excise taxes, where the state receipts must be multiplied by some proportion to determine the local share (this proportion may depend upon the gasoline sales, and hence, indirectly depends on the personal income in the area). Local data should replace the extrapolated state data where available. Since the gasoline tax receipts from some future year (under the assumption that the activity has not been undertaken) is the basis for the measurement; at a minimum, tax receipts for at least two past years should be linearly extrapolated to arrive at the desired figure.

Changes in payments to the local government or its agencies by individuals, businesses, and other agencies for particular goods or services (e.g., water and other public utilities) should be included on a specific basis. Transfer payments from outside sources that are direct or indirect compensations for incurred expenses should be based on the specific changes in these costs caused by the project activity, following standard reimbursement procedures. For example, compensation for increased educational expenses for military families is a transfer payment to the local area. Total changes in receipts from taxes, subsidies, and transfer payments due to the project activity should be summed to arrive at an aggregate figure.

Changes in local public expenses due to the activity may be assumed to be proportional to changes in the total personal income in the area, reflecting both the number of consumers of a public good or service and the per capita level of consumption. The *Statistical Abstract of the United States* provides information on direct expenditures for state and local governments, and gives figures for public sector expenditures for $1000 of personal income. These ratios must be multiplied by the proportion of expenses accruing to the local government for each category: education, highways, and health. When available, these ratios should be calculated from local information for all types of public goods and services that change in the same proportion as total personal income. Some expenses, such as welfare payments, do not change proportionally and must be calculated through independent analyses. Among these expenditures are damages to public facilities or any other temporary or permanent costs identified as resulting from the project activity, but not through social or economic changes within the population. The percentage change in personal income in the impact area is determined from the per capita consumption attribute, and this proportion must be multiplied by the public sector expenses per unit of income. The resulting figures are the changes in public expenditures if the project occurs, and summing them gives the total change in expenses.

Evaluation and interpretation of data.   The changes in public sector revenues and expenditures must be compared to determine whether or not there is a net gain or loss to the public sector subsequent to the project. The severity of the impact (either positive or negative) would remain a matter of individual judgment and would be partly subject to considerations of indebtedness of the community.

Special conditions.   The measurement can be improved if a more accurate estimate of future revenue and expenditure levels without the project can be determined. A detailed analysis, perhaps using multiple regression techniques, would improve these projections as well as help identify and evaluate more precisely the causal relationships between public sector revenues and costs and the direct impacts of the proposed activity.

Geographical and temporal limitations.   In general, the same geographical and temporal limitations that exist for the per capita consumption attribute are applicable in this measurement situation. The geographical range of local governments and civilian public agencies with respect to both the revenues and expenditures must be determined in a manner similar to an analysis of the market and supply areas of a private sector business enterprise.

Mitigation of impact.   A negative impact can be mitigated if the project activities are designed either to reduce costs to the local community (e.g., demands for public sector goods and services, physical or economic damages to existing infrastructure) or to increase the direct or indirect payments to the local government.

Secondary effects.   See the description under regional economic stability.

Per capita consumption

Definition of the attribute.   Annual per capita consumption is the yearly use of goods and services by each person, derived by dividing the quantity of use by the number of people. This variable can serve as a direct measure of personal economic well-being.

Activities that affect the attribute.   Increases (or decreases) in local employment, industrial expansion (or reduction or deletion), and construction all have the potential for affecting per capita consumption.

Source of effects.   A change in demand for local goods or services results in increased or decreased money available for purchase of goods and services (disposable income). As another example, disposable income and, therefore, consumption may be affected by a changed tax base resulting from government project acquisition of formerly taxable land.

Variables to be measured.   The baseline measure is the average amount that will be spent in each future year throughout the life of the project by each resident of the affected area for goods or services meant for personal consumption, assuming the project has not been undertaken. The variable indicating change is that same calculation, but under the assumption that the project or activity has been undertaken (with everything else exactly the same) minus the baseline measure.

How variables are measured.   Assuming that businesses are not at full capacity, a change in final output (in dollars) will be reflected in a change in all short-run costs, including labor wages. With constant returns to scale for inputs, the change is completely proportional, and output revenue and all costs will change proportionately to the change in production, based on the current ratio of these values. In addition to labor costs, profits which accrue to the owners of a business may change. A determination of how a profit change affects personal income in the region must be based on an individual analysis of each business, with consideration of the amount of profit per dollar of output and the location of the owners (where the changed income of nonlocal residents is not included). Thus, a coefficient for a particular industry may be determined, showing the ratio of local personal income (wages, salaries, profits) to the dollar output of the industry.

Changes in output due to project activity may be approximately determined by first noting all industries, firms, or individuals who supply some needed input to the activity and the amount of this input in dollars. Included as inputs are such goods and services as local raw materials, retail goods and services bought by project personnel and their families, and contributions to local charities.

Changes in the inputs (which are the outputs of the supplying firms) must be calculated or estimated with as much accuracy as possible. Assuming a constant, linear production function (constant input mix), the change in a supplying firm's output can be approximated by first determining the ratio of the activity's current requirements for the firm's output (in dollar terms) to the current total requirements for that activity (which need not be measurable in dollar terms). Multiplying this ratio by the change in activity, the change in the firm's output is determined. This is multiplied by the previously calculated personal income–output ratio to produce the desired figure.

Prices are assumed to be constant, but if a price change is expected to result from the activity, then the input-output ratio has to be recomputed based on the new price before being used. Direct employment changes, changes in the average wage rate (perhaps due to a change in the size of the labor force), or other changes that are directly caused by the proposed activity should be examined. Any additional information indicating how the total wage bill changes with a change in an activity should be used, if possible. For example, business failures or disruptions caused by the activity and resulting in employment changes should be included. Other determinants of personal income, such as proprietor's income, dividends, interest, transfer payments, and other personal costs and revenues may be assumed to be changes in the same proportion as output revenue unless specific information indicates otherwise. Attempts should be made to assess these ratios whenever possible.

The change in disposable income equals the change in personal income minus the change in personal tax payments. Assuming a constant effective income tax rate (due to small incremental changes in income), this rate times the change in personal income gives the total income tax change. The tax rate may be obtained from the tables of information pertaining to rates of state personal income taxes for different adjusted gross income levels, using the same filing basis (such as married couple with two dependants) for both before and after. Changes in taxable property, together with the pertinent rate, give the property tax alteration.

The change in personal consumption is determined by a rough calculation of the coefficient of consumption applied to the change in disposable income. Thus, the proportion of disposable personal income spent on personal consumption expenditures, calculated from national data if local information is missing, may be assumed to apply to local disposable income. The *Statistical Abstract of the United States,* in the table on personal income and disposition of income, gives the pertinent data from which the necessary coefficient can be calculated for the appropriate data (approximately 0.9 for all years). Multiplying this by the change in disposable incomes provides an estimate of the initial change in consumption.

The most difficult data requirements involve the identification of all activities linked to the proposed activity through an input-output relationship, and the determination of each coefficient indicating the dollar change in the supplying firm's (or individual's) output due to a unit change in the project activity (where this output relates to the particular activity being investigated). This information can come only from a detailed examination of project activity.

Evaluation and interpretation of data. The interpretation of these data must be based on exercised individual judgment. Judgments regarding high or low impacts should be made by persons performing the assessment. The reason for the judgment should be stated also.

Special conditions. The analysis can be improved if a complete input-output analysis is completed together with a detailed economic analysis of the change in personal income (and then in personal disposable income) that results from a change in the output of economic activities linked to the project. Where data are uncertain, an attempt should be made to use expected values, if possible.

Geographical and temporal limitations. The geographical area within which the change in consumption occurs must be determined *a priori*, but it should be defined by the spatial distribution of the affected labor force. Where project activities affect consumption outside the area (and hence, would not normally be included in the analysis), efforts should be made to separate locally important effects from the effects that are far removed from the project's impact.

The attribute measurement methodology presented assumes an average of the total annual changes over the lifetime of the project or activity. This is an arbitrary procedure, and temporal trade-offs (time discounting) can be applied if desired. Calculations can be made for different years in the future with or without project changes, and the separate figures are aggregated by first multiplying them by arbitrarily assigned normalized weights. Another simple alternative is to choose a single future year to compare with and without project changes, implicitly weighting all other years as zero.

Mitigation of impact.   Any detrimental impacts can be mitigated best if direct linkages are established with area industries, businesses, or other economic activities, encouraging an inflow of money into the local economy.

Secondary effects.   See description given under regional economic stability.

References   Bureau of the Census, Department of Commerce, *Statistical Abstraction of the United States.* http:www.ntis.gov/product/statistical-abstract.htm.
Economics and Statistics Administration. Department of Commerce, *State and Metropolitan Area Data Book, 1997-98.* http://www.ntis.gov/product/sma.htm.

Resources

Resources include assets that can take many forms—natural, cultural, economic, historic, etc. In many NEPA review situations, cultural resources may be of particular significance. These resources are discussed in Chapter 13. As used here, *resources* refers only to natural resources.
   Natural resources include the land, air, water, vegetation, animal, and mineral resources which constitute our natural environment and provide the raw materials and spatial settings which are utilized in developing our familiar human-modified environment. These resources may be nonrenewable, such as metals and fuels, or renewable, such as water. Nonrenewable resources are of particular interest, since their consumption or utilization represents a commitment that is potentially irreversible or irretrievable, and constitute a special NEPA responsibility.
   Since fuel resources hold a position of extreme importance, they are treated as a separate attribute. Also, since many of the other natural resources are discussed through other attributes (ecology, air, water, and land), another attribute emphasizes the remaining nonfuel resources which are utilized in either a natural or transformed state for products and materials in the development of the human environment. A third attribute considers the aesthetic qualities of natural and human-modified environments—modified through the use of natural resources.
   These attributes are summarized as follows:

Fuel resources

Nonfuel resources

Aesthetics

Fuel resources

Definition of the attribute. Fuel resources include all basic fuel supplies utilized for heating, electrical production, transportation, and other forms of energy requirements. These resources may take the form of fossil fuels (oil, coal, gas, etc.), radioactive materials used in nuclear power plants, or miscellaneous fuels such as wood, solid waste, or other combustible materials. Solar, wind, and hydroelectric energy resources or other energy sources in a current developmental state are not addressed in this text.

Activities that affect the attribute. Since energy consumption relies almost entirely upon fuel resources, it is probable that almost any activity that consumes energy consumes substantial fuel resources as well. Actions requiring consumption of energy can be categorized into (1) residential, (2) commercial, (3) industrial, or (4) transportation activities.

Residential activities include space heating, water heating, cooking, clothes drying, refrigeration, and air conditioning associated with the operation of housing facilities. Also included is the operation of energy-intensive appliances such as hair dryers and toasters.

Commercial activities include space heating, water heating, cooking, refrigeration, air conditioning, feedstock, and other energy-consuming aspects of building or physical plant operation. Facilities which consume particularly significant amounts of energy include bakeries, laundries, and hospital services.

Industrial activities which require large amounts of fuel resources include power plants, boiler and heating plants, and cold storage and air-conditioning plants. Other industrial operations that require process steam, electric drivers, electrolytic processes, direct heat, or feedstock may also have a heavy impact upon fuel resources.

Transportation activities involving the movement of equipment, materials or personnel require the consumption of fuel resources. The mode of transportation may include aircraft, automobile, bus, truck, train, pipeline, or watercraft.

Source of effects. Most presently utilized fuel resources are limited to the supplies of existing fossil and nonfossil fuels at or beneath the earth's surface. The demand for these fuels in the United States far outstrips the production rates of domestic supplies; hence, much of the fuel resources consumed daily in the United States comes from foreign sources. This places a dependence upon these foreign sources which bears heavily upon economic stability and has obvious strategic implications. Furthermore, known reserves of certain fuels—particularly natural gas—are limited to the extent that unless conservation measures are effected immediately, these supplies will be consumed in the foreseeable future.

Variables to be measured.   The most important variables to be considered in determining impacts on fuel resources are the rate of fuel consumption for the particular activity being considered and the useful energy output derived from the fuel being consumed. Various units may be utilized in describing consumption rates: Miles per gallon, cubic feet per minute, and tons per day are commonly used in describing the consumption of gasoline, natural gas, and coal, respectively. Similarly, the energy output of various fuel- and energy-consuming equipment and facilities may be described in many different units—horsepower, kilowatt-hours, and tons of cooling are a few examples.

A common unit of heat, the Btu, may be applied to most cases involving fuel or energy consumption. The Btu is the quantity of heat required to raise the temperature of 1 pound of water 1 Fahrenheit degree. In the evaluation of transportation systems, for example, alternatives may be compared on a Btu per ton-mile basis.

Other variables of concern include the availability (short- and long-term) of fuel alternatives, cost factors involved, and transportation distribution and storage system features required for each alternative.

Data on the consumption of fuel resources may be applied to almost any environmental impact analysis, but the depth and degree to which data are required depend upon the nature of the project under consideration. For an analysis of existing facilities or operations, sufficient information should be available from existing records and reference sources. Where alternative fuels or transportation systems are the focus of the proposed action, additional background information may be necessary to evaluate not only efficiencies but cost-effectiveness and long-term reliability.

How variables are measured.   Because of the complexities in the nature of the variables discussed above, most are measured by engineering, resource, and other professionals, although the results may be applied by most individuals with a technical background.

Once the heat contents of fuels are known, comparisons may be made on the bases of the heat content of each required to achieve a given performance. An energy ratio can be established as the tool for comparison. The ratio is defined as the number of Btus of one fuel equivalent to 1 Btu of another fuel supplying the same amount of useful heat. Determination of energy ratios requires careful testing in laboratory or field comparisons and usually yields reliable results when conducted under impartial and competently supervised conditions. These ratios have been determined in various tests and are summarized in such publications as the *Gas Engineer's Handbook.*

The consumption of fuels on an installation may be determined from procurement and operational records. Measurements may be made by using conventional meters, gauges, and other devices.

Evaluation and interpretation of data. The fuel resource data will more than likely be used in an environmental impact analysis for the benefit of planners and decision makers for either (1) evaluating the alternatives where either fuel consumption or fuel-consuming equipment or facilities are involved or (2) determining the aspects of fuel and energy conservation that exist with regard to ongoing actions. These aspects include the irreversible and irretrievable commitments of resources resulting from the action, the short- and long-term trade-offs, and the identification of areas for potential conservation and mitigation of unnecessary waste.

As previously discussed, the analysis should include aspects of efficiency, availability, cost of fuel and support facilities (transportation distribution, storage, etc.), and projected changes in these values which might occur in the future. Secondary effects should also be considered, such as environmental effects due to fuel production (mining, refining, etc.) and impacts on air and water quality from combustion and related pollution control measures.

Special conditions. If the activity results in additional demands or waste of fuels already in short supply, or nuclear fuel, expect public controversy to follow. Natural gas supplies, presently limited or unavailable in some areas, should be considered with special emphasis. Electric consumption, in most cases, bears directly upon fuel resources, the effects of which should be included in the analysis.

Geographical and temporal limitations. Concern for fuel resources typically peaks during summer (when air conditioning loads are high) and winter (when heating loads are high). Thus, projects in northern climates would be expected to have the greatest concern for heating fuels, while southern facilities would be more concerned with heavy cooling requirements in the summer, although exceptions to this general trend may occur due to localized demands or geographical or climatic effects. Proximity to natural supplies also may play an important role in fuel selection, since transportation may affect the availability and economic desirability of certain fuels.

Mitigation of impact. Mitigation of impacts directly and indirectly attributable to fuel resources falls into two categories. The first pertains to mitigation by alternate fuel selection and is based on a number of complex variables—availability, cost, environmental effects, and pollution control requirements, to name a few. Other factors to be considered in the selection are the short- and long-term effects of a particular choice, and the irreversible and irretrievable commitment of resources associated with the selection.

The second category of mitigation is associated with the conservation of fuel resources, regardless of the type or types of fuel being consumed. Such measures can be applied to new construction in the form of additional insulation and design to incorporate energy conservation features related to color, orientation, shape, lighting, etc. The conservation of energy can be applied to existing facilities in the form of added insulation and programs to reduce loads on heating, cooling, and other utility consumption. Likewise, in the operation and maintenance of equipment, steps may be taken to further reduce fuel consumption by increasing efficiencies through proper equipment maintenance, reducing transportation requirements, and scheduling replacement of old equipment with newer, highly efficient models.

Secondary effects.   Conversion of fossil and nuclear fuels into useful energy can lead to secondary effects on the biophysical and socioeconomic environment. Air emissions occur during extraction, processing, and combustion processes. Water quality may be affected by spills, acid mine drainage, and thermal discharges. Land-use impacts include loss of habitat, land disturbance, erosion, and aesthetic blight. Solid waste problems resulting from mining and production activities include leachates, radioactive wastes, slags, and tailings. Chapter 12 discusses many of these environmental considerations.

Nonfuel resources

Definition of the attribute.   This attribute considers the nonfuel resources which are utilized in either a natural or transformed state for products and materials in the development of the human environment. Various nonfuel products are manufactured from fuel resources and are included in the definition. Specific examples include wood and wood products, metals, plastics, and nonmetallic minerals and materials.

Activities that affect the attribute.   Few, if any, activities do not depend on natural resources in some way. Any activity that consumes materials and supplies, requires equipment and machinery, utilizes land, or produces waste products may have an effect on natural resources. Various materials—lumber, aggregates, cement, steel, asphalt, etc.—are utilized in construction and repair activities. Operation of facilities depends on equipment that is manufactured and requires metallic and nonmetallic parts and components. Land use may deny access to minerals or other resources. Disposal of some waste may result in loss of valuable resources that could effectively be recycled, reclaimed, or reused.

Source of effects.   In order to develop and maintain our present lifestyles, many nonrenewable resources are being consumed at rates which indicate depletion of many critical materials within the century, or, in some cases, within a few decades. Furthermore, some of these materials in short supply are controlled by foreign powers, which results in even further complications, and dependency and strategic implications become important.

Variables to be measured. For the impact assessment procedure, a study should be made that will (1) identify the activities or points of consumption of natural resources, (2) indicate the consumption rates, and (3) reveal the quantities and content of wastes resulting from those activities.

How variables are measured. Qualitative determinations relating specific activities and resource consumption may be made on the basis of firsthand knowledge of the activities and their mode of accomplishment, and a general knowledge of resources and resource management. Once these relationships are identified, the information may be utilized repeatedly, as it will remain valid until changes in the activity or its mode of accomplishment occur.

Consumption rates are somewhat more difficult to quantify, and technical expertise may be required. Depending on the kind of activity and the type of resource, the rates may be reported in such various terms as pounds per year or tons per day. Based upon purchasing data and other records, input-output models may be constructed which depict the total effect of the resource utilization.

Content and quantities of waste products resulting from an activity may be determined from field studies during which actual waste samples are classified and analyzed, or may be estimated on the basis of the same input-output models discussed above or by simpler procedures (e.g., emission factors).

Evaluation and interpretation of data. After determination of the points of resource consumption, quantities involved, and waste products produced, an evaluation may be made of the total impact by considering each of the resources being consumed in light of its individual status—abundance, importance, availability, economics, origin, energy to produce, recycle potential, etc. Life-cycle thinking should be incorporated (i.e., looking at an activity with regard to the resource requirements for the life of a project from origin to completion, operation, and eventual disposal).

Special conditions. Special conditions may arise due to resource availability and price that can affect natural resources markets. Natural scarcities do exist for many resources, and these availabilities can be further jeopardized by embargoes or other supply interruptions (e.g., strikes). Although prices can actually assist in resource allocation in a free market supply-and-demand situation, efforts to artificially increase prices through such means as cartels, price-gouging, or cartel-like actions may occasionally place specific resources in a position of increased importance.

Geographical and temporal limitations.   Specific geographicalal considerations include the origin of specific resources and the strategic implications associated with resource control. Also, transportation consumes fuel resources and should be considered in choosing alternatives (e.g., specification of a particular type of wood or building product that is unavailable locally). Seasonal aspects affect some resources (vegetation, mining, etc.), but most temporal limitations on resources are artificially produced.

Mitigation of impact.   Adverse impacts on natural resources and resource consumption can be minimized by economizing on resource requirements, development and use of substitutes, and recycling of scrap materials. These mitigations all can be considered as forms of conservation resulting in the use of less raw material per unit of output. Specific programs might include recycling of tires, glass, paper, metals, petroleum waste, construction and demolition debris, and general solid waste. These areas not only provide potential for conservation of materials, but some may be used for energy conversion, resulting in fuel conservation as well.

Secondary effects.   In addition to energy consumption, other environmental effects may be related to the consumption of resources. Activities associated with the extraction, transportation, and processing of materials to produce the finished products may have an impact on air, water, land, and ecology. Other social and economic factors may be affected as well.

Aesthetics

Definition of the attribute.   The aesthetic attribute may be used to describe impacts on the environment which are apprehended through the senses—sight, taste, smell, hearing, and touch. Although treated in part in other attributes (e.g., odors in air and the entire category of noise), tolerance levels based on aesthetic criteria are often somewhat different, in addition to the fact that aesthetic perceptions generally require the consideration of all the senses simultaneously. Visual perception is perhaps the most familiar of the areas, and the ensuing discussion will emphasize visual aesthetics and natural and human-modified landscapes.

Activities that affect the attribute. Generally, any activity that will alter the quality or distinguishable characteristic of the perceived environment can be considered as having an effect on aesthetics. Visual perception may be altered by activities involving construction, forestry and recreation management, transportation, water resource and land-use planning, and other activities involving landscape and scenic vista modification. Other aesthetic perceptions (hearing, smell, etc.) may be affected by industrial activities, burning, aircraft operations, waste discharges, and various facility operation and maintenance activities.

Source of effects. The activities that affect aesthetics do so by creating changes in the aesthetic characteristics of the environment as they are perceived by individuals (examples of characteristics include color, texture, scale, harmony, etc.). These perceptions are explained more fully in the following section.

Variables to be measured. Individual perceptions and values for defining beauty make it difficult to quantify aesthetic impacts. Perception of ugliness, however, is more nearly agreed upon. In most cases, aesthetic criteria can be formulated by persons who have had experience in design and have acquired a sensitivity to the characteristics of the natural setting and structures that make them pleasing or displeasing to the human senses. Measurement techniques for identifying and describing aesthetic impacts are basically of two types:

1. Subjective: The qualitative analysis procedures based on the developer's best knowledge of design characteristics.
2. Objective: The quantitative analysis procedures based on established thresholds. The essence of this methodology includes design standards, architectural controls, sign ordinances, and landscape criteria. As an example, natural landscape aesthetics may be analyzed using the variables as follows (Litton, 1971, Bagley, 1973):
   a. *Landscape character* in terms of the landscape setting
      (1) Boundary definition: physical, vegetative, topographic, etc.
      (2) General form and terrain pattern.
      (3) Vegetational patterns.
      (4) Features: hills, valleys, cliffs, promontories.
      (5) Water and land interfaces: conditions and quality.
      (6) Weather patterns.
      (7) Cultural interfaces: artificial objects, transportation facilities, structures, etc.
      (8) Natural and human-made acoustical features: sound absorption, falling water, birds.

b. *Macro (major) components*
   (1) Unity: the cohesion of the parts into a single harmonious unit, described by the presence or absence of a single dominant factor and complementing subordinate elements, contributing to a pleasant total composition.
   (2) Variety: diversity without confusion, more than one element contributing richness; the maximum opportunity for visual stimulus.
   (3) Vividness: quality lending to sharp visual impression-distinction.
c. *Micro (minor) descriptive elements*
   (1) Texture: identifying quality or disposition of the vista (e.g., rocks, trees, grass, and cultivated crop patterns), soft, sharp, flowing, rough.
   (2) Color: may be described in terms of hue, lightness, and saturation.
   (3) Contrast: diversity of adjacent parts in color, shape, or texture.
   (4) Uniformity: similarity between features.
   (5) Scale: proportion of one object compared to another, particularly important in considering modified landscapes.
d. *Changing qualities*
   (1) Distance: proximity to components in the vista.
   (2) Observer position: aesthetic qualities of a given area may vary with viewer location.
   (3) Speed of observation: duration of viewer's observance.
   (4) Time: daily and seasonal changes.
   (5) Observer's state of mind: expectations, values, mood.

How variables are measured; evaluation and interpretation of data. Due to the nature of aesthetics and human perception, significant features are often difficult to quantify. Many methods, however, have been developed in an attempt to establish standards of comparison, such as to arrive at a basis for determining which type of landscape (for most persons) is more desirable than another. These methods take two general forms:

1. A relative numerical weighting of each of the various intrinsic and extrinsic landscape resources as individual components and as a composition reflecting the presence and relationships of the descriptive elements listed above. These procedures attempt to quantify visual relationships, place a value on aesthetic resources, and describe the implications of changes on the landscape in terms of scenic quality, as ranked with other environmental changes.
2. The nonnumerical methodologies tend to place emphasis on ranking of visual attributes according to the same elements as the numerical scheme, but evaluate the aesthetic elements in terms of

comparative analysis based on established criteria. They do not assign numerical weights but may, in some cases, assign a position on negative value. In addition, most studies can be categorized as

a. The *visual methodologies:* Visual components of the environment are inventoried and assessed by the planning staff, decision makers, or consultants.

b. The *user-analysis methodologies:* Designed for attempting to find out how the general public feels about various aesthetic and potential impacts. Used as inputs to above assessments.

Special conditions. Since the value, importance, or expression of beauty is relative to the variable of perception, it is important to note that the following conditions bear significantly on the degree of aesthetic impact:

- The observer's state of mind: Factors of current perceptual setting and environmental lifestyle, coupled with past experiences and future expectations, can produce varying impressions of aesthetic quality.

- The observer's background: Cultural, economic, ethnic, and social background can determine perceived aesthetic qualities.

- Context of the observation: The setting of an observation may bear upon its acceptability (e.g., is a structure otherwise acceptable, but "out of place"?).

Mitigation of impact. Aesthetic impacts are frequently controversial. While it is generally agreed that everyone would like to enjoy clean air, pristine waters, scenic vistas, and serenity in their everyday living, economics and other "facts of life" do not always make this possible. However, many adverse aesthetic impacts may be minimized once an aesthetic inventory is provided to planners and designers, so that desirable features associated with a project might be maintained and enhanced or incorporated into the project, and undesirable features of the project redesigned or eliminated.

Secondary effects. Aesthetic qualities may be associated closely with land use characteristics—an association leading to potential secondary impacts on almost any other biophysical or socioeconomic attribute. Aesthetic impacts not only reflect upon psychological needs, but frequently may be related to land prices, economic security, and community needs.

References

Bagley, M. D., et al. *Aesthetics for Environmental Planning.* Environmental Protection Agency, Report No. EPA-600/5-73-009, November 1973.

Litton, R. Burton, et al. *An Aesthetic Overview of the Role of Water in the Landscape.* Prepared for the National Water Commission by the Department of Landscape Architecture, University of California, Berkeley, July 1971.

Gas Engineer's Handbook, Chap. 22, "Fuel Comparisons." New York: The Industrial Press, 1969.

Proshansky, Harold M., et al., eds. *Environmental Psychology: Man and His Physical Setting.* New York: Holt, Rinehart and Winston, 1973.

A Step-by-Step Procedure
for Preparing
Environmental Assessments
and Statements

This appendix presents a simplified flowchart of the steps which will be completed if a systematic approach to preparation of environmental documentation is to be followed (Fig. C.1). The major steps are presented in a sequential fashion, using hypothetical examples.

Many different methodologies for impact assessment exist, both in theory and in practice, as discussed in Chapter 6. This appendix presents one simplified, systematic approach to the development of an environmental impact analysis, utilizing the matrix approach and the multidisciplinary attribute descriptor package from Appendix B. Although the matrix method is employed, similar procedures using environmental attributes could be developed, and the reader's attention is directed to those places where variation is normally found.

The following steps, as shown in Fig. C.1, detail the procedure to be used in preparing environmental assessment documentation. The degree of consideration to be exercised within some of the steps may vary with project scope and magnitude, but the basic algorithm is applicable in all cases

Step 1. Define the Action

The first step in the assessment process for either an environmental impact statement or an environmental assessment is to determine what the proposed action is, and identify the purpose and need that it

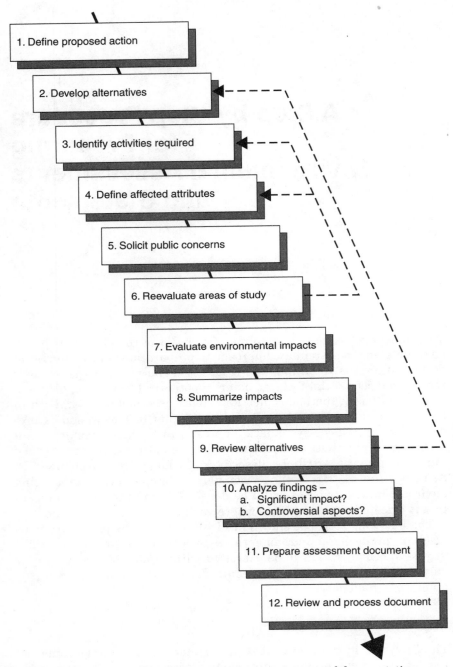

Figure C.1 A step-by-step procedure for preparing environmental documentation.

addresses. At first glance, this step may seem unnecessary. However, often this is the most difficult step of the process, and may mean the difference between agency success or failure if a court or administrative law judge reviews the assessment.

If the proponent is a typical government agency, a "proposed action" may begin as an idea for change. Often these "good ideas" are rather nebulous in nature, at least at first. An agency manager may want to build a new facility but may not have thought through why this would be a good idea—in other words, What is the underlying purpose and need for the action? The agency staff may take "because the boss said so" as a reason, but this approach will rarely stand up to the test of agency or public scrutiny. The NEPA assessor must think through the underlying reasons why a new building would be needed. Is the agency growing and more space needed? Is the old building no longer functional or in disrepair? Is the old building ill suited to modern technology? Would the staff find the workplace more inviting if housed in new surroundings? By answering these types of questions, the assessor will delve into the purpose and need for the proposed action, which will in turn affect what alternative courses of action are considered.

A NEPA analysis relates to physical actions proposed to take place, and how they would affect the human environment. Even an environmental impact statement prepared on a new agency policy or proposed legislation must be couched in terms of the physical actions that would be expected to ensue and how these might affect the environment. Although NEPA is most useful if applied early in the planning process for a new proposal, meaningful assessment is not possible until the agency can identify firm descriptions of the physical actions expected under the proposal.

Once the physical activities required to implement a proposal are defined, the assessor must determine whether or not environmental documentation, under NEPA or other regulations, will be required. All federal agencies, and many state and local agencies, are required to consider the environmental effects of implementing their major programs or actions. The Council on Environmental Quality (CEQ) regulations require federal agencies to develop specific criteria for, and identification of, three typical classes of action (see Chapter 4):

- Those normally requiring an environmental impact statement (EIS)

- Those normally not requiring either an environmental impact statement or an environmental assessment (i.e., a categorical exclusion)

- Those normally requiring an environmental assessment (EA) but not necessarily an EIS

This guidance usually takes the form of lists of types of actions or criteria for inclusion or exclusion. Multiple criteria and/or flowcharts may also be employed. These requirements are publicly available from each agency.

First, the person preparing the NEPA assessment should compare the project under consideration with the agency's lists of typical classes of actions.

- If the project appears on the list of actions that *normally require an environmental impact statement,* unless there are extenuating circumstances approved by the agency, the assessor should prepare an EIS, and should proceed with steps 2 through 12 (Fig. C.1).

- If the project appears on the list of actions that the agency has *categorically excluded* from preparation of either an environmental impact statement or an environmental assessment, the assessor should briefly check to see if there are extraordinary circumstances, such as the presence of an endangered species, that would negate the use of a categorical exclusion. Many agencies have documentation requirements for all or some types of categorical exclusions, and the assessor should become familiar with and follow the agency requirements. Categorical exclusions are not transferable from agency to agency; one agency may have listed a categorical exclusion for drilling water wells, for example, while another agency may require a more stringent documentation of impact of this same activity through an environmental assessment.

- For *all other actions,* in other words, those that are not known to require an environmental impact statement *and* are not listed for consideration as a categorical exclusion, the assessor should proceed with preparation of an EA, and should proceed with steps 2 through 10 (Fig. C.1). Many agencies have published lists of types of actions normally requiring an environmental assessment at the initial stage of NEPA review; however, experience has shown that these lists are not all-inclusive and provide at best an indication of general agency expectations.

- For any action, an agency may *opt* to prepare a full environmental impact statement even though not required, "to further the purposes of NEPA," regardless of the significance of the impacts of the action. This is generally done in order to provide a fuller public disclosure of environmental impacts, to accommodate the preference of another federal or state agency, or to assess the impacts of on-going activities where there is no proposal for change.

As an example, assume that an agency receives a proposal involving the redevelopment of a tract of suburban land where a developer wants to construct 500 housing units. The cognizant state or federal approving agency has placed such an action on its list of actions which normally require preparation of an environmental impact statement. This project, unless some extraordinary element is present, will require the preparation of an EIS. The advice which follows is applicable to either an agency assessment preparer or a contractor employed by the agency or the developer. The requirements are the same regardless of who prepares the document.

Step 2.   Develop Alternatives

The CEQ considers the development of alternatives, including the proposed action, as "the heart" of the environmental analysis. Why is this so important? At times it seems as if the engineering or economic studies prepared by the agency "prove" that there is only one "best" way to proceed with a proposed action, in part because these types of studies are generally focused on optimizing the agency's preferred approach rather than exploring options. In fact, every agency is aiming for that "best" plan. NEPA and the CEQ regulations specifically require that an agency consider alternative ways to reach a desired goal, and this is done to ensure that the agency does not foreclose reasonable options too early in the planning process. Alternative approaches may mean that the agency consider different sites for a new facility, different designs for a proposed structure, different seasonal timing for construction, alternative construction techniques, or even an alternative approach to meet agency needs without new construction. This is done so that the agency will consider a reasonable range of alternative approaches to meeting its underlying purpose and need for action, and disclose to the interested public the full range of options under consideration. Considering a full range of alternatives will help the agency make better decisions. Remember, also, that the "best" alternative may be one which can be implemented at reasonable cost without unreasonable delays. Agency goals may be better served in this manner than by fighting for many years to carry out the technically superior choice.

For major federal actions, the agency is required to identify its "preferred alternative" as soon as is reasonable, at least by the time that the final EIS is circulated (but before a final agency decision has been made). If the agency has invested in engineering or economic studies, these may serve to support the agency's identification of its preferred course of action. To facilitate pubic disclosure, the agency should identify its preference as soon as possible. If the agency has a concrete engineering proposal, it may wish to identify this as its preferred course of action in the first public notifications (Notice of Intent, published in the

Federal Register, or similar public notification). However, the CEQ regulations caution against going too far and committing too many resources to an engineering solution until the NEPA review is complete. In other situations, the agency may not be in a position to identify its preferred course of action until it has identified environmental impacts and considered public input through the environmental analysis process; in this case, the preferred alternative may first be identified in the final EIS.

Just as the definition of the proposed action requires some thought (see Step 1), the identification of alternatives also requires that the assessor think broadly to identify alternative ways of meeting the agency's stated purpose and need to take the action. The agency must have accurately identified the need to which it is responding, or else the alternatives will not be responsive to the appropriate issue. For example, a forest manager may wish to cut trees to thin a forest tract. If the underlying purpose of this proposal is to improve the stand of trees for silvicultural purposes (timber harvest), the alternatives considered by the agency will be very different from those developed if the underlying purpose is to improve forest health or in response to an order to create habitat for an endangered species. In all three cases the proposed action might be essentially the same, but without knowing what is at issue, the forest manager may look at an inappropriate suite of alternatives (for example, alternative forest management practices to increase timber harvest, such as thinning or limbing to produce tall, straight, evenly spaced tree trunks, may be counterproductive to wildlife habitat management, which may require irregular stands of mixed tree species and some variation in understory habitat).

Must a NEPA reviewer consider endless arrays of possible alternatives? No. The agency must consider a "reasonable" range of alternatives. As this term is somewhat subjective, the NEPA practitioner may make use of the public and agency scoping process (Steps 5 and 6, Fig. C.1) to help decide what is a "reasonable" range. The agency may identify and reject alternative approaches that are very similar to alternatives that will be analyzed, alternatives that are not considered to be "reasonable" (such as those that are too expensive, or where necessary technology is not fully developed), or alternatives that are not responsive to the agency's stated purpose and need for the action. Agencies are encouraged to disclose the alternatives considered and dismissed in order to assist the public in discerning this aspect of the decision-making process. Remember that the analysis of alternatives is a way to help the decision maker consider the environmental consequences of the proposal and alternative approaches; consideration of very similar alternatives, or a large array of alternatives, may not be useful to help focus on those matters that are "ripe for decision." In practice, most environmental impact statements consider between three and six alternatives, and most environmental assessments consider somewhat fewer. Agencies are expected to flesh out and analyze all alternatives in an EIS to a "comparable" degree; however, in an EA, an agency may focus its description and analysis on the preferred alternative. What is a realistic alternative? This is a tough call. Broadly, it is one which accomplishes most of or all of the agency's purposes, as discussed above. It should not be a frivolous action, which would serve no purpose *if implemented.*

A NEPA analysis is a comparative analysis: The environmental impacts of taking an action are compared to the impacts of the different alternatives analyzed. An impact can be thought of as the degree of change that would be expected to occur over time. In order to determine the degree of change, the assessor must first determine what the baseline condition is; in other words, What would happen to the ambient environment if the action were not taken? The description of the environmental conditions that would be expected over time if no action were taken is often referred to as the "no-action" alternative. By law and regulation, the agency must determine and consider the effects of *not* taking the proposed action (or any action) alongside the consideration of the impacts of the proposed action and the alternatives analyzed. See also Section 5.3.

In some cases, an agency may want to quit performing an ongoing activity. For example, an agency may want to cease operation of a military base, a federal facility, or ongoing forest harvest. Some agencies take the position that ceasing an activity is not subject to NEPA review; however, taking action to restore the site, such as decommissioning or tearing down a facility, or cleanup of hazardous waste, would be subject to NEPA review.

Step 3.   Identify Relevant Project Activities

In order to conduct an impact assessment, the assessor must know what activities would be expected to occur. To identify detailed activities associated with implementing the project or the program, agency activities may be categorized into functional areas. For each functional area, detailed activities associated with implementing projects or programs may be developed. The user should supplement these activities with project-specific activities.

In the case of the construction of 500 dwelling units described in Step 1, the 63 construction activities shown in Table C.1 may be used as a starting point for the analysis. Those activities not applicable to the project should be crossed off, and supplemental activities should be added to encompass the project-specific requirements.

TABLE C.1   Construction Activities Typical of a Major Development

Site access / delivery
  Railroad
  Road
  Water
  Air
  Pipeline
Support facilities operation
  Asphalt plant
  Aggregate production
  Concrete operations
  Foundry and metal shop
  Fuel storage and dispensing
  Material storage
  Personnel support
  Utilities provision
  Solid waste disposal
  Sewage disposal
Site preparation
  Clearing and grubbing
  Tree removal
  Existing structure removal
  Demolition debris disposal
  Grading
Excavation
  Topsoil stripping
  Excavation
  Backfill
  Channeling and dredging
  Hauling
Quarrying and subsurface excavation
  Cutting and drilling
  Loosening
  Hauling
  Drainage
Foundations (buildings and roads)
  Base course
  Footings
  Compaction
  Piling
  Foundation mats
  Groundwater control

Bituminous construction
  Hauling
  Mixing
  Placing and spreading
  Compaction
  Curing and sealing
Concrete construction
  Hauling
  Mixing
  Placing
  Finishing
Masonry construction
  Hauling
  Forming
  Mortar mixing
  Placing
  Finishing
Steel construction
  Hauling
  Erecting
  Finishing
Timber construction
  Hauling
  Pest/insect protection
  Cutting and shaping
  Erecting
  Finishing
Finishing—general
  HVAC (heating, ventilation,
    and air conditioning)
  Electrical
  Plumbing
  Cleanup operations
  Landscaping
  Painting

If the agency has experienced construction project managers, they may be able to describe the phases of a typical agency project in some detail, and the assessor may build a similar table of relevant activities. Remember that "Begin Phase II" is not inherently meaningful in environmental assessment terms, but "clear the site" and "lay drainage tile" can be related to real environmental effects.

Step 4.   Examine Attributes Likely to Be Affected

The user should examine the environmental attributes in Appendix B and become familiar with the general nature of the individual attributes and the kinds of activities that may have an impact on them. In addition, the descriptor packages may be used to identify areas where available technical expertise is deficient and additional assistance may be required.

Step 5.   Solicit Public Concerns

Throughout the NEPA review process, the agency is expected to listen to, identify, and address environmental concerns raised by the public or external parties. (Refer also to Chapter 11.) In particular, an agency with regulatory expertise is a vitally important party here.

One of the purposes of NEPA is to lay federal decision making open to public scrutiny; a corollary to this is the challenge established by NEPA and the CEQ regulations to encourage and facilitate public involvement in public decisions. The CEQ regulations require that an agency solicit public involvement at specific points when it prepares an EIS. Although the CEQ regulations are generally silent on public involvement when an EA is prepared, many federal agencies or offices have made similar provisions for public involvement in this type of NEPA review as well. Means for conducting an effective public participation program are discussed in Chapter 11.

Under the process for preparing an EIS, the regulations provide for public input at the following specific points:

- *Public scoping process.* After announcing its intent to prepare an EIS, an agency must invite participation from affected federal, state, and local governments, American Indian tribes, other interested parties, and the general public to help the agency determine the scope of the analysis.

- *Review and comment on draft impact statement.* Before preparing a final EIS, an agency must make a draft version available to the public for review and comment. The review and comment period must be at least 45 days long, but is often longer.

- *Review and comment on final impact statement.* An agency must make a final EIS available to the public prior to taking any action on the proposal analyzed. At the agency's option, it may solicit agency and public comment on a final EIS prior to making a final decision. For all EISs there must be at least 30 days between issuing the final statement and taking the action decided upon; some agencies issue a Record of Decision (ROD) and then wait at least 30 days before it becomes effective, while other agencies wait for at least 30 days before issuing the ROD.

The CEQ regulations are not as specific in requiring public input into an environmental assessment. The regulations require that if an agency reaches a finding of no significant impact (FONSI), the finding must be made available to the public. In some limited situations (if the proposed action is close to that which would normally require an EIS, or if the proposed action is one without precedent) the agency must make the FONSI available for public review for at least 30 days prior to making a final determination. While the CEQ regulations do not impose specific requirements for the EA process, some federal agencies or offices have adopted a "mini-EIS" approach to public involvement for environmental assessments. Agencies may include a scoping process to help determine the proposed action, affected environment, and alternatives or solicit public and agency review of a draft EA or the completed EA prior to reaching a determination of significance.

As part of their own regulations or guidance on how to conduct the NEPA review process, many agencies have adopted provisions for giving notice to the public and detailed public participation procedures. Some agencies have statutory or administrative processes that must be considered alongside the NEPA review process. An agency can go beyond the requirements of law or regulation and involve the public or other parties at additional points in the assessment process. For example, the agency may circulate a plan on how it will approach preparing the environmental review; seek public input into a public participation plan; provide advance notice of its intent to prepare a future environmental review; conduct a prescoping exercise to determine if a suggested proposed action is warranted; circulate more than one draft assessment for review and comment; prepare and circulate ancillary documents such as a technology review of alternatives, a rationale for the preferred alternative, or an economic feasibility report; or circulate a draft decision document for review. At some point, however, the agency must forge ahead with the analysis in order to support a meaningful agency decision and action.

Step 6.   Reevaluate Identification of Activities and Attributes

As a result of consulting with other agencies, the public, and interest groups, the originally proposed identification of activities which will be required and attributes which may be affected should be reexamined. The dashed lines in Fig. C.1 which lead from Step 6 back to Steps 3 and 4 represent this reexamination and reformulation following the public involvement stage. The members of the various publics may have additional concerns which had not been brought to light. They may also not be particularly concerned about some elements which were believed to be important. It is at this stage that the plans for the agency may first be modified to become more acceptable, or the planning for preparation of the EA or EIS may be directed toward those elements of greatest concern. The agency should not avoid mention of a known attribute just because the public is not concerned. There is still a responsibility to cover all relevant attributes. It is just that the focus may be directed most strongly toward those aspects which are of the most public concern. If the comments are from an agency with regulatory expertise, the comments should be given great deference.

Step 7.   Evaluate Impacts Using Descriptor Package and Worksheets

Using the activities (as developed in Step 3) and the attribute list, the assessor may find it useful to construct a matrix worksheet, with activities on one axis and environmental attributes on the other. Figure C.2 indicates an example format. The attribute descriptor packages (Appendix B) may be used to identify environmental attributes.

The matrix in Fig. C.2 can be used to

- Identify potential impacts on the environment by placing an X at the appropriate element of the worksheet

- Collect baseline data on the affected attributes

- Quantify the impact where possible using an analytic approach

For instance, construction of the 500-unit housing development discussed in Step 1 will almost certainly require large-scale excavation, which might later cause erosion, which might result in increased suspended solids in the receiving waters of a nearby stream and cause a decrease in dissolved oxygen. An X is marked on the worksheet for all such negative potential impacts, and a + marks any positive potential impacts. It should be emphasized that this evaluation is to be done on an interdisciplinary basis.

It is important to note that the CEQ regulations require EISs to be analytic rather than encyclopedic. The purpose, of course, is to reduce the bulk of the EIS and also to make the document useful to the decision maker. In order to comply with the regulations, it is therefore essential that the EIS include analytic and quantitative information for environmental impacts where possible. Lengthy subjective discussion of the impacts and general "boilerplating" of the document should

	Environmental Attributes																	
	Air									Water								
Project	Diffusion factor	Particulates	Sulfur oxides	Hydrocarbons	Nitrogen oxide	Carbon monoxide	Photochemical oxidants	Hazardous toxicants	Odor	Aquifer safe yield	Flow variations	Oil	Radioactivity	Suspended solids	Thermal pollution	Acid and alkali	Biochem. oxygen demand	Dissolved oxygen (DO)	Dissolved solids
Site Access/Delivery																		
Railroad																		
Road																		
Water																		
Air																		
Pipeline																		
Support Facilities Operation																		
Asphalt plant																		
Aggregate production																		
Concrete production																		
Foundry & metal shop																		
Fuel storage & dispensing																		
Material storage																		
Personnel support																		
Utilities provision																		
Solid waste disposal																		
Sewage disposal																		
Site Preparation																		
Clearing & grubbing																		
Tree removal																		
Existing structure removal																		
Demolition debris disposal																		

(Left margin label: Activities)

Figure C.2 An example impact matrix for a major construction project. *Note:* The original matrix used in this example included 63 activities (see Table C.1) and 49 environmental attributes (see Fig. 5.1 and Appendix B).

be avoided. Standard lists of endangered species found somewhere in the state are of little value, for example. The decision maker will need to know if there are any endangered species or critical habitats on the site or in an area affected by the action.

In the housing example used above, the analysis started with a consideration of the proposed action, that is, building the 500 housing units. However, in some cases, it may be more useful to start with an analysis of the existing case as described in the no-action alternative. This is applicable in situations where the agency is considering alternative approaches but has not yet formulated a proposed action or preferred course of action. Examples would be considerations of changes to existing forestry practices, development of a resource management plan, and using the NEPA process to develop options to existing patterns of land management.

Step 8.   Summarize Impacts

For potential impacts marked with X or + on the worksheet (Fig. C.2), summarize the impacts using Fig. C.3. Shade the areas of net positive or net negative impacts, using the shading intensity to indicate the significance of the impact. For example, for impacts on erosion, suspended solids, and dissolved oxygen, the magnitude of the project is evaluated along with the site characteristics, and the scientific information provided in Appendix B, and the degree of severity of the impact on each attribute is determined. Finally, the impacts on each attribute from all project activities are summarized by using the key shown in Fig. C.3. In practice, most agencies concentrate on identifying and quantifying adverse impacts instead of positive impacts.

Step 9.   Review Other Alternatives

Repeat the procedure for each of the alternatives considered. Examples of alternatives for the housing development project example may be

- No-action alternative
- Alternatives related to different designs and/or projects, such as a high-rise apartment building, and different activity sites
- Buying existing vacant housing stock and renting to prospective occupants
- Alternative measures to provide for mitigation of fish and wildlife losses discovered to be associated with the soil erosion problem

The dashed line in Fig. C.1, which flows from Step 9 back to Step 2, represents this reevaluation of the alternatives possible. Remember to be imaginative. It may be necessary to evaluate alternatives which are not within the prerogative of the proposing agency. This is specifically *required* by the NEPA regulations if the alternative is otherwise reasonable.

Step 10.   Analyze Findings

Determine whether significant adverse environmental impacts would be expected, and identify mitigation measures. Among other

Category	Attribute	Net Positive Impact +	Attr. #	Net Negative Impact X
Economic Resources	Aesthetics		49	
	Nonfuel resources		48	
	Fuel resources		47	
	Per capita consumption		46	
	Public sector revenue		45	
	Regional economic stability		44	
Human	Community needs		43	
	Physiological systems		42	
	Psychological needs		41	
	Lifestyles		40	
Sound	Social behavior effects		39	
	Performance effects		38	
	Communications effects		37	
	Psychological effects		36	
	Physiological effects		35	
Ecology	Aquatic plants		34	
	Natural land vegetation		33	
	Threatened species		32	
	Field crops		31	
	Fish, shellfish, and waterfowl		30	
	Small game		29	
	Predatory birds		28	
	Large animals		27	
Land	Land-use patterns		26	
	Natural hazard		25	
	Erosion		24	
Water	Fecal coliform		23	
	Aquatic life		22	
	Toxic compounds		21	
	Nutrients		20	
	Dissolved solids		19	
	Dissolved oxygen		18	
	Biochemical oxygen demands		17	
	Acid and alkali		16	
	Thermal pollution		15	
	Suspended solids		14	
	Radioactivity		13	
	Oil		12	
	Flow variations		11	
	Aquifer safe yield		10	
Air	Odor		9	
	Hazardous toxicants		8	
	Photochemical oxidants		7	
	Carbon monoxide		6	
	Nitrogen oxide		5	
	Hydrocarbons		4	
	Sulfur oxides		3	
	Particles		2	
	Diffusion factor		1	
		Net Positive Impact +	Attr. #	Net Negative Impact X

☐ No significant impact

▨ Moderate impact

■ Significant impact

Project name _____

Project number _____

Alternative _____

Figure C.3 Summary of impacts.

requirements, a NEPA review is used to do two things: disclose to the decision maker and public whether the environmental impacts of a proposed action would be significant, and assist in developing mitigation measures to lessen adverse impacts. Although academically speaking NEPA does not distinguish between positive and negative impacts, in practice most agencies do not prepare NEPA reviews to identify significant environmental impacts when those impacts are solely positive.

If the agency is preparing an EA, it must determine whether the adverse impacts of the proposed action, or any of the alternatives considered, are "significant" within the meaning of NEPA. If the impacts are found to be "not significant," the agency may come to a FONSI and proceed with the proposed action. If the impacts are deemed "significant," in order to carry out the proposed action the agency must prepare a full EIS. (If the agency is already preparing an EIS impact statement, it is irrelevant whether the environmental impacts identified and analyzed are "significant" or "not significant" within the definitions of NEPA, although the severity of adverse impact may be of interest to the agency or the public for other reasons.)

What determines significance? NEPA requires that an EIS be prepared for "major federal actions," defined as those actions having significant environmental impacts. To determine significance, answer the following questions concerning the action:

- Will the implementation of the action or program have a significant adverse effect on the quality of the human environment?

- Will the action be deemed *environmentally* controversial? (Controversy caused by other considerations, such as local politics, does not, by itself, trigger this requirement.)

- Are the possible adverse environmental impacts highly uncertain or do they involve unique or unknown risks? (In other words, is the assessor unable to determine if impacts would be significant or not?)

Note: While the attribute descriptor package (see Appendix B), impact analysis worksheet (see Step 7), and summary sheet (see Step 8) assist in identifying environmental impacts of a scientific or technological nature, determining the potential for public controversy requires a more subjective approach. (Refer to Public Participation in Chapter 11.)

If the answer to any of the above three questions is Yes, the agency cannot take the proposed action without preparing an EIS. Similarly, if the impacts, while individually not significant, cross a threshold of significance in the cumulative, an EIS would be required. In some cases, the proponent may decide that it is preferable to abandon or delay the project until uncertainties can be dealt with or the project can be reconfigured, rather than proceed with preparing an EIS.

If the answer to all of the above three questions is No, then the agency can complete the EA, prepare a FONSI, and proceed with the action without preparing a full EIS. The finding must be made available to the public. Many agencies have requirements to publish the FONSI, though this is not a universal requirement.

Often, a look at the summary sheets from Step 8 will demonstrate that many impacts are the same for all alternatives. In this case, the assessor may document in the environmental review those impacts that are common to all alternatives (including mitigation measures that may be applied), and eliminate these from further discussion or analysis. This allows the agency to concentrate its environmental review on the impacts that vary by alternative, thereby assisting in bringing into sharp focus those environmental considerations that bear on the decision to be made.

A second use of the analysis of findings is identifying where mitigating measures would be useful in eliminating an adverse impact or reducing its degree of adversity. Some adverse impacts cannot be mitigated. Sometimes the mitigation itself might create problems to other aspects of the environment (for example, mitigations to assist wildlife might have an adverse impact on cultural resources). Sometimes it is not clear that a suggested mitigation measure would substantially ameliorate the adverse impact. For these reasons, the mitigation measures, once identified, are then evaluated under Step 7, and the benefits or detriments are captured in the summary sheets prepared under Step 8. Step 10 is then revisited to address the effect of the mitigation measures under each alternative.

We note here that *mitigation* has two different meanings in the NEPA context. It may refer either to impact *avoidance* measures or to *compensation* measures to "atone" for some damage or loss. The latter context is regularly seen when one deals with habitat losses, particularly of wetlands.

Mitigation measures can be included in both an EIS and an EA. Some agencies require additional agency or public involvement if the mitigation measures included in an EA would reduce the severity of an impact from "significant" to "not significant," because the issuance of a FONSI (instead of preparation of an EIS) then becomes dependent upon proper implementation of the mitigation measures. This promise has been interpreted as having the weight of law in some jurisdictions. A postproject mitigation-monitoring plan may be required to assure that the promised mitigations are implemented.

Step 11.   Prepare Assessment Document

The agency must document its NEPA review in plain, easy-to-understand language. The type of documentation depends upon the type of review (see Chapter 4). Completed NEPA documents are publicly available under the provisions of NEPA and the Freedom of Information Act (subject to the exclusions of those laws regarding public disclosure of certain types of information, such as classified or proprietary material), although most agencies have policies or methods to ease the process for making NEPA documents readily available. Recent NEPA documents of many agencies are routinely placed on-line.

- *Categorical exclusions.* Some agencies have requirements to document all or certain types of categorical exclusions, often with some sort of standard form. Sometimes the documentation includes a checklist or other annotation that extraordinary circumstances, such as the presence of threatened or endangered species, were considered and did not affect the review.

- *Environmental assessments.* EAs are always documented. Some agencies prepare a draft EA for public review, similar to the EIS process. Most agencies follow the format for EISs, although some agencies follow other formats or options for the EA. If the EA leads to a FONSI, that finding must be documented. Some agencies also prepare a decision record for actions assessed through an EA and FONSI.

- *Environmental impact statements.* The CEQ regulations prescribe the requirements for an EIS and the ensuing ROD. Although exceptions may be granted, most agencies closely follow the recommended format.

- *Other environmental reviews.* Some agencies prepare other types of NEPA reviews, such as special environmental analyses, supplemental analyses, assessment memoranda, NEPA strategies, or implementation plans. Many agencies combine NEPA reviews with other types of environmental reviews, such as floodplain and wetland assessments or hazardous waste–related permitting processes. The agency's requirements for special types of reviews or combined reviews are generally spelled out in their regulations, handbooks, guidance, or policy, and should be followed closely.

Step 12.   Review and Process Document

The last step of the NEPA process outlined in Fig. C.1 is to make the NEPA document available for review, and process the document through proper administrative channels. Most agencies have specific directives for this step, which vary depending upon the type of NEPA document being prepared. Each agency has specific ways in which NEPA documents are reviewed internally and approved for public release, and these should be followed closely.

An agency must print sufficient copies of the NEPA document for agency and public distribution. For large documents, such as a multi-volume EIS, most agencies follow the U.S. Government Printing Office requirements, although there are some exceptions. Depending on the complexity of the document, printing may take some time. More and more agencies are making documents available on-line as an alternative to printing; however, to date, the agency must make printed copies available as well. For greatest efficiency and to allow for ease in searching and accessing document information, new documents that are put on-line should be prepared in specific electronic formats; the document preparer should anticipate this requirement before the document is finalized to avoid costly delays at the end of the process. Documents can be electronically scanned, but this can limit the ease of accessing information in the document.

The Environmental Protection Agency (EPA) provides public notice of the availability of any draft or final EIS; these notices are published weekly in the *Federal Register*. The notification processes of the proponent agency, the EPA, and the *Federal Register* must be followed exactly or the agency may find that its public notification is not published at the time anticipated. The completed draft or final EIS must be filed with the EPA and must already be in distribution to agencies and the public before the Notice of Availability is published. An agency may publish, on its own behalf, notice of other types of NEPA documents in the *Federal Register*, such as a notice of availability of an EA, ROD, or FONSI, and some agencies require that these types of notices or documents be so published.

Once the NEPA review is complete, the agency must archive NEPA documents. Time frames vary among agencies and by type of document, but for most NEPA documents the agency must keep an archived copy indefinitely. Some agencies also require that the complete administrative record (including, for example, background reports or analyses) be kept as well. Some agencies allow for disposal of certain kinds of NEPA documents after a fixed period, often several years.

Appendix

D

Regulations for Implementing Procedural Provisions of the National Environmental Policy Act

This appendix presents the full text of provisions of the NEPA Regulations, found at 40 CFR 1500-1508. It contains the changes made in 1986.

PART 1500—PURPOSE, POLICY, AND MANDATE

Sec.
1500.1 Purpose.
1500.2 Policy.
1500.3 Mandate.
1500.4 Reducing paperwork.
1500.5 Reducing delay.
1500.6 Agency authority.

AUTHORITY: NEPA, the Environmental Quality Improvement Act of 1970, as amended (42 U.S.C. 4371 *et seq.*), sec. 309 of the Clean Air Act, as amended (42 U.S.C. 7609) and E.O. 11514, Mar. 5, 1970, as amended by E.O. 11991, May 24, 1977).

SOURCE: 43 FR 55990, Nov. 28, 1978, unless otherwise noted.

§ 1500.1 Purpose.

(a) The National Environmental Policy Act (NEPA) is our basic national charter for protection of the environment. It establishes policy, sets goals (section 101), and provides means (section 102) for carrying out the policy. Section 102(2) contains "action-forcing" provisions to make sure that federal agencies act according to the letter and spirit of the Act. The regulations that follow implement section 102(2). Their purpose is to tell federal agencies what they must do to comply with the procedures and achieve the goals of the Act. The President, the federal agencies, and the courts share responsibility for enforcing the Act so as to achieve the substantive requirements of section 101.

(b) NEPA procedures must insure that environmental information is available to public officials and citizens before decisions are made and before actions are taken. The information must be of high quality. Accurate scientific analysis, expert agency comments, and public scrutiny are essential to implementing NEPA. Most important, NEPA documents must concentrate on the issues that are truly significant to the action in question, rather than amassing needless detail.

(c) Ultimately, of course, it is not better documents but better decisions that count. NEPA's purpose is not to generate paperwork—even excellent paperwork—but to foster excellent action. The NEPA process is intended to help public officials make decisions that are based on understanding of environmental consequences, and take actions that protect, restore, and enhance the environment. These regulations provide the direction to achieve this purpose.

§ 1500.2 Policy.

Federal agencies shall to the fullest extent possible:

(a) Interpret and administer the policies, regulations, and public laws of the United States in accordance with the policies set forth in the Act and in these regulations.

(b) Implement procedures to make the NEPA process more useful to decisionmakers and the public; to reduce paperwork and the accumulation of extraneous background data; and to emphasize real environmental issues and alternatives. Environmental impact statements shall be concise, clear, and to the point, and shall be supported by evidence that agencies have made the necessary environmental analyses.

(c) Integrate the requirements of NEPA with other planning and environmental review procedures required by law or by agency practice so that all such procedures run concurrently rather than consecutively.

(d) Encourage and facilitate public involvement in decisions which affect the quality of the human environment.

(e) Use the NEPA process to identify and assess the reasonable alternatives to proposed actions that will avoid or minimize adverse effects of these actions upon the quality of the human environment.

(f) Use all practicable means, consistent with the requirements of the Act and other essential considerations of national policy, to restore and enhance the quality of the human environment and avoid or minimize any possible adverse effects of their actions upon the quality of the human environment.

§ 1500.3 Mandate.

Parts 1500 through 1508 of this title provide regulations applicable to and binding on all Federal agencies for implementing the procedural provisions of the National Environmental Policy Act of 1969, as amended (Pub. L. 91-190, 42 U.S.C. 4321 *et seq.*) (NEPA or the Act) except where compliance would be inconsistent with other statutory requirements. These regulations are issued pursuant to NEPA, the Environmental Quality Improvement Act of 1970, as amended (42 U.S.C. 4371 *et seq.*) section 309 of the Clean Air Act, as amended (42 U.S.C. 7609) and Executive Order 11514, Protection and Enhancement of Environmental Quality (March 5, 1970, as amended by Executive Order 11991, May 24, 1977). These regulations, unlike the predecessor guidelines, are not confined to sec. 102(2)(C) (environmental impact statements). The regulations apply to the whole of section 102(2). The provisions of the Act and of these regulations must be read together as a whole in order to comply with the spirit and letter of the law. It is the Council's intention that judicial review of agency compliance with these regulations not occur before an agency has filed the final environmental impact statement, or has made a final finding of no significant impact (when such a finding will result in action affecting the environment), or takes action that will result in irreparable injury. Furthermore, it is the Council's intention that any trivial violation of these regulations not give rise to any independent cause of action.

§ 1500.4 Reducing paperwork.

Agencies shall reduce excessive paperwork by:

(a) Reducing the length of environmental impact statements (§ 1502.2(c)), by means such as setting appropriate page limits (§§ 1501.7(b)(1) and 1502.7).

(b) Preparing analytic rather than encyclopedic environmental impact statements (§ 1502.2(a)).

(c) Discussing only briefly issues other than significant ones (§ 1502.2(b)).

(d) Writing environmental impact statements in plain language (§ 1502.8).

(e) Following a clear format for environmental impact statements (§ 1502.10).

(f) Emphasizing the portions of the environmental impact statement that are useful to decisionmakers and the public (§§ 1502.14 and 1502.15) and reducing emphasis on background material (§ 1502.16).

(g) Using the scoping process, not only to identify significant environmental issues deserving of study, but also to deemphasize insignificant issues, narrowing the scope of the environmental impact statement process accordingly (§ 1501.7).

(h) Summarizing the environmental impact statement (§ 1502.12) and circulating the summary instead of the entire environmental impact statement if the latter is unusually long (§ 1502.19).

(i) Using program, policy, or plan environmental impact statements and tiering from statements of broad scope to those of narrower scope, to eliminate repetitive discussions of the same issues (§§ 1502.4 and 1502.20).

(j) Incorporating by reference (§ 1502.21).

(k) Integrating NEPA requirements with other environmental review and consultation requirements (§ 1502.25).

(l) Requiring comments to be as specific as possible (§ 1503.3).

(m) Attaching and circulating only changes to the draft environmental impact statement, rather than rewriting and circulating the entire statement when changes are minor (§ 1503.4(c)).

(n) Eliminating duplication with State and local procedures, by providing for joint preparation (§ 1506.2), and with other Federal procedures, by providing that an agency may adopt appropriate environmental documents prepared by another agency (§ 1506.3).

(o) Combining environmental documents with other documents (§ 1506.4).

(p) Using categorical exclusions to define categories of actions which do not individually or cumulatively have a significant effect on the human environment and which are therefore exempt from requirements to prepare an environmental impact statement (§ 1508.4).

(q) Using a finding of no significant impact when an action not otherwise excluded will not have a significant effect on the human environment and is therefore exempt from requirements to prepare an environmental impact statement (§ 1508.13).

[43 FR 55990, Nov. 29, 1978; 44 FR 873, Jan. 3, 1979]

§ 1500.5 Reducing delay.

Agencies shall reduce delay by:

(a) Integrating the NEPA process into early planning (§ 1501.2).

(b) Emphasizing interagency cooperation before the environmental impact statement is prepared, rather than submission of adversary comments on a completed document (§ 1501.6).

(c) Insuring the swift and fair resolution of lead agency disputes (§ 1501.5).

(d) Using the scoping process for an early identification of what are and what are not the real issues (§ 1501.7).

(e) Establishing appropriate time limits for the environmental impact statement process (§§ 1501.7(b)(2) and 1501.8).

(f) Preparing environmental impact statements early in the process (§ 1502.5).

(g) Integrating NEPA requirements with other environmental review and consultation requirements (§ 1502.25).

(h) Eliminating duplication with State and local procedures by providing for joint preparation (§ 1506.2) and with other Federal procedures by providing that an agency may adopt appropriate environmental documents prepared by another agency (§ 1506.3).

(i) Combining environmental documents with other documents (§ 1506.4).

(j) Using accelerated procedures for proposals for legislation (§ 1506.8).

(k) Using categorical exclusions to define categories of actions which do not individually or cumulatively have a significant effect on the human environment (§ 1508.4) and which are therefore exempt from requirements to prepare an environmental impact statement.

(l) Using a finding of no significant impact when an action not otherwise excluded will not have a significant effect on the human environment (§ 1508.13) and is therefore exempt from requirements to prepare an environmental impact statement.

§ 1500.6 Agency authority.

Each agency shall interpret the provisions of the Act as a supplement to its existing authority and as a mandate to view traditional policies and missions in the light of the Act's national environmental objectives. Agencies shall review their policies, procedures, and regulations accordingly and revise them as necessary to insure full compliance with the purposes and provisions of the Act. The phrase "to the fullest extent possible" in section 102 means that each agency of the Federal Government shall comply with that section unless existing law applicable to the agency's operations expressly prohibits or makes compliance impossible.

PART 1501—NEPA AND AGENCY PLANNING

Sec.
1501.1 Purpose.
1501.2 Apply NEPA early in the process.
1501.3 When to prepare an environmental assessment.
1501.4 Whether to prepare an environmental impact statement.
1501.5 Lead agencies.
1501.6 Cooperating agencies.
1501.7 Scoping.
1501.8 Time limits.

AUTHORITY: NEPA, the Environmental Quality Improvement Act of 1970, as amended (42 U.S.C. 4371 *et seq.*), sec. 309 of the Clean Air Act, as amended (42 U.S.C. 7609, and E.O. 11514 (Mar. 5, 1970, as amended by E.O. 11991, May 24, 1977).

SOURCE: 43 FR 55992, Nov. 29, 1978, unless otherwise noted.

§ 1501.1 Purpose.

The purposes of this part include:

(a) Integrating the NEPA process into early planning to insure appropriate consideration of NEPA's policies and to eliminate delay.

(b) Emphasizing cooperative consultation among agencies before the environmental impact statement is prepared rather than submission of adversary comments on a completed document.

(c) Providing for the swift and fair resolution of lead agency disputes.

(d) Identifying at an early stage the significant environmental issues deserving of study and deemphasizing insignificant issues, narrowing the scope of the environmental impact statement accordingly.

(e) Providing a mechanism for putting appropriate time limits on the environmental impact statement process.

§ 1501.2 Apply NEPA early in the process.

Agencies shall integrate the NEPA process with other planning at the earliest possible time to insure that planning and decisions reflect environmental values, to avoid delays later in the process, and to head off potential conflicts. Each agency shall:

(a) Comply with the mandate of section 102(2)(A) to "utilize a systematic, interdisciplinary approach which will insure the integrated use of the natural and social sciences and the environmental design arts in planning and in decisionmaking which may have an impact on man's environment," as specified by § 1507.2.

(b) Identify environmental effects and values in adequate detail so they can be compared to economic and technical analyses. Environmental documents and appropriate analyses shall be circulated and reviewed at the same time as other planning documents.

(c) Study, develop, and describe appropriate alternatives to recommended courses of action in any proposal which involves unresolved conflicts concerning alternative uses of available resources as provided by section 102(2)(E) of the Act.

(d) Provide for cases where actions are planned by private applicants or other non-Federal entities before Federal involvement so that:

(1) Policies or designated staff are available to advise potential applicants of studies or other information foreseeably required for later Federal action.

(2) The Federal agency consults early with appropriate State and local agencies and Indian tribes and with interested private persons and organizations when its own involvement is reasonably foreseeable.

(3) The Federal agency commences its NEPA process at the earliest possible time.

§ 1501.3 When to prepare an environmental assessment.

(a) Agencies shall prepare an environmental assessment (§ 1508.9) when necessary under the procedures adopted by individual agencies to supplement these regulations as described in § 1507.3. An assessment is not necessary if the agency has decided to prepare an environmental impact statement.

(b) Agencies may prepare an environmental assessment on any action at any time in order to assist agency planning and decisionmaking.

§ 1501.4 Whether to prepare an environmental impact statement.

In determining whether to prepare an environmental impact statement the Federal agency shall:

(a) Determine under its procedures supplementing these regulations (described in § 1507.3) whether the proposal is one which:

(1) Normally requires an environmental impact statement, or

(2) Normally does not require either an environmental impact statement or an environmental assessment (categorical exclusion).

(b) If the proposed action is not covered by paragraph (a) of this section, prepare an environmental assessment (§ 1508.9). The agency shall involve environmental agencies, applicants, and the public, to the extent practicable, in preparing assessments required by § 1508.9(a)(1).

(c) Based on the environmental assessment make its determination whether to prepare an environmental impact statement.

(d) Commence the scoping process (§ 1501.7), if the agency will prepare an environmental impact statement.

(e) Prepare a finding of no significant impact (§ 1508.13), if the agency determines on the basis of the environmental assessment not to prepare a statement.

(1) The agency shall make the finding of no significant impact available to the affected public as specified in § 1506.6.

(2) In certain limited circumstances, which the agency may cover in its procedures under § 1507.3, the agency shall make the finding of no significant impact available for public review (including State and areawide clearinghouses) for 30 days before the agency makes its final determination whether to prepare an environmental impact statement and before the action may begin. The circumstances are:

(i) The proposed action is, or is closely similar to, one which normally requires the preparation of an environmental impact statement under the procedures adopted by the agency pursuant to § 1507.3, or

(ii) The nature of the proposed action is one without precedent.

§ 1501.5 Lead agencies.

(a) A lead agency shall supervise the preparation of an environmental impact statement if more than one Federal agency either:

(1) Proposes or is involved in the same action; or

(2) Is involved in a group of actions directly related to each other because of their functional interdependence or geographical proximity.

(b) Federal, State, or local agencies, including at least one Federal agency, may act as joint lead agencies to prepare an environmental impact statement (§ 1506.2).

(c) If an action falls within the provisions of paragraph (a) of this section the potential lead agencies shall determine by letter or memorandum which agency shall be the lead agency and which shall be cooperating agencies. The agencies shall resolve the lead agency question so as not to cause delay. If there is disagreement among the agencies, the following factors (which are listed in order of descending importance) shall determine lead agency designation:

(1) Magnitude of agency's involvement.

(2) Project approval/disapproval authority.

(3) Expertise concerning the action's environmental effects.

(4) Duration of agency's involvement.

(5) Sequence of agency's involvement.

(d) Any Federal agency, or any State or local agency or private person substantially affected by the absence of lead agency designation, may make a written request to the potential lead agencies that a lead agency be designated.

(e) If Federal agencies are unable to agree on which agency will be the lead agency or if the procedure described in paragraph (c) of this section has not resulted within 45 days in a lead agency

designation, any of the agencies or persons concerned may file a request with the Council asking it to determine which Federal agency shall be the lead agency.

A copy of the request shall be transmitted to each potential lead agency. The request shall consist of:

(1) A precise description of the nature and extent of the proposed action.

(2) A detailed statement of why each potential lead agency should or should not be the lead agency under the criteria specified in paragraph (c) of this section.

(f) A response may be filed by any potential lead agency concerned within 20 days after a request is filed with the Council. The Council shall determine as soon as possible but not later than 20 days after receiving the request and all responses to it which Federal agency shall be the lead agency and which other Federal agencies shall be cooperating agencies.

[43 FR 55992, Nov. 29, 1978; 44 FR 873, Jan. 3, 1979]

§ 1501.6 Cooperating agencies.

The purpose of this section is to emphasize agency cooperation early in the NEPA process. Upon request of the lead agency, any other Federal agency which has jurisdiction by law shall be a cooperating agency. In addition any other Federal agency which has special expertise with respect to any environmental issue, which should be addressed in the statement may be a cooperating agency upon request of the lead agency. An agency may request the lead agency to designate it a cooperating agency.

(a) The lead agency shall:

(1) Request the participation of each cooperating agency in the NEPA process at the earliest possible time.

(2) Use the environmental analysis and proposals of cooperating agencies with jurisdiction by law or special expertise, to the maximum extent possible consistent with its responsibility as lead agency.

(3) Meet with a cooperating agency at the latter's request.

(b) Each cooperating agency shall:

(1) Participate in the NEPA process at the earliest possible time.

(2) Participate in the scoping process (described below in § 1501.7).

(3) Assume on request of the lead agency responsibility for developing information and preparing environmental analyses including portions of the environmental impact statement concerning which the cooperating agency has special expertise.

(4) Make available staff support at the lead agency's request to enhance the latter's interdisciplinary capability.

(5) Normally use its own funds. The lead agency shall, to the extent available funds permit, fund those major activities or analyses it requests from cooperating agencies. Potential lead agencies shall include such funding requirements in their budget requests.

(c) A cooperating agency may in response to a lead agency's request for assistance in preparing the environmental impact statement (described in paragraph (b)(3), (4), or (5) of this section) reply that other program commitments preclude any involvement or the degree of involvement requested in the action that is the subject of the environmental impact statement. A copy of this reply shall be submitted to the Council.

§ 1501.7 Scoping.

There shall be an early and open process for determining the scope of issues to be addressed and for identifying the significant issues related to a proposed action. This process shall be termed scoping. As soon as practicable after its decision to prepare an environmental impact statement and before the scoping process the lead agency shall publish a notice of intent (§ 1508.22) in the FEDERAL REGISTER except as provided in § 1507.3(e).

(a) As part of the scoping process the lead agency shall:

(1) Invite the participation of affected Federal, State, and local agencies, any affected Indian tribe, the proponent of the action, and other interested persons (including those who might not be in accord with the action on environmental grounds), unless there is a limited exception under § 1507.3(c). An agency may give notice in accordance with § 1506.6.

(2) Determine the scope (§ 1508.25) and the significant issues to be analyzed in depth in the environmental impact statement.

(3) Identify and eliminate from detailed study the issues which are not significant or which have been covered by prior environmental review (§ 1506.3), narrowing the discussion of these issues in the statement to a brief presentation of why they will not have a significant effect on the human environment or providing a reference to their coverage elsewhere.

(4) Allocate assignments for preparation of the environmental impact statement among the lead and cooperating agencies, with the lead agency retaining responsibility for the statement.

(5) Indicate any public environmental assessments and other environmental impact statements which are being or will be prepared that are related to but are not part of the scope of the impact statement under consideration.

(6) Identify other environmental review and consultation requirements so the lead and cooperating agencies may prepare other required analyses and studies concurrently with, and integrated with, the environmental impact statement as provided in § 1502.25.

(7) Indicate the relationship between the timing of the preparation of environmental analyses and the agency's tentative planning and decisionmaking schedule.

(b) As part of the scoping process the lead agency may:

(1) Set page limits on environmental documents (§ 1502.7).

(2) Set time limits (§ 1501.8).

(3) Adopt procedures under § 1507.3 to combine its environmental assessment process with its scoping process.

(4) Hold an early scoping meeting or meetings which may be integrated with any other early planning meeting the agency has. Such a scoping meeting will often be appropriate when the impacts of a particular action are confined to specific sites.

(c) An agency shall revise the determinations made under paragraphs (a) and (b) of this section if substantial changes are made later in the proposed

Council on Environmental Quality   **§ 1502.1**

action, or if significant new circumstances or information arise which bear on the proposal or its impacts.

§ 1501.8 Time limits.

Although the Council has decided that prescribed universal time limits for the entire NEPA process are too inflexible, Federal agencies are encouraged to set time limits appropriate to individual actions (consistent with the time intervals required by § 1506.10). When multiple agencies are involved the reference to agency below means lead agency.

(a) The agency shall set time limits if an applicant for the proposed action requests them: *Provided,* That the limits are consistent with the purposes of NEPA and other essential considerations of national policy.

(b) The agency may:

(1) Consider the following factors in determining time limits:

(i) Potential for environmental harm.

(ii) Size of the proposed action.

(iii) State of the art of analytic techniques.

(iv) Degree of public need for the proposed action, including the consequences of delay.

(v) Number of persons and agencies affected.

(vi) Degree to which relevant information is known and if not known the time required for obtaining it.

(vii) Degree to which the action is controversial.

(viii) Other time limits imposed on the agency by law, regulations, or executive order.

(2) Set overall time limits or limits for each constituent part of the NEPA process, which may include:

(i) Decision on whether to prepare an environmental impact statement (if not already decided).

(ii) Determination of the scope of the environmental impact statement.

(iii) Preparation of the draft environmental impact statement.

(iv) Review of any comments on the draft environmental impact statement from the public and agencies.

(v) Preparation of the final environmental impact statement.

(vi) Review of any comments on the final environmental impact statement.

(vii) Decision on the action based in part on the environmental impact statement.

(3) Designate a person (such as the project manager or a person in the agency's office with NEPA responsibilities) to expedite the NEPA process.

(c) State or local agencies or members of the public may request a Federal Agency to set time limits.

PART 1502—ENVIRONMENTAL IMPACT STATEMENT

Sec.
1502.1 Purpose.
1502.2 Implementation.
1502.3 Statutory requirements for statements.
1502.4 Major Federal actions requiring the preparation of environmental impact statements.
1502.5 Timing.
1502.6 Interdisciplinary preparation.
1502.7 Page limits.
1502.8 Writing.
1502.9 Draft, final, and supplemental statements.
1502.10 Recommended format.
1502.11 Cover sheet.
1502.12 Summary.
1502.13 Purpose and need.
1502.14 Alternatives including the proposed action.
1502.15 Affected environment.
1502.16 Environmental consequences.
1502.17 List of preparers.
1502.18 Appendix.
1502.19 Circulation of the environmental impact statement.
1502.20 Tiering.
1502.21 Incorporation by reference.
1502.22 Incomplete or unavailable information.
1502.23 Cost-benefit analysis.
1502.24 Methodology and scientific accuracy.
1502.25 Environmental review and consultation requirements.

AUTHORITY: NEPA, the Environmental Quality Improvement Act of 1970, as amended (42 U.S.C. 4371 *et seq.*), sec. 309 of the Clean Air Act, as amended (42 U.S.C. 7609), and E.O. 11514 (Mar. 5, 1970, as amended by E.O. 11991, May 24, 1977).

SOURCE: 43 FR 55994, Nov. 29, 1978, unless otherwise noted.

§ 1502.1 Purpose.

The primary purpose of an environmental impact statement is to serve as an action-forcing device to insure that the policies and goals defined in the

Act are infused into the ongoing programs and actions of the Federal Government. It shall provide full and fair discussion of significant environmental impacts and shall inform decisionmakers and the public of the reasonable alternatives which would avoid or minimize adverse impacts or enhance the quality of the human environment. Agencies shall focus on significant environmental issues and alternatives and shall reduce paperwork and the accumulation of extraneous background data. Statements shall be concise, clear, and to the point, and shall be supported by evidence that the agency has made the necessary environmental analyses. An environmental impact statement is more than a disclosure document. It shall be used by Federal officials in conjunction with other relevant material to plan actions and make decisions.

§ 1502.2 Implementation.

To achieve the purposes set forth in § 1502.1 agencies shall prepare environmental impact statements in the following manner:

(a) Environmental impact statements shall be analytic rather than encyclopedic.

(b) Impacts shall be discussed in proportion to their significance. There shall be only brief discussion of other than significant issues. As in a finding of no significant impact, there should be only enough discussion to show why more study is not warranted.

(c) Environmental impact statements shall be kept concise and shall be no longer than absolutely necessary to comply with NEPA and with these regulations. Length should vary first with potential environmental problems and then with project size.

(d) Environmental impact statements shall state how alternatives considered in it and decisions based on it will or will not achieve the requirements of sections 101 and 102(1) of the Act and other environmental laws and policies.

(e) The range of alternatives discussed in environmental impact statements shall encompass those to be considered by the ultimate agency decisionmaker.

(f) Agencies shall not commit resources prejudicing selection of alternatives before making a final decision (§ 1506.1).

(g) Environmental impact statements shall serve as the means of assessing the environmental impact of proposed agency actions, rather than justifying decisions already made.

§ 1502.3 Statutory requirements for statements.

As required by sec. 102(2)(C) of NEPA environmental impact statements (§ 1508.11) are to be included in every recommendation or report.

On proposals (§ 1508.23).

For legislation and (§ 1508.17).

Other major Federal actions (§ 1508.18).

Significantly (§ 1508.27).

Affecting (§§ 1508.3, 1508.8).

The quality of the human environment (§ 1508.14).

§ 1502.4 Major Federal actions requiring the preparation of environmental impact statements.

(a) Agencies shall make sure the proposal which is the subject of an environmental impact statement is properly defined. Agencies shall use the criteria for scope (§ 1508.25) to determine which proposal(s) shall be the subject of a particular statement. Proposals or parts of proposals which are related to each other closely enough to be, in effect, a single course of action shall be evaluated in a single impact statement.

(b) Environmental impact statements may be prepared, and are sometimes required, for broad Federal actions such as the adoption of new agency programs or regulations (§ 1508.18). Agencies shall prepare statements on broad actions so that they are relevant to policy and are timed to coincide with meaningful points in agency planning and decisionmaking.

(c) When preparing statements on broad actions (including proposals by more than one agency), agencies may find it useful to evaluate the proposal(s) in one of the following ways:

(1) Geographically, including actions occurring in the same general location, such as body of water, region, or metropolitan area.

(2) Generically, including actions which have relevant similarities, such

as common timing, impacts, alternatives, methods of implementation, media, or subject matter.

(3) By stage of technological development including federal or federally assisted research, development or demonstration programs for new technologies which, if applied, could significantly affect the quality of the human environment. Statements shall be prepared on such programs and shall be available before the program has reached a stage of investment or commitment to implementation likely to determine subsequent development or restrict later alternatives.

(d) Agencies shall as appropriate employ scoping (§ 1501.7), tiering (§ 1502.20), and other methods listed in §§ 1500.4 and 1500.5 to relate broad and narrow actions and to avoid duplication and delay.

§ 1502.5 Timing.

An agency shall commence preparation of an environmental impact statement as close as possible to the time the agency is developing or is presented with a proposal (§ 1508.23) so that preparation can be completed in time for the final statement to be included in any recommendation or report on the proposal. The statement shall be prepared early enough so that it can serve practically as an important contribution to the decisionmaking process and will not be used to rationalize or justify decisions already made (§§ 1500.2(c), 1501.2, and 1502.2). For instance:

(a) For projects directly undertaken by Federal agencies the environmental impact statement shall be prepared at the feasibility analysis (go-no go) stage and may be supplemented at a later stage if necessary.

(b) For applications to the agency appropriate environmental assessments or statements shall be commenced no later than immediately after the application is received. Federal agencies are encouraged to begin preparation of such assessments or statements earlier, preferably jointly with applicable State or local agencies.

(c) For adjudication, the final environmental impact statement shall normally precede the final staff recommendation and that portion of the public hearing related to the impact study. In appropriate circumstances the statement may follow preliminary hearings designed to gather information for use in the statements.

(d) For informal rulemaking the draft environmental impact statement shall normally accompany the proposed rule.

§ 1502.6 Interdisciplinary preparation.

Environmental impact statements shall be prepared using an inter-disciplinary approach which will insure the integrated use of the natural and social sciences and the environmental design arts (section 102(2)(A) of the Act). The disciplines of the preparers shall be appropriate to the scope and issues identified in the scoping process (§ 1501.7).

§ 1502.7 Page limits.

The text of final environmental impact statements (e.g., paragraphs (d) through (g) of § 1502.10) shall normally be less than 150 pages and for proposals of unusual scope or complexity shall normally be less than 300 pages.

§ 1502.8 Writing.

Environmental impact statements shall be written in plain language and may use appropriate graphics so that decisionmakers and the public can readily understand them. Agencies should employ writers of clear prose or editors to write, review, or edit statements, which will be based upon the analysis and supporting data from the natural and social sciences and the environmental design arts.

§ 1502.9 Draft, final, and supplemental statements.

Except for proposals for legislation as provided in § 1506.8 environmental impact statements shall be prepared in two stages and may be supplemented.

(a) Draft environmental impact statements shall be prepared in accordance with the scope decided upon in the scoping process. The lead agency shall work with the cooperating agencies and shall obtain comments as required in part 1503 of this chapter. The draft statement must fulfill and satisfy to the fullest extent possible the requirements established for final statements in section 102(2)(C) of the Act. If a draft statement is so inadequate as to preclude meaningful analysis, the agency shall prepare and circulate a revised draft of the appropriate portion. The agency shall make every effort to disclose and discuss at appropriate points in the draft statement all major points of view on the environmental impacts of the alternatives including the proposed action.

(b) Final environmental impact statements shall respond to comments as required in part 1503 of this chapter. The agency shall discuss at appropriate points in the final statement any responsible opposing view which was not adequately discussed in the draft statement and shall indicate the agency's response to the issues raised.

(c) Agencies:

(1) Shall prepare supplements to either draft or final environmental impact statements if:

(i) The agency makes substantial changes in the proposed action that are relevant to environmental concerns; or

(ii) There are significant new circumstances or information relevant to environmental concerns and bearing on the proposed action or its impacts.

(2) May also prepare supplements when the agency determines that the purposes of the Act will be furthered by doing so.

(3) Shall adopt procedures for introducing a supplement into its formal administrative record, if such a record exists.

(4) Shall prepare, circulate, and file a supplement to a statement in the same fashion (exclusive of scoping) as a draft and final statement unless alternative procedures are approved by the Council.

§ 1502.10   Recommended format.

Agencies shall use a format for environmental impact statements which will encourage good analysis and clear presentation of the alternatives including the proposed action. The following standard format for environmental impact statements should be followed unless the agency determines that there is a compelling reason to do otherwise:

(a) Cover sheet.
(b) Summary.
(c) Table of contents.

(d) Purpose of and need for action.

(e) Alternatives including proposed action (sections 102(2)(C)(iii) and 102(2)(E) of the Act).

(f) Affected environment.

(g) Environmental consequences (especially sections 102(2)(C)(i), (ii), (iv), and (v) of the Act).

(h) List of preparers.

(i) List of Agencies, Organizations, and persons to whom copies of the statement are sent.

(j) Index.

(k) Appendices (if any).

If a different format is used, it shall include paragraphs (a), (b), (c), (h), (i), and (j), of this section and shall include the substance of paragraphs (d), (e), (f), (g), and (k) of this section, as further described in §§ 1502.11 through 1502.18, in any appropriate format.

§ 1502.11   Cover sheet.

The cover sheet shall not exceed one page. It shall include:

(a) A list of the responsible agencies including the lead agency and any cooperating agencies.

(b) The title of the proposed action that is the subject of the statement (and if appropriate the titles of related cooperating agency actions), together with the State(s) and county(ies) (or other jurisdiction if applicable) where the action is located.

(c) The name, address, and telephone number of the person at the agency who can supply further information.

(d) A designation of the statement as a draft, final, or draft or final supplement.

(e) A one paragraph abstract of the statement.

(f) The date by which comments must be received (computed in cooperation with EPA under § 1506.10).

The information required by this section may be entered on Standard Form 424 (in items 4, 6, 7, 10, and 18).

§ 1502.12   Summary.

Each environmental impact statement shall contain a summary which adequately and accurately summarizes the statement. The summary shall stress the major conclusions, areas of controversy (including issues raised by agencies and the public), and the issues to be resolved (including the choice among alternatives). The summary will normally not exceed 15 pages.

§ 1502.13 Purpose and need.

The statement shall briefly specify the underlying purpose and need to which the agency is responding in proposing the alternatives including the proposed action.

§ 1502.14 Alternatives including the proposed action.

This section is the heart of the environmental impact statement. Based on the information and analysis presented in the sections on the Affected Environment (§ 1502.15) and the Environmental Consequences (§ 1502.16), it should present the environmental impacts of the proposal and the alternatives in comparative form, thus sharply defining the issues and providing a clear basis for choice among options by the decisionmaker and the public. In this section agencies shall:

(a) Rigorously explore and objectively evaluate all reasonable alternatives, and for alternatives which were eliminated from detailed study, briefly discuss the reasons for their having been eliminated.

(b) Devote substantial treatment to each alternative considered in detail including the proposed action so that reviewers may evaluate their comparative merits.

(c) Include reasonable alternatives not within the jurisdiction of the lead agency.

(d) Include the alternative of no action.

(e) Identify the agency's preferred alternative or alternatives, if one or more exists, in the draft statement and identify such alternative in the final statement unless another law prohibits the expression of such a preference.

(f) Include appropriate mitigation measures not already included in the proposed action or alternatives.

§ 1502.15 Affected environment.

The environmental impact statement shall succinctly describe the environment of the area(s) to be affected or created by the alternatives under consideration. The descriptions shall be no longer than is necessary to understand the effects of the alternatives. Data

and analyses in a statement shall be commensurate with the importance of the impact, with less important material summarized, consolidated, or simply referenced. Agencies shall avoid useless bulk in statements and shall concentrate effort and attention on important issues. Verbose descriptions of the affected environment are themselves no measure of the adequacy of an environmental impact statement.

§ 1502.16 Environmental consequences.

This section forms the scientific and analytic basis for the comparisons under § 1502.14. It shall consolidate the discussions of those elements required by sections 102(2)(C)(i), (ii), (iv), and (v) of NEPA which are within the scope of the statement and as much of section 102(2)(C)(iii) as is necessary to support the comparisons. The discussion will include the environmental impacts of the alternatives including the proposed action, any adverse environmental effects which cannot be avoided should the proposal be implemented, the relationship between short-term uses of man's environment and the maintenance and enhancement of long-term productivity, and any irreversible or irretrievable commitments of resources which would be involved in the proposal should it be implemented. This section should not duplicate discussions in § 1502.14. It shall include discussions of:

(a) Direct effects and their significance (§ 1508.8).

(b) Indirect effects and their significance (§ 1508.8).

(c) Possible conflicts between the proposed action and the objectives of Federal, regional, State, and local (and in the case of a reservation, Indian tribe) land use plans, policies and controls for the area concerned. (See § 1506.2(d).)

(d) The environmental effects of alternatives including the proposed action. The comparisons under § 1502.14 will be based on this discussion.

(e) Energy requirements and conservation potential of various alternatives and mitigation measures.

(f) Natural or depletable resource requirements and conservation potential of various alternatives and mitigation measures.

(g) Urban quality, historic and cultural resources, and the design of the built environment, including the reuse and conservation potential of various alternatives and mitigation measures.

(h) Means to mitigate adverse environmental impacts (if not fully covered under § 1502.14(f)).

[43 FR 55994, Nov. 29, 1978; 44 FR 873, Jan. 3, 1979]

§ 1502.17 List of preparers.

The environmental impact statement shall list the names, together with their qualifications (expertise, experience, professional disciplines), of the persons who were primarily responsible for preparing the environmental impact statement or significant background papers, including basic components of the statement (§§ 1502.6 and 1502.8). Where possible the persons who are responsible for a particular analysis, including analyses in background papers, shall be identified. Normally the list will not exceed two pages.

§ 1502.18 Appendix.

If an agency prepares an appendix to an environmental impact statement the appendix shall:

(a) Consist of material prepared in connection with an environmental impact statement (as distinct from material which is not so prepared and which is incorporated by reference (§ 1502.21)).

(b) Normally consist of material which substantiates any analysis fundamental to the impact statement.

(c) Normally be analytic and relevant to the decision to be made.

(d) Be circulated with the environmental impact statement or be readily available on request.

§ 1502.19 Circulation of the environmental impact statement.

Agencies shall circulate the entire draft and final environmental impact statements except for certain appendices as provided in § 1502.18(d) and unchanged statements as provided in § 1503.4(c). However, if the statement is unusually long, the agency may circulate the summary instead, except that the entire statement shall be furnished to:

(a) Any Federal agency which has jurisdiction by law or special expertise with respect to any environmental impact involved and any appropriate Federal, State or local agency authorized to develop and enforce environmental standards.

(b) The applicant, if any.

(c) Any person, organization, or agency requesting the entire environmental impact statement.

(d) In the case of a final environmental impact statement any person, organization, or agency which submitted substantive comments on the draft.

If the agency circulates the summary and thereafter receives a timely request for the entire statement and for additional time to comment, the time for that requestor only shall be extended by at least 15 days beyond the minimum period.

§ 1502.20 Tiering.

Agencies are encouraged to tier their environmental impact statements to eliminate repetitive discussions of the same issues and to focus on the actual issues ripe for decision at each level of environmental review (§ 1508.28). Whenever a broad environmental impact statement has been prepared (such as a program or policy statement) and a subsequent statement or environmental assessment is then prepared on an action included within the entire program or policy (such as a site specific action) the subsequent statement or environmental assessment need only summarize the issues discussed in the broader statement and incorporate discussions from the broader statement by reference and shall concentrate on the issues specific to the subsequent action. The subsequent document shall state where the earlier document is available. Tiering may also be appropriate for different stages of actions. (Section 1508.28).

§ 1502.21 Incorporation by reference.

Agencies shall incorporate material into an environmental impact statement by reference when the effect will be to cut down on bulk without impeding agency and public review of the action. The incorporated material shall be cited in the statement and its content briefly described. No material

Council on Environmental Quality   § 1502.24

may be incorporated by reference unless it is reasonably available for inspection by potentially interested persons within the time allowed for comment. Material based on proprietary data which is itself not available for review and comment shall not be incorporated by reference.

§ 1502.22 Incomplete or unavailable information.

When an agency is evaluating reasonably foreseeable significant adverse effects on the human environment in an environmental impact statement and there is incomplete or unavailable information, the agency shall always make clear that such information is lacking.

(a) If the incomplete information relevant to reasonably foreseeable significant adverse impacts is essential to a reasoned choice among alternatives and the overall costs of obtaining it are not exorbitant, the agency shall include the information in the environmental impact statement.

(b) If the information relevant to reasonably foreseeable significant adverse impacts cannot be obtained because the overall costs of obtaining it are exorbitant or the means to obtain it are not known, the agency shall include within the environmental impact statement:

(1) A statement that such information is incomplete or unavailable; (2) a statement of the relevance of the incomplete or unavailable information to evaluating reasonably foreseeable significant adverse impacts on the human environment; (3) a summary of existing credible scientific evidence which is relevant to evaluating the reasonably foreseeable significant adverse impacts on the human environment, and (4) the agency's evaluation of such impacts based upon theoretical approaches or research methods generally accepted in the scientific community. For the purposes of this section, "reasonably foreseeable" includes impacts which have catastrophic consequences, even if their probability of occurrence is low, provided that the analysis of the impacts is supported by credible scientific

evidence, is not based on pure conjecture, and is within the rule of reason.

(c) The amended regulation will be applicable to all environmental impact statements for which a Notice of Intent (40 CFR 1508.22) is published in the FEDERAL REGISTER on or after May 27, 1986. For environmental impact statements in progress, agencies may choose to comply with the requirements of either the original or amended regulation.

[51 FR 15625, Apr. 25, 1986]

§ 1502.23 Cost-benefit analysis.

If a cost-benefit analysis relevant to the choice among environmentally different alternatives is being considered for the proposed action, it shall be incorporated by reference or appended to the statement as an aid in evaluating the environmental consequences. To assess the adequacy of compliance with section 102(2)(B) of the Act the statement shall, when a cost-benefit analysis is prepared, discuss the relationship between that analysis and any analyses of unquantified environmental impacts, values, and amenities. For purposes of complying with the Act, the weighing of the merits and drawbacks of the various alternatives need not be displayed in a monetary cost-benefit analysis and should not be when there are important qualitative considerations. In any event, an environmental impact statement should at least indicate those considerations, including factors not related to environmental quality, which are likely to be relevant and important to a decision.

§ 1502.24 Methodology and scientific accuracy.

Agencies shall insure the professional integrity, including scientific integrity, of the discussions and analyses in environmental impact statements. They shall identify any methodologies used and shall make explicit reference by footnote to the scientific and other sources relied upon for conclusions in the statement. An agency may place discussion of methodology in an appendix.

§ 1502.25 Environmental review and consultation requirements.

(a) To the fullest extent possible, agencies shall prepare draft environmental impact statements concurrently with and integrated with environmental impact analyses and related surveys and studies required by the Fish and Wildlife Coordination Act (16 U.S.C. 661 *et seq.*), the National Historic Preservation Act of 1966 (16 U.S.C. 470 *et seq.*), the Endangered Species Act of 1973 (16 U.S.C. 1531 *et seq.*), and other environmental review laws and executive orders.

(b) The draft environmental impact statement shall list all Federal permits, licenses, and other entitlements which must be obtained in implementing the proposal. If it is uncertain whether a Federal permit, license, or other entitlement is necessary, the draft environmental impact statement shall so indicate.

PART 1503—COMMENTING

Sec.
1503.1 Inviting comments.
1503.2 Duty to comment.
1503.3 Specificity of comments.
1503.4 Response to comments.

AUTHORITY: NEPA, the Environmental Quality Improvement Act of 1970, as amended (42 U.S.C. 4371 *et seq.*), sec. 309 of the Clean Air Act, as amended (42 U.S.C. 7609), and E.O. 11514 (Mar. 5, 1970, as amended by E.O. 11991, May 24, 1977).

SOURCE: 43 FR 55997, Nov. 29, 1978, unless otherwise noted.

§ 1503.1 Inviting comments.

(a) After preparing a draft environmental impact statement and before preparing a final environmental impact statement the agency shall:

(1) Obtain the comments of any Federal agency which has jurisdiction by law or special expertise with respect to any environmental impact involved or which is authorized to develop and enforce environmental standards.

(2) Request the comments of:

(i) Appropriate State and local agencies which are authorized to develop and enforce environmental standards;

(ii) Indian tribes, when the effects may be on a reservation; and

(iii) Any agency which has requested that it receive statements on actions of the kind proposed.

Office of Management and Budget Circular A-95 (Revised), through its system of clearinghouses, provides a means of securing the views of State and local environmental agencies. The clearinghouses may be used, by mutual agreement of the lead agency and the clearinghouse, for securing State and local reviews of the draft environmental impact statements.

(3) Request comments from the applicant, if any.

(4) Request comments from the public, affirmatively soliciting comments from those persons or organizations who may be interested or affected.

(b) An agency may request comments on a final environmental impact statement before the decision is finally made. In any case other agencies or persons may make comments before the final decision unless a different time is provided under § 1506.10.

§ 1503.2 Duty to comment.

Federal agencies with jurisdiction by law or special expertise with respect to any environmental impact involved and agencies which are authorized to develop and enforce environmental standards shall comment on statements within their jurisdiction, expertise, or authority. Agencies shall comment within the time period specified for comment in § 1506.10. A Federal agency may reply that it has no comment. If a cooperating agency is satisfied that its views are adequately reflected in the environmental impact statement, it should reply that it has no comment.

§ 1503.3 Specificity of comments.

(a) Comments on an environmental impact statement or on a proposed action shall be as specific as possible and may address either the adequacy of the statement or the merits of the alternatives discussed or both.

(b) When a commenting agency criticizes a lead agency's predictive methodology, the commenting agency should describe the alternative methodology which it prefers and why.

(c) A cooperating agency shall specify in its comments whether it needs additional information to fulfill other applicable environmental reviews or consultation requirements and what information it needs. In particular, it shall specify any additional information it needs to comment adequately on the draft statement's analysis of significant site-specific effects associated with the granting or approving by that cooperating agency of necessary Federal permits, licenses, or entitlements.

(d) When a cooperating agency with jurisdiction by law objects to or expresses reservations about the proposal on grounds of environmental impacts, the agency expressing the objection or reservation shall specify the mitigation measures it considers necessary to allow the agency to grant or approve applicable permit, license, or related requirements or concurrences.

§ 1503.4 Response to comments.

(a) An agency preparing a final environmental impact statement shall assess and consider comments both individually and collectively, and shall respond by one or more of the means listed below, stating its response in the final statement. Possible responses are to:

(1) Modify alternatives including the proposed action.

(2) Develop and evaluate alternatives not previously given serious consideration by the agency.

(3) Supplement, improve, or modify its analyses.

(4) Make factual corrections.

(5) Explain why the comments do not warrant further agency response, citing the sources, authorities, or reasons which support the agency's position and, if appropriate, indicate those circumstances which would trigger agency reappraisal or further response.

(b) All substantive comments received on the draft statement (or summaries thereof where the response has been exceptionally voluminous), should be attached to the final statement whether or not the comment is thought to merit individual discussion by the agency in the text of the statement.

(c) If changes in response to comments are minor and are confined to the responses described in paragraphs (a)(4) and (5) of this section, agencies may write them on errata sheets and attach them to the statement instead of rewriting the draft statement. In such cases only the comments, the responses, and the changes and not the final statement need be circulated (§1502.19). The entire document with a new cover sheet shall be filed as the final statement (§1506.9).

PART 1504—PREDECISION REFERRALS TO THE COUNCIL OF PROPOSED FEDERAL ACTIONS DETERMINED TO BE ENVIRONMENTALLY UNSATISFACTORY

Sec.
1504.1   Purpose.
1504.2   Criteria for referral.
1504.3   Procedure for referrals and response.

AUTHORITY: NEPA, the Environmental Quality Improvement Act of 1970, as amended (42 U.S.C. 4371 et seq.), sec. 309 of the Clean Air Act, as amended (42 U.S.C. 7609), and E.O. 11514 (Mar. 5, 1970, as amended by E.O. 11991, May 24, 1977).

§ 1504.1 Purpose.

(a) This part establishes procedures for referring to the Council Federal interagency disagreements concerning proposed major Federal actions that might cause unsatisfactory environmental effects. It provides means for early resolution of such disagreements.

(b) Under section 309 of the Clean Air Act (42 U.S.C. 7609), the Administrator of the Environmental Protection Agency is directed to review and comment publicly on the environmental impacts of Federal activities, including actions for which environmental impact statements are prepared. If after this review the Administrator determines that the matter is "unsatisfactory from the standpoint of public health or welfare or environmental quality," section 309 directs that the matter be referred to the Council (hereafter "environmental referrals").

(c) Under section 102(2)(C) of the Act other Federal agencies may make similar reviews of environmental impact statements, including judgments on the acceptability of anticipated environmental impacts. These reviews must be made available to the President, the Council and the public.

[43 FR 55998, Nov. 29, 1978]

§ 1504.2 Criteria for referral.

Environmental referrals should be made to the Council only after concerted, timely (as early as possible in the process), but unsuccessful attempts to resolve differences with the lead agency. In determining what environmental objections to the matter are appropriate to refer to the Council, an agency should weigh potential adverse environmental impacts, considering:

(a) Possible violation of national environmental standards or policies.

(b) Severity.

(c) Geographical scope.

(d) Duration.

(e) Importance as precedents.

(f) Availability of environmentally preferable alternatives.

[43 FR 55998, Nov. 29, 1978]

§ 1504.3 Procedure for referrals and response.

(a) A Federal agency making the referral to the Council shall:

(1) Advise the lead agency at the earliest possible time that it intends to refer a matter to the Council unless a satisfactory agreement is reached.

(2) Include such advice in the referring agency's comments on the draft environmental impact statement, except when the statement does not contain adequate information to permit an assessment of the matter's environmental acceptability.

(3) Identify any essential information that is lacking and request that it be made available at the earliest possible time.

(4) Send copies of such advice to the Council.

(b) The referring agency shall deliver its referral to the Council not later than twenty-five (25) days after the final environmental impact statement has been made available to the Environmental Protection Agency, commenting agencies, and the public. Except when an extension of this period has been granted by the lead agency, the Council will not accept a referral after that date.

(c) The referral shall consist of:

(1) A copy of the letter signed by the head of the referring agency and delivered to the lead agency informing the lead agency of the referral and the reasons for it, and requesting that no action be taken to implement the matter until the Council acts upon the referral. The letter shall include a copy of the statement referred to in (c)(2) of this section.

(2) A statement supported by factual evidence leading to the conclusion that the matter is unsatisfactory from the standpoint of public health or welfare or environmental quality. The statement shall:

(i) Identify any material facts in controversy and incorporate (by reference if appropriate) agreed upon facts,

(ii) Identify any existing environmental requirements or policies which would be violated by the matter,

(iii) Present the reasons why the referring agency believes the matter is environmentally unsatisfactory,

(iv) Contain a finding by the agency whether the issue raised is of national importance because of the threat to national environmental resources or policies or for some other reason,

(v) Review the steps taken by the referring agency to bring its concerns to the attention of the lead agency at the earliest possible time, and

(vi) Give the referring agency's recommendations as to what mitigation alternative, further study, or other course of action (including abandonment of the matter) are necessary to remedy the situation.

(d) Not later than twenty-five (25) days after the referral to the Council the lead agency may deliver a response to the Council, and the referring agency. If the lead agency requests more time and gives assurance that the matter will not go forward in the interim, the Council may grant an extension. The response shall:

(1) Address fully the issues raised in the referral.

(2) Be supported by evidence.

(3) Give the lead agency's response to the referring agency's recommendations.

(e) Interested persons (including the applicant) may deliver their views in writing to the Council. Views in support of the referral should be delivered not later than the referral. Views in support of the response shall be delivered not later than the response.

(f) Not later than twenty-five (25) days after receipt of both the referral and any response or upon being informed that there will be no response (unless the lead agency agrees to a longer time), the Council may take one or more of the following actions:

(1) Conclude that the process of referral and response has successfully resolved the problem.

(2) Initiate discussions with the agencies with the objective of mediation with referring and lead agencies.

(3) Hold public meetings or hearings to obtain additional views and information.

(4) Determine that the issue is not one of national importance and request the referring and lead agencies to pursue their decision process.

(5) Determine that the issue should be further negotiated by the referring and lead agencies and is not appropriate for Council consideration until one or more heads of agencies report to the Council that the agencies' disagreements are irreconcilable.

(6) Publish its findings and recommendations (including where appropriate a finding that the submitted evidence does not support the position of an agency).

(7) When appropriate, submit the referral and the response together with the Council's recommendation to the President for action.

(g) The Council shall take no longer than 60 days to complete the actions specified in paragraph (f)(2), (3), or (5) of this section.

(h) When the referral involves an action required by statute to be determined on the record after opportunity for agency hearing, the referral shall be conducted in a manner consistent with 5 U.S.C. 557(d) (Administrative Procedure Act).

[43 FR 55998, Nov. 29, 1978; 44 FR 873, Jan. 3, 1979]

PART 1505—NEPA AND AGENCY DECISIONMAKING

Sec.
1505.1 Agency decisionmaking procedures.
1505.2 Record of decision in cases requiring environmental impact statements.
1505.3 Implementing the decision.

AUTHORITY: NEPA, the Environmental Quality Improvement Act of 1970, as amended (42 U.S.C. 4371 et seq.), sec. 309 of the Clean Air Act, as amended (42 U.S.C. 7609), and E.O. 11514 (Mar. 5, 1970, as amended by E.O. 11991, May 24, 1977).

SOURCE: 43 FR 55999, Nov. 29, 1978, unless otherwise noted.

§ 1505.1 Agency decisionmaking procedures.

Agencies shall adopt procedures (§ 1507.3) to ensure that decisions are made in accordance with the policies and purposes of the Act. Such procedures shall include but not be limited to:

(a) Implementing procedures under section 102(2) to achieve the requirements of sections 101 and 102(1).

(b) Designating the major decision points for the agency's principal programs likely to have a significant effect on the human environment and assuring that the NEPA process corresponds with them.

(c) Requiring that relevant environmental documents, comments, and responses be part of the record in formal rulemaking or adjudicatory proceedings.

(d) Requiring that relevant environmental documents, comments, and responses accompany the proposal through existing agency review processes so that agency officials use the statement in making decisions.

(e) Requiring that the alternatives considered by the decisionmaker are encompassed by the range of alternatives discussed in the relevant environmental documents and that the decisionmaker consider the alternatives described in the environmental impact statement. If another decision document accompanies the relevant environmental documents to the decisionmaker, agencies are encouraged to make available to the public before the decision is made any part of that document that relates to the comparison of alternatives.

§ 1505.2

§ 1505.2   Record of decision in cases requiring environmental impact statements.

At the time of its decision (§ 1506.10) or, if appropriate, its recommendation to Congress, each agency shall prepare a concise public record of decision. The record, which may be integrated into any other record prepared by the agency, including that required by OMB Circular A-95 (Revised), part I, sections 6(c) and (d), and part II, section 5(b)(4), shall:

(a) State what the decision was.

(b) Identify all alternatives considered by the agency in reaching its decision, specifying the alternative or alternatives which were considered to be environmentally preferable. An agency may discuss preferences among alternatives based on relevant factors including economic and technical considerations and agency statutory missions. An agency shall identify and discuss all such factors including any essential considerations of national policy which were balanced by the agency in making its decision and state how those considerations entered into its decision.

(c) State whether all practicable means to avoid or minimize environmental harm from the alternative selected have been adopted, and if not, why they were not. A monitoring and enforcement program shall be adopted and summarized where applicable for any mitigation.

§ 1505.3   Implementing the decision.

Agencies may provide for monitoring to assure that their decisions are carried out and should do so in important cases. Mitigation (§ 1505.2(c)) and other conditions established in the environmental impact statement or during its review and committed as part of the decision shall be implemented by the lead agency or other appropriate consenting agency. The lead agency shall:

(a) Include appropriate conditions in grants, permits or other approvals.

(b) Condition funding of actions on mitigation.

(c) Upon request, inform cooperating or commenting agencies on progress in carrying out mitigation measures which they have proposed and which

were adopted by the agency making the decision.

(d) Upon request, make available to the public the results of relevant monitoring.

PART 1506—OTHER REQUIREMENTS OF NEPA

Sec.

1506.1 Limitations on actions during NEPA process.
1506.2 Elimination of duplication with State and local procedures.
1506.3 Adoption.
1506.4 Combining documents.
1506.5 Agency responsibility.
1506.6 Public involvement.
1506.7 Further guidance.
1506.8 Proposals for legislation.
1506.9 Filing requirements.
1506.10 Timing of agency action.
1506.11 Emergencies.
1506.12 Effective date.

AUTHORITY: NEPA, the Environmental Quality Improvement Act of 1970, as amended (42 U.S.C. 4371 *et seq.*), sec. 309 of the Clean Air Act, as amended (42 U.S.C. 7609), and E.O. 11514 (Mar. 5, 1970, as amended by E.O. 11991, May 24, 1977).

SOURCE: 43 FR 56000, Nov. 29, 1978, unless otherwise noted.

§ 1506.1   Limitations on actions during NEPA process.

(a) Until an agency issues a record of decision as provided in § 1505.2 (except as provided in paragraph (c) of this section), no action concerning the proposal shall be taken which would:

(1) Have an adverse environmental impact; or

(2) Limit the choice of reasonable alternatives.

(b) If any agency is considering an application from a non-Federal entity, and is aware that the applicant is about to take an action within the agency's jurisdiction that would meet either of the criteria in paragraph (a) of this section, then the agency shall promptly notify the applicant that the agency will take appropriate action to insure that the objectives and procedures of NEPA are achieved.

(c) While work on a required program environmental impact statement is in progress and the action is not covered by an existing program statement, agencies shall not undertake in the interim any major Federal action covered by the program which may significantly affect the quality of the human environment unless such action:

(1) Is justified independently of the program;

(2) Is itself accompanied by an adequate environmental impact statement; and

(3) Will not prejudice the ultimate decision on the program. Interim action prejudices the ultimate decision on the program when it tends to determine subsequent development or limit alternatives.

(d) This section does not preclude development by applicants of plans or designs or performance of other work necessary to support an application for Federal, State or local permits or assistance. Nothing in this section shall preclude Rural Electrification Administration approval of minimal expenditures not affecting the environment (*e.g.* long leadtime equipment and purchase options) made by non-governmental entities seeking loan guarantees from the Administration.

§ 1506.2 Elimination of duplication with State and local procedures.

(a) Agencies authorized by law to cooperate with State agencies of statewide jurisdiction pursuant to section 102(2)(D) of the Act may do so.

(b) Agencies shall cooperate with State and local agencies to the fullest extent possible to reduce duplication between NEPA and State and local requirements, unless the agencies are specifically barred from doing so by some other law. Except for cases covered by paragraph (a) of this section, such cooperation shall to the fullest extent possible include:

(1) Joint planning processes.

(2) Joint environmental research and studies.

(3) Joint public hearings (except where otherwise provided by statute).

(4) Joint environmental assessments.

(c) Agencies shall cooperate with State and local agencies to the fullest extent possible to reduce duplication between NEPA and comparable State and local requirements, unless the agencies are specifically barred from doing so by some other law. Except for

cases covered by paragraph (a) of this section, such cooperation shall to the fullest extent possible include joint environmental impact statements. In such cases one or more Federal agencies and one or more State or local agencies shall be joint lead agencies. Where State laws or local ordinances have environmental impact statement requirements in addition to but not in conflict with those in NEPA, Federal agencies shall cooperate in fulfilling these requirements as well as those of Federal laws so that one document will comply with all applicable laws.

(d) To better integrate environmental impact statements into State or local planning processes, statements shall discuss any inconsistency of a proposed action with any approved State or local plan and laws (whether or not federally sanctioned). Where an inconsistency exists, the statement should describe the extent to which the agency would reconcile its proposed action with the plan or law.

§ 1506.3 Adoption.

(a) An agency may adopt a Federal draft or final environmental impact statement or portion thereof provided that the statement or portion thereof meets the standards for an adequate statement under these regulations.

(b) If the actions covered by the original environmental impact statement and the proposed action are substantially the same, the agency adopting another agency's statement is not required to recirculate it except as a final statement. Otherwise the adopting agency shall treat the statement as a draft and recirculate it (except as provided in paragraph (c) of this section).

(c) A cooperating agency may adopt without recirculating the environmental impact statement of a lead agency when, after an independent review of the statement, the cooperating agency concludes that its comments and suggestions have been satisfied.

(d) When an agency adopts a statement which is not final within the agency that prepared it, or when the action it assesses is the subject of a referral under part 1504, or when the statement's adequacy is the subject of a judicial action which is not final, the agency shall so specify.

§ 1506.4 Combining documents.

Any environmental document in compliance with NEPA may be combined with any other agency document to reduce duplication and paperwork.

§ 1506.5 Agency responsibility.

(a) *Information.* If an agency requires an applicant to submit environmental information for possible use by the agency in preparing an environmental impact statement, then the agency should assist the applicant by outlining the types of information required. The agency shall independently evaluate the information submitted and shall be responsible for its accuracy. If the agency chooses to use the information submitted by the applicant in the environmental impact statement, either directly or by reference, then the names of the persons responsible for the independent evaluation shall be included in the list of preparers (§ 1502.17). It is the intent of this paragraph that acceptable work not be redone, but that it be verified by the agency.

(b) *Environmental assessments.* If an agency permits an applicant to prepare an environmental assessment, the agency, besides fulfilling the requirements of paragraph (a) of this section, shall make its own evaluation of the environmental issues and take responsibility for the scope and content of the environmental assessment.

(c) *Environmental impact statements.* Except as provided in §§ 1506.2 and 1506.3 any environmental impact statement prepared pursuant to the requirements of NEPA shall be prepared directly by or by a contractor selected by the lead agency or where appropriate under § 1501.6(b), a cooperating agency. It is the intent of these regulations that the contractor be chosen solely by the lead agency, or by the lead agency in cooperation with cooperating agencies, or where appropriate by a cooperating agency to avoid any conflict of interest. Contractors shall execute a disclosure statement prepared by the lead agency, or where appropriate the cooperating agency, specifying that they have no financial or other interest in the outcome of the project. If the document is prepared by contract, the responsible Federal official shall furnish guidance and participate in the preparation and shall independently evaluate the statement prior to its approval and take responsibility for its scope and contents. Nothing in this section is intended to prohibit any agency from requesting any person to submit information to it or to prohibit any person from submitting information to any agency.

§ 1506.6 Public involvement.

Agencies shall:

(a) Make diligent efforts to involve the public in preparing and implementing their NEPA procedures.

(b) Provide public notice of NEPA-related hearings, public meetings, and the availability of environmental documents so as to inform those persons and agencies who may be interested or affected.

(1) In all cases the agency shall mail notice to those who have requested it on an individual action.

(2) In the case of an action with effects of national concern notice shall include publication in the FEDERAL REGISTER and notice by mail to national organizations reasonably expected to be interested in the matter and may include listing in the *102 Monitor.* An agency engaged in rulemaking may provide notice by mail to national organizations who have requested that notice regularly be provided. Agencies shall maintain a list of such organizations.

(3) In the case of an action with effects primarily of local concern the notice may include:

(i) Notice to State and areawide clearinghouses pursuant to OMB Circular A–95 (Revised).

(ii) Notice to Indian tribes when effects may occur on reservations.

(iii) Following the affected State's public notice procedures for comparable actions.

(iv) Publication in local newspapers (in papers of general circulation rather than legal papers).

(v) Notice through other local media.

(vi) Notice to potentially interested community organizations including small business associations.

(vii) Publication in newsletters that may be expected to reach potentially interested persons.

(viii) Direct mailing to owners and occupants of nearby or affected property.

(ix) Posting of notice on and off site in the area where the action is to be located.

(c) Hold or sponsor public hearings or public meetings whenever appropriate or in accordance with statutory requirements applicable to the agency. Criteria shall include whether there is:

(1) Substantial environmental controversy concerning the proposed action or substantial interest in holding the hearing.

(2) A request for a hearing by another agency with jurisdiction over the action supported by reasons why a hearing will be helpful. If a draft environmental impact statement is to be considered at a public hearing, the agency should make the statement available to the public at least 15 days in advance (unless the purpose of the hearing is to provide information for the draft environmental impact statement).

(d) Solicit appropriate information from the public.

(e) Explain in its procedures where interested persons can get information or status reports on environmental impact statements and other elements of the NEPA process.

(f) Make environmental impact statements, the comments received, and any underlying documents available to the public pursuant to the provisions of the Freedom of Information Act (5 U.S.C. 552), without regard to the exclusion for interagency memoranda where such memoranda transmit comments of Federal agencies on the environmental impact of the proposed action. Materials to be made available to the public shall be provided to the public without charge to the extent practicable, or at a fee which is not more than the actual costs of reproducing copies required to be sent to other Federal agencies, including the Council.

§ 1506.7 Further guidance.

The Council may provide further guidance concerning NEPA and its procedures including:

(a) A handbook which the Council may supplement from time to time, which shall in plain language provide guidance and instructions concerning the application of NEPA and these regulations.

(b) Publication of the Council's Memoranda to Heads of Agencies.

(c) In conjunction with the Environmental Protection Agency and the publication of the 102 Monitor, notice of:

(1) Research activities;

(2) Meetings and conferences related to NEPA; and

(3) Successful and innovative procedures used by agencies to implement NEPA.

§ 1506.8 Proposals for legislation.

(a) The NEPA process for proposals for legislation (§ 1508.17) significantly affecting the quality of the human environment shall be integrated with the legislative process of the Congress. A legislative environmental impact statement is the detailed statement required by law to be included in a recommendation or report on a legislative proposal to Congress. A legislative environmental impact statement shall be considered part of the formal transmittal of a legislative proposal to Congress; however, it may be transmitted to Congress up to 30 days later in order to allow time for completion of an accurate statement which can serve as the basis for public and Congressional debate. The statement must be available in time for Congressional hearings and deliberations.

(b) Preparation of a legislative environmental impact statement shall conform to the requirements of these regulations except as follows:

(1) There need not be a scoping process.

(2) The legislative statement shall be prepared in the same manner as a draft statement, but shall be considered the "detailed statement" required by statute; *Provided*, That when any of the following conditions exist both the draft and final environmental impact statement on the legislative proposal shall be prepared and circulated as provided by §§ 1503.1 and 1506.10.

(i) A Congressional Committee with jurisdiction over the proposal has a rule requiring both draft and final environmental impact statements.

(ii) The proposal results from a study process required by statute (such as those required by the Wild and Scenic Rivers Act (16 U.S.C. 1271 *et seq.*) and the Wilderness Act (16 U.S.C. 1131 *et seq.*)).

(iii) Legislative approval is sought for Federal or federally assisted construction or other projects which the agency recommends be located at specific geographic locations. For proposals requiring an environmental impact statement for the acquisition of space by the General Services Administration, a draft statement shall accompany the Prospectus or the 11(b) Report of Building Project Surveys to the Congress, and a final statement shall be completed before site acquisition.

(iv) The agency decides to prepare draft and final statements.

(c) Comments on the legislative statement shall be given to the lead agency which shall forward them along with its own responses to the Congressional committees with jurisdiction.

§ 1506.9 **Filing requirements.**

Environmental impact statements together with comments and responses shall be filed with the Environmental Protection Agency, attention Office of Federal Activities (A-104), 401 M Street SW., Washington, DC 20460. Statements shall be filed with EPA no earlier than they are also transmitted to commenting agencies and made available to the public. EPA shall deliver one copy of each statement to the Council, which shall satisfy the requirement of availability to the President. EPA may issue guidelines to agencies to implement its responsibilities under this section and § 1506.10.

§ 1506.10 **Timing of agency action.**

(a) The Environmental Protection Agency shall publish a notice in the FEDERAL REGISTER each week of the environmental impact statements filed during the preceding week. The minimum time periods set forth in this section shall be calculated from the date of publication of this notice.

(b) No decision on the proposed action shall be made or recorded under § 1505.2 by a Federal agency until the later of the following dates:

(1) Ninety (90) days after publication of the notice described above in paragraph (a) of this section for a draft environmental impact statement.

(2) Thirty (30) days after publication of the notice described above in paragraph (a) of this section for a final environmental impact statement.

An exception to the rules on timing may be made in the case of an agency decision which is subject to a formal internal appeal. Some agencies have a formally established appeal process which allows other agencies or the public to take appeals on a decision and make their views known, after publication of the final environmental impact statement. In such cases, where a real opportunity exists to alter the decision, the decision may be made and recorded at the same time the environmental impact statement is published. This means that the period for appeal of the decision and the 30-day period prescribed in paragraph (b)(2) of this section may run concurrently. In such cases the environmental impact statement shall explain the timing and the public's right of appeal. An agency engaged in rulemaking under the Administrative Procedure Act or other statute for the purpose of protecting public health or safety, may waive the time period in paragraph (b)(2) of this section and publish a decision on the final rule simultaneously with publication of the notice of the availability of the final environmental impact statement as described in paragraph (a) of this section.

(c) If the final environmental impact statement is filed within ninety (90) days after a draft environmental impact statement is filed with the Environmental Protection Agency, the minimum thirty (30) day period and the minimum ninety (90) day period may run concurrently. However, subject to paragraph (d) of this section agencies shall allow not less than 45 days for comments on draft statements.

(d) The lead agency may extend prescribed periods. The Environmental Protection Agency may upon a showing by the lead agency of compelling reasons of national policy reduce the prescribed periods and may upon a showing by any other Federal agency of compelling reasons of national policy also extend prescribed periods, but only after consultation with the lead agency. (Also see § 1507.3(d).) Failure to file timely comments shall not be a sufficient reason for extending a period. If the lead agency does not concur with the extension of time, EPA may not extend it for more than 30 days. When the Environmental Protection Agency reduces or extends any period of time it shall notify the Council.

[43 FR 56000, Nov. 29, 1978; 44 FR 874, Jan. 3, 1979]

§ 1506.11 Emergencies.

Where emergency circumstances make it necessary to take an action with significant environmental impact without observing the provisions of these regulations, the Federal agency taking the action should consult with the Council about alternative arrangements. Agencies and the Council will limit such arrangements to actions necessary to control the immediate impacts of the emergency. Other actions remain subject to NEPA review.

§ 1506.12 Effective date.

The effective date of these regulations is July 30, 1979, except that for agencies that administer programs that qualify under section 102(2)(D) of the Act or under section 104(h) of the Housing and Community Development Act of 1974 an additional four months shall be allowed for the State or local agencies to adopt their implementing procedures.

(a) These regulations shall apply to the fullest extent practicable to ongoing activities and environmental documents begun before the effective date. These regulations do not apply to an environmental impact statement or supplement if the draft statement was filed before the effective date of these regulations. No completed environmental documents need be redone by reasons of these regulations. Until these regulations are applicable, the Council's guidelines published in the FEDERAL REGISTER of August 1, 1973, shall continue to be applicable. In cases where these regulations are applicable the guidelines are superseded. However, nothing shall prevent an agency from proceeding under these regulations at an earlier time.

(b) NEPA shall continue to be applicable to actions begun before January 1, 1970, to the fullest extent possible.

PART 1507—AGENCY COMPLIANCE

Sec.
1507.1 Compliance.
1507.2 Agency capability to comply.
1507.3 Agency procedures.

AUTHORITY: NEPA, the Environmental Quality Improvement Act of 1970, as amended (42 U.S.C. 4371 *et seq.*), sec. 309 of the Clean Air Act, as amended (42 U.S.C. 7609), and E.O. 11514 (Mar. 5, 1970, as amended by E.O. 11991, May 24, 1977).

SOURCE: 43 FR 56002, Nov. 29, 1978, unless otherwise noted.

§ 1507.1 Compliance.

All agencies of the Federal Government shall comply with these regulations. It is the intent of these regulations to allow each agency flexibility in adapting its implementing procedures authorized by § 1507.3 to the requirements of other applicable laws.

§ 1507.2 Agency capability to comply.

Each agency shall be capable (in terms of personnel and other resources) of complying with the requirements enumerated below. Such compliance may include use of other's resources, but the using agency shall itself have sufficient capability to evaluate what others do for it. Agencies shall:

(a) Fulfill the requirements of section 102(2)(A) of the Act to utilize a systematic, interdisciplinary approach which will insure the integrated use of the natural and social sciences and the environmental design arts in planning and in decisionmaking which may have an impact on the human environment. Agencies shall designate a person to be responsible for overall review of agency NEPA compliance.

(b) Identify methods and procedures required by section 102(2)(B) to insure that presently unquantified environmental amenities and values may be given appropriate consideration.

(c) Prepare adequate environmental impact statements pursuant to section 102(2)(C) and comment on statements in the areas where the agency has jurisdiction by law or special expertise or is authorized to develop and enforce environmental standards.

(d) Study, develop, and describe alternatives to recommended courses of action in any proposal which involves unresolved conflicts concerning alternative uses of available resources. This requirement of section 102(2)(E) extends to all such proposals, not just the more limited scope of section 102(2)(C)(iii) where the discussion of alternatives is confined to impact statements.

(e) Comply with the requirements of section 102(2)(H) that the agency initiate and utilize ecological information in the planning and development of resource-oriented projects.

(f) Fulfill the requirements of sections 102(2)(F), 102(2)(G), and 102(2)(I), of the Act and of Executive Order 11514, Protection and Enhancement of Environmental Quality, Sec. 2.

§ 1507.3 Agency procedures.

(a) Not later than eight months after publication of these regulations as finally adopted in the FEDERAL REGISTER, or five months after the establishment of an agency, whichever shall come later, each agency shall as necessary adopt procedures to supplement these regulations. When the agency is a department, major subunits are encouraged (with the consent of the department) to adopt their own procedures. Such procedures shall not paraphrase these regulations. They shall confine themselves to implementing procedures. Each agency shall consult with the Council while developing its procedures and before publishing them in the FEDERAL REGISTER for comment. Agencies with similar programs should consult with each other and the Council to coordinate their procedures, especially for programs requesting similar information from applicants. The procedures shall be adopted only after an opportunity for public review and after review by the Council for conformity with the Act and these regulations. The Council shall complete its review within 30 days. Once in effect they shall be filed with the Council and made readily available to the public. Agencies are encouraged to publish ex-

planatory guidance for these regulations and their own procedures. Agencies shall continue to review their policies and procedures and in consultation with the Council to revise them as necessary to ensure full compliance with the purposes and provisions of the Act.

(b) Agency procedures shall comply with these regulations except where compliance would be inconsistent with statutory requirements and shall include:

(1) Those procedures required by §§ 1501.2(d), 1502.9(c)(3), 1505.1, 1506.6(e), and 1508.4.

(2) Specific criteria for and identification of those typical classes of action:

(i) Which normally do require environmental impact statements.

(ii) Which normally do not require either an environmental impact statement or an environmental assessment (categorical exclusions (§ 1508.4)).

(iii) Which normally require environmental assessments but not necessarily environmental impact statements.

(c) Agency procedures may include specific criteria for providing limited exceptions to the provisions of these regulations for classified proposals. They are proposed actions which are specifically authorized under criteria established by an Executive Order or statute to be kept secret in the interest of national defense or foreign policy and are in fact properly classified pursuant to such Executive Order or statute. Environmental assessments and environmental impact statements which address classified proposals may be safeguarded and restricted from public dissemination in accordance with agencies' own regulations applicable to classified information. These documents may be organized so that classified portions can be included as annexes, in order that the unclassified portions can be made available to the public.

(d) Agency procedures may provide for periods of time other than those presented in § 1506.10 when necessary to comply with other specific statutory requirements.

(e) Agency procedures may provide that where there is a lengthy period between the agency's decision to prepare an environmental impact statement and the time of actual preparation, the notice of intent required by § 1501.7 may be published at a reasonable time in advance of preparation of the draft statement.

PART 1508—TERMINOLOGY AND INDEX

Sec.
1508.1 Terminology.
1508.2 Act.
1508.3 Affecting.
1508.4 Categorical exclusion.
1508.5 Cooperating agency.
1508.6 Council.
1508.7 Cumulative impact.
1508.8 Effects.
1508.9 Environmental assessment.
1508.10 Environmental document.
1508.11 Environmental impact statement.
1508.12 Federal agency.
1508.13 Finding of no significant impact.
1508.14 Human environment.
1508.15 Jurisdiction by law.
1508.16 Lead agency.
1508.17 Legislation.
1508.18 Major Federal action.
1508.19 Matter.
1508.20 Mitigation.
1508.21 NEPA process.
1508.22 Notice of intent.
1508.23 Proposal.
1508.24 Referring agency.
1508.25 Scope.
1508.26 Special expertise.
1508.27 Significantly.
1508.28 Tiering.

AUTHORITY: NEPA, the Environmental Quality Improvement Act of 1970, as amended (42 U.S.C. 4371 et seq.), sec. 309 of the Clean Air Act, as amended (42 U.S.C. 7609), and E.O. 11514 (Mar. 5, 1970, as amended by E.O. 11991, May 24, 1977).

SOURCE: 43 FR 56003, Nov. 29, 1978, unless otherwise noted.

§ 1508.1 Terminology.

The terminology of this part shall be uniform throughout the Federal Government.

§ 1508.2 Act.

Act means the National Environmental Policy Act, as amended (42 U.S.C. 4321, et seq.) which is also referred to as "NEPA."

§ 1508.3 Affecting.

Affecting means will or may have an effect on.

§ 1508.4 Categorical exclusion.

Categorical exclusion means a category of actions which do not individually or cumulatively have a significant effect on the human environment and which have been found to have no such effect in procedures adopted by a Federal agency in implementation of these regulations (§ 1507.3) and for which, therefore, neither an environmental assessment nor an environmental impact statement is required. An agency may decide in its procedures or otherwise, to prepare environmental assessments for the reasons stated in § 1508.9 even though it is not required to do so. Any procedures under this section shall provide for extraordinary circumstances in which a normally excluded action may have a significant environmental effect.

§ 1508.5 Cooperating agency.

Cooperating agency means any Federal agency other than a lead agency which has jurisdiction by law or special expertise with respect to any environmental impact involved in a proposal (or a reasonable alternative) for legislation or other major Federal action significantly affecting the quality of the human environment. The selection and responsibilities of a cooperating agency are described in § 1501.6. A State or local agency of similar qualifications or, when the effects are on a reservation, an Indian Tribe, may by agreement with the lead agency become a cooperating agency.

§ 1508.6 Council.

Council means the Council on Environmental Quality established by title II of the Act.

§ 1508.7 Cumulative impact.

Cumulative impact is the impact on the environment which results from the incremental impact of the action when added to other past, present, and reasonably foreseeable future actions regardless of what agency (Federal or non-Federal) or person undertakes such other actions. Cumulative impacts can result from individually minor but collectively significant actions taking place over a period of time.

§ 1508.8 Effects.

Effects include:

(a) Direct effects, which are caused by the action and occur at the same time and place.

(b) Indirect effects, which are caused by the action and are later in time or farther removed in distance, but are still reasonably foreseeable. Indirect effects may include growth inducing effects and other effects related to induced changes in the pattern of land use, population density or growth rate, and related effects on air and water and other natural systems, including ecosystems.

Effects and impacts as used in these regulations are synonymous. Effects includes ecological (such as the effects on natural resources and on the components, structures, and functioning of affected ecosystems), aesthetic, historic, cultural, economic, social, or health, whether direct, indirect, or cumulative. Effects may also include those resulting from actions which may have both beneficial and detrimental effects, even if on balance the agency believes that the effect will be beneficial.

§ 1508.9 Environmental assessment.

Environmental assessment:

(a) Means a concise public document for which a Federal agency is responsible that serves to:

(1) Briefly provide sufficient evidence and analysis for determining whether to prepare an environmental impact statement or a finding of no significant impact.

(2) Aid an agency's compliance with the Act when no environmental impact statement is necessary.

(3) Facilitate preparation of a statement when one is necessary.

(b) Shall include brief discussions of the need for the proposal, of alternatives as required by section 102(2)(E), of the environmental impacts of the proposed action and alternatives, and a listing of agencies and persons consulted.

§ 1508.10 Environmental document.

Environmental document includes the documents specified in § 1508.9 (environmental assessment), § 1508.11 (environ-

mental impact statement), § 1508.13 (finding of no significant impact), and § 1508.22 (notice of intent).

§ 1508.11 Environmental impact statement.

Environmental impact statement means a detailed written statement as required by section 102(2)(C) of the Act.

§ 1508.12 Federal agency.

Federal agency means all agencies of the Federal Government. It does not mean the Congress, the Judiciary, or the President, including the performance of staff functions for the President in his Executive Office. It also includes for purposes of these regulations States and units of general local government and Indian tribes assuming NEPA responsibilities under section 104(h) of the Housing and Community Development Act of 1974.

§ 1508.13 Finding of no significant impact.

Finding of no significant impact means a document by a Federal agency briefly presenting the reasons why an action, not otherwise excluded (§ 1508.4), will not have a significant effect on the human environment and for which an environmental impact statement therefore will not be prepared. It shall include the environmental assessment or a summary of it and shall note any other environmental documents related to it (§ 1501.7(a)(5)). If the assessment is included, the finding need not repeat any of the discussion in the assessment but may incorporate it by reference.

§ 1508.14 Human environment.

Human environment shall be interpreted comprehensively to include the natural and physical environment and the relationship of people with that environment. (See the definition of "effects" (§ 1508.8).) This means that economic or social effects are not intended by themselves to require preparation of an environmental impact statement. When an environmental impact statement is prepared and economic or social and natural or physical environmental effects are interrelated, then the environmental impact statement will discuss all of these effects on the human environment.

§ 1508.15 Jurisdiction by law.

Jurisdiction by law means agency authority to approve, veto, or finance all or part of the proposal.

§ 1508.16 Lead agency.

Lead agency means the agency or agencies preparing or having taken primary responsibility for preparing the environmental impact statement.

§ 1508.17 Legislation.

Legislation includes a bill or legislative proposal to Congress developed by or with the significant cooperation and support of a Federal agency, but does not include requests for appropriations. The test for significant cooperation is whether the proposal is in fact predominantly that of the agency rather than another source. Drafting does not by itself constitute significant cooperation. Proposals for legislation include requests for ratification of treaties. Only the agency which has primary responsibility for the subject matter involved will prepare a legislative environmental impact statement.

§ 1508.18 Major Federal action.

Major Federal action includes actions with effects that may be major and which are potentially subject to Federal control and responsibility. Major reinforces but does not have a meaning independent of significantly (§ 1508.27). Actions include the circumstance where the responsible officials fail to act and that failure to act is reviewable by courts or administrative tribunals under the Administrative Procedure Act or other applicable law as agency action.

(a) Actions include new and continuing activities, including projects and programs entirely or partly financed, assisted, conducted, regulated, or approved by federal agencies; new or revised agency rules, regulations, plans, policies, or procedures; and legislative proposals (§§ 1506.8, 1508.17). Actions do not include funding assistance solely in the form of general revenue sharing funds, distributed under the State and Local Fiscal Assistance Act of 1972, 31 U.S.C. 1221 *et seq.*, with no

Federal agency control over the subsequent use of such funds. Actions do not include bringing judicial or administrative civil or criminal enforcement actions.

(b) Federal actions tend to fall within one of the following categories:

(1) Adoption of official policy, such as rules, regulations, and interpretations adopted pursuant to the Administrative Procedure Act, 5 U.S.C. 551 *et seq.;* treaties and international conventions or agreements; formal documents establishing an agency's policies which will result in or substantially alter agency programs.

(2) Adoption of formal plans, such as official documents prepared or approved by federal agencies which guide or prescribe alternative uses of Federal resources, upon which future agency actions will be based.

(3) Adoption of programs, such as a group of concerted actions to implement a specific policy or plan; systematic and connected agency decisions allocating agency resources to implement a specific statutory program or executive directive.

(4) Approval of specific projects, such as construction or management activities located in a defined geographic area. Projects include actions approved by permit or other regulatory decision as well as federal and federally assisted activities.

§ 1508.19 Matter.

Matter includes for purposes of part 1504:

(a) With respect to the Environmental Protection Agency, any proposed legislation, project, action or regulation as those terms are used in section 309(a) of the Clean Air Act (42 U.S.C. 7609).

(b) With respect to all other agencies, any proposed major federal action to which section 102(2)(C) of NEPA applies.

§ 1508.20 Mitigation.

Mitigation includes:

(a) Avoiding the impact altogether by not taking a certain action or parts of an action.

(b) Minimizing impacts by limiting the degree or magnitude of the action and its implementation.

(c) Rectifying the impact by repairing, rehabilitating, or restoring the affected environment.

(d) Reducing or eliminating the impact over time by preservation and maintenance operations during the life of the action.

(e) Compensating for the impact by replacing or providing substitute resources or environments.

§ 1508.21   NEPA process.

NEPA process means all measures necessary for compliance with the requirements of section 2 and title I of NEPA.

§ 1508.22   Notice of intent.

Notice of intent means a notice that an environmental impact statement will be prepared and considered. The notice shall briefly:

(a) Describe the proposed action and possible alternatives.

(b) Describe the agency's proposed scoping process including whether, when, and where any scoping meeting will be held.

(c) State the name and address of a person within the agency who can answer questions about the proposed action and the environmental impact statement.

§ 1508.23   Proposal.

Proposal exists at that stage in the development of an action when an agency subject to the Act has a goal and is actively preparing to make a decision on one or more alternative means of accomplishing that goal and the effects can be meaningfully evaluated. Preparation of an environmental impact statement on a proposal should be timed (§ 1502.5) so that the final statement may be completed in time for the statement to be included in any recommendation or report on the proposal. A proposal may exist in fact as well as by agency declaration that one exists.

§ 1508.24   Referring agency.

Referring agency means the federal agency which has referred any matter to the Council after a determination that the matter is unsatisfactory from the standpoint of public health or welfare or environmental quality.

§ 1508.25   Scope.

Scope consists of the range of actions, alternatives, and impacts to be considered in an environmental impact statement. The scope of an individual statement may depend on its relationships to other statements (§§ 1502.20 and 1508.28). To determine the scope of environmental impact statements, agencies shall consider 3 types of actions, 3 types of alternatives, and 3 types of impacts. They include:

(a) Actions (other than unconnected single actions) which may be:

(1) Connected actions, which means that they are closely related and therefore should be discussed in the same impact statement. Actions are connected if they:

(i) Automatically trigger other actions which may require environmental impact statements.

(ii) Cannot or will not proceed unless other actions are taken previously or simultaneously.

(iii) Are interdependent parts of a larger action and depend on the larger action for their justification.

(2) Cumulative actions, which when viewed with other proposed actions have cumulatively significant impacts and should therefore be discussed in the same impact statement.

(3) Similar actions, which when viewed with other reasonably foreseeable or proposed agency actions, have similarities that provide a basis for evaluating their environmental consequencies together, such as common timing or geography. An agency may wish to analyze these actions in the same impact statement. It should do so when the best way to assess adequately the combined impacts of similar actions or reasonable alternatives to such actions is to treat them in a single impact statement.

(b) Alternatives, which include:

(1) No action alternative.

(2) Other reasonable courses of actions.

(3) Mitigation measures (not in the proposed action).

(c) Impacts, which may be: (1) Direct; (2) indirect; (3) cumulative.

§ 1508.26 Special expertise.

Special expertise means statutory responsibility, agency mission, or related program experience.

§ 1508.27 Significantly.

Significantly as used in NEPA requires considerations of both context and intensity:

(a) *Context*. This means that the significance of an action must be analyzed in several contexts such as society as a whole (human, national), the affected region, the affected interests, and the locality. Significance varies with the setting of the proposed action. For instance, in the case of a site-specific action, significance would usually depend upon the effects in the locale rather than in the world as a whole. Both short- and long-term effects are relevant.

(b) *Intensity*. This refers to the severity of impact. Responsible officials must bear in mind that more than one agency may make decisions about partial aspects of a major action. The following should be considered in evaluating intensity:

(1) Impacts that may be both beneficial and adverse. A significant effect may exist even if the Federal agency believes that on balance the effect will be beneficial.

(2) The degree to which the proposed action affects public health or safety.

(3) Unique characteristics of the geographic area such as proximity to historic or cultural resources, park lands, prime farmlands, wetlands, wild and scenic rivers, or ecologically critical areas.

(4) The degree to which the effects on the quality of the human environment are likely to be highly controversial.

(5) The degree to which the possible effects on the human environment are highly uncertain or involve unique or unknown risks.

(6) The degree to which the action may establish a precedent for future actions with significant effects or represents a decision in principle about a future consideration.

(7) Whether the action is related to other actions with individually insignificant but cumulatively significant impacts. Significance exists if it is reasonable to anticipate a cumulatively

significant impact on the environment. Significance cannot be avoided by terming an action temporary or by breaking it down into small component parts.

(8) The degree to which the action may adversely affect districts, sites, highways, structures, or objects listed in or eligible for listing in the National Register of Historic Places or may cause loss or destruction of significant scientific, cultural, or historical resources.

(9) The degree to which the action may adversely affect an endangered or threatened species or its habitat that has been determined to be critical under the Endangered Species Act of 1973.

(10) Whether the action threatens a violation of Federal, State, or local law or requirements imposed for the protection of the environment.

[43 FR 56003, Nov. 29, 1978; 44 FR 874, Jan. 3, 1979]

§ 1508.28 Tiering.

Tiering refers to the coverage of general matters in broader environmental impact statements (such as national program or policy statements) with subsequent narrower statements or environmental analyses (such as regional or basinwide program statements or ultimately site-specific statements) incorporating by reference the general discussions and concentrating solely on the issues specific to the statement subsequently prepared. Tiering is appropriate when the sequence of statements or analyses is:

(a) From a program, plan, or policy environmental impact statement to a program, plan, or policy statement or analysis of lesser scope or to a site-specific statement or analysis.

(b) From an environmental impact statement on a specific action at an early stage (such as need and site selection) to a supplement (which is preferred) or a subsequent statement or analysis at a later stage (such as environmental mitigation). Tiering in such cases is appropriate when it helps the lead agency to focus on the issues which are ripe for decision and exclude from consideration issues already decided or not yet ripe.

Act	1508.2
Action	1508.18, 1508.25
Action-forcing	1500.1, 1502.1
Adoption	1500.4(n), 1500.5(h), 1506.3
Affected Environment	1502.10(f), 1502.15
Affecting	1502.3, 1508.3.
Agency Authority	1500.6
Agency Capability	1501.2(a), 1507.2
Agency Compliance	1507.1
Agency Procedures	1505.1, 1507.3
Agency Responsibility	1506.5
Alternatives	1501.2(c), 1502.2, 1502.10(e), 1502.14, 1505.1(e), 1505.2, 1507.2(d), 1508.25(b)
Appendices	1502.10(k), 1502.18, 1502.24
Applicant	1501.2(d)(1), 1501.4(b) 1501.8(a), 1502.19 (b), 1503.1(a)(3), 1504.3(e), 1506.1 (d), 1506.5(a), 1506.5(b)
Apply NEPA Early in the process.	1501.2
Categorical Exclusion	1500.4(p), 1500.5(k), 1501.4(a), 1507.3(b), 1508.4
Circulating of Environmental Impact Statement	1502.19, 1506.3
Classified Information	1507.3(c)
Clean Air Act	1504.1, 1508.19(a)
Combining Documents	1500.4(o), 1500.5(i), 1506.4
Commenting	1502.19, 1503.1, 1503.2, 1503.3, 1503.4, 1506.6(f)
Consultation Requirement	1500.4(k), 1500.5(g), 1501.7(a)(6), 1502.25
Context	1508.27(a)
Cooperating Agency	1500.5(b), 1501.1(b), 1501.5(c), 1501.5(f), 1501.6, 1503.1(a)(1), 1503.2, 1503.3,1506.3(c), 1506.5(a), 1508.5.
Cost-Benefit	1502.23
Council on Environmental Quality	1500.3, 1501.5(e), 1501.5(f), 1501.6(c), 1502.9(c)(4), 1504.1, 1504.2, 1504.3, 1506.6(f), 1506.9, 1506.10(d), 1506.11, 1507.3, 1508.6, 1508.24
Cover Sheet	1502.10(a), 1502.11
Cumulative Impact	1508.7, 1508.25(a), 1508.25(c)
Decisionmaking	1505.1, 1506.1
Decision points	1505.1(b)
Dependent	1508.25(a)
Draft Environmental Impact Statement	1502.9(a)
Early Application of NEPA	1501.2
Economic Effects	1508.8
Effective Date	1506.12
Effects	1502.16, 1508.8
Emergencies	1506.11
Endangered Species Act	1502.25, 1508.27(b)(9)
Energy	1502.16(e)
Environmental Assessment	1501.3, 1501.4(b), 1501.4(c), 1501.7(b)(3), 1506.2(b)(4), 1506.5(b), 1508.4, 1508.9, 1508.10, 1508.13
Environmental Consequences	1502.10(g), 1502.16
Environmental Consultation Requirements	1500.4(k), 1500.5(g), 1501.7(a)(6), 1502.25, 1503.3(c)
Environmental Documents	1508.10
Environmental Impact Statement	1500.4, 1501.4(c), 1501.7, 1501.3, 1502.1, 1502.2, 1502.3, 1502.4, 1502.5, 1502.6, 1502.7, 1502.8, 1502.9, 1502.10, 1502.11, 1502.12, 1502.13, 1502.14, 1502.15, 1502.16,

Environmental | 1502.17, 1502.18,
Impact Statement | 1502.19, 1502.20,
(Cont.): | 1502.21, 1502.22,
  | 1502.23, 1502.24,
  | 1502.25,
  | 1506.2(b)(4),
  | 1506.3, 1506.8,
  | 1508.11
Environmental | 1502.11(f), 1504.1,
Protection | 1504.3, 1506.7(c),
Agency | 1506.9, 1506.10,
  | 1508.19(a)
Environmental | 1500.4(k), 1500.5(g),
Review | 1501.7(a)(6), 1502.25,
Requirements | 1503.3(c)
Expediter | 1501.8(b)(2)
Federal Agency | 1508.12
Filing | 1506.9
Final | 1502.9(b), 1503.1,
Environmental | 1503.4(b)
Impact Statement
Finding of No | 1500.3, 1500.4(q),
Significant Impact | 1500.5(1),
  | 1501.4(e), 1508.13
Fish and Wildlife | 1502.25
Coordination Act
Format for | 1502.10
Environmental
Impact Statement
Freedom of | 1506.6(f)
Information Act
Further Guidance | 1506.7
Generic | 1502.4(c)(2)
General Services | 1506.8(b)(5)
Administration
Geographic | 1502.4(c)(1)
Graphics | 1502.8
Handbook | 1506.7(a)
Housing and | 1506.12, 1508.12
Community
Development Act
Human | 1502.3, 1502.22,
Environment | 1508.14
Impacts | 1508.8, 1508.25(c)
Implementing | 1505.3
the Decision
Incomplete or | 1502.22
Unavailable
Information
Incorporation by | 1500.4(j), 1502.21
Reference
Index | 1502.10(j)

Indian Tribes | 1501.2(d)(2),
  | 1501.7(a)(1),
  | 1502.15(c),
  | 1503.1(a)(2)(ii),
  | 1506.6(b)(3)(ii),
  | 1508.5, 1508.12
Intensity | 1508.27(b)
Interdisciplinary | 1502.6, 1502.17
Preparation
Interim Actions | 1506.1
Joint Lead Agency | 1501.5(b), 1506.2
Judicial Review | 1500.3
Jurisdiction by Law | 1508.15
Lead Agency | 1500.5(c), 1501.1(c),
  | 1501.5, 1501.6,
  | 1501.7, 1501.8,
  | 1504.3, 1506.2
  | (b)(4), 1506.8(a),
  | 1506.10(e), 1508.16
Legislation | 1500.5(j), 1502.3,
  | 1506.8, 1508.17,
  | 1508.18(a)
Limitation on | 1506.1
Action During
NEPA Process
List of Preparers | 1502.10(h), 1502.17
Local or State | 1500.4(n), 1500.5(h),
  | 1501.2(d)(2),
  | 1501.5(b),
  | 1501.5(d),
  | 1501.7(a)(1),
  | 1501.8(c),
  | 1502.16(c),
  | 1503.1(a)(2),
  | 1506.2(b),
  | 1506.6(b)(3),
  | 1508.5, 1502.12,
  | 1508.18
Major Federal | 1502.3, 1508.18
Action
Mandate | 1500.3
Matter | 1504.1, 1504.2,
  | 1504.3, 1508.19
Methodology | 1502.24
Mitigation | 1502.14(f),
  | 1502.16(h),
  | 1503.3(d),
  | 1505.2(c), 1505.3,
  | 1508.20
Monitoring | 1505.2(c), 1505.3
National Historic | 1502.25
Preservation Act

National Register 1508.27(b)(8)
of Historical Places
Natural or 1502.16(f)
Depletable Resource
Requirements
Need for Action 1502.10(d), 1502.13
NEPA Process 1508.21
Non-Federal 1501.2(d)
Sponsor
Notice of Intent 1501.7, 1507.3(e),
1508.22
OMB Circular A-95 1503.1(a)(2)(iii),
1505.2,
1506.6(b)(3)(i)
102 Monitor 1506.6(b)(2),
1506.7(c)
Ongoing Activities 1506.12
Page Limits 1500.4(a), 1501.7(b),
1502.7
Planning 1500.5(a),1501.2(b),
1502.4(a), 1508.18
Policy 1500.2, 1502.4(b),
1508.18(a)
Program 1500.4(i), 1502.4,
1502.20, 1508.18
Environmental
Impact Statement
Programs 1502.4, 1508.18(b)
Projects 1508.18
Proposal 1502.4, 1502.5,
1506.8, 1508.23
Proposed Action 1502.10(e), 1502.14,
1506.2(c)
Public Health 1504.1
and Welfare
Public 1501.4(e),
Involvement 1503.1(a)(3), 1506.6
Purpose 1500.1, 1501.1,
1502.1, 1504.1
Purpose of Action 1502.10(d), 1502.13
Record of Decision 1505.2, 1506.1
Referrals 1504.1, 1504.2,
1504.3, 1506.3(d)
Referring Agency 1504.1, 1504.2,
1504.3
Response to 1503.4
Comments
Rural 1506.1(d)
Electrification
Administration
Scientific Accuracy 1502.24
Scope 1502.4(a), 1502.9(a),
1508.25

Scoping 1500.4(b), 1501.1(d),
1501.4(d), 1501.7,
1502.9(a), 1506.8(a)
Significantly 1502.3, 1508.27
Similar 1508.25
Small Business 1506.6(b)(3)(vi)
Association
Social Effects 1508.8
Special Expertise 1508.26
Specificity of 1500.4(l), 1503.3
Comments
State and 1501.4(e)(1), 1503.3,
Areawide Clearing- 1501.4(e)(2),
houses 1503.1(a)(2)(iii),
1506.6(b)(3)(i)
State and Local 1500.4(n), 1500.5(h),
1501.2(d)(2),
1501.5(b),
1501.5(d),
1501.7(a)(1),
1501.8(c),
1502.16(c),
1503.1(a)(2),
1506.2(b),
1506.6(b)(3),
1508.5,
1508.12, 1508.18
State and Local 1508.18(a)
Fiscal Assistance
Act
Summary 1500.4(h),
1502.10(b), 1502.12
Supplements to 1502.9(c)
Environmental
Impact Statements
Table of Contents 1502.10(c)
Technological 1502.4(c)(3)
Development
Terminology 1506.1
Tiering 1500.4(i), 1502.4(d),
1502.20, 1508.28
Time Limits 1500.5(e), 1501.1(e),
1501.7(b)(2), 1501.8
Timing 1502.4, 1502.5,
1506.10
Treaties 1508.17
When to Prepare 1501.3
an Environmental
Impact Statement
Wild and Scenic 1506.8(b)(2)(ii)
Rivers Act
Wilderness Act 1506.8(b)(2)(ii)
Writing 1502.8

Abbreviations and Acronyms

ACHP	Advisory Council on Historic Preservation
AID	Agency for International Development
ASCS	Agricultural Stabilization and Conservation Service, former agency of USDA; now Farm Service Agency
BLM	Bureau of Land Management
BOD	biochemical oxygen demand
Btu	British thermal unit
C	Celsius
CA 90	Clean Air Act Amendments of 1990 (*See also* CAAA)
CAA	Clean Air Act of 1970, as amended
CAAA	Clean Air Act Amendments of 1990 (*See also* CA 90)
CARE	Cooperative for Assistance and Relief Everywhere, Inc.
CEA	Cumulative Effects Analysis
CEC	Commission of the European Communities
CEO	Chief Executive Officer
CEQ	Council on Environmental Quality
CERCLA	Comprehensive Environmental Response, Compensation and Liability Act
CFC	chlorofluorocarbon
CFR	Code of Federal Regulations
CH$_4$	methane
Ci	curie
CITES	Convention on International Trade in Endangered Species of Wild Fauna and Flora
CO	carbon monoxide
CO$_2$	carbon dioxide

COD	chemical oxygen demand
COE	Corps of Engineers, U.S. Army
CX	Categorical Exclusion
D.C.	District of Columbia
dB(A)	decibels (A-weighted)
DEIS	Draft Environmental Impact Statement
DO	dissolved oxygen
DOD	U.S. Department of Defense
DOE	U.S. Department of Energy
DOT	U.S. Department of Transportation
EA	Environmental Assessment
EEA	European Environmental Agency
EEC	European Economic Community
EEG	electroencephalogram
EEU	European Economic Union
EIA	environmental impact analysis process
EIS	Environmental Impact Statement
EO	Executive Order
EPA	Environmental Protection Agency
EQ	environmental quality
ER	environmental reservations, EPA rating system for EISs
ESA	Endangered Species Act
ESCAP	Economic and Social Commission for Asia and the Pacific, United Nations
EU	environmentally unsatisfactory, EPA rating system for EISs
F	Fahrenheit
FAA	Federal Aviation Act
FAA	Federal Aviation Administration
FAO	Food and Agriculture Organization, United Nations
FEA	Federal Energy Administration, former agency; now part of DOE
FEIS	Final Environmental Impact Statement
FID	flame ionization detector
FIFRA	Federal Insecticide, Fungicide, and Rodenticide Act

Fig.	Figure
FLITE	Federal Legal Information Through Electronics
FONSI	Finding of No Significant Impact
FR	*Federal Register*
ft^2	square feet
FWS	Fish and Wildlife Service (*See also* USFWS)
gal	gallon, U.S.
GAO	General Accounting Office
GHG	greenhouse gas
GIS	geographic information system
GNP	gross national product
GPO	Government Printing Office
GSA	General Services Administration
h	hour
ha	hectare
HR	House Report
HTC	hazardous toxicant concentration
HUD	Housing and Urban Development Agency
HVAC	heating, ventilation, and air conditioning
ICC	Interstate Commerce Commission
IMPROVE	Interagency Monitoring of Protected Visual Environments
IPCC	Intergovernmental Panel on Climate Change
IUCN	International Union for Conservation of Nature and Natural Resources, United Nations
km^2	square kilometers
l	liter
lb	pound
LEXIS	a proprietary legal search system
LO	lack of objections, EPA rating system for EISs
MCL	maximum contaminant level
MCLG	maximum contaminant level goal
mg	milligram
Mha	million hectares

mi^2	square miles
min	minute
ml	milliliter
MOA	Memorandum of Agreement
MOU	Memorandum of Understanding
MPN	most probable number
MSDS	Material Safety Data Sheet
N/A	not applicable
N$_2$O	nitrous oxide
NAAQS	National Ambient Air Quality Standards
NAGPRA	Native American Graves Protection and Repatriation Act
NEDA	a chemical solution used as a chelating or reducing agent
NEPA	National Environmental Policy Act
NEXIS	a proprietary legal search system
NGO	Non-Governmental Organization
NHPA	National Historic Preservation Act
NIOSH	National Institute for Occupational Health and Safety
NMFS	National Marine Fisheries Service
NO	nitric oxide
NO$_2$	nitrogen dioxide
NOAA	National Oceanic and Atmospheric Administration
NOI	Notice of Intent
NO$_x$	nitrogen oxides
NPDES	National Pollutant Discharge Elimination System
NRC	National Research Council
NRC	Nuclear Regulatory Commission
NRDA	Natural Resource Damage Assessment
OCS	off-coast survey
OCS	outer continental shelf
OCZM	Office of Coastal Zone Management
OECD	Organization for Economic Co-Operation and Development
OMB	Office of Management and Budget
ONWI	Office of Nuclear Waste Isolation, DOE

OPEC Organization of Petroleum Exporting Countries

OSHA Occupational Safety and Health Administration

OTA Office of Technology Assessment

P phosphorus

PAN peroxyacetyl nitrate

PCB polychlorinated biphenyl

pCi picocurie

pH measure of acidity

PL Public Law (*See also* Pub.L.)

PPA Pollution Prevention Act

ppm parts per million

Pub.L. Public law (*See also* PL)

RCRA Resource Conservation and Recovery Act

RCW red cockaded woodpecker

ROD Record of Decision

SARA Superfund Amendments and Reauthorization Act

SB Senate Bill

SCS Soil Conservation Service, former agency of the USDA; Natural Resources Conservation Service since 1994

SEPA State Environmental Policy Act

SETAC Society of Environmental Toxicology and Chemistry

SHPO State Historic Preservation Officer or Office

SIA Social Impact Assessment

SIC Standard Industrial Classification

SO$_2$ sulfur dioxide

TCM tetrachloromercurate

TDS total dissolved solids

THC total hydrocarbon concentration

TNC The Nature Conservancy (a private organization)

TOC total organic carbon

TSCA Toxic Substances Control Act

TVA Tennessee Valley Authority

U.S. United States

UNESCO	United Nations Educational, Scientific and Cultural Organization
USA	United States of America
USA	U.S. Army
USAID	U.S. Agency for International Development (*See also* AID)
USC	United States Code
USDA	United States Department of Agriculture
USFWS	United States Fish and Wildlife Service (See also FWS)
USGS	United States Geological Survey
V	volt
WESTLAW	a proprietary legal search system
WHO	World Health Organization
yd^3	cubic yard
μCi	microcurie
μg	microgram
μm	micrometer

Bibliography

The following legislation, regulations, and publications have been cited in the text and appendices of this book.

7 U.S.C. 136 et seq., *Federal Insecticide, Fungicide, and Rodenticide Act* (1972–1978).

15 U.S.C. 2601 et seq., *Toxic Substances Control Act* (1976–1986).

16 U.S.C. 470-470t et seq., *National Historic Preservation Act* (1966–1980).

16 U.S.C. 661 et seq., *Fish and Wildlife Coordination Act* (1958–1965).

16 U.S.C. 1271 et seq., *Wild and Scenic Rivers Act* (1968–1987).

16 U.S.C. 1451 et seq., *Coastal Zone Management Act* (1972–1986).

16 U.S.C. 1531–1542 et seq., *Endangered Species Act* (1973–1984).

33 U.S.C. 1251 et seq., *Clean Water Act* (1948–1987).

33 U.S.C. 1401 et seq., *Marine Protection, Research and Sanctuaries Act* (1972–1980).

42 U.S.C. 300f et seq., *Safe Drinking Water Act* (1974–1986).

42 U.S.C. 4321 et seq., *National Environmental Policy Act,* (PL 91-190; 83 Stat 852) (1970–1975).

42 U.S.C. 4901 et seq., *Noise Control Act* (1972–1978).

42 U.S.C. 6901 et seq., *Resource Conservation and Recovery Act* (1976–1986).

42 U.S.C. 7401 et seq., *Clean Air Act* (1963–1990).

42 U.S.C. 9601 et seq., *Comprehensive Environmental Response, Compensation and Liability Act* (1980–1987).

42 U.S.C. 11001 et seq., *Superfund Amendments and Reauthorization Act* (1986).

"Agenda for the 21st Century: Managing Earth's Resources," *Business Week,* June 18, 1990.

"Is the Endangered Species Act Losing Its Bite?" *Audubon,* July/August 1991, p. 8.

Biological Diversity Hearing before the Subcommittee on Fisheries and Wildlife Conservation and the Environment. House of Representatives, May 22, 1991.

Calvert Cliffs Coordinating Committee versus Atomic Energy Commission. 499 F. 2d 2ERC1779, 1ELR2036, D.C. Cir 1971 (West, 1971).

CF Letter. The Conservation Foundation, May 1972.

Control of Erosion and Sediment Deposition from Construction of Highways and Land Development. U.S. Environmental Protection Agency, 1971.

Environmental Guidelines. Western Systems Coordinating Council, Environmental Committee, 1971.

Environmental Law Reporter. 7 ELR 10182, October 1977.

Environmental Law Reporter. 8 ELR 10010, January 1978.

Environmental Quality. Annual Reports of the Council on Environmental Quality, 1971–1991. Washington D.C: U.S. Government Printing Office.

Environmental Resources Management. Prepared for the Department of Housing and Urban Development, Central New York Regional Planning and Development Board, October 1972.

Gas Engineer's Handbook, Chap. 22, "Fuel Comparisons." New York: The Industrial Press, 1969.

Handbook of Forecasting Techniques. IWR Report 75-7, U.S. Army Engineer Institute for Water Resources, Ft. Belvoir, Virginia, December 1975.

IUCN Red List of Threatened Animals. IUCN Conservation Monitoring Centre, Cambridge, U.K., 1988.

JAG Law Review XIV, No. 1: 10–13, 25–34, 35–67, 1972.

Legislation, Programs and Organization. U.S. Environmental Protection Agency, January 1979, p. 28.

Matrix Analysis of Alternatives for Water Resource Development. U.S. Army Corps of Engineers, Tulsa District, July 31, 1972.

Noise and Conservation of Hearing. U.S. Army, Technical Bulletin Med. 251, March 1972.

Noise Effects Handbook: A Desk Reference to Health and Welfare Effects of Noise. U.S. Environmental Protection Agency, EPA 550/9-82/106, 1981.

Official World Wildlife Fund Guide to Endangered Species of North America. Washington, D.C.: Beecham Publishing, 1990.

Rainfall Erosion Losses for Cropland. Agricultural Research Service, USDA Agricultural Handbook No. 282.

Review of Federal Actions Impacting the Environment. Environmental Protection Agency (EPA), Washington, D.C., March 1975.

U.S. Environmental Initiatives in Eastern Europe. Hearing Before the Subcommittee on Transportation and Hazardous Materials. House of Representatives. April 23, 1990.

U.S. Environmental Protection Agency. "Report to the President and Congress on Noise," December 1971.

Abrahamson, Dean Edwin. "Global Warming: The Issue, Impacts, Responses," in *The Challenge of Global Warming.* Washington, D.C.: Island Press, 1989.

Adkins, William G., and Burke Dock, Jr. *Interim Report: Social Economic, and Environmental Factors in Highway Decision Making.* Research conducted for the Texas Highway Department in cooperation with the U.S. Department of Transportation, Federal Highway Administration, Texas Transportation Institute, Texas A & M University, October 1971.

Advisory Council on Historic Preservation (ACHP). "Section 106, Step-by-Step." 1986.

Advisory Council on Historic Preservation (ACHP). "A Five-Minute Look at Section 106 Review." 1989.

Allenby, Brad R. "Earth Systems Engineering: The World as Human Artifact," *The Bridge,* **30**:5–13, 2000.

American Conference of Government Industrial Hygienists (ACGIH), 1973. http://www.acgih.org/home.htm.

American National Standards. *Methods for Measurement of Sound Pressure Levels.* American National Standards Institute, New York, ANSI S1.13-1971.

American Planning Association, Chapter Presidents Council. *A Study Manual for the Comprehensive Planning Examination of the American Institute of Certified Planners (AICP).* Memphis, Tenn.: Graduate Program in City and Regional Planning, Memphis State University, 1990.

Anderson, Terry L., and Donald R. Leal. *Free Market Environmentalism.* San Francisco: Pacific Research Institute for Public Policy, 1991.

Argonne National Laboratory (ANL). *Summary Description of SEAM: The Social and Economic Assessment Model.* ANL/IAPE/TM78-9, April 1978.

Arrandale, Tom. "Endangered Species," *CQ Researcher,* 1:7 (June 21, 1991), pp. 395–415.

Arrhenius, S. "On the Influence of Carbonic Acid in the Air upon the Temperature of the Ground," *Philosophical Magazine,* 1896, 41:237.

Ayres, R.V., and A.V. Kneese, "Production, Consumption, and Externalities," *American Economic Review,* 59: 3-14, 1970.

Bagley, M. D., et al. *Aesthetics for Environmental Planning.* Environmental Protection Agency, Report No. EPA-600/5-73-009, November 1973.

Bandow, Doug, ed. *Protecting the Environment: A Free Market Strategy.* Washington, D.C.: The Heritage Foundation, 1986.

Bartell, Steven M., Robert H. Gardner, and Robert V. O'Neill. *Ecological Risk Estimation.* Chelsea, Mich.: Lewis Publishers, 1992.

Bass, Ronald. "Evaluating Environmental Justice under the National Environmental Policy Act," *Environmental Impact Assessment Review,* **18**:83–92, 1998.

Battelle-Columbus Division. *Socioeconomic Effects of the DOE Gas Centrifuge Enrichment Plant.* ORO-EP-111-P1, May 1979.

Beland, R. D., and T. P. Mann. *Aircraft Noise Impact Planning Guidelines for Local Agencies.* U.S. Department of Housing and Urban Development, November 1972, NTIS-PB-213020.

Beranek, L. L. *Noise Reduction.* New York: McGraw-Hill, 1960.

Berglund, Birgitta, and Thomas Lindvall, eds. "Community Noise." Archives of the Center for Sensory Research, 2(1):1–195, 1995.

Bragdon, Clifford R. "Community Noise: A Status Report," paper presented at the 84th meeting of the Acoustical Society of America, Miami, Florida, November 1972.

Bragdon, Clifford R. "Noise Control in Urban Planning," *Journal of Urban Planning and Development Division*, Vol. 99, No. 1, ASCE, 1973.

Bragdon, Clifford R. *Noise Pollution: The Unquiet Crisis*. Philadelphia: University of Pennsylvania Press, 1971.

Brown, Lester R., Christopher Flavin, and Sandra Postel. "Picturing a Sustainable Society," in *State of the World 1990: A Worldwatch Institute Report on Progress Toward a Sustainable Society*. New York: Norton, 1990.

Bullard, Robert D. "Ecological Inequities and the New South: Black Communities Under Siege," *Journal of Ethnic Studies*, 17(4):101–115, 1990.

Bureau of Land Management. Manual 1601—Land Use Planning. Rel. 1-1666, November 22, 2000. Washington, D.C.: Bureau of Land Management.

Bureau of Reclamation. Multiagency Task Force, *Guidelines for Implementing Principles and Standards for Multiobjective Planning of Water Resources*, U.S. Bureau of Reclamation, U.S. Department of the Interior, 1972 [draft].

Calabrese, Edward J., and Linda A. Baldwin. *Performing Ecological Risk Assessments*. Chelsea, Mich.: Lewis Publishers, 1993.

Caldwell, Lynton H. *The National Environmental Policy Act: An Agenda for the Future*. Indiana University Press, 1998.

California Environmental Protection Agency, Air Resources Board. *Transportation-Related Land Use Strategies to Minimize Motor Vehicle Emissions: An Indirect Source Research Study*. Final Report, June 1995.

Canter, Larry, and Ray Clark, eds. *Environmental Policy and NEPA: Past, Present, and Future*. Boca Raton, Fla.: St. Lucie Press, 1997.

Carley, Michael. *Social Measurement and Social Indicators: Issues of Policy and Theory*. London: George Allen and Unwin, 1981.

Carroll, John E. "Acid Rain—Acid Diplomacy," in *The Acid Rain Debate: Scientific, Economic and Political Dimensions*. Edited by Ernest J. Yanarella and Randall H. Ihara. Boulder, Colo.: Westview Press, 1985.

Carter, F. W., and David Turnock. *Environmental Problems in Eastern Europe*. New York: Routledge, 1997.

Cavanagh, Ralph, David Goldstein, and Robert Watson. "One Last Chance for a National Energy Policy," in *The Challenge of Global Warming*. Edited by D. E. Abrahamson. Washington, D.C.: Island Press, 1989.

CELDS: United States Army Corps of Engineers, Construction Engineering Research Laboratory, Champaign, Ill.

Central New York Regional Planning and Development Board. *Environmental Resources Management*, Department of Housing and Urban Development, Central New York Regional Planning and Development Board, 1972.

CHABA. *Guidelines for Preparing Environmental Impact Statements on Noise*. Report of Working Group No. 69. National Research Council, Assembly of Behavioral and Social Sciences, Committee on Hearing, Bioacoustics, and Biomechanics (CHABA). February 1977.

Ciborowski, Peter. "Sources, Sinks, Trends and Opportunities," in *The Challenge of Global Warming*. Edited by D. E. Abrahamson. Washington, D.C.: Island Press, 1989.

Clark, Ray, and Larry Canter, eds. *Environmental Policy and NEPA: Past, Present, and Future*. Boca Raton, Fla.: CRC Press LLC, 1997.

Clinton, William Jefferson. Memorandum for the heads of all departments and agencies; Subject: Executive Order on federal actions to address environmental justice in minority populations and low-income populations. Comprehensive Presidential Documents No. 279. Washington, D.C.: The White House, 1994.

Cohn, Jeffrey. "The Politics of Extinction," *Government Executive*, Oct. 1990, pp. 18–23.

Cohrssen, John J., and Vincent T. Covello. *Risk Analysis: A Guide to Principles and Methods for Analyzing Health and Environmental Risks*. Washington, D.C.: Council on Environmental Quality, Executive Office of the President, 1985.

Committee on Environment and Natural Resources (CENR), National Science and Technology Council. *Ecological Risk Assessment in the Federal Government.* EPA 600R99058 (CENR/5-99/001). Washington, D.C.: Executive Office of the President, 1999. See also www.nnic.noaa.gov/CENR/cenr.html.

Congressional Budget Office. *Environmental Regulation and Economic Efficiency.* Washington, D.C.: CBO, 1985.

Corbitt, R. A. *Standard Handbook of Environmental Engineering.* New York: McGraw-Hill, 1999.

Council on Environmental Quality. *Annual Report of the Council on Environmental Quality.* Washington, D.C., 1993.

Council on Environmental Quality. *Considering Cumulative Effects under the National Environmental Policy Act,* January 1997.

Council on Environmental Quality. *Environmental Justice: Guidance under the National Environmental Policy Act.* Washington, D.C, 1998.

Council on Environmental Quality. *Environmental Quality, Fourth Annual Report of the Council on Environmental Quality.* U.S. Government Printing Office, September 1973.

Council on Environmental Quality. *Environmental Quality: 20th Annual Report.* Washington, D.C.: GPO, 1990.

Council on Environmental Quality. *Environmental Quality: 21st Annual Report.* Washington, D.C.: GPO, 1990.

Council on Environmental Quality. *Guidance for Addressing Environmental Justice under the National Environmental Policy Act.* Washington, D.C.: Executive Office of the President, 1996.

Council on Environmental Quality. "Preparation of Environmental Impact Statements: Guidelines," *Federal Register,* Vol. 38, No. 147, Part II:20550–20562, August 1, 1973.

Council on Environmental Quality. *The 1997 Report of the Council on Environmental Quality.* Washington, D.C.

Council on Environmental Quality. *The National Environmental Policy Act: A Study of Its Effectiveness after 25 Years.* Washington, D.C.: Executive Office of the President, January, 1997. http://ceq.eh.doe.gov/nepa/nepanet.htm.

Crampton, R. C. Testimony before Joint Meeting of the Interior and Public Works Committees of U.S. Senate, March 7, 1972.

Culhane, P.J., H. P. Friesema, and J. A. Beecher. *Forecasts and Environmental Decision-making: the Content and Predictive Accuracy of Environmental Impact Statements.* Social Impact Assessment Series N. 14, Boulder Col.: Westview Press, 1987.

Cumberland, J.H. *Regional Development Experience in the United States of America.* The Hague: Mouton, 1971.

Davis, R. M., G. S. Stacey. G. I. Nebman, and F. K. Goodman, "Development of an Economic-Environmental Trade-Off Model for Industrial Land Use Planning, " *Rev. Regional Studies,* 4:1, Spring, 1974.

Dee, Norbert, et al. *Environmental Evaluation System for Water Resources Planning.* Reports to the U.S. Bureau of Reclamation, Battelle Memorial Institute, January 1972.

Dee, Norbert, et al. *Planning Methodology for Water Quality Management: Environmental Evaluation System.* Battelle Memorial Institute, July 1973.

Defenders of Wildlife. "Southwest Wolf Reintroduction Given Another Chance: Defenders Applauds Decision to Translocate Mexican Wolves." News Release, March 21, 2000.

Department of Energy, Office of Environmental Policy and Assistance, RCRA/CERCLA Division, EH-413. *Natural Resources Damage Assessment Implementation Project; Savannah River Site.* DOE/EH-0510. Washington, D.C.: Department of Energy, 1995.

Department of Energy. *Dual Axis Radiographic Hydrodynamic Test Facility Final Environmental Impact Statement,* DOE/EIS-0228, August 1995. Los Alamos, N.Mex.: Los Alamos Area Office, Department of Energy.

Department of Energy. *Special Environmental Analysis for the Department of Energy, National Nuclear Security Administration: Actions Taken in Response to the Cerro Grande Fire at Los Alamos National Laboratory, Los Alamos, New Mexico,* DOE/SEA-03, September 2000. Los Alamos, N.Mex.: Los Alamos Area Office, Department of Energy.

Department of Energy; National Nuclear Security Administration; "Emergency Activities Conducted at Los Alamos National Laboratory, Los Alamos County, New Mexico in Response to Major Disaster Conditions Associated with the Cerro Grande Fire," *Federal Register,* Vol. 65, No. 120, pp. 38522–38527, June 21, 2000.

Diersing, Victor E., William D. Severinghaus, and Edward W. Novak. *Annotated Directory of Endangered Wildlife on Selected US Army Installations West of the Mississippi River.* Champaign, Ill.: U.S. Army Corps of Engineers, Construction Engineering Research Laboratory, 1985.

Dodson, Roy D. *Storm Water Pollution Control: Municipal, Industrial and Construction NPDES Compliance,* Second Ed. New York: McGraw-Hill, September 1998.

Drucker, Peter F. *Management: Tasks, Responsibilities, Practices.* New York: Harper & Row, 1973.

Ehrlich, P. "Dodging the Crisis," *Saturday Review,* Vol. 53, No. 11, 1970, p. 73.

Ellis, William B., and Turner T. Smith Jr. "The Limits of Federal Environmental Responsibility and Control Under the National Environmental Policy Act," *Environmental Law Reporter,* Vol. 18, 1988, pp. 10055–10061.

European Environmental Agency (EEA). Website policy statement, 2001. http://org.eea.eu.int/.

Executive Order 11991, 42 FR 26967 (1977), Amending Executive Order 11514 (1970).

Executive Order 12898, Federal Actions to Address Environmental Justice in Minority Populations and Low-Income Populations, 59 FR 7630 (February 11, 1994). Washington, D.C.

Faludi, Andreas. *Planning Theory.* Oxford: Pergamon Press, 1973.

Fay, Thomas H., ed. *Noise & Health.* New York: New York Academy of Medicine, 1991.

Federal Energy Administration. "Energy Conservation in the Manufacturing Sector 1954–1980," *Project Independence Blueprint Final Task Force Report 3,* November 1974.

Findley, Roger W., and Daniel A. Faber. *Cases and Materials on Environmental Law.* St. Paul, Minn.: West Group, 1999.

Flavin, Christopher. *Reassessing Nuclear Power: The Fallout from Chernobyl.* Worldwatch Paper 75. Worldwatch Institute, 1987.

Flavin, Christopher. "Slowing Global Warming," in *State of the World 1990: A Worldwatch Institute Report on Progress Toward a Sustainable Society.* New York: Norton, 1990.

FLITE: United States Air Force Office, Denver, Colorado.

Foreman, John E. K. *Sound Analysis and Noise Control.* New York: Van Nostrand Reinhold, 1990.

French, Hilary. "Clearing the Air," in *State of the World 1990: A Worldwatch Institute Report on Progress Toward a Sustainable Society.* New York: Norton, 1990.

French, Hilary F. *Green Revolutions: Environmental Reconstruction in Eastern Europe and the Soviet Union.* Worldwatch Paper 99, Worldwatch Institute, 1990.

Fresquez, P. R., D. R. Armstrong, and L. Naranjo, Jr. *Radionuclides and Heavy Metals in Rainbow Trout from Tsichomo, Nana Ka, Wen Povi, and Pin De Lakes in Santa Clara Canyon.* LA-13441-MS. Los Alamos, N.Mex.: Los Alamos National Laboratory, 1998a.

Fresquez, P. R., D. R. Armstrong, M. A. Mullen, and L. Naranjo, Jr. "The Uptake of Radionuclides by Beans, Squash, and Corn Growing in Contaminated Alluvial Soils at Los Alamos National Laboratory." *Journal of Environmental Science and Health,* **B33**(1):99–122. New York: Marcel Dekker, Inc., 1998.

Gallegos, Anthony F., and Gilbert J. Gonzales. *Documentation of the Ecological Risk Assessment Computer Model ECORSK.5.* LA-13571-MS. Los Alamos, N.Mex.: Los Alamos National Laboratory, 1999.

General Accounting Office. *An Analysis of Issues Concerning 'Acid Rain.'* Report to the Congress of the United States by the Comptroller General. Washington, D.C.: GAO, 1984.

Gibbs, M. J. "Economic Analysis of Sea Level Rise: Methods and Results," in *Greenhouse Effect and Sea Level Rise, A Challenge for This Generation.* Edited by M. C. Barth and J. G. Titus. New York: Van Nostrand Reinhold, 1984.

Gilmore, J.S. and Duff, M.K. *Boom Town Growth Management: A Case Study of Rock Springs – Green River, Wyoming.* Boulder Col.: Westview Press, 1975.

Gilpin, Alan. *Environmental Impact Assessment: Cutting Edge for the Twenty-First Century.* New York: Cambridge University Press, 1995.

Giroult, E. "WHO Interest in Environmental Health Impact Assessment," in *Environmental Impact Assessment: Theory and Practice.* Edited by Peter Wathern. Winchester, Mass.: Allen & Unwin, 1988, pp. 259–271.

Goff, R. J., and E. W. Novak. *Environmental Noise Impact Analysis for Army Military Activities: User Manual.* USA-CERL Technical Report N-30, 1977.

Goldfarb, Joan R. "Extraterritorial Compliance with NEPA Amid the Current Wave of Environmental Alarm," *Boston College Environmental Affairs Law Review,* 18:543, 1991, pp. 543–603.

Gonzales, Gilbert J., et al. *A Spatially-Dynamic Preliminary Risk Assessment of the Bald Eagle at the Los Alamos National Laboratory.* LA-13399-MS. Los Alamos, N.Mex.: Los Alamos National Laboratory, 1998.

Goodman, William I., and Eric C. Freund, eds. *Principles and Practice of Urban Planning.* Washington, D.C.: Institute for Training in Municipal Administration, International City Managers' Association, 1968.

Gradwohl, Judith, and Michael Greensberg. *Saving the Tropical Forests.* Washington, D.C.: Island Press, 1988.

Grove, Noel. "Quietly Conserving Nature," *National Geographic,* Dec. 1988, pp. 818–844.

Hahn, Robert W. "Economic Prescriptions for Environmental Problems: How the Patient Followed the Doctor's Orders," *Journal of Economic Perspectives,* Vol. 3, No. 2, 1990, pp. 94–114.

Haklay, Mordechay, et al. "The Potential of a GIS-Based Scoping System: An Israeli Proposal and Case Study," *Environmental Impact Assessment Review,* 18:439–459, 1998.

Hallstrom, Lars K. "Industry Versus Ecology: Environment in the New Europe." *Futures,* 31:25–38, 1999.

Hannon, B. M. "Bottles, Cans, Energy," *Environment,* 14, No. 2, March, 1972.

Harris, David A., ed. *Noise Control Manual: Guidelines for Problem-Solving in the Industrial/Commercial Acoustical Environment.* New York: Van Nostrand Reinhold, 1991.

Hayes, A. C., P. R. Fresquez, and W. F. Whicker. *Uranium Uptake Study, Nambe, New Mexico: Source Document.* LA-13614-MS. Los Alamos, N. Mex.: Los Alamos National Laboratory, 2000.

Hoeller, Peter, Andrew Dean, and Jon Nicolaisen. "Macroeconomic Implications of Reducing Greenhouse Gas Emissions: A Survey of Empirical Studies," *OECD Economic Studies* No. 16, 1991.

Horberry, John. "International Organization and EIA in Developing Countries," *Environmental Impact Assessment Review,* Vol. 5, 1985, pp. 207–222.

Htun, Nay. "The EIA Process in Asia and the Pacific Region," in *Environmental Impact Assessment: Theory and Practice.* Edited by Peter Wathern. Winchester, Mass.: Allen & Unwin, 1988, pp. 223–238.

Hudson, Barclay M. "Comparison of Current Planning Theories: Counterparts and Contradictions." *Journal of the American Planning Association,* 45:(4) (October 1979), p. 387. Chicago: American Planning Association, *Journal of the American Planning Association.*

Ihara, Randall H. "An Overview of the Acid Rain Debate: Politics, Science and the Search for Consensus," in *The Acid Rain Debate: Scientific, Economic and Political Dimensions.* Edited by Ernest J. Yanarella and Randall H. Ihara. Boulder, Colo.: Westview Press, 1985.

Institute of Medicine, Committee of Environmental Justice, Health Sciences Policy Program, Health Sciences Section. *Toward Environmental Justice: Research, Education and Health Policy Needs.* Washington, D.C.: National Academy Press, 1999. See also www.nap.edu.

International Rubber Study Group (IRSG). http://www.rubberstudy.com/STATS.htm, 2001.

Isard, W. *Ecologic-Economic Analysis for Regional Development.* New York: The Free Press, 1972.

Jacobson, Jodi. "Holding Back the Sea," in *State of the World 1990: A Worldwatch Institute Report on Progress toward a Sustainable Society.* New York: Norton, 1990.

Jain, R. K., et al. *Environmental Impact Assessment Study for Army Military Programs.* Interim Report D-13/AD 771062, U.S. Army Construction Engineering Research Laboratory [CERL], December 1973.

Jain, R. K., et al. *Handbook for Environmental Impact Analysis.* Technical Report E-59/ADA006241, CERL, September 1974.

Jansen, H. M. A., et al. "Impacts of Sea Level Rise: An Economic Approach," in *Climate Change: Evaluating the Socio-Economic Impacts.* Paris: OEDC, 1991.

Jessee, Lee. "The National Environmental Policy Act Net (NEPAnet) and DOE NEPA Web: What They Bring to Environmental Impact Assessment," *Environmental Impact Assessment Review,* **18**:73–82, 1998.

Jimenez-Beltran, D. *The role of the European Environment Agency in delivering information to support the policy process.* Address to Green Week, Brussels, April, 2001. http://org.eea.eu.int/documents/speeches/brussels20010424.

Job, R. F. S. "The Influence of Subjective Reactions to Noise on Health Effects of the Noise," *Environmental International,* **22**(1):93–104, 1996.

Joint Services Planning Manual. *Environmental Protection for Planning in the Noise Environment.* (AFM 19-10; NAVFAC P-970). U.S. Army TM 5-803-2, June 1978.

Jones, Murray G. "Canadian Federal and Ontario Provincial Environmental Assessment Procedures," in *Perspectives on Environmental Impact Assessment.* Boston: Reidel, 1984, pp. 35–50.

Jones, Murray G. "The Evolving EIA Procedure in the Netherlands," in *Perspectives on Environmental Impact Assessment.* Boston: Reidel, 1984, pp. 57–68.

JURIS: United States Department of Justice, Justice Retrieval and Inquiry System, Room 4016 Chester Arthur Bldg, Washington, D.C.

Karimi, Hassan A., et al. "Evaluating Strategies for Integrating Environmental Models with GIS: Current Trends and Future Needs," *Computer Environment and Urban Systems,* **20**(6):413–425, 1996.

Kennedy, W. V. "Environmental Impact Assessment and Bilateral Development Aid: An Overview," in *Environmental Impact Assessment: Theory and Practice.* Edited by Peter Wathern. Winchester, Mass.: Allen & Unwin, 1988, pp. 272–285.

Ketcham, David. "How Does the Scoping Process Affect the Substance of an EIS?" Edited by Nicholas A. Robinson. *Environmental Impact Assessment—Proceedings of a Conference on the Preparation and Review of Environmental Impact Statements.* New York State Bar Association, 1988, pp. 84–89.

King, Thomas F. *Cultural Resource Laws and Practice: An Introductory Guide.* Walnut Creek, Calif.: AltaMira Press, Rowman and Littlefield Publishers, Inc., 1998.

Kiser, Jonathan. In *In Defense of Garbage* by Judd H. Alexander. Westport Conn.: Praeger Publishers, 1993.

Krauskopf, Thomas M., and Dennis C. Bunde. *Evaluation of Environmental Impact Through a Computer Modelling Process, Environmental Impact Analysis: Philosophy and Methods.* Edited by Robert Ditton and Thomas Goodale. University of Wisconsin Sea Grant Program, 1972, pp. 107–125.

Kryter, K. D. *The Effects of Noise on Man.* New York: Academic Press, 1985.

Land Use Planning Reports. *A Summary of State Land Use Controls,* September 1973.

Laurent, E.A. and J.C. Hite. *Economic-Ecologic Analysis in the Charleston Metropolitan Region: An Input-Output Study.* Water Resources Research Institute, Clemson University, April 1971.

Lee, E. Y. S., et al. *Environmental Impact Computer System.* Technical Report E-37 (CERL, September 1974).

Leontief, W. "Environmental Repercussions and the Economic Structure: An Input-Output Approach." *A Challenge to Social Studies,* Shiegeto Tsuru, Ed. Tokyo: Asahi, 1970.

Leopold, Luna B., et al. *A Procedure for Evaluating Environmental Impact.* Geological Survey Circular 645, Government Printing Office, 1971.

LEXIS: Mead Data Central, 200 Park Ave., New York, N.Y. 10166.

Lim, Gill-Chin. "Theory and Practice of EIA Implementation: A Comparative Study of Three Developing Countries," *Environmental Impact Assessment Review,* Vol. 5, 1985, pp. 133–153.

Liroff, Richard. "Eastern Europe: Restoring a Damaged Environment," *EPA Journal*, July–August 1990, pp. 50–55.

Litton, R. Burton, et al., *An Aesthetic Overview of the Role of Water in the Landscape*. National Water Commission by the Department of Landscape Architecture, University of California, Berkeley, July 1971.

Lohani, B. N. "Status of Environmental Impact Assessment in the Asian and Pacific Region," *Environmental Impact Assessment Worldletter*, Nov./Dec. 1986.

Lowry, Kem, and Richard A. Carpenter. "Institutionalizing Sustainable Development: Experiences in Five Countries," *Environmental Impact Assessment Review*, Vol. 5, 1985, pp. 239–254.

Lynch, Kevin, and Gary Hack. *Site Planning*. Cambridge, Mass.: The MIT Press, 1984.

Mahar, Dennis J. "Policies Affecting Land Use in the Brazilian Amazon: Impact on the Rainforest," *Land Use Policy* 7, January 1990, pp. 59–69.

Marcoux, Alain. *Population and the Environment: A Review and Concepts for Population Programmes. Part III: Population and Deforestation*. Food and Agriculture Organization, June 2000.

McHarg, Ian. *A Comprehensive Highway Route-Selection Method*. Highway Research Record No. 246, 1968, pp. 1–15.

McHarg, Ian L. *Design with Nature*. New York: John Wiley & Sons, 1991 (originally published in 1969).

Miernyk, W. H. "Long Range Forecasting With a Regional Input-Output Model," *Western Econ. Journal*, VI, No. 3, June 1968.

Miller, J. D. *Effects of Noise on People*. U.S. Environmental Protection Agency, Washington, D.C., December 31, 1971, PB-206723.

Monbailliu, Xavier. "EIA Procedures in France," in *Perspectives on Environmental Impact Assessment*. Boston: Reidel, 1984, pp. 51–56.

Moomaw, William R. "Near-Term Congressional Options for Responding to Global Climate Change," in *State of the World 1990: A Worldwatch Institute Report on Progress toward a Sustainable Society*. New York: Norton, 1990.

Moore, John L., et al. *A Methodology for Evaluating Manufacturing Environmental Impact Statements for Delaware's Coastal Zone*. Report to the State of Delaware. Battelle Memorial Institute, June 1973.

Moreira, Verocai. "EIA in Latin America," in *Environmental Impact Assessment: Theory and Practice*. Edited by Peter Wathern. Winchester, Mass.: Allen & Unwin, 1988, pp. 239–253.

Morgan, Richard K. *Environmental Impact Assessment*. Boston: Kluwer Academic Publishers, 1998.

Multiagency Task Force. *Guidelines for Implementing Principles and Standards for Multiobjective Planning of Water Resources*. U.S. Bureau of Reclamation, December 1972 [draft].

Munn, R. E. *Environmental Impact Assessment: Principles and Procedures, United Nations*. SCOPE, Secretariat, 51, Boulevard de Montmorency 75016 Paris, France. (Report 5, Toronto, Canada) 1975.

National Association of Regional Councils. "2000 Directory of Regional Councils in the United States." http://www.narc.org/.

National Council for Science and the Environment. "Recommendations on Improving the Scientific Basis for Environmental Decisionmaking. *First National Conference on Science, Policy and the Environment*, Washington, D.C., December 2000. Washington, D.C.: National Council on Science, Policy and the Environment. See also www.cnie.org.

National Research Council. *Science and the Endangered Species Act*. Washington, D.C.: National Academy Press, 1995.

National Wildlife Federation. *The Gray Wolf Returns: Reintroduction Fact Sheet*, January 1995. http://www.nwf.org/endangered/learn/wolfrein.html.

Nelson, Marj. "Habitat Conservation Planning," *Endangered Species Bulletin*, 6:12–13, 1999.

Nemerow, N. L. *Stream, Estuary, Lake and Ocean Pollution*. New York: Van Nostrand Reinhold, 1991.

Newton, David E. *Environmental Justice: A Reference Handbook*. Santa Barbara, Calif.: ABC-CLIO, Inc., 1996.

Nicolaisen, John, Andrew Dean, and Peter Hoeller. "Economics and the Environment: A Survey of Issues and Policy Options," *OECD Economic Studies*, No. 16, 1991, pp. 7–43.

Nordhaus, W. D. "A Survey of Estimates of the Cost of Reduction of Greenhouse Gas Emissions," unpublished paper, 1990.

Nordhaus, W. D. "The Economics of the Greenhouse Effect," paper prepared for the 1989 Meetings of the International Energy Workshop and the MIT Symposium on Environment and Energy, August 1989.

Nordhaus, W. D. "To Slow or Not to Slow," revision of the above paper, November 1990.

O'Brien, Mary H. "The Importance of Scoping," in *Environmental Impact Assessment— Proceedings of a Conference on the Preparation and Review of Environmental Impact Statements*. Edited by Nicholas A. Robinson. New York State Bar Association, 1988, pp. 90–92.

O'Brien, Stephen J., and Ernst Mayr. "Bureaucratic Mischief: Recognizing Endangered Species and Subspecies," *Science*, Vol. 251, March 8, 1991, pp. 1187–1188.

Odum, E. P. *Fundamentals of Ecology*. Philadelphia: W. B. Saunders, 1971.

Office of Technology Assessment. *Changing by Degrees: Steps to Reduce Greenhouse Gases*. Washington, D.C.: OTA, 1991.

Office of Technology Assessment. *Technologies to Maintain Biological Diversity*. Washington, D.C.: OTA, 1987.

Organization for Economic Cooperation and Development. *Fighting Noise: Strengthening Noise Abatement Policies*. Paris: OECD, 1986.

Organization for Economic Co-Operation and Development. *The Macro-Economic Impact of Environmental Expenditure*. Paris: OECD, 1985.

Park, Chris. *Acid Rain*. New York; Methuen, 1987.

Park, R. A, T. V. Armentano, and C. L. Cloonan. "Predicting the Effect of Sea Level Rise on Coastal Wetlands," in *Effects of Changes in Stratospheric Ozone and Global Climate*. Edited by J. G. Titus. Washington, D.C.: NOAA, 1986.

Pasztor, Janos. "What Role Can Nuclear Power Play in Mitigating Global Warming?" *Energy Policy*, 19, March 1991, pp. 98–109.

Pearce, David W. *Environmental Policy Benefits: Monetary Valuation*. Paris: OECD, 1989.

Pearce, David W. "Evaluating the Socio-Economic Impacts of Climate Change: An Introduction," in *Climate Change: Evaluating the Socio-Economic Impacts*. Paris: OECD, 1991.

Peterson, A. P. G., and E. E. Gross. *Handbook of Noise Measurement*. West Concord, Mass.: General Radio Company, 1980.

Polichtchouk, Y. "Geoinformation Systems and Regional Environmental Prediction," *Safety Science*, 30:63–70, 1998.

Post, N. K. "Odor Control of Wastes," *Industrial Waste Disposal*, R.D. Ross (ed.). New York: Van Nostrand Reinhold, 1968.

Postel, Sandra. "A Green Fix to the Global Warm-up," *World Watch*, Vol. 1, Sept./Oct. 1988, pp. 29–36.

President's Council on Sustainable Development. *PCSD Final Report*. Washington, D.C., 2000.

Proshansky, Harold M., et al., eds. *Environmental Psychology—Man and His Physical Setting*. New York: Holt, Rinehart and Winston, 1973.

Renard, M.L. "A Review of Comparative Energy Use in Materials Potentially Recovered from Municipal Solid Waste," prepared for the U.S. Department of Energy. Washington D.C.: National Center for Resources Recovery, Inc. March 1982.

Renner, Michael G. "War on Nature," *World Watch*, May–June 1991, pp. 18–25.

Rickson, R.E., R.J. Burdge, T. Hundloe, G.T. McDonald. "Institutional constraints to adoption of social impact assessment as a decision making and planning tool," in *EIA Review*, 10, 233-243, 1990.

Roberts, John, and Diane Fairhall, eds. *Noise Control in the Built Environment*. Brookfield, Vt.: Gower Technical, 1988.

Robinson, D.P., J.W. Hamilton, R.D. Webster, and M.J. Olson. "Economic Impact Forecast System (EIFS) II: User's Manual, Updated Edition." U.S. Army Construction

Engineering Research Laboratory Technical Report N-69 (Revised). Champaign, Ill., May 1984.

Rodgers, W. H., Jr. *Environmental Law.* St. Paul, Minn.: West Publishing Co., 1977.

Rohlf, Daniel J. *The Endangered Species Act: A Guide to Its Protection and Implementation.* Stanford Environmental Law Society, 1989.

Roque, Celso R. "Environmental Impact Assessment in the Association of Southeast Asian Nations," *Environmental Impact Assessment Review,* Vol. 5, 1985, pp. 257–263.

Satchell, Michael. "The Endangered Logger," *US News and World Report,* June 25, 1990, pp. 27–30.

Sax, Joseph. *Defending the Environment: A Handbook for Citizen Action.* New York: Vintage Books, 1970.

Schomer, P. D. "Assessment of Community Response to Impulsive Noise," *Journal of the Acoustical Society of America,* Vol. 77, No. 2, Feb. 1985, pp. 520–535.

Schomer, P. D. "Evaluation of C-weighted Ldn for Assessment of Impulse Noise," *Journal of the Acoustical Society of America,* Vol. 62, No. 2, 1977, pp. 386–399.

Schomer, P. D. *Handbook of Acoustical Measurements and Noise Control,* Chap. 50, "Community Noise Measurements," 2d ed. New York: John Wiley and Sons, 1991.

Schomer, P. D. "High-Energy Impulsive Noise Assessment," *Journal of the Acoustical Society of America,* Vol. 79, No. 1, Jan. 1986, pp. 182–186.

Schreiber, Helmut. "The Threat from Environmental Destruction in Eastern Europe," *Journal of International Affairs,* Vol. 44, No. 2, 1991, pp. 359–391.

Schultz, Cynthia B., and Tamara Raye Crockett. "Economic Development, Democratization, and Environmental Protection in Eastern Europe," *Boston College Environmental Affairs Law Review,* Vol. 18, No. 1, 1990, pp. 53–84.

Schultze, C. L. *The Public Use of Private Interest.* Washington, D.C.: The Brookings Institution, 1977, p. 32.

Sedjo, R. A. "Forests to Offset the Greenhouse Effect," *Journal of Forestry,* July 1990, pp. 12–15.

Severinghaus, W. D., H. E. Balbach, and L. L. Radke. *Endangered Species on U.S. Army Installations.* Champaign, Ill.: U.S. Army Corps of Engineers, Construction Engineering Research Laboratory, 1982.

Silveire, Vicente Fernando, et al. "An Information Management System for Forecasting Environmental Change," *Computer Industrial Engineering,* 31(1–2):289–292, 1996.

Simons, Marlise. "Europe Environment Plan Hits Snag: France," *The New York Times,* Feb. 16, 1992, p. 4.

Smith, William L. *Quantifying the Environmental Impacts of Transportation Systems.* Van Doer-Hazard-Stallings-Schnacke, undated.

Society of Environmental Toxicology and Chemistry. *Ecological Risk Assessment.* Technical issue paper. Pensacola, Fla.: SETAC, 1997.

Solbrig, Otto T. "The Origin and Function of Biodiversity," *Environment,* June 1991, pp. 17–38.

Solow, R. M. "The Economics of Resources or the Resources of Economics," in *Economics of Environment,* 2d ed. Edited by R. Dorfman. New York: W.W. Norton and Co., 1977, p. 368.

Sorensen, Jens. *A Framework for Identification and Control of Resource Degradation and Conflict in the Multiple Use of the Costal Zone,* University of California, Berkeley, Department of Landscape Architecture, 1970.

Sorensen, Jens, and James E. Pepper. *Procedures for Regional Clearinghouse Review of Environmental Impact Statements – Phase Two,* Report to the Association of Bay Area Governments, April 1973.

Sriwattanatamma, Patra, and Patrick Breysse. "Comparison of NIOSH Criteria and OSHA Hearing Conservation Criteria," *American Journal of Industrial Medicine,* 37:334–338, 2000.

Stacey G.S. "Proposed nomination of sites for site characterization and recommendation of issues for environmental assessments and site characterization plans." ONWI-505. Columbus Ohio: Battelle Memorial Institute, 1985.

State and Metropolitan Area Data Book, 1997-98. U.S. Department of Commerce, Economics and Statistics Administration, National Technical Information Service, http://www.ntis.gov/product/sma.htm, February 2000.

Statistical Abstract of the United States. U.S. Department of Commerce, Bureau of the Census, National Technical Information Service, http://www.ntis.gov/product/statistical-abstract.htm, June 2001.

Stein-Hudson, Kathleen E. "How Local and Regional Public Participation Affect the EIS Process," in *Environmental Impact Assessment—Proceedings of a Conference on the Preparation and Review of Environmental Impact Statements.* New York: New York State Bar Association, 1988, pp. 102–105.

Stover, Lloyd V. *Environmental Impact Assessment: A Procedure.* Miami, Fla.: Sanders and Thomas, 1972.

Sudara, Suraphol. "EIA Procedures in Developing Countries," in *Perspectives on Environmental Impact Assessment.* Edited by Brian D. Clark et al. Dordrecht, Netherlands: Reidel, 1984, pp. 81–90.

Suter, Alice H. "Noise and Its Effects," prepared for the Administrative Conference of the United States, November 1991.

Suter, Alice H., and John R. Franks, eds. *A Practical Guide to Effective Hearing Conservation Programs in the Workplace.* U.S. Dept. of Health and Human Services, Public Health Service, Centers for Disease Control, National Institute for Occupational Safety and Health, 1990.

Suter, G. W., II, et al. *Approach and Strategy for Performing Ecological Risk Assessments for the U.S. Department of Energy's Oak Ridge Reservation: 1995 revision.* ES/ER/TM-33/R2 (ON: DE96007746). Oak Ridge, Tenn.: Oak Ridge National Laboratory, 1995.

Suter, G. W., II. *Ecological Risk Assessment.* Chelsea, Mich.: Lewis Publishers, 1993.

Tennessee Valley Authority (TVA). *Belefonte Nuclear Plant Construction Employer Survey, May 1976.* Tennessee Valley Authority, Regional Planning Staff, Knoxville, Tenn., November 1976.

The Nature Conservancy (TNC). "About Us: Quick Facts." http://nature.org/aboutus/, 2001.

Thomann, R. V., and J. A. Mueller. *Principles of Surface Water Quality Modeling and Control.* New York: Harper & Row, 1987.

Titus, James G. "Greenhouse Effect, Sea Level Rise, and Barrier Islands: Case Study of Long Beach Island, New Jersey," *Coastal Management,* Vol. 18, 1986, pp. 65–89.

Titus, James G. "Strategies for Adapting to the Greenhouse Effect," *APA Journal,* Summer 1990, pp. 311–332.

Toronto Conference. "The Changing Atmosphere: Implications for Global Security," *The Challenge of Global Warming.* Edited by D. E. Abrahamson. Washington, D.C.: Island Press, 1989.

Trudeau, Rebecca L., and Olexa, M. T. "The Pollution Prevention Act," SS-FRE-13, Food and Resource Economics, Florida Cooperative Extension Service, Institute of Food and Agricultural Sciences, University of Florida, September 1994 http://www.ifas.ufl.edu/www/extension/ces.htm.

University of Georgia. *Optimum Pathway Matrix Analysis Approach to the Environmental Decision Making Process: Test Case: Relative Impact of Proposed Highway Alternatives.* University of Georgia, Institute of Ecology, 1971.

U.S. Army Corps of Engineers (USACOE). *Matrix Analysis of Alternatives for Water Resource Development.* U.S. Army Corps of Engineers, Tulsa District, July 1972.

U.S. Army Medical Research and Development Command, Department of the Army, Department of Defense. *Biological Defense Research Program, Final Environmental Impact Statement.* U.S. Army Medical Research and Development Command, Ft. Detrick, MD. U.S. Government Printing Office, 1989.

U.S. Army Military Standard, *Noise Limits for Army Materiel,* MIL-STD-1474 (MI), 1 March 1973.

U.S. Department of Commerce, Panel on Noise Abatement. *The Noise Around Us,* September 1970, COM 71-00147.

U.S. Department of Energy, Energy Information Administration. Web page: www.eia.doe.gov/aer/enduse.html.

U.S. Department of Energy. *Carbon Sequestration: State of the Science.* Washington, D.C., April 1999.

U.S. Department of Interior (USDI). Resource Management Systems, "A Description of Potential Socioeconomic Impacts from Energy-Related Developments on Campbell

County, Wyoming," Report prepared for the U.S. Department of the Interior, September, 1975.

U.S. Department of Transportation (DOT). *Transportation and Environment: Synthesis for Action: Impact of National Environmental Policy Act of 1969 on the Department of Transportation 3,* Office of the Secretary, U.S. Department of Transportation, 1969.

U.S. Department of Transportation. *Transportation Statistics Annual Report,* 1999.

U.S. Environmental Protection Agency. *33/55 Program: The Final Record, March 1999.* EPA-745-R-99-004. Washington, D.C.: Environmental Protection Agency, Office of Pollution Prevention and Toxics, 1999.

U.S. Environmental Protection Agency, *Community Noise,* NTID 300.3, 1971.

U.S. Environmental Protection Agency. *Community Relations in Superfund: A Handbook.* EPA/540/G-88/002. Washington, D.C.: Office of Emergency and Remedial Response, USEPA, 1988.

U.S. Environmental Protection Agency. *Environmental Equity: Reducing Risk for All Communities.* EPA230R92008. Washington, D.C.: Environmental Protection Agency, 1992b.

U.S. Environmental Protection Agency. *Filing System Guidance for Implementing 1506.9 and 1506.10 of the CEQ Regulations,* March 7, 1989.

U.S. Environmental Protection Agency. *Final Guidance for Incorporating Environmental Justice Concerns in EPA's NEPA Compliance Analysis.* Washington, D.C.: Office of Federal Activities, 1998.

U.S. Environmental Protection Agency. *Guidelines for Ecological Risk Assessment; Notice.* 63 FR 26846 (Thursday, May 14, 1998). EPA 630R95002F. Washington, D.C.: Environmental Protection Agency, 1998a. See also: www.epa.gov/ncea.

U.S. Environmental Protection Agency. *Innovative Management Strategies: Environmental Equity.* EPA 220B9224. Washington, D.C.: Office of Environmental Justice, 1992a.

U.S. Environmental Protection Agency. *Legislation, Programs and Organization.* January 1979, p. 28.

U.S. Environmental Protection Agency. *Noise Pollution,* August 1972, Washington, D.C.: USGPO, 5500-0072.

U.S. Environmental Protection Agency. *Policy Options to Stabilize Global Climate.* Washington, D.C.: EPA, 1990.

U.S. Environmental Protection Agency. *Progress Report on the EPA Acid Rain Program.* EPA430-R-99-011. Washington, D.C., November 1999.

U.S. Environmental Protection Agency. *Protective Noise Limits,* EPA550/9-79-100, 1998.

U.S. Environmental Protection Agency. *Public Health and Welfare Criteria for Noise,* July 27, 1973.

U.S. Environmental Protection Agency. *Report to the President and Congress on Noise,* December 1971, Washington, D.C.: USGPO, 1972.

U.S. Environmental Protection Agency. *Report to the President and Congress on Noise,* February 1972, Washington, D.C.: USGPO, 5500-0040.

U.S. Environmental Protection Agency. *Terms of Environment,* rev. ed. EPA 175B97001. Washington, D.C.: Environmental Protection Agency, 1998b. See also www.epa.gov/ OCEPAterms.

U.S. Environmental Protection Agency. From EPA web page; maintained by Jeff Kelley, Office of Public Affairs, 1997.

U.S. Environmental Protection Agency, Office of Water, Drinking Water Standards Program. http://www.epa.gov/safewater/standards.html, May 2001.

U.S. Fish & Wildlife Service. "Mexican Wolf Frequently Asked Questions." Southwest Region Ecological Services, July 1999. http://mexicanwolf.fws.gov/FAQ.cfm.

U.S. Fish & Wildlife Service. "Tenth Circuit Court Rules Wolf Introduction Legal; Wolves Are Here to Stay." News Release, January 13, 2000.

U.S. Fish and Wildlife Service. "Threatened and Endangered Species System." http://endangered.fws.gov/wildlife.html.

U.S. Fish & Wildlife Service. *ESA Basics: Over 25 Years of Protecting Endangered Species.* Washington, D.C., June 1998.

U.S. General Services Association. *Environmental Justice Fact Sheet,* 1998 http://www.gsa.gov/pbs/pt/call-in/factshet/0298b/02_98_6.htm.

U.S. Interagency Burned Area Emergency Rehabilitation (BAER) Team. *Cerro Grande Rehabilitation Report, Cerro Grande Burned Area Emergency Rehabilitation Team,* June 2000. Santa Fe, N.Mex.: National Park Service.

University of Tennessee. "A Study of Payments In-lieu-of-Taxes by the Atomic Energy Commission to Anderson and Roane Counties, Tennessee, under the Special Burdens Provisions of Section 168 of the Atomic Energy Act of 1954." Center for Business and Economic Research, Bureau of Public Administration, The University of Tennessee. Knoxville, Tenn.: June 1973.

Unpublished Data—CEQ: All EISs Filed 1973–1999. http://ceq.eh.doe.gov/nepa/ nepanet.htm.

Unpublished Data—CEQ: Total EISs Filed by Year & Selected Agencies 1992–1998. http://ceq.eh.doe.gov/nepa/nepanet.htm.

Unpublished Data—EPA, Acid Rain Program: Allowance Trading System Fact Sheet. http://www.epa.gov/acidrain/allsys.html.

Unpublished Data—EPA, Acid Rain Program: Program Overview. http://www.epa.gov/ acidrain/overview.html.

Unpublished Data—EPA, Office of Pesticide Programs: About the Endangered Species Protection Program. http://www.epa.gov/espp/aboutespp.htm.

Unpublished Data—EPA, Office of Pesticide Programs: Endangered and Threatened Species. http://www.epa.gov/espp/coloring/especies.htm.

Unpublished Data—U.S. Department of Energy, Fossil Energy: Carbon Sequestration. http://www.fe.doe.gov/coal_power/sequestration/index.html.

Urban, L. V., et al. *User Manual—Computer-Aided Environmental Impact Analysis for Construction Activities.* Technical Report E-50, CERL, March 1975.

Verity, C. William, Jr. "You Should Make the First Move Toward Working with Government," *Area Development: Sites and Facility Planning.* Amy, 1977.

Villach Conference. "The Scientific Consensus," in *The Challenge of Global Warming.* Edited by D. E. Abrahamson. Washington, D.C.: Island Press, 1989.

von Moltke, Konrad. "Impact Assessment in the United States and Europe," in *Perspectives on Environmental Impact Assessment.* Edited by Brian D. Clark et al. Dordrecht, Netherlands: Reidel, 1984, pp. 25–34.

Walton, L. Ellis, Jr., and James E. Lewis. *A Manual for Conducting Environmental Impact Studies,* Virginia Highway Research Council, January 1971.

Wandesforde-Smith, Geoffrey, et al. "EIA in Developing Countries: An Introduction," *Environmental Impact Assessment Review,* Vol. 5, 1985, pp. 201–206.

Warner, M. L., et al. *An Assessment Methodology for the Environmental Impact of Water Resources Projects* (600/5-75-016). Washington, D.C.: Environmental Protection Agency, July 1974.

Wathern, P. "The EIA Directive of the European Community," in *Environmental Impact Assessment: Theory and Practice.* Edited by Peter Wathern. Winchester, Mass.: Allen & Unwin, 1988, pp. 192–209.

Webb, M. Diana. "DARHT—an 'Adequate' EIS: A NEPA Case Study." *Proceedings of the 22d Annual Conference of the National Association of Environmental Professionals.* Boulder, Colo.: National Association of Environmental Professionals, 1997, pp. 1014–1027.

Webber, David J. "Equitably Reducing Transboundary Causes of Acid Rain: An Economic Incentive Regulatory Approach," in *The Acid Rain Debate: Scientific, Economic and Political Dimensions.* Edited by Ernest J. Yanarella and Randall H. Ihara. Boulder, Colo.: Westview Press, 1985.

Webster, R. D., et al. *Development of the Environmental Technical Information.* Interim Report E-52, CERL, April 1975.

Weisburd, N. I. "Field Operations and Enforcement Normal for Air Pollution Control 3," U.S. Environmental Protection Agency, August 1972.

Welsh, R. L. "User Manual for the Computer-Aided Environmental Legislative Data System," Technical Report E-78, Construction Engineering Research Laboratory, Champaign, Ill., November 1975.

Wertenbaker, T. J. *The Puritan Oligarchy—The Founding of American Civilization.* New York, Scribner, 1947.

Western Systems Coordinating Council. *Environmental Guidelines.* Western Systems Coordinating Council, Environmental Committee, 1971.

WESTLAW: West Publishing Co., P.O. Box 3526, St. Paul, Minnesota 55165.

Wilcoxen, P. J. "Coast Erosion and Sea Level Rise: Implications for Ocean Beach and San Francisco's West Side Transport Project," *Coastal Zone Management Journal,* Vol. 14, No. 3, 1986, pp. 173–191.

Wilson, Edward O. *Diversity of Life.* Cambridge, Mass.: Harvard University Press, 1993.

Wirth, Timothy E., and John Heinz. *Project 88—Round II: Incentives for Action: Designing Market-Based Environmental Strategies.* Washington, D.C., 1991.

Wood, Christopher. "The European Directive on Environmental Impact Assessment: Implementation at Last?" *The Environmentalist,* Vol. 8, No. 3, 1988, pp. 177–186.

Wood, Christopher. "The Genesis and Implementation of Environmental Impact Assessment in Europe," in *The Role of Environmental Impact Assessment in the Planning Process.* London: Mansell, 1988, pp. 88–102.

World Resources Institute. *A Directory of Impact Assessment Guidelines,* 2d ed. IIED Environmental Planning Group, WRI, 1998.

Index

A

Acid rain, 340–347, 391
  causes of, 340–341
  Clean Air Act provisions related to,
    20–21
  damages from, 342–344
  environmental assessment implications
    of, 347
  international efforts related to,
    346–347
  policy options for dealing with,
    344–346
  uncertainties about, 341
Acid Rain Program, 340–341
Actions, 76
  alternatives to, 100–102
  categorical exclusions, 93, 95
  classes of, 92–93
  emergency, 120–122
  federal, 65–66, 72
  and level of NEPA review, 93–97
  private vs. federal, 65–66
Activities
  categorization of, 193–197
  commercial, 310–311
  dredge and fill, 30
  identifying relevant, 573–574
  industrial, 311
  reevaluating identification of, 575–576
  residential, 310
  transportation, 311, 313–315
Ad hoc methodologies, 166
Ad hoc review (EIS), 219, 223
Adaptation strategies (global warming),
  339
Administrative compliance review (EIS),
  212
Administrative constraints, 166
Advisory Council on Historic
  Preservation web site, 17

Aesthetics, 148, 562–565
Affected environment, 5, 102–103, 149,
  277
Afghanistan, 232
Africa, 348
Aggregation, 155
Air, 127–130, 407–439
  carbon monoxide in, 426–428
  Clean Air Act provisions related to, 20
  diffusion factor of, 408–410
  environmental effects of energy con-
    sumption on, 313
  hazardous toxicants in, 430–433
  hydrocarbons in, 418–422
  National Ambient Air Quality
    Standards, 20
  nitrogen oxides in, 422–426
  odors in, 434–439
  particulates in, 410–414
  photochemical oxidants in, 428–430
  sulfur oxides in, 414–418
Allowance trading, 345
Alternative actions
  EIS section on, 100–102
  as methodological consideration, 164
  no-action alternative, 119–120, 148–149
Alternative fuels, 316–318
Aluminum recycling, 324
Animals (see Wildlife)
Antarctica, 228
Appendices to EIS, 104
Aquatic life, 132, 473–474
Aquatic plants, 514–517
Aquifer safe yield, 442–443, 446–455
Argentina, 232
Arrhenius, Svante, 327
Asia, 232, 348
Asian Development Bank, 235
Association for Biodiversity Information,
  366

Attributes, environmental, 126–148,
    407–565
  air, 127–130, 407–439
    carbon monoxide in, 426–428
    Clean Air Act provisions, 20
    diffusion factor of, 408–410
    hazardous toxicants in, 430–433
    hydrocarbons in, 418–422
    National Ambient Air Quality
        Standards, 20
    nitrogen oxides in, 422–426
    odors in, 434–439
    particulates in, 410–414
    photochemical oxidants in, 428–430
    sulfur oxides in, 414–418
  categorization of, 197–200
  comparison among, 152–155
  definition of, 126
  ecology, 136–138, 492–517
    aquatic plants, 514–517
    field crops, 506–509
    fish/shellfish/waterfowl, 501–506
    large animals, 493–496
    listed species (threatened/endan-
        gered), 509–511
    natural land vegetation, 511–514
    predatory birds, 496–499
    small game, 499–501
  economics, 145–147, 547–556
    per capita consumption, 553–556
    public sector revenue/expenditures,
        550–553
    regional economic stability, 548–549
  human aspects, 143–145, 539–547
    community needs, 545–547
    lifestyles, 540–542
    physiological systems, 544–545
    psychological needs, 542–544
  impacts on, 151
  land, 132–136, 476–492
    erosion of, 476–481
    and natural hazards, 481–487
    patterns, land-use, 487–492
  reevaluating identification of, 575–576
  resources, 147–148, 556–565
    aesthetics, 562–565
    fuel, 556–560
    nonfuel, 560–562
  sound, 138–143, 517–538
    communication effects of, 529–533
    performance effects of, 533–535
    physiological effects of, 517–525
    psychological effects of, 525–529
    social behavior effects of, 535–538

Attributes, environmental (*Cont.*)
  water, 129–132, 440–475
    aquatic life in, 473–474
    aquifer safe yield, 442–443, 446–455
    biochemical oxygen demand of,
        464–466
    dissolved oxygen in, 466–467
    dissolved solids in, 467–469
    EPA primary drinking water regula-
        tions, 446–454
    fecal coliforms in, 474–475
    flow variations in, 455–456
    groundwater systems, 441
    nutrients in, 469–471
    oil in/on, 457–458
    pH of, 463–464
    radioactivity in, 458–459
    self-purification of natural, 440
    surface systems, 440–441
    suspended solids in, 459–461
    thermal discharge in, 461–463
    toxic compounds in, 471–473
Australia, 232, 240
Aviation noise, 22, 140

B

Bangladesh, 232
Baseline characteristics (impact assess-
    ment), 150–151
Biochemical oxygen demand (BOD),
    464–466
Biodiversity, 360–366, 391
  and Endangered Species Act, 362–363
  environmental assessment implications
      of, 365–366
  and extinction of species, 362
  in "healthy" ecosystems, 360–362
  international efforts related to,
      363–364
  nongovernmental efforts promoting,
      363
  uncertainties about, 364–365
Birds
  Division of Migratory Bird
      Management web site, 18
  and Migratory Bird Treaty Act of 1918,
      18
  Neotropical Migratory Bird
      Conservation Program, 364
  predatory, 496–499
BOD (*see* Biochemical oxygen demand)
Boomtown development, 258–262
Brazil, 230–232, 348–349
British thermal units (Btus), 311

Bryson, Reid, 329
Btus (British thermal units), 311
Buildings, acid rain damage to, 344
Bureau of Land Management, 81–82
Burger v. County of Mendocino, 65

C

CA 90 (*see* Clean Air Act Amendments of 1990)
California, 74, 283, 340
California Air Resources Board, 314
Calvert Cliffs, 68–69
Camp Shelby, Miss., 119–120
Canada, 232–233, 236, 342, 347
Carbon dioxide, 332
Carbon monoxide, 426–428
Carbon sequestration, 333–335
Case study method (SIA), 252, 254, 255
Categorical exclusions, 93
Cat-X (CX) actions, 93
CEA (*see* Cumulative Effects Analysis)
CEC (*see* Commission of the European Communities)
CEQ (*see* Council on Environmental Quality)
CERCLA (*see* Comprehensive Environmental Response, Compensation and Liability Act)
Cerro Grande Fire, 121–122
CFCs (*see* Chlorofluorocarbons)
Checklists, 167, 252, 253
Chlorofluorocarbons (CFCs), 330–332
CITES (Convention on International Trade in Endangered Species of Wild Fauna and Flora), 351
City of Carmel-By-The-Sea v. U.S. DOT, 64
Clean Air Act Amendments of 1990 (CA 90), 20, 129, 340–341
Clean Air Act of 1970, 19–22, 71, 128
Clean Air Amendments of 1977, 128–129
Clean Water Act, 27–31, 130
Climate(s), 133 (*See also* Global warming)
Coal, 316, 345–346
Coastal Zone Management Act, 47–49
Code of Federal Regulations, 17
Colombia, 227–228, 232, 348
Colorado, 283–284, 344
Commercial activities, 310–311
Commission of the European Communities (CEC), 233, 234
Committee for the National Institutes for the Environment web site, 161
Communication effects of sound, 529–533

Communication techniques, 297–301
Community needs, 545–547
Community relations (*see* Public participation)
Comprehensive Environmental Response, Compensation and Liability Act (CERCLA), 34–37, 386–389 (*See also* Superfund Amendments and Reauthorization Act)
A Comprehensive Highway Route-Selection Method, and Design with Nature, 184
Computer-aided combination methodologies, 167
Computer-Aided Environmental Impact Analysis for Construction Activities: User Manual, 180–181
Computer-based EA systems, 201–205
Connecticut, 350
Consequences, environmental, 5
Conservation
of energy, 318–319
of resources, 1
Conservation Reserve Program, 349
Consumption
energy, 310–311
per capita, 553–556
Content analysis method (SIA), 255
Controversy (methodological consideration), 165–166
Convention on International Trade in Endangered Species of Wild Fauna and Flora (CITES), 351
Cook Island, 232
Corporate environmentalism, 4
Costa Rica, 360
Cost-benefit analyses, 244, 338–339
Costs
economic *vs.* environmental, 14, 55–56
environmental, 68
of pollution control, 319
transportation, 314
Council on Environmental Quality (CEQ), 44, 70, 85
classes of actions recognized by, 92–93
and data access, 159–160
draft environmental impact statements required by, 91
EIS format prescribed by (*see under* Environmental impact statements)
electric power alternatives study by, 316
environmental assessments required by, 90

Council on Environmental Quality (CEQ)
(*Cont.*)
environmental documents required by,
90–92
and environmental justice, 272–273,
277–278
final environmental impact statements
required by, 91
findings of no significant impact
required by, 90–91
law establishing, 401–404
and level of NEPA review, 92–97
notices of intent required by, 91
Record of Decisions required by, 91–92
regulations issued by, 88–90
supplemental reviews under, 112–113
web site of, 16
Cover sheet (EIS), 97
Crampton, Roger C., 69
Critical habitat, 353, 359
Cultural resources, 366–374, 391
definition of, 367–368
Native American, 371–373
and NEPA, 373–374
responsibility under NHPA for,
368–371
Cumulative Effects Analysis (CEA), 175,
177
Cumulative impacts, 156–157, 173–176
Czechoslovakia, 3

D

DARHT Facility (*see* Dual Axis
Radiographic Hydrodynamic Test
Facility)
Debt-for-nature swaps, 349–350
Decision making
EIS review for, 212
public participation in, 288
risk-based, 384
Deforestation, 347–350, 391
in developing countries, 349–349
in industrialized countries, 349
policy options for dealing with, 349–350
DEIS (*see* Draft environmental impact
statement)
Delphi method (SIA), 256
Delphi technique, 152–155
Developing countries
deforestation in, 349–349
EA in, 230–231
Diffusion factor (air), 408–410
Dissolved oxygen (in water), 466–467
Dissolved solids (in water), 467–469

District of Columbia, 73–74
Division of Migratory Bird Management
web site, 18
Documents, environmental, 7–9, 85–86,
88, 90–122, 567–584
and application of documentation
process, 92–97
case studies related to, 116–122
draft environmental impact statements, 91, 109–110
environmental assessments, 90,
104–107
environmental impact statements,
97–104
affected environment section of,
102–103
alternatives including the proposed
action section of, 100–102
appendices to, 104
cover sheet of, 97
environmental consequences section
of, 103
index of, 104
list of preparers in, 103–104
outline of, 98–99
preparation of, 116
purpose and need statement in,
99–100
scoping process with, 107–109
summary in, 97
supplements to, 112–113
table of contents in, 97, 99
time frames for, 113–114
final environmental impact statement,
91, 111–112
finding of no significant impact, 90–91
and function/purpose of assessment
process, 85–88
mitigation statements in, 115–116
notice of intent, 91
preparation of, 567–584
analyzing findings (step 10), 580–582
defining the action (step 1), 567,
569–571
developing alternatives (step 2),
571–573
evaluating impacts using descriptor
package and worksheets (step 7),
576–579
examining attributes likely to be
affected (step 4), 574
flow chart for, 567–568
identifying relevant project activities
(step 3), 573–574

Documents, environmental (*Cont.*)
  preparing assessment document
      (step 11), 583
  reevaluating identification of activi-
      ties/attributes (step 6), 575–576
  reviewing other alternatives (step 9),
      580
  reviewing/processing document (step
      12), 583–584
  soliciting public concerns (step 5),
      574–575
  summarizing impacts (step 8),
      579–580
  Record of Decision, 91–92
  related to listed species, 357
  tiering of, 115
  web site for downloading of, 18
DOD (*see* U.S. Department of Defense)
DOE (*see* U.S. Department of Energy)
Draft environmental impact statement
    (DEIS), 91, 109–112, 209, 291–292
Dredge and fill activities, 30, 40–41
Drinking water
  EPA regulations for, 446–454
  Safe Drinking Water Act, 25–27
Drinking Water Act of 1974, 26
Dual Axis Radiographic Hydrodynamic
    Test (DARHT) Facility, 117–119
Dumping, ocean, 40–41
Dynamic simulation method (SIA),
    257–258

E

EA documents (*see* Environmental
    Assessment documents)
EA process (*see* Environmental assess-
    ment process)
Earth Summit, 229, 330
Eastern Europe, 2–3, 225
Ecological biodiversity, 363
Ecological risk assessment (ecorisk),
    374–391
  applications of, 376–378
  definition of, 375
  description of, 374–376
  federal laws requiring/facilitated by,
      386–390
  implementation of decisions based on,
      384–386
  limitations of, 390
  phases in, 378–382
    analysis, 379–380
    problem formulation, 379, 380
    risk characterization, 380–381

Ecological risk assessment (ecorisk)
    (*Cont.*)
  as resource management tool, 390
  uncertainties about, 382–383
  value of, 391
Ecology, 136–138, 492–517
  aquatic plants, 514–517
  field crops, 506–509
  fish/shellfish/waterfowl, 501–506
  large animals, 493–496
  listed species, 509–511
  natural land vegetation, 511–514
  predatory birds, 496–499
  small game, 499–501
Econometric models, 240
Economic base models, 246–251
Economic costs, environmental costs *vs.*,
    14, 55–56
Economic impact analysis, 145–147,
    242–251, 547–556
  economic base models for, 246–251
  future direction for, 251
  input-output model for, 244–246
  and per capita consumption, 553–556
  public sector revenue/expenditures in,
      550–553
  and regional economic stability,
      548–549
Economic Impact Forecast System,
    246–251
Ecorisk (*see* Ecological risk assessment)
Ecosystem(s)
  carbon sequestration in,
      334–335
  diversity, ecosystem, 360
  healthy, 360–362
Ehrlich, Paul, 68
EISs (*see* Environmental impact state-
    ments)
Electronic journals, 160–161
Emergency actions case study,
    120–122
Emergency planning, 38
Emergency Planning and Community
    Right-to-Know Act of 1986, 37
Emissions
  acid rain from, 340–341
  Clean Air Act provisions related to, 20
  contributing to greenhouse effect, 330
  sulfur dioxide, 340
  technologies for reduction of, 332–335
Employment, 146, 269
Endangered species, 350–353, 359, 391
  critical habitat, 353, 359

Endangered species (*Cont.*)
  listed species, 351–353, 359
Endangered Species Act (ESA), 49–50,
    351–359
  compliance with Section 7 of, 354
  compliance with Section 9 of, 354–355
  conservation of ecosystems under, 389
  NEPA compliance with, 355–359
Endangered Species Conservation Act of
    1969, 350
Endangered Species Preservation Act of
    1966, 350
Energy conservation, 318–319
Energy considerations, 309–325
  categories of energy consumption,
      310–311
  with commercial activities, 310–311
  conservation of energy, 318–319
  with fuel alternatives, 316–318
  impacts on energy resources, 310–311
  impacts on fuel resources, 311–313
  with industrial activities, 311
  in operation of pollution control sys-
      tems, 319
  in recycling of materials, 319–325
    glass, 321–323
    metals, 324
    oil wastes, 324–325
    paper, 323
    plastics, 324
    solid waste, 325
    tires/rubber products, 323
  with residential activities, 310
  with transportation activities, 311,
      313–315
Energy ratio, 312
Energy resources, 147–148
Eneweitak v. Laird, 227
Environment
  definition of, 4–5
  social, 143–145
Environmental Assessment (EA) docu-
    ments, 7, 90
  as initial level of NEPA review,
      95–96
  mitigation in, 115–116
  preparation/processing of, 104–107
  standard *vs.* alternative formats for,
      106
  tiering of, 115
Environmental assessment (EA) process,
    5–10, 125–161
  and activities associated with project
    implementation, 125–126

Environmental assessment (EA) process
    (*Cont.*)
  categorization of attributes in, 126–148
      (*See also specific headings*)
    air, 127–130
    ecology, 136–138
    economics, 145–147
    human aspects, 143–145
    land, 132–136
    resources, 147–148
    sound, 138–143
    water, 129–132
  contemporary issues in, 391 (*See also
    specific topics*)
  determination of environmental impact
      in, 148–157
    aggregation, 155
    baseline characteristics, 150–151
    comparison among attributes,
        152–155
    cumulative impacts, 156–157
    and identification of impacts, 150
    measurement of impacts,
        151–152
    role of attributes, 151
    secondary impacts, 155–156
    with and without the project,
        148–150
  generalized approach for, 193–205
    categorization of agency activities,
        193–197
    categorization of environmental
        attributes, 197–200
    computer-based system for, 201–205
    environmental setting, 201
    institutional constraints, 200
    output from, 201
    system for, 201
  implications of acid rain for, 347
  implications of biodiversity for,
      365–366
  implications of global warming for,
      339
  information technology aids for,
      158–161
  integrating art and science of, 8
  methodologies for (*see* Methodologies
      for assessment)
  necessity for, 6–7
  reporting findings of, 157–158
  steps in, 5–6
Environmental consequences section
    (EIS), 103

Environmental Defense Fund v. Massey,
67
Environmental Defense Fund v. USAID,
227
Environmental Evaluation System for
Water Resources Planning, 178–179
Environmental Guidelines methodology,
190
Environmental impact analysis, 1
and long-term planning process, 86
reasons for undertaking, 86–87
in Asia and the Pacific, 232
in Canada, 232–233
in developing countries, 230–231
in Europe, 233–235
and extraterritorial application of
NEPA, 226–229
international issues related to,
225–236
in Asia and Pacific, 232
in Canada, 232–233
in developing countries, 230–231
in Europe, 233–235
extraterritorial NEPA application,
226–229
in Latin America, 232
organizations, international, 235–236
in Latin America, 232
Environmental Impact Assessment: A
Procedure, 187
Environmental Impact Assessment Study
for Army Military Programs,
180–181
Environmental impact statements (EISs),
7–10, 85–86, 97–104
appendices to, 104
CEQ-prescribed outline for content of,
98–99
cover sheet of, 97
draft (DEIS), 91, 109–110
environmental consequences section of,
103
EPA categories of, 209–211
final (FEIS), 91, 111–112
format of, 97–104
affected environment section,
102–103
alternatives including the proposed
action section, 100–102
appendices to, 104
cover sheet, 97
environmental consequences section,
103
index of, 104

Environmental impact statements (EISs)
(*Cont.*)
list of preparers, 103–104
outline of, 98–99
purpose and need statement, 99–100
summary, 97
table of contents, 97, 99
index of, 104
as initial level of NEPA review, 93–95
list of preparers in, 103–104
mitigation in, 115–116
preparation of, 116
primary purpose of, 86
purpose and need statement of, 99–100
review of, 207–223
ad hoc review, 219, 223
administrative compliance review,
212
for decision making, 212
EPA review, 209–211
general document review, 212–213
independent analysis, 215–216
interagency review, 208–209
internal review, 208
predetermined evaluation criteria
for, 216–222
procedures for, 213–214
public review, 211
screening questions for, 217–219
systematic approaches to, 214–223
technical review, 213
scoping process with, 107–109
summary in, 97
supplements to, 112–113
table of contents in, 97, 99
tiering of, 115
time frames for, 113–114
Environmental justice, 269–278
CEQ guidance on, 277–278
definition of, 270–271
determination of issues in, 271–272
disproportionate effects in, 274–275
and Executive Order 12898, 72–73
populations affected by, 272–274
public involvement in, 276
requirement to improve research on,
276–277
Environmental Law (William H.
Rodgers), 65
Environmental management, 73
Environmental Quality Improvement Act,
amended, 404–406
Environmental Resources Management,
185

Environmental setting, 201
EPA (*see* U.S. Environmental Protection
    Agency)
Equity, environmental (*see*
    Environmental justice)
Erosion, 476–481
ESA (*see* Endangered Species Act)
Europe, 233–235, 342, 343, 346
European Commission, 235
European Community, noise emission
    regulation by, 22–23
European Investment Bank, 235
Evaluation criteria (EIS review), 216–222
Evaluation of Environmental Impact
    Through a Computer Modeling
    Process, Environmental Impact
    Analysis: Philosophy and Methods,
    182–183
Executive Orders, environmental, 15,
    70–73, 228, 271, 276
Expenditures
    estimation of, 266–267
    public sector, 550–553
External scoping process, 107, 108
Extinction of species, 362
Extraterritorial NEPA application, 66–67,
    226–229

F

FAA (*see* Federal Aviation
    Administration)
Facility plans, 79
Failure, market, 14
Fecal coliforms, 474–475
Federal actions, 65–66, 72
Federal agencies
    document preparation in, 7–10
    impact mandate in, 6–7
    NEPA responses of, 73
    planning process for, 75–77
    web site for, 16
Federal Aviation Act, 23
Federal Aviation Administration (FAA),
    22–24
Federal Aviation Noise Policy, 22
Federal Insecticide, Fungicide, and
    Rodenticide Act (FIFRA), 39–40, 389
Federal Land Policy and Management
    Act, 81
Federal Legal Information Through
    Electronics (FLITE) web site, 19
Federal Register, 17, 89, 93, 108, 113
Federal Water Pollution Control Act
    Amendments of 1972, 130

FEIS (*see* Final environmental impact
    statement)
Field crops, 506–509
FIFRA (*see* Federal Insecticide,
    Fungicide, and Rodenticide Act)
Fiji, 232
Final environmental impact statement
    (FEIS), 111–112, 211
Finding of no significant impact (FONSI),
    90–91, 94, 95, 106–107
First National People-of-Color
    Environmental Leadership Summit,
    270
Fish, 501–506
Fish and Wildlife Coordination Act,
    50–51
Fish and Wildlife Service, 17–18, 49–51
Flexibility (methodology), 172–173
FLITE (Federal Legal Information
    Through Electronics) web site, 19
Flow variations (water), 455–456
FONSI (*see* Finding of no significant
    impact)
Foreign Assistance Act of 1983, 364
Framework Convention on Climate
    Change, 330
A Framework for Identification and
    Control of Resource Degradation and
    Conflict in the Multiple Use of the
    Coastal Zone, 186–187
France, 233–235, 252
Fritiofsen v. Alexander, 174
Fuel(s), 311–313, 556–560
    alternative, 316–318
    consumption by end-use sectors, 320
    and greenhouse effect, 328

G

Game (wildlife), 499–501
General document review (EIS), 212–213
Genetic diversity, 360
Geographic characteristics (impact
    assessment), 150
Geographic Information System (GIS),
    174, 176, 203–205
Germany, 236, 336, 341, 342
Gillham Dam, 64
GIS (*see* Geographic information system)
Glass recycling, 321–323
Global warming, 327–339, 391
    effects of, 329–330
    and emission reduction technologies,
        332–335
    emissions contributing to, 331–332

Global warming (*Cont.*)
  environmental assessment implica-
    tions of, 339
  international efforts to mitigate, 330
  policy options for dealing with,
    335–339
  prevention of, 330
  process of, 329–330
  uncertainties about, 328–329
Goals, objectives *vs.,* 76
Government Printing Office web site, 17,
  160
Great Britain, 341
Greenhouse effect (*see* Global warming)
Greenpeace USA v. Stone, 227
Greenpeace v. Stone, 66–67
Groundwater, 54, 130–131, 441
Guatemala, 350
Guidelines for Implementing Principles
  and Standards for Multiobjective
  Planning of Water Resources,
  187–188

H

Habitat, critical, 353, 359
Handbook for Environmental Impact
  Analysis, 181–182
Hazardous air pollutants, 20, 430–433
Hazardous Substance Superfund, 35, 36
Hazardous wastes, 32–34
  and Comprehensive Environmental
    Response, Compensation and
    Liability Act, 36–37
  definition classes of, 32
  land disposal limitations for, 54
  and Resource Conservation and
    Recovery Act, 32–34
  and Superfund Amendments and
    Reauthorization Act, 37–39
Hazards, natural, 481–487
Health effects
  of acid rain, 344
  and environmental justice, 276–277
  of noise, 140–142
Hearings, public, 303–304
Historic preservation (*see* Cultural
  resources)
Hong Kong, 232
Human development, 1
Human ecology, 145
Human impact, 143–145, 539–547
  community needs, 545–547
  lifestyles, 540–542
  physiological systems, 544–545

Human impact (*Cont.*)
  psychological needs, 542–544
Hydrocarbons, 418–422

I

Illinois, 340
Impact(s), 5
  change *vs.,* 148
  communication of, 169
  cumulative, 156–157
  on energy resources, 310
  on fuel resources, 311–313
  identification of, 150, 168
  interpretation of, 168–169
  measurement of, 151–152, 168
  possible, 6
  secondary, 155–156
  (*See also* Human impact)
IMPROVE (Interagency Monitoring of
  Protected Visual Environments), 344
Income, 146
Independent analysis (EIS review),
  215–216
Index (EIS), 104
India, 232
Indiana, 340
Indonesia, 232, 348
Industrial activities, 311
Industrialized countries, deforestation in,
  349
Information technology, 158–161
Input-output economic impact model,
  244–246
Institutional constraints, 200
Interagency Monitoring of Protected
  Visual Environments (IMPROVE),
  344
Interagency review (EIS), 208–209
Inter-American Development Bank, 235
Interference from theory method (SIA),
  258
Intergovernmental Panel on Climate
  Change (IPCC), 328
Interim Report: Social, Economic, and
  Environmental Factors in Highway
  Decision Making, 176, 178
Internal review (EIS), 208
Internal scoping process, 107–108
International EA issues, 225–236
  acid rain, 346–347
  in Asia and Pacific, 232
  biodiversity goals, 363–364
  in Canada, 232–233

International EA issues (*Cont.*)
  in developing countries, 230–231
  in Europe, 233–235
  extraterritorial NEPA application,
    66–67, 226–229
International EA issues
  global warming, 330
  and *Greenpeace v. Stone,* 66–67
  in Latin America, 232
International organizations promoting
  EA, 235–236
International Union for Conservation of
  Nature and Natural Resources
  (IUCN), 364
Internet
  data resources on, 159–160
  laws, regulations, guidance resources
    on, 16–19, 159
  models resources on, 160
  and public participation, 304–306
  reference resources on, 160–161
IPCC (Intergovernmental Panel on
  Climate Change), 328
Iran, 232
Isaak Walton League, 289
Isard model, 244
IUCN (International Union for
  Conservation of Nature and Natural
  Resources), 364
Ivory Coast, 348

J

Japan, 232, 336
Johnston Atoll, 67
Journals, electronic, 160–161
Justice, environmental (*see*
  Environmental justice)

K

Kentucky, 340
Korea, 230, 232
Kyoto Protocol, 330

L

Labor force assessment (SIA),
  263–266
Lacey Act, 364
Lakes, acid rain effects on, 342–343
Land, 132–136, 476–492
  environmental effects of energy con-
    sumption on, 313
  erosion of, 476–481
  and natural hazards, 481–487
  and wetlands regulation, 54–55

Land use, 135–136
  elements producing, 144
  matrix comparing, 245, 246
  patterns of, 487–492
  plans, land use, 78–79, 81
Landforms, 133–135
Latin America, EA in, 232
Legislation and regulation, 13–57
  Clean Air Act, 19–22
  Clean Water Act, 27–31
  Coastal Zone Management Act,
    47–49
  Comprehensive Environmental
    Response, Compensation and
    Liability Act, 34–37
  concerns related to, 15
  data systems for current information
    on, 15–19
  economic *vs.* environmental costs in,
    14, 55–56
  Endangered Species Act, 49–50
  federal, 19–54
  Federal Insecticide, Fungicide, and
    Rodenticide Act, 39–40
  federal/nonfederal roles in, 55
  Fish and Wildlife Coordination Act,
    50–51
  Marine, Protection, Research, and
    Sanctuaries Act, 40–41
  National Environmental Policy Act,
    44–45
  National Historic Preservation Act,
    45–46
  Noise Control Act, 22–25
  Pollution Prevention Act, 51–54
  rationale for, 14–15
  Resource Conservation and Recovery
    Act, 31–34
  Safe Drinking Water Act, 25–27
  Superfund Amendments and
    Reauthorization Act, 37–39
  Toxic Substances Control Act, 41–43
  trends in, 54–57
  Wild and Scenic Rivers Act, 46–47
  (*See also* National Environmental
    Policy Act [NEPA])
Leontief's general equilibrium model,
  244–245
LEXIS®-NEXIS® web site, 17–18
Libraries, on-line, 160
Lifestyles, 540–542
Life-support systems, 14
Listed species, 351–353, 509–511
Litigation, NEPA, 63–67

Location quotients, 249, 250
Lovejoy, Thomas, 350
Low-income populations, 272–274

M

Madison, James, 282
Maine, 74
Malaysia, 232
Managed species, 137
A Manual for Conducting Environmental
    Impact Studies, 189–190
Marine, Protection, Research, and
    Sanctuaries Act, 40–41
Marine Mammal Protection Act, 350–351
Marine Sanctuaries Program, 41
Market failure, 14
Massachusetts, 282
Material safety data sheets (MSDSs), 38
Matrices, 167
Matrix Analysis of Alternatives for Water
    Resource Development, 188–189
Measurements, impact, 151–152
Metals recycling, 324
Methane, 331
Methodologies for assessment, 163–191
    ad hoc, 166
    CEQ requirements for, 191
    checklists, 167
    criteria for choosing, 163–166
    development of agency-specific,
        190–191
    matrices, 167
    networks, 167
    overlays, 166
    review criteria for analyzing, 167–177
    review of, 176–190 (See also specific
        methodologies)
A Methodology for Evaluating
    Manufacturing Environmental
    Impact Statements for Delaware's
    Coastal Zone, 184–185
Mexico, 232
Michigan, 74
Migratory Bird Treaty Act of 1918, 18
Minority populations, 272–274
Mississippi Army National Guard,
    119–120
Missouri, 340
Missouri Botanical Garden, 366
Mitigation, 115–116 (See also specific
    headings)
Models, access to, 160
Montana, 74
Montreal Protocol, 331–332

MSDSs (material safety data sheets), 38
Muir, John, 283
Multiplier effect, 248, 249, 251
Myers, Norman, 362

N

NAAQS (see National Ambient Air
    Quality Standards)
NAGPRA (see Native American Graves
    Protection and Repatriation Act of
    1990)
National Ambient Air Quality Standards
    (NAAQS), 20, 21
National Audubon Society, 289
National Environmental Policy Act
    (NEPA), 6–8, 61–75, 77–82, 397–398
    CEQ regulations for compliance with,
        85–104
    classes of actions recognized by,
        92–93
    draft environmental impact state-
        ment, 91
    EIS format prescribed by, 97–104
    environmental assessments (EA)s, 90
    environmental documents required
        by, 90–92
    final environmental impact state-
        ment, 91
    and level of NEPA review, 92–97
    notice of intent (NOI), 91
    Record of Decision (ROD), 91–92
    Council on Environmental Quality
        established by, 70
    court cases resulting from, 63–67
    energy implications of, 309–310
    Environmental Quality Improvement
        Act, amended, 404–406
    executive orders related to, 70–73
    extraterritorial application of, 66–67,
        226–229
    federal agency responses to, 73
    function/purpose of assessment process
        under, 85–88
    and Greenpeace v. Stone, 66–67
    as model for state NEPA programs,
        73–74
    parallels between planning and review
        process of, 77–78
    as planning tool, 74, 77–82
        relationship between NEPA and
            planning, 77–78
        and timing of NEPA reviews, 79
        and types of plans, 78–79
    purpose of, 397

National Environmental Policy Act
(NEPA) (*Cont.*)
  regulations, NEPA
    public involvement introduced by,
      284–285
    public participation in development
      of, 296
    text of, 585–614
  reviews, NEPA, 74–75, 77–82
    concurrent planning with, 81–82
    determining level of, 92–97
    EA as initial level of, 95–96
    EIS as initial level of, 93–95
    planning process parallels with,
      77–78
    post-planning, 79–80
    pre-planning, 80–81
    timing of, 79–82
    types of plans subject to, 78–79
  timing of reviews under, 79–82
  Title I of (Declaration of National
    Environmental Policy), 62–63,
    398–401
  Title II of (Council on Environmental
    Quality), 63, 70, 401–404
  types of plans subject to review under,
    78–79
National Historic Preservation Act of
  1966 (NHPA), 17, 45–46, 367–371,
  373
National Marine Fisheries Service,
  17–18, 49
National Oceanic and Atmospheric
  Administration (NOAA), 41, 47–48
National Pollutant Discharge
  Elimination System (NPDES), 27–29
National Register of Historic Places, 45,
  46, 367, 368
National Science Foundation (NSF), 67
Native American cultural resources,
  371–374
Native American Graves Protection and
  Repatriation Act of 1990 (NAGPRA),
  372–374
Natural gas, 317
Natural hazards, 481–487
Natural land vegetation, 511–514
Natural resources (*see* Resources)
*Natural Resources Defense Council, Inc.
  v. Nuclear Regulatory Commission,*
  227
The Nature Conservancy (TNC), 363–365
Neotropical Migratory Bird Conservation
  Program, 364

NEPA (*see* National Environmental
  Policy Act)
Nepal, 232
Netherlands, 234–236
Networks, 167
New Jersey, 74, 342
New Mexico, 121–122
New York, 74, 342
New Zealand, 232
NGOs (nongovernmental organizations),
  306
NHPA (*see* National Historic Preservation
  Act of 1966)
Nitrogen oxides, 346, 422–426
Nitrous oxide, 331
NOAA (*see* National Oceanic and
  Atmospheric Administration)
No-action alternative, 148–149
NOI (*see* Notice of intent)
Noise (*see* Sound)
Noise Control Act, 22–25
Nonattainment areas, 20
Nonfuel resources, 560–562
North American Waterfowl Management
  Plan, 364
North Carolina, 270
Northern spotted owls, 358
Norway, 236, 331, 333, 341
Notice of intent (NOI), 91, 113
NPDES (*see* National Pollutant
  Discharge Elimination System)
NSF (National Science Foundation), 67
Nuclear energy, 317–318
Nuclear Regulatory Commission,
  227–228
Nuclear waste management, 55
Nutrients (in water), 469–471

O

Objectives, goals *vs.,* 76
Oceans
  carbon sequestration in, 334
  dumping in, 40–41
OCS activities (*see* Off-coast survey activities)
Odors (in air), 434–439
Off-coast survey (OCS) activities, 48–49
Office of Technology Assessment (OTA),
  365
Ohio, 340
Oil
  as fuel, 316, 317
  in/on water, 457–458
  waste recycling, 324–325

Oil Pollution Act, 387–388
On-line libraries, 160
Optimum Pathway Matrix Analysis
    Approach to the Environmental
    Decision-Making Process: Test Case:
    Relative Impact of Proposed
    Highway Alternatives, 179–180
Organization of American States, 235
OTA (Office of Technology Assessment),
    365
Outline of EIS, 98–99
Output, economic, 147
Overlays, 166
Ozone protection, 21

P

Pacific, EA in, 232
Pakistan, 232
Panama, 227–228
Paper recycling, 323
Papua New Guinea, 232
Participant observation method (SIA),
    256–257
Particulates (in air), 410–414
Patterns, land-use, 487–492
PCBs, 43
Pennsylvania, 340
Per capita consumption, 553–556
Perception of effects, 143
Performance effects of sound, 533–535
Permits, air, 21
Peru, 232
Pesticides, 39–40
Philippines, 230, 232
Photochemical oxidants, 428–430
Physiological effects of sound, 140,
    517–525
Physiological systems, 544–545
Planning Methodology for Water Quality
    Management: Environmental
    System, 179
Planning process
    and environmental impact analysis,
        86
    NEPA parallels with, 77–78
    steps in, 75–77
    timing of NEPA reviews in, 79–82
Plans, types of, 78–79
Plastics recycling, 324
Poland, 2, 3
Pollution (see specific topics, e.g.: Air)
Pollution control, 71–72, 319
Pollution Prevention Act (PPA) of 1990, 4,
    51–54

Possible impacts, 6
Poverty, 3–4
PPA (see Pollution Prevention Act of
    1990)
Predatory birds, 496–499
Preparers, list of (EIS), 103–104
A Procedure for Evaluating
    Environmental Impact, 183
Procedures for Regional Clearinghouse
    Review of Environmental Impact
    Statements— Phase Two, 186–187
Programmatic plans, 78
Proponent of an action, 7
Proposed action, 5
Psychological effects of sound, 525–529
Psychological needs, 542–544
Public hearings, 303–304
Public participation, 281–306
    application of input from, 295–296
    case studies of, 292–295
    commenting on EIS as special form of,
        291–292
    in decision-making process, 288
    definition of, 285–287
    in development of regulations, 296
    in early America, 292–294
    in environmental justice, 276
    input from, 290–291
    Internet's effect on, 304–306
    level of, 288
    means of informing public for, 289–290
    as methodological consideration, 164
    NEPA requirements for, 284–285
    objectives for effective, 296–302
    in public hearings, 303–304
    terms related to, 294
    value of, 302–303
Public review (EIS), 211
Public sector revenue/expenditures,
    550–553
Public works projects, 240
The Puritan Oligarchy—The Founding of
    American Civilization (Wertenbaker),
    282
Purpose and need statement (EIS),
    99–100

Q

Qualitative impact measurements,
    151–152
Quantifying the Environmental Impact of
    Transportation System, 186
Quantitative impact measurements, 151
Quiet Communities Act, 23–24

R

Radioactivity (in water), 458–459
RCRA (*see* Resource Conservation and
    Recovery Act)
Record of Decision (ROD), 91–92, 94
Recycling of materials, 319–325
    glass, 321–323
    metals, 324
Recycling of materials
    oil wastes, 324–325
    paper, 323
    plastics, 324
    solid waste, 325
    tires/rubber products, 323
Regional analysis model, 245
Regional economic stability, 548–549
Regulations, environmental, 13 (*See also*
    Legislation and regulation)
Reliability (methodology), 172
Reports of toxic chemical release, 38
Residential activities, 310
Resource Conservation and Recovery Act
    (RCRA), 31–34, 388
Resource management plans, 78
Resources, 556–565
    aesthetics, 562–565
    and biodiversity, 361
    energy, 147–148, 310–311
    environment as, 14
    fuel, 311–313, 556–560
    groundwater, 54
    as methodological consideration, 165
    nonfuel, 560–562
Results, reporting, 6
Revenue
    estimation of, 266–267
    public sector, 550–553
Reviews
    of environmental impact statements,
        207–223
    ad hoc review, 219, 223
    administrative compliance review,
        212
    for decision making, 212
    EPA review, 209–211
    general document review, 212–213
    independent analysis, 215–216
    interagency review, 208–209
    internal review, 208
    predetermined evaluation criteria
        for, 216–222
    procedures for, 213–214
    public review, 211
    screening questions for, 217–219

Reviews (*Cont.*)
    systematic approaches to, 214–223
    technical review, 213
    NEPA, 74–75, 77–82
        concurrent planning with, 81–82
        determining level of, 92–97
        EA as initial level of, 95–96
        EIS as initial level of, 93–95
        planning process parallels with,
            77–78
        post-planning, 79–80
        pre-planning, 80–81
        timing of, 79–82
        types of plans subject to, 78–79
Ringsted v. Duluth, 66
Risk management, 375
Risk-based decision making, 384
Rivers
    acid rain effects on, 342–343
    Wild and Scenic Rivers Act, 46–47
Road traffic, 138, 140
*Robertson v. Methow Valley Citizen
    Council,* 66
ROD (*see* Record of Decision)
Rodgers, William H., 65
Rubber products recycling, 323

S

Safe Drinking Water Act, 25–27
SARA (*see* Superfund Amendments and
    Reauthorization Act)
Scandinavia, 342
Scenarios, SIA, 259
Scoping, 107–109
    with EIS, 107–109
    environmental justice concerns during,
        277
    scope creep case study, 117–119
Secondary impacts, 155–156
Self-purification of natural waters, 440
SEPAs (*see* State environmental policy
    acts)
Setting, environmental, 201
Shellfish, 501–506
SHPOs (*see* State Historic Preservation
    Officers)
SIA (*see* Social impact assessment)
Sierra Club, 283, 289
Sierra Club v. Adams, 227
Sierra Club v. Froehlke, 64
Similarity method (SIA), 257
Snail darters, 356
Social behavior effects of sound, 535–538
Social environment, 143–145

Social impact assessment (SIA), 239,
    251–259
  case study method for, 255
  content analysis method for, 255
  Delphi method for, 256
  dynamic simulation method for,
      257–258
  interference from theory method for,
      258
Social impact assessment (SIA)
  participant observation method for,
      256–257
  scenarios for, 259
  similarity method for, 257
  surveys for, 258
  trend analysis method for, 255
Socioeconomic impact assessment,
    239–278
  economic impact analysis, 242–251
    economic base models for, 246–251
    future direction for, 251
    input-output model for, 244–246
  effects covered in, 259–262
  and environmental justice, 269–278
    affected populations, 272–274
    CEQ guidance on, 277–278
    definition of, 270–271
    determination of issues in, 271–272
    disproportionate effects in, 274–275
    public involvement in, 276
    requirement to improve research on,
        276–277
  estimation of revenues/expenditures in,
      266–267
  labor force in, 263–266
  within NEPA, 240–242
  problem areas in, 267–269
  social impact assessment, 251–259
    case study method for, 255
    content analysis method for, 255
    Delphi method for, 256
    dynamic simulation method for,
        257–258
    interference from theory method for,
        258
    participant observation method for,
        256–257
    scenarios for, 259
    similarity method for, 257
    surveys for, 258
    trend analysis method for, 255
  time dimension in, 262–263
Soil(s), 132–133
Solid Waste Disposal Act, 31, 35

Solid waste(s)
  disposal sites for, 33
  environmental effects of energy con-
      sumption on, 313
  recycling of, 325
  and Resource Conservation and
      Recovery Act, 33–34
Sound, 138–143, 517–538
  communication effects of, 529–533
Sound
  and Noise Control Act, 22–25
  performance effects of, 533–535
  physiological effects of, 517–525
  psychological effects of, 525–529
  social behavior effects of, 535–538
Species diversity, 137, 360
Sri Lanka, 232
State environmental policy acts (SEPAs),
    73–74
State Historic Preservation Officers
    (SHPOs), 368–370, 373, 374
Steel recycling, 324
Sulfur oxides, 414–418
Summary (EIS), 97
Superfund (see Comprehensive
    Environmental Response,
    Compensation and Liability Act)
Superfund Amendments and
    Reauthorization Act (SARA), 37–39
Supplemental reviews, 112–113
Surface water, 130–132, 440–441
Surveys, SIA, 258
Suspended solids (in water), 459–461
Sustainable development and production,
    4
Sustainable economic growth, 3–4
Swain v. Brinegar, 64
Sweden, 236, 331, 344
Synoptic planning, 75–76
System stability (ecology), 137

T
Table of contents (EIS), 97, 99
Target groups (community activities), 286
Technical review (EIS), 213
Tellico Dam, 356–358
Temporal characteristics (impact assess-
    ment), 150–151
Tennessee, 340, 378
Tennessee Valley Authority, 356
Texas, 340
Thailand, 232, 348
Thermal discharge (water), 461–463
Thoreau, Henry David, 283

Threatened species (*see* Listed species)
Tiering of environmental assessments, 115
Time
  as methodological consideration, 165
  in socioeconomic impact assessment, 262–263, 267–268
Timing of NEPA reviews, 79–82
Tires, recycling of, 323
TNC (*see* The Nature Conservancy)
Toxic chemical release reporting, 38
Toxic compounds (in water), 471–473
Toxic Substances Control Act (TSCA), 41–43, 389, 390
Transportation activities, 311, 313–315
Transportation and Environment: Synthesis for Action: Impact of National Environmental Policy Act of 1969 on the Department of Transportation, 183–184
Trend analysis method (SIA), 255
Tropical deforestation, 348–349
Trust Territory of the Pacific Islands, 232
TSCA (*see* Toxic Substances Control Act)
Tuvalu, 232

U

Unemployment, 269
United Kingdom, 346
United Nations Development Programme, 235
United States, 6–7, 252
  acid rain issues in, 341, 342, 346–347
  carbon dioxide emissions in, 336
  CFC use in, 331
  deforestation in, 347
  ecosystem types in, 364–365
  energy crisis in, 309
  history of public participation in, 282–284
  sulfur dioxide emissions in, 340
  transportation system in, 313–314
  (*See also specific topics*)
The University of Michigan School of Natural Resources Conference on Race and the Incidence of Environmental Hazards, 270
Uruguay, 232
U.S. Army, 194, 196, 227–228
U.S. Army Corps of Engineers, 28–31, 41, 66, 285, 292–293
U.S. Department of Defense (DOD), 294–295

U.S. Department of Energy (DOE), 117–119, 121–122, 160, 333–334
U.S. Environmental Protection Agency (EPA)
  Acid Rain Program of, 345
  court cases against, 65–66
  and drinking water regulation, 25, 26
  ecorisk guidelines of, 374, 375
  EIS review by, 209–211
U.S. Environmental Protection Agency (EPA)
  environmental justice defined by, 271–272
  hazardous waste regulation by, 32–34, 36–37
  and noise emissions, 23, 24
  ocean dumping regulation by, 40, 41
  pesticide regulation by, 39–40
  and Pollution Prevention Act of 1990, 51–54
  primary drinking water regulations of, 446–454
  risk assessment process identified by, 378–379
  solid waste regulation by, 33, 34
  toxic substances regulation by, 41–43
  water pollution regulation by, 27–31
  web site of, 16, 160
U.S. Fish and Wildlife Service, 17–18
U.S. Forest Service, 119–120, 302
U.S. Supreme Court decisions, web site for, 19
USAID, 227, 236

V

Vegetation, 511–514
Venezuela, 232
Visibility, acid rain and, 343–344

W

Wallisville Dam Project, 64
Washington state, 74
Water Pollution Control Act of 1972, 30
Water Pollution Control amendments of 1981, 31
Water Quality Act of 1987, 30
Waterfowl, 501–506
Water(s), 129–132, 440–475
  acid rain effects on, 342–343
  aquatic life in, 473–474
  aquifer safe yield, 442–443, 446–455
  biochemical oxygen demand of, 464–466

Water(s) (*Cont.*)
  Clean Water Act, 27–31
  dissolved oxygen in, 466–467
  dissolved solids in, 467–469
  environmental effects of energy con-
    sumption on, 313
  EPA primary drinking water regula-
    tions, 446–454
  fecal coliforms in, 474–475
  flow variations in, 455–456
  groundwater, 54, 441
  and Marine, Protection, Research, and
    Sanctuaries Act, 40–41
  nutrients in, 469–471
  oil in/on, 457–458
  pH of, 463–464
  quality standards for, 29–30
  radioactivity in, 458–459
  Safe Drinking Water Act, 25–27
  self-purification of natural, 440
  surface water, 130–132, 440–441
  suspended solids in, 459–461
  thermal discharge in, 461–463
  toxic compounds in, 471–473

  and wetlands, 54–55
  Wild and Scenic Rivers Act, 46–47
Watt, Kenneth, 329
Weighting (attributes), 152–153
West Germany, 227–228
West Virginia, 340
Westlaw web site, 18
Wetlands, 54–55, 71
WHO (World Health Organization), 235
Wild and Scenic Rivers Act, 46–47
Wildlife
  and Endangered Species Act, 49–50
  and Fish and Wildlife Coordination Act,
    50–51
  and Fish and Wildlife Service, 17–18
  large animals, 493–496
  managed species, 137
  small game, 499–501
  (*See also* Birds)
Wilson, Edward O., 362
Woolard, Edgar J., Jr., 4
World Bank, 235, 306
World Health Organization (WHO), 235
World Resources Institute, 366

ABOUT THE AUTHORS

R. K. Jain, Ph.D., is Dean of the School of Engineering and Professor of Engineering at the University of the Pacific. He has served on numerous National Task Forces and Advisory Councils for the Department of Defense, NSF, Navy, Army, EPA and NAS. He is a fellow ASCE and Diplomate American Academy of Environmental Engineers. Along with his prior appointments, Dr. Jain has published ten books and over 100 journal articles, book chapters, and technical reports. He is Editor-In-Chief of the international journal *Environmental Engineering and Policy*.

L. V. URBAN is Director of the Water Resources Center and Professor of Civil Engineering at Texas Tech University. He consults to industry on water resources, environmental impact analysis, and environmental engineering. He is also an active member of the American Society of Civil Engineers and registered professional engineer in Texas.

GARY S. STACEY was formerly a senior economist at the Battelle-Europe Centres de Recherche in Geneva, Switzerland, and has more than 30 years' experience in economic impact assessment. He has helped develop strategies for many U.S. government agencies, including the EPA and the Department of Energy. Dr. Stacey is the author or co-author of more than a dozen books and scores of professional articles and client reports in these fields.

HAROLD BALBACH is a senior biologist at the U.S. Army Engineer Research and Development Center, Champaign, IL. He is a Fellow of the Society of American Military Engineers, and serves on their Board of Directors in addition to representing the Military Installations Land Management Division on the Board of Directors of the American Society of Agronomy. He has co-authored or edited six books, scores of agency reports, and numerous NEPA documents.

M. DIANA WEBB is Group Leader of the Ecology Group, Los Alamos National Laboratory, New Mexico, operated by the University of California. Ms. Webb has over 30 years experience in environmental compliance issues with the Laboratory, Department of Energy, Bureau of Land Management, Army Corps of Engineers, and private consulting firms. She is the recipient of numerous awards and honors, including a Secretarial NEPA Quality Award.